Epigenetics of Aging

Trygve O. Tollefsbol
Editor

Epigenetics of Aging

Springer

Editor
Trygve O. Tollefsbol
Department of Biology
University of Alabama at Birmingham
1300 University Blvd.
Birmingham, AL 35294-1170
USA
trygve@uab.edu

ISBN 978-1-4419-0638-0 e-ISBN 978-1-4419-0639-7
DOI 10.1007/978-1-4419-0639-7
Springer New York Dordrecht Heidelberg London

Library of Congress Control Number: 2009932905

Printed on acid-free paper

Springer is part of Springer Science+Business Media (www.springer.com)

Preface

Numerous studies have indicated that epigenetic mechanisms may play a major role in both cellular and organismal aging. These epigenetic processes not only include DNA methylation and histone modifications but also extend to many other epigenetic mediators such as the polycomb group proteins, chromosomal position effects, and noncoding RNA. The topics of this seminal book on aging epigenetics range from fundamental changes in DNA methylation in aging to the most recent research on intervention into epigenetic modifications to modulate the aging process and age-associated disorders. The major topics of aging epigenetics covered in this book are (1) DNA methylation and histone modifications in aging, (2) other epigenetic processes and aging, (3) impact of epigenetics on aging, (4) epigenetics of age-related diseases, (5) epigenetic interventions and aging, and (6) future directions/perspectives in aging epigenetics.

The most studied of epigenetic processes, DNA methylation, has been associated with cellular aging and aging of organisms for many years. It is now apparent that both global and gene-specific alterations occur not only in DNA methylation during aging but also in several types of histone modifications. Many epigenetic aberrations may have an impact on aging processes through control of telomerase, modifications of telomeres, and epigenetic drift. The latter is evident in the recent studies of aging monozygotic twins.

Numerous age-related diseases are affected by epigenetic mechanisms. For example, recent studies have shown that DNA methylation is altered in Alzheimer's disease and autoimmunity. Other prevalent diseases that have been associated with age-related epigenetic changes include cancer and osteoarthritis. Epigenetic alterations appear to have an effect on several of the progeroid syndromes of premature aging as well. Moreover, the impact of dietary or drug intervention into epigenetic processes as they affect normal aging or age-related diseases is becoming increasingly feasible.

This book is intended for those with interests ranging from the fundamental basis of aging to interventions in slowing the aging process or treating age-related disorders. The study of epigenetics as it relates to aging and age-related diseases is a relatively new field that is showing considerable promise in revolutionizing how the aging process is viewed. The purpose of this book on aging epigenetics is to provide coverage of not only established aspects of epigenetics as applied to the aging process but also new approaches and perceptions in this important area of research.

Trygve O. Tollefsbol
Birmingham, AL, USA

Contents

Contributors

Peter D. Adams Fox Chase Cancer Center, Philadelphia, PA 19111, USA, peter.adams@fccc.edu

Ruben Agrelo Research Institute of Molecular Pathology (IMP), Dr. Bohr-Gasse 7, 1030 Vienna, Austria, agrelo@imp.univie.ac.at, ragrelo@adinet.com.uy

Adebayo D. Akintola Systems Biology and Translational Medicine, Texas A&M Health Science Center, College of Medicine, College Station, TX 77847-1114, USA

Gerald L. Arthur Department of Pathology and Anatomical Sciences, Ellis Fischel Cancer Center, University of Missouri School of Medicine, Columbia, MO 65212, USA

Gowrishankar Banumathy Fox Chase Cancer Center, Philadelphia, PA 19111, USA

Lynda B. Bennett Department of Pathology and Anatomical Sciences, Ellis Fischel Cancer Center, University of Missouri School of Medicine, Columbia, MO 65212, USA

Pinaki Bose Department of Biochemistry and Molecular Biology, Southern Alberta Cancer Research Institute, University of Calgary, 3330 Hospital Drive NW, Calgary, Alberta, Canada 42N 4N1

Charles W. Caldwell Department of Pathology and Anatomical Sciences, Ellis Fischel Cancer Center, University of Missouri School of Medicine, Columbia, MO 65212, USA, caldwellc@health.missouri.edu

Vincenzo Calvanese Department of Immunology and Oncology, National Center for Biotechnology, CNB-CSIC, Cantoblanco, Madrid E-28049, Spain

Huaping Chen Department of Biology, 175 CH, 1300 University Boulevard, University of Alabama, Birmingham, AL 35294-1170, USA

J. Tyson DeAngelis Department of Biology, 175 CH, 1300 University Boulevard, University of Alabama, Birmingham, AL 35294-1170, USA

Yuqing Dong Department of Biological Sciences, Clemson University, Clemson, SC 29634, USA,

Mario F. Fraga Department of Immunology and Oncology, National Center for Biotechnology, CNB-CSIC, Cantoblanco, Madrid E-28049, Spain; Laboratorio de Epigenética del Cáncer, Instituto Universitario de Oncología del Principado de Asturias (IUOPA), Universidad de Oviedo, Oviedo, Spain, mffraga@cnio.es

Amudha Ganapathy Department of Biochemistry and Molecular Biology, Southern Alberta Cancer Research Institute, University of Calgary, 3330 Hospital Drive NW, Calgary, Alberta, Canada 42N 4N1

Eric Gilson Laboratoire de Biologie Moleculaire de la Cellule, CNRS UMR 5239, Ecole Normale Superieure de Lyon, UCBL1, IFR 128, 46 allee d-Italie, 69364 Lyon Cedex 07, France

Jorg J. Goronzy Department of Medicine, Kathleen B. and Mason I. Lowance Center for Human Immunology, Emory University School of Medicine, Woodruff Memorial Research Building, Atlanta, GA, USA, jgoronz@emory.edu

Robin Holliday Australian Academy of Science, Canberra, Australia, randl.holliday@bigpond.com

Inga Kadish Department of Cell Biology, THT 912, University of Alabama, Birmingham, AL 35294-0006, USA; Center for Aging, University of Alabama, Birmingham, AL, USA, ikadisha@uab.edu

Ester Lara Department of Immunology and Oncology, National Center for Biotechnology, CNB-CSIC, Cantoblanco, Madrid E-28049, Spain

Guangjin Li Department of Medicine, Kathleen B. and Mason I. Lowance Center for Human Immunology, Emory University School of Medicine, Woodruff Memorial Research Building, Atlanta, GA, USA

Yuanyuan Li Department of Biology, 175 CH, 1300 University Boulevard, University of Alabama, Birmingham, AL 35294-1170, USA

Wil A.M. Loenen Department of Toxicogenetics, Leiden University Medical Centre, Einthovenweg 20, Leiden, The Netherlands, w.a.m.loenen@lumc.nl

J. Arturo Loñdono-Vallejo UMR7147, Institut Curie-UPMC-CNRS, 26, rue d'Ulm, 75248 Paris, France, arturo.londono@curie.fr

Frederique Magdinier Laboratoire de Biologie Moleculaire de la Cellule, CNRS UMR 5239, Ecole Normale Superieure de Lyon, UCBL1, IFR 128, 46 allee d-Italie, 69364 Lyon Cedex 07, France, frederique.magdinier@ens-lyon.fr

Ulrich Mahlknecht Department of Internal Medicine, Saarland University Medical Center, Division of Immunotherapy and Gene Therapy, Jose Carreras Center for Immunotherapy and Gene Therapy, D-66421 Homburg/Saar, Germany, ulrich.mahlknecht@uks.eu

Ashley McCollum Alabama State University, Montgomery, AL 36104, USA

Krishnaveni Mishra Department of Biochemistry, School of Life Sciences, University of Hyderabad, Hyderabad, India, kmsl@uohyd.ernet.in

Rakesh K. Mishra Centre for Cellular and Molecular Biology, Uppal Road, Hyderabad 500007, India, mishra@ccmb.res.in

Alan R. Parrish Systems Biology and Translational Medicine, Texas A&M Health Science Center, College of Medicine, College Station, TX 77847-1114, USA, parrish@medicine.tamhsc.edu

Igor P. Pogribny Division of Biochemical Toxicology, National Center for Toxicological Research, Jefferson, AR 72079, USA, igor.pogribny@fda.hhs.gov

Karl Riabowol Departments of Biochemistry & Molecular Biology, Southern Alberta Cancer Research Institute, University of Calgary, 3330 Hospital Drive NW, Calgary, Alberta, Canada 42 N 4N1; Departments of Oncology, Southern Alberta Cancer Research Institute, University of Calgary, 3330 Hospital Drive NW, Calgary, Alberta, Canada 42 N 4N1, karl@ucalgary.ca

Helmtrud I. Roach Bone and Joint Research Group, Southampton General Hospital, University of Southampton, Southampton, SO16 6YD, UK, h.roach@southampton.ac.uk

Sabita N. Saldanha Department of Biology, 175 CH, 1300 University Boulevard, University of Alabama, Birmingham, AL 35294-1170, USA

Huidong Shi Department of Pathology and Anatomical Sciences, Ellis Fischel Cancer Center, University of Missouri School of Medicine, Columbia, MO 65212, USA

Kristen H. Taylor Department of Pathology and Anatomical Sciences, Ellis Fischel Cancer Center, University of Missouri School of Medicine, Columbia, MO 65212, USA

Trygve O. Tollefsbol Department of Biology, University of Alabama, Birmingham, AL 35294, USA; Center for Aging, University of Alabama, Birmingham, AL 35294, USA; Comprehensive Cancer Center, University of Alabama, Birmingham, AL 35294, USA; Clinical Nutrition Research Center, University of Alabama, Birmingham, AL 35294, USA, trygve@uab.edu

Thomas van Groen Department of Cell Biology, THT 912, University of Alabama, Birmingham, AL 35294-0006, USA; Department of Neurobiology, University of Alabama, Birmingham, AL. USA, vangroen@uab.edu

Boris F. Vanyushin Belozersky Institute of Physico-Chemical Biology, Lomonosov Moscow State University, Moscow 119899, Russia, vanyush@belozersky.msu.ru

Eugenia Wang Department of Biochemistry, , University of Louisville School of Medicine, Louisville, KY 40202, USA; Gheens Center on Aging, University of Louisville School of Medicine, Louisville, KY 40202, USA, eugenia.wang@louisville.edu

Cornelia M. Weyand Department of Medicine, Kathleen B. and Mason I. Lowance Center for Human Immunology, Emory University School of Medicine, Woodruff Memorial Research Building, Atlanta, GA, USA

Sige Zou Functional Genomics Unit, Laboratory of Experimental Gerontology, National Institute on Aging, Baltimore, MD 21224, USA, zous@grc.nia.nih.gov

Barbara Zschoernig Department of Internal Medicine, Saarland University Medical Center, Division of Immunotherapy and Gene Therapy, Jose Carreras Center for Immunotherapy and Gene Therapy, D-66421 Homburg/Saar, Germany

Epigenetics and the Aging Process

Trygve O. Tollefsbol

Abstract Although there has been considerable interest in the aging process for quite some time and in epigenetics for over two decades, it is relatively recent that we have witnessed a surge in interest in aging epigenetics. Epigenetic processes are vast and influence many genes and other processes related to aging that impact the physiological decline that characterizes the aging process. This ranges from changes in DNA methylation and histone modifications that control genes such as the telomerase regulatory gene, *hTERT*, to modes of modulating environmental factors that affect epigenetic processes and aging. The potential for advances in aging epigenetics seems limitless and the purpose of this book is to convey the cutting-edge research that is rapidly driving the field of aging epigenetics to the forefront of medical science.

Keywords Epigenetic · Aging · DNA methylation · Histone modification

Introduction

Epigenetics involves effects carried out by a vast array of chromatin modifications, DNA methylation, noncoding RNA and other processes that are heritable and reversible. There is no change in DNA or protein sequence that occurs during epigenetic processes. Aging of course is an even broader field that encompasses the decline in the physiological capacity of an organism over time due to a number of changes at the molecular, cellular and organismal levels. While both the study of epigenetics and aging have developed rapidly in the past two decades, the field of aging epigenetics is relatively new. Despite this, however, it has become

T.O. Tollefsbol (✉)
Center for Aging and Department of Biology, 175 Campbell Hall, 1300 University of Alabama at Birmingham, AL 35294-1170, USA
e-mail: trygve@uab.edu

T.O. Tollefsbol (ed.), *Epigenetics of Aging*, DOI 10.1007/978-1-4419-0639-7_1,

increasingly clear that epigenetic processes play a major role in aging (Liu et al., 2003; Sedivy et al., 2008; Fraga and Esteller, 2007).

DNA methylation is the mainstay of epigenetics and in most eukaryotes is characterized by the addition of methyl groups to the cytosine-5 position and this usually occurs in CpG dinucleotides. The DNA methyltransferases (DNMTs) catalyze DNA methylation and three DNMTs are responsible for most of the DNA methylation that occurs in the genome. The patterns of methylation that are inherited with each DNA replication are carried out by DNMT1 and this represents the vast majority of DNA methylation. However, methylation patterns are constantly subject to modifications and DNMT3A and DNMT3B catalyze most of the *de novo* methylation in the cell that involves the addition of new methyl moities at cytosines that were not previously methylated. DNA methylation influences numerous biological processes such as genomic imprinting, cellular differentiation and X-chromosome inactivation. However, perhaps the most significant biological effect due to DNA methylation is the control of gene expression where hypermethylation in a gene regulatory region is usually associated with transcriptional repression and hypomethylation is generally associated with transcriptional activity with some notable exceptions (Lai et al., 2005).

Gene expression can also be controlled by another major epigenetic process, chromatin changes, that are characterized by posttranslational modifications such as acetylation and methylation of histones. These site-specific modifications result in regulation of the binding and activities of other proteins that interact with DNA and the histones. The major enzymes that carry out histone modifications are the histone acetyltransferases (HATs) and histone deacetylases (HDACs) and these enzymes allow for dynamic changes in chromatin while also serving to maintain chromatin configurations to allow stable inheritance of epigenetic traits. Most histone modifications occur in the amino terminal regions of these nucleosomal proteins and histone acetylation is generally associated with increased transcriptional activity while histone methylation has variable effects on gene activity depending upon the specific residue that is methylated and the nature of the methylation. For example, methylation of lysine 4 (K4) on histone H3 is often found in active chromatin regions while methylation of H3-K9 usually leads to transcriptional repression (Groth et al., 2007). The processes of DNA methylation and chromatin modifications often interact together to impact epigenetic changes. Methylation at K3-K9 can promote cytosine methylation while methylation of cytosines can foster changes at H3-K9 involving histone methylation and deacetylation (Fuks et al., 2000).

Recent advances have merged the broad fields of epigenetics and aging. For instance, global hypomethylation of the genome and regional hypermethylation of specific genes occur during the aging of cells and tissues (Liu et al., 2003). May age-related diseases such as cancer, Alzheimer's disease, autoimmunity, and osteoarthritis are associated with epigenetic alterations and changes in the epigenetic machinery have also been reported in premature aging diseases such as progeria. While epigenetics impacts many different biological processes other than aging and aging involves numerous mechanisms other than epigenetics, it is clear that these two processes are linked and it seems likely that epigenetics will be proven to have a major role in aging not only in the aging of cells, but also in organismal aging.

DNA Methylation and Histone Modifications in Aging

Hypomethylation of aging cells and tissues has been documented in numerous studies and likely plays a major role in the aging process. As described in Chapter 2 of this book, although many different mechanisms have been proposed to lead to the genomic hypomethylation that occurs during aging such as a decline in DNMT1, this reliable characteristic of aging remains an enigma and the causes for decreased methylation during aging are still unresolved. DNA hypomethylation is observed in many different age-related diseases such as cancer, atherosclerosis, Alzheimer's disease, autoimmunity and macular degeneration. Therefore, genomic hypomethylation may not only be involved in the basic mechanisms of aging, but it also is a major risk factor in the development of chronic age-associated pathologies. Besides genomic hypomethylation, DNA methylation is also important in aging due to gene- or region-specific hypermethylation (Chapter 3). Many different genes have been shown to undergo hypermethylation of their control regions during aging such as the tumor suppressor genes that play an important role in cancer. Moreover, age-related hypermethylation has been estimated to involve up to half of the genes that are hypermethylated in colon cancer and other tissues such as breast, kidney and lung have been shown to undergo age-related hypermethylation of key genes such as those encoding estrogen receptor α, E-cadherin, and p16. Therefore, hypermethylation of genes may also play a role in the mechanism of aging and serve as a risk factor for selected pathologies that are associated with aging.

Histone modifications change during the aging process and often occur together with changes in DNA methylation. For example, with increasing age of an organism, gene expression is altered in part due to an aberrant balance between the activity of the HATs and HDACs (Chapter 4). The altered histone acetylation in concert with aberrant DNA methylation can lead to changes in gene expression that contribute to the progression of aging and processes such as neurodegeneration. Moreover, the sirtuins are a large group of protein deacetylases that have been extensively studied during the aging process (Chapter 5). These enzymes regulate lifespan in part through modulating calorie restriction pathways and the sirtuins have therefore been the subject of intense interest for the discovery of prolongevity compounds such as resveratrol. In Chapter 6, the role of chromatin structure in senescent cells is analyzed, especially as it relates to Wnt-signaling. The authors suggest that elevated Wnt-signaling can suppress some aspects of tissue aging through its ability to regulate chromatin structure as cells approach senescence.

One of the most versatile compounds in a cell is S-adenosylmethionine (SAM) that serves as the methyl donor for most of the methyltransferases. In Chapter 7, a role of SAM in aging is proposed whereby deregulation of its metabolism can occur during the aging process through environmental or hereditary factors, folate or dietary deficiency, alcohol abuse and other factors. The collective effects of aberrations in SAM metabolism may lead to cancer, autoimmune disease, neurological disorders and other processes that contribute to longevity and aging.

Therefore, many alterations in DNA methylation and histone modifications occur during the aging process that are expressed at the cellular level and ultimately in aging of the organism itself. Gene-specific alterations occur in DNA methylation and histone modifications as well as genome-wide changes such as the extensively documented decline in DNA methylation of the genome in aging cells, tissues and organisms. These aberrations in epigenetic processes during aging contribute to the development of a number of age-associated diseases and intervention into these age-related epigenetic aberrations through prolongevity compounds that act on DNA methylation or histone modifications may have considerable utility in not only extending longevity, but also in combating the many diseases that accompany the aging process.

Other Epigenetic Processes and Aging

Besides changes in DNA methylation and histone modifications during aging, other epigenetic mediators such as the polycomb group of genes, chromosomal position effects and noncoding RNA also contribute to aging. In Chapter 8, the age-related alterations in the polycomb group proteins that are involved in regulation of a number of loci and thereby influence aging are discussed. The polycomb group proteins can repress loci involved in cell cycle control that facilitates normal cell mitosis. The polycomb group of genes may undergo changes with aging that influence longevity and they have also been proposed to be responsive to environmental signaling as well as calorie restriction. Likewise, DNA rearrangements can occur during aging in response to exogenous stress or other factors leading to a redistribution of epigenetic factors. This redistribution process has been proposed to alter the chromatin environment thereby affecting gene expression through chromosomal position effect (Chapter 9). In fact, a global reprogramming of gene expression associated with position effect mechanisms may contribute to aging as well as the premature aging characteristic of the Werner syndrome.

Noncoding RNAs have recently attracted considerable attention and may have a role in aging and age-dependent diseases (Chapter 10). For example, micro-RNAs are important epigenetic factors that control the expression of numerous genes through a negative regulatory mechanism. The microRNAs can degrade the mRNA of specific genes by binding to their coding regions or they can inhibit translation of genes by binding at the 3'-untranslated region. Recent developments have suggested the existence of longevity microRNAs since these noncoding RNAs control DNA repair, oxidative defense and other processes that are likely important in determining the lifespan of individuals. Slowing or reducing mid-life dysregulation of microRNAs may have an important impact on the molecular degeneration processes that increase especially late in life. Thus, molecular aberrations in non-traditional epigenetic processes such as the polycomb group genes, chromosomal position effects and noncoding RNAs as well as potentially other epigenetic processes may have a major role in aging and age-associated disorders.

Impact of Epigenetics on Aging

Besides the underlying fundamental basis of epigenetics and aging, there is also considerable interest in the impact of these processes. In Chapter 11, the role of epigenetics in controlling telomerase, the enzyme that maintains chromosomal ends, is discussed especially as this pertains to aging. Since the absence of telomerase leads to telomeric attrition in normal cells which contributes to cellular senescence (Lai et al., 2005), the regulation of the major gene of telomerase, *TERT (telomerase reverse transcriptase)*, has been of major interest to biogerontologists. Although the epigenetic control of the *TERT* gene is not yet fully resolved, it is clear that cellular modifications especially in DNA methylation and histone modifications affect the *TERT* promoter region which lead in part to regulation of this important gene in cellular aging. If the regulation of the *TERT* gene can be controlled, this could have a tremendous impact on the aging process. Epigenetic mechanisms are also important in the control of the telomeric ends of chromosomes. For example, age-related changes occur in DNA methylation and histone modifications during cancer development that affect telomere length (Chapter 12). Conversely, telomere shortening during aging may impact the heterochromatic state of many telomeres in the cells and influence gene expression which is often altered in aging cells.

The tumor suppressors impact chromatin structure and epigenetic processes during aging through various mechanisms that affect enzymes such as the HATs and HDACs (Chapter 13). These effects on heterochromatin may lead to the formation of heterochromatic foci that contribute to cellular senescence. These tumor suppressors, such as p53 and Rb, appear to provide a crucial link between aging and cancer not only through tumor suppressive mechanisms, but also through their control of chromatin modifications.

Many of the epigenetic changes that occur during aging are stochastic or dependent upon environmental factors that lead to a phenomenon referred to as epigenetic drift that may have a major impact on the progression of aging (Chapter 14). The loss of DNA methylation during aging is an example of epigenetic drift and changes in histone modifications, such as trimethylation of H4-K20, occur during aging as well. In fact, studies of monozygotic twins have illustrated epigenetic drift during aging where young monozygotic twin pairs had few epigenetic changes while some elderly monozygotic twin pairs had substantial variations in DNA methylation and chromatin modifications of many genes in a number of tissues.

Epigenetic mechanisms also likely impact neurological functions such as long-term memory. The loss of DNMT1 expression during aging could have a major effect on long-term memory formation given that DNMT1 is highly expressed in brain tissue (Chapter 15). Hypomethylation of key genes such as the gene that encodes protein phosphatase 1 (PP1) could lead to an up-regulation of PP1 expression and decrease long-term potentiation involved in the memory process during aging. Therefore, many epigenetic mechanisms are thought to have an impact in the aging process and age-related diseases. The control of the telomerase regulatory gene (*TERT*) and telomere length may be under epigenetic control and the tumor suppressors appear to influence epigenetic processes and aging. Epigenetic drift may play a major role in impacting the expression of aging and epigenetic mechanisms have also been implicated in affecting features of aging such as the decline in long-term memory capacity.

Epigenetics of Age-Related Diseases

Epigenetic processes affect many of the diseases that accompany aging. Chief among these age-related diseases is cancer. It has long been known that alterations in DNA methylation and chromatin contribute to cancer and Chapter 16 details a number of mechanisms involved in this process as well as model diseases for age-related epigenetic changes and cancer such as acute lymphoblastic leukemia. This disease covers the entire lifespan and a number of age-related differences are observed indicating that acute lymphoblastic leukemia may serve as an excellent model to decipher age-related impacts on disease. As mentioned above, epigenetic processes such as DNA methylation play a role in Alzheimer's disease. The decreased DNA methylation status with aging appears to affect several genes related to the expression of Alzheimer's disease such as the *amyloid precursor protein (APP)* gene leading to higher levels of amyloid β protein in Alzheimer's disease (Chapter 17). Other genes important to Alzheimer's disease such as the *presenilin 1 (PS1)* gene may also undergo epigenetic changes that contribute to Alzheimer's disease. DNA methylation has also been proposed to have a major role in the epigenetic regulation of adaptive responses of the immune system and genes important in T-cell differentiation have been found to be differentially expressed through epigenetic processes during aging (Chapter 18). Moreover, age-related demethylation appears to impair immune competence and the frequency of autoimmunity in the elderly and may contribute to diseases such as rheumatoid arthritis.

Epigenetic mechanisms have also been implicated in the premature aging diseases such as the Werner syndrome and progeria. For example, the *WRN* gene that leads to the Werner syndrome is epigenetically silenced in a wide variety of tumor types leading to speculation that epigenetic control of this gene may play a role in the pathogenic features of the Werner syndrome (Chapter 19). Lastly, it is likely that DNA methylation is important in osteoarthritis, a debilitating disease with a strong age-associated expression. The aberrations in DNA methylation that occur with increasing age may affect the expression of a number of genes that contribute to osteoarthritis (Chapter 20). Thus, epigenetic mechanisms appear to not only influence the fundamental aspects of aging, but to also contribute significantly to a number of age-associated diseases such as cancer, Alzheimer's disease, autoimmunity and osteoarthritis as well as potentially premature aging diseases and many other disorders.

Epigenetic Interventions and Aging

Besides learning more about the role of epigenetics in aging and age-related diseases, we are also very interested in elucidating modes through which we can exploit this knowledge to intervene into aging or age-related diseases by modulating epigenetic mechanisms. For instance, modulation of SIRT1, an important sirtuin histone deacetylase involved in the aging process, has been shown to mimic calorie restriction which is the most consistent and currently the most effective mode of influencing longevity (Chapter 21). Additional studies are underway to assess the utility of other members of the sirtuin family in potential intervention into the aging process.

Dietary factors are also receiving considerable attention in terms of their control of epigenetic factors and their influence on aging and age-associated diseases (Chapter 22). Several lines of study are currently underway but especially promising are studies on caloric restriction, epigenetics and aging as mentioned above. Of equal interest are the nutritional modulation of DNA methylation by the green tea component, EGCG (epigallocatechin-3-gallate) and the soybean product, genistein. The use of histone deacetylase inhibitors such as butyrate, diallyl disulfide and sulforaphane found in many foods is also gaining considerable interest in intervening into aging and age-related disorders. Moreover, environmental factors such as stress and toxic metal exposure as well as maternal nurturing can influence epigenetic modifications that change with aging (Chapter 23). It is also likely that environmental factors have a major impact on epigenetic drift that is gaining increasing attention in the field of biogerontology. Therefore, there is optimism that we may eventually be able to intervene into the epigenetic processes so important to aging and the diseases that accompany aging.

Future Directions/Perspectives

With the exponential rise in interest in epigenetics in general and epigenetics as it applies to aging and age-related diseases such as cancer, the question that now comes to the forefront is "Where do we go from here?". Clearly there are many fruitful avenues of investigation that will address not only the role of epigenetic processes in aging mechanisms but also the role of epigenetics in the many disorders associated with aging. Further improvements in high-throughput technologies, bioinformatics, genome-wide analyses of epigenetic patterns, treatment strategies, and new epigenetic modifiers as applied to the epigenetics of aging are several areas identified for future research in aging epigenetics in Chapter 24. The development of new NIH initiatives in these areas will almost certainly stimulate advancements in the study of the epigenetics of aging.

Lastly, in Chapter 25, a pioneer not only in epigenetics but also in aging epigenetics, Robin Holliday, provides numerous interesting perspectives into the epigenetics of aging. He begins with a discussion of the term "epigenetics" and analyzes the multiple causes of aging and the biological reasons for aging. Other topics of considerable interest to aging epigenetics that are covered are the role of development in aging, DNA methylation, cell-signaling, X-chromosome reactivation and epigenetic and non-epigenetic events during aging. His perspectives provide a fascinating conclusion to a topic that is gaining increasing attention among many investigators ranging from basic molecular biologists to physicians.

Conclusion

For quite some time it has been thought that genetic causes for cancer were much more important than epigenetic process that not so long ago were given a relatively minor role in cancer etiology and progression. However, investigations into cancer epigenetics have now overturned this dogma and we are currently in a revolutionary period of understanding cancer epigenetics and its importance. Many now feel that cancer may be more an epigenetic than a genetic disease. We may see a similar future for aging epigenetics. For decades aging has been largely associated with traditional genetic mechanisms. However, the contents of this book suggest that epigenetic processes also play an essential role in aging. It cannot be said at this point whether we will view aging as largely an epigenetic condition as has occurred with cancer, but mounting evidence indicates that this could be the case. Future investigations will reveal the true role of epigenetics in aging but for now it is indeed an exciting and rapidly emerging field that deserves considerable more attention as we attempt to fully understand and assume better control over aging and the many diseases the accompany aging.

References

Liu, L., Wylie, R.C., Andrews, L.G., and Tollefsbol, T.O. 2003. Aging, cancer and nutrition: the DNA methylation connection. Mech. Ageing Dev. 124: 989–98.

Sedivy, J.M., Banumathy, G., and Adams, P.D. 2008. Aging by epigenetics–a consequence of chromatin damage? Exp. Cell Res. 314:1909–1917.

Fraga, M.F. and Esteller, M. 2007. Epigenetics and aging: the targets and the marks. Trends Genet. 23: 413–418.

Lai, S.R., Phipps, S.M., Liu, L., Andrews, L.G., Tollefsbol, T.O. 2005. Epigenetic control of telomerase and modes of telomere maintenance in aging and abnormal systems. Front. Biosci. 10: 1779–1796.

Groth, A., Rocha, W., Verreault, A., Almouzni, G. 2007. Chromatin challenges during DNA replication and repair. Cell 128: 721–733.

Fuks F, Burgers WA, Brehm A, Hughes-Davies L, Kouzarides T. 2000. DNA methyltransferase Dnmt1 associates with histone deacetylase activity. Nat. Genet. 24: 88–91.

Part I
DNA Methylation and Histone Modifications in Aging

Age-Related Genomic Hypomethylation

Igor P. Pogribny and Boris F. Vanyushin

Abstract Aging is a multi-factorial process of the progressive gradual decline of cellular functions with the passage of time. It is clear that aging affects the mammalian epigenome, including hypomethylation of DNA. DNA methylation is a crucial biological process that controls maintenance of genomic integrity and an accurate expression of genetic information. The accurate status of DNA methylation is balanced in mature cells, but with age this balance is strongly shifted in favor of DNA demethylation. Therefore, DNA hypomethylation that occurs during normal aging appears to be a critical risk factor contributing to the development of chronic age-related human pathological states. This review describes the involvement of DNA hypomethylation in the pathogenesis of several major age-related human diseases, including cancer, atherosclerosis, Alzheimer's disease, psychiatric disorders, and autoimmune pathologies.

Introduction

Aging can be defined as a multifactorial process of the accumulation of molecular alterations driven by genetic and epigenetic events in the organism that lead to a loss of phenotypic plasticity over time. Epigenetic modifications, including DNA methylation and histone modification, are essential for normal development and for the maintenance of cellular functions in an adult organism (Allis et al. 2007; Tost 2008). The significance of epigenetic alterations, especially changes in DNA methylation, in the pathogenesis of a number of human diseases has been well established and studied extensively (Egger et al. 2004; Feinberg 2007; Jones and Baylin 2007; Feinberg 2008). Unfortunately, their role in aging has received less attention

---

The views expressed in this chapter do not necessarily represent those of the US Food and Drug Administration.

I.P. Pogribny (✉)
Division of Biochemical Toxicology, National Center for Toxicological Research, Jefferson, AR 72079, USA
e-mail: igor.pogribny@fda.hhs.gov

T.O. Tollefsbol (ed.), *Epigenetics of Aging*, DOI 10.1007/978-1-4419-0639-7_2,

(Fraga and Esteller 2007; Feinberg 2008). There is, however, growing evidence that DNA methylation changes, both hypo- and hypermethylation of DNA, arise with age (Holliday 1985, 1987; Ahuja and Issa 2000; Issa 2003; Bandyopadhyay and Medrano 2003; Fraga and Esteller 2007).

Aging and Genomic Hypomethylation

DNA hypomethylation signifies one of the major DNA methylation states and in most cases refers to a relative situation that indicates a decrease from the "normal" methylation level (Dunn 2003). The first experimental evidence of the age-dependent loss of genomic methylation was provided by Berdishev et al. (1967) and Vanyushin et al. (1969), who found that the content of 5-methylcytosine (5meC) in DNA isolated from the various organs of humpback salmon was significantly decreased during ontogenesis (Vanyushin et al. 1970). These findings were confirmed in later studies that documented age-dependent DNA hypomethylation in many mammalian tissues (Vanyushin et al. 1973; Romanov and Vanyushin 1981). The maximal amount of 5meC was observed in DNA isolated from tissues of embryos and newborn animals and gradually decreased upon aging. For instance, Fig. 1 shows the age-associated changes in the DNA methylation status in the brain, heart, spleen, and liver tissues of rats. Interestingly, the brain and heart tissues are characterized by the most notable loss of genomic methylation, especially in old animals.

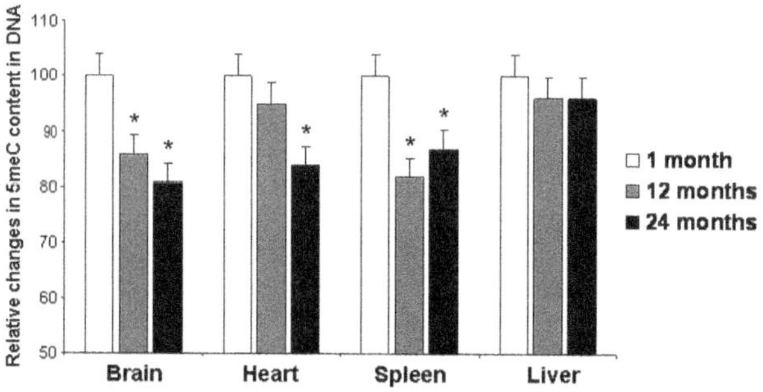

Fig. 1 Age-related DNA methylation changes in the brain, heart, spleen, and liver tissues of rats. The 5meC contents in DNA of 1-month-old rats are presented as 100%
* Significantly different from 1-month-old rats.

Further evidence confirming the loss of DNA methylation during the aging process was obtained in in vitro studies that demonstrated that a marked decrease in the 5meC content in DNA is associated with a number of cell divisions in normal diploid mouse, hamster, and human cells, in contrast to the immortal cell lines (Wilson and Jones 1983). Based on these findings, it has been proposed by Holliday (1985) that changes in DNA methylation patterns may have significance in the aging process. Since that time, a number of in vivo and in vitro findings have established that normal aging mammalian cells (mice, rats, and humans) show a progressive loss of 5meC content in DNA (Wilson et al. 1987; Singhal et al. 1987; Ono et al. 1993; Drinkwater et al. 1989; Richardson 2002). Furthermore, the measurement of the 5meC content in DNA isolated from the tissues of various ages of mice characterized by different durations of natural life has demonstrated that the rate of loss of 5meC content in DNA is inversely related to life span (Wilson et al. 1987).

Another piece of evidence clearly indicating the intimate link between DNA hypomethylation and the aging process was obtained in the studies on effects of the demethylating agent 5-aza-2′-deoxycytidine (5-aza-dC) on the life span of normal MRC-5 human diploid cells in vitro. Treatment of human fibroblasts with 5-aza-dC induced demethylation of DNA and substantially shortened the cell life span (Holliday 1986; Fairweather et al. 1987).

It is well known that the majority of cytosine methylation resides within GC-rich repetitive DNA sequences (Gama-Sosa et al. 1983b; Yoder et al. 1997; Rollins et al. 2006). Because of this, a progressive loss of DNA methylation during the aging process is primarily associated with demethylation of those DNA repetitive elements (Mays-Hoopes et al. 1983; Ono et al. 1989; Hornsby et al. 1992; Barbot et al. 2002). However, in addition to demethylation of repetitive DNA sequences, the methylation status of numerous single-copy genes decreased with age in the brain, liver, spleen, mammary glands, and other tissues (Ono et al. 1989a; Yenbutr et al. 1998; Zhang et al. 2002; Ravindran and Ticku 2005). Specifically, age-related hypomethylation and activation of *c-myc* and *NR2B* genes were detected in the mouse spleen and brain, respectively, as well as an increased loss of estrogen receptor I gene methylation in normal rat mammary glands.

Mechanism of DNA Hypomethylation During Aging

The mechanism of DNA hypomethylation is still unclear, and there is very likely no universal mechanism that ascribes demethylation of DNA alone. However, it is well established that upon aging, several factors including the activity and expression of DNA methyltransferases, the status of one-carbon metabolism, and the integrity of the genome may trigger and contribute to the loss of genomic methylation.

DNA Methyltrasferases and Age-Dependent DNA Hypomethylation

The addition of methyl groups from the universal methyl donor, *S*-adenosyl-l-methionine (SAM), to cytosines at CpG sites in mammalian cells is catalyzed by the enzymatic activity of the DNA methyltransferase (DNMT) family, which includes the maintenance DNA methyltransferase DNMT1 and the de novo DNA methyltransferases DNMT3A and DNMT3B (Bestor 2000; Goll and Bestor 2005). DNA methylation is initiated and established during embryonic development by means of DNMT3A and DNMT3B (Okano et al. 1999) in the presence of their regulator DNMT3L (Gowher et al. 2005) and maintained during DNA replication by a complex cooperative interplay of the maintenance DNMT1 with methyl-CpG-binding protein MeCP2 (Kimura and Shiota 2003) and the ubiquitin-like, containing PHD and RING finger domains 1 (UHRF1) protein (Arita et al. 2008; Avvakumov et al. 2008; Hashimoto et al. 2008). Therefore, the function of DNMT1 is considered essential for maintaining genomic methylation patterns. Indeed, a large body of evidence clearly demonstrates the importance of DNMT1's proper function in the maintenance of the faithful status of DNA methylation (Li et al. 1992; Loriot et al. 2006). For instance, the reduction of *Dnmt1* expression by gene targeting to 10% in mice caused a significant hypomethylation of centromeric and endogenous retroviral intracesternal A particle (IAP) repetitive sequences (Gaudet et al. 2003). Similarly, the loss of DNA methyltransferase through the inhibition of its activity by 5-aza-dC (Creusot et al. 1982), homocysteine, or its metabolite *S*-adenosyl-l-homocysteine (SAH) resulted in rapid demethylation of DNA (Yi et al. 2000).

The results of several studies have demonstrated that the activity of DNMT1 substantially decreases with aging (Lopatina et al. 2002; Casillas et al. 2003). In light of this, it is believed that genomic hypomethylation during aging is a result of passive demethylation, especially of highly methylated GC-rich DNA domains caused by the functional inability of DNMT1 to maintain a normal DNA methylation pattern. This observation was further supported by the results of in vivo and in vitro studies demonstrating that expression of *Dnmt1* and *Dnmt3a* decreased with aging (Zhang, et al. 2002; Ray et al. 2006) and by the fact that the deletion of *Dnmt3a* gene substantially shortens the mouse life span (Nguyen et al. 2007). Proper DNMT activity is also crucial for the maintenance of the normal methylation pattern and the function of single-copy genes during aging. For example, methylation and activation of the killer Ig-like receptor (*KIR2DL4*) gene, whose upregulation contributes to age-related diseases, is sensitive to the inhibition of DNMT activity (Li et al. 2008).

The accurate maintenance of DNA methylation, in addition to DNMTs, depends on function and cooperation with other factors. Specifically, recent studies have demonstrated that lymphocyte-specific helicase (LSH) and proliferation-associated SNF-2-like gene (PASG) are major regulators of DNA methylation (Dennis et al. 2001; Sun et al. 2004; Myant and Stancheva 2008). The disruption of LSH or PASG not only causes genomic hypomethylation in mice but also leads to their premature aging (Sun et al. 2004; Sun and Arceci 2005).

However, several lines of evidence have demonstrated that loss of DNA methylation is not always associated with altered DNMT function and may be independent of DNMT levels (Arnaud et al. 1985; Ehrlich et al. 2006). This may be explained by the fact that DNMT1 is incapable by itself of maintaining the normal level of DNA methylation status, especially of GC-rich domains when the DNA methylation status is compromised (Liang et al. 2002).

One-Carbon Metabolism and Age-Dependent DNA Hypomethylation

One-carbon metabolism, a network of cellular interrelated biochemical reactions, is essential for de novo nucleotide biosynthesis, methionine biosynthesis, and cellular methylation reactions (Mason 2003). The methyl groups that are needed for all cellular biological methylation reactions, including DNA methylation, are acquired from SAM, the primary universal donor of methyl groups in mammals derived from methionine in the one-carbon metabolic pathway (Chiang et al. 1996). This indispensably connects the status of epigenetic modifications to the functioning of the one-carbon metabolic pathway. Indeed, it is well established that the level of cellular one-carbon metabolism has a great impact on the DNA methylation status (Niculescu and Zeisel 2002; Ulrey et al. 2005).

During recent years, interest in the one-carbon metabolism noticeably expanded, driven by the growing evidence of the significance of its improper functioning upon aging and in the pathogenesis of a number of age-related human pathologies (Selhub 2002; Morris 2003; Refsum et al. 2006; Schulz 2007). The fundamental effect of a compromised one-carbon metabolic pathway is hypomethylation of DNA that is associated with an accumulation of homocysteine and SAH, two major inhibitors of cellular methylation reactions (Yi et al. 2000; James et al. 2002). There are two types of risk factors that may compromise the normal functioning of the one-carbon metabolic pathway and subsequently alter the cellular epigenetic profile. The first group consists of nonmodifiable genetic risk factors, such as genetic variations in genes encoding enzymes involved in the cellular one-carbon metabolism (Rozen 2000; Chen et al. 2001; Refsum et al. 2006). The second group consists of potentially modifiable factors, specifically essential nutrients (methionine, choline, folic acid, and vitamin B_{12}) involved in the metabolism of methyl groups (Oomen et al. 2005; Nijhout et al. 2006), and the supply and the availability of methyl groups are considered the most significant factors affecting DNA methylation (Chiang et al. 1996; Niculescu and Zeisel 2002; Ulrey et al. 2005). Therefore, the proper maintenance of the one-carbon metabolism is critical for healthy aging and may prevent or slow the development of age-related pathologies (Kuo et al. 2005). However, a number of studies have demonstrated that loss of DNA methylation may occur in the environment of the uncompromised status of one-carbon metabolism and the normal SAM and SAH levels (Arnaud et al. 1985; Seivwright et al. 1993; Pogribny et al. 2008).

DNA Integrity and Age-Dependent DNA Hypomethylation

The integrity of the genome is another critical factor that impacts the normal status of DNA methylation. Every living organism is exposed to various genomic insults on a daily basis caused by many endogenous and exogenous factors (Chen et al. 2007; Vijg 2008). The results of several studies have demonstrated that the presence of unrepaired lesions in DNA substantially alters the methylation capacity of DNA methyltransferases, leading to DNA hypomethylation (Cerda and Weitzman 1997; Valinluck and Sowers 2007; Pogribny et al. 2008). The significance of this mechanism of DNA hypomethylation progressively increases with age due to an age-dependent decrease in DNA repair proficiency (Gorbunova et al. 2007; Spry et al. 2007; Bertram and Hass 2008).

The above-provided evidence emphasized the importance and interdependence between the function of DNA methyltransferases, the status of one-carbon metabolism, and the integrity of the genome in the maintenance of a normal level of DNA methylation. Thus, disruption of any of these mechanisms during aging may result in a subsequent loss of DNA methylation.

Age-Related Diseases and DNA Hypomethylation

The loss of DNA methylation upon aging correlates with a number of adverse outcomes, among which the link of DNA hypomethylation to the development of age-related pathologies such as cancer, atherosclerosis, neurodegeneration, and autoimmune diseases is the most significant.

Cancer

Advancing age is the most potent risk factor for human tumors, especially for epithelial breast, prostate, lung, colon, and basal and squamous cell carcinomas (DePinho 2000). Importantly, global DNA hypomethylation is a central feature, one of the most common molecular alterations, and the first epigenetic abnormality identified in human cancer cells (Gama-Sosa et al. 1983; Feinberg and Vogelstein 1983; Flatau et al. 1983; Bedford and van Helden 1987; Feinberg et al. 1988; Fraga et al. 2004; Wilson et al. 2007; Agrawal et al. 2007). A substantial loss of DNA methylation has been demonstrated at very early stages of human cancer and even at premalignant stages of cancer development. Because of that, the loss of DNA methylation has been suggested to be an important event in carcinogenesis (Feinberg and Tycko 2004; Hoffman and Schulz 2005; Ehrlich 2006; Wilson et al. 2007). Furthermore, recent evidence has documented a causative role of DNA hypomethylation in cancer development (Gaudet et al. 2003).

The mechanistic significance of DNA hypomethylation in cancer is associated with the instability of the genome induced by the loss of methylation. Several lines of evidence have demonstrated that activation and transposition of repetitive DNA sequences (Howard et al. 2008), increased mutation rate (Chen et al. 1998), and chromosomal aberration (Eden et al. 2003; Karpf and Matsui 2005; Rodriguez et al. 2006) are causally linked to hypomethylation of DNA. The fundamental role of these events, as an integral part of neoplastic cell transformation, is now commonly accepted (Coleman and Tsongalis 2006).

In addition to global genomic hypomethylation, cancer-linked promoter hypomethylation has been shown in many primary human cancers (Sato et al. 2003; Laner et al. 2005; Wu et al. 2005; Grunau et al. 2008). More importantly, detection of cancer-linked genome- and gene-specific hypomethylation can be used as sensitive molecular markers for cancer detection (Laner et al. 2005; Guerrero-Preston et al. 2007; Moore et al. 2008).

Atherosclerosis

Atherosclerosis and its complications are a major cause of mortality, morbidity, and disability in developed Western countries (Worthley et al. 2001; Mensah and Brown 2007). The disease is characterized by infiltration of lipid particles into the arterial wall, accompanied by the recruitment of inflammatory and immune cells, migration and proliferation of smooth muscle cells, synthesis of the extracellular matrix, and development of fibrocellular lesions (Raines and Ross 1997; Ross 1999). In contrast to cancer research, the involvement of DNA hypomethylation in the context of atherosclerosis was first formulated by Newman (1999), which is only less than a decade ago. The hypothesis was based on evidence suggesting that elevated plasma homocysteine is a risk factor for atherosclerosis (Nehler et al. 1997; Fallest-Strobl et al. 1997) and the fact that homocysteine and SAH efficiently inhibit DNA methyltransferases, causing hypomethylation of DNA (Yi et al. 2000). Currently, the significance of the loss of DNA methylation in atherosclerosis is widely documented (Dong et al. 2002; Hiltunen and Ylä-Herttuala 2003; Zaina et al. 2005). Substantial global DNA hypomethylation has been found in peripheral white blood cells (Castro et al. 2003), smooth muscle cells (Hiltunen and Ylä-Herttuala 2003; Yideng et al. 2007), and atherosclerotic lesions (Hiltunen et al. 2002) in patients with atherosclerosis. These correlative studies, without undermining the underlying role of homocysteine as a risk factor for atherosclerosis, suggested that hypomethylation during atherosclerosis may be a consequential secondary passive event induced by elevated homocysteine. However, the result of a recent study has clearly demonstrated the significance of DNA hypomethylation in the pathogenesis of atherosclerosis and in the susceptibility to the disease. This was evidenced by the occurrence of global DNA hypomethylation prior to the formation of atherosclerotic lesions in genetically atherosclerosis-prone *Apoe*$^{-/-}$ mice (Lund et al. 2004). Furthermore, transcriptional upregulation of *5-ipoxygenase* and *15-lypooxygenase* genes, key enzymes implicated in the pathogenesis of atherosclerosis (Zhao and Funk 2004), is mediated by promoter hypomethylation (Uhl et al. 2002; Liu et al. 2004).

Alzheimer's Disease

Alzheimer's disease is an age-related progressive neurodegenerative disorder characterized by the presence of amyloid plaques and intracellular tangles in the brain (Hardy and Selkoe 2002; Pastorino and Lu 2006). The biogenesis and accumulation of amyloid plaques, which primarily consist of 40- to 42-residue β-amyloid peptides (Aβ40 and Aβ42) derived from amyloid precursor protein (APP) mediated by sequential proteolic processing by β-secretase (BACE1) and γ-secretase complex (Wolfe 2007; Verdile et al. 2007; Das 2008), is a key event in Alzheimer's disease. The fact that the brain along with the heart is characterized by the most pronounced age-associated loss of DNA methylation (Fig. 1) suggests that DNA hypomethylation in the brain upon aging may have significance in the pathogenesis of the disease. Indeed, this suggestion is supported by findings showing hypomethylation of the *APP* gene in an Alzheimer's brain (West et al. 1995) and by substantial age-dependent *APP* promoter demethylation in the human cortex (Tohgi et al. 1999). Specifically, the frequency of methylation of cytosine residues at −207, −204, −200, and −182 in the *APP* promoter region in subjects younger than 70 years was substantially greater (55%) compared to subjects older than 70 years (5%) (Tohgi et al. 1999). Additionally, expression of the *presenilin 1* (*PS1*) gene, a key component of the γ-secretase complex, is regulated by methylation (Fuso et al. 2005, 2007). In light of these considerations, we propose the following model of pathogenesis of Alzheimer's disease driven by the age-related DNA hypomethylation events (Fig. 2). First, the age-related hypomethylation of *APP* promoter provokes an overexpression of *APP* gene, leading to greater levels of APP. Second, the age-associated hypomethylation and upregulation of the *PS1* gene consequently induces the activity of γ-secretase complex and stimulates the proteolytic cleavage of APP, leading to the accumulation of Aβ40 and Aβ42. Importantly, this model allows us to bring together the two most widely accepted, the amyloid and the presilin, hypotheses of Alzheimer's disease into one mechanism.

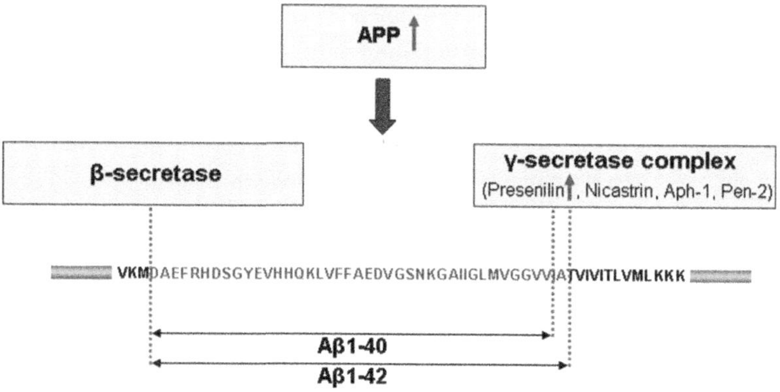

Fig. 2 Proteolytic processing of the amyloid precursor protein (APP) driven by DNA hypomethylation events. The age-related hypomethylation of the *APP* and *PS1* promoters (*arrows*) provokes an overexpression of *APP* and *PS1* genes, leading to greater levels of APP and enhanced proteolytic cleavage of APP, respectively, resulting in the accumulation of Aβ40 and Aβ42. Interestingly, genes encoding nicastrin, Aph-1, and Pen-2, three other components of γ-secretase complex, contain CpG island according to http://cpgislands.usc.edu

Autoimmunity

The age-associated changes in DNA methylation, in particular, DNA hypomethylation, have emerged as an attractive mechanistic link between aging and autoimmunity (Prelog 2006; Yung and Julius 2008). Evidence from in vitro and in vivo models has indicated that DNA hypomethylation might be implicated in the development of autoimmunity (Richardson 2002; Richardson 2003a). The treatment of T cells with 5-aza-dC demethylated DNA induced autoreactivity that correlated with overexpression of lymphocyte function-associated antigen-1 (LFA-1) (Richardson 2003a). Likewise, the T cells from patients with active lupus are characterized by decreased DNA methylation and increased expression of LFA-1. It has been demonstrated that overexpression of LFA-1 is mediated by promoter hypomethylation-induced upregulation of *ITGAL (CD11a)* gene, which encodes CD11a – a subunit of LFA-1 (Lu et al. 2002). Interestingly, the *ITGAL* promoter hypomethylation and increased expression of LFA-1 has been documented upon aging (Zhang et al. 2002). Additionally, the results of recent studies have documented that the age-related overexpression of *perforin* and *KIR2DL4* genes in T cells is associated with promoter demethylation (Kaplan et al. 2004; Li et al. 2008). The upregulation of *perforin* and *KIR2DL4* genes has been implicated in a number of autoimmune diseases and, therefore, may explain the increasing incidence of these inflammatory diseases with age (Goronzy and Weyand 2003).

Age-Related Macular Degeneration

Age-related macular degeneration (AMD) is the leading cause of an irreversible blindness in people 50 years and older and has no effective cure (Jager et al. 2008). The disease is characterized by the focal deposition of acellular polymorphous debris, called drusen, between retinal pigment epithelium (RPE) and Bruch's membrane (Jager et al. 2008). One of the major proteins accumulated in drusen is clusterin, whose expression is regulated by promoter methylation. Recent findings showing that treatment of the RPE cell line, ARPE-19, with 5-aza-dC potentiates the clusterin induction and secretion to medium provided an additional evidence regarding the involvement of DNA hypomethylation in the pathogenesis of AMD (Suuronen et al. 2007).

Concluding Remarks

The mechanism of aging is a complex multifactorial process characterized by many biologically significant changes. One of these changes is the alteration of DNA methylation. It is clear that aging can impact DNA methylation, including hypomethylation of DNA. However, the loss of DNA methylation can also have an impact on aging (Richardson 2003b). Thus, aging and DNA undermethylation go hand in hand.

DNA methylation is a crucial biological process that programs an accurate expression of genetic information in vertebrates. Recent evidence indicates that DNA methylation in cells is very dynamic and depends on a combination and precise balance of methylation and demethylation reactions (Métivier et al. 2008; Kangaspeska et al. 2008). These processes are nicely balanced in mature cells, but with age this balance is strongly shifted in favor of DNA undermethylaion and demethylation. This may be due to the passive age-dependent decline in DNA methylation, the increase of the active DNA demethylation, or both. In aging cells, DNA undermethylation may be essentially linked to the age-dependent decrease in the activity of DNA methyltransferases and/or the decreased methylation capacity of DNA methyltransferases caused by compromised genome integrity. Additionally, global loss of DNA methylation may be a result of an active demethylation caused by the age-specific activation of specific enzymes, demethylases, described recently (Hamm et al. 2008).

Therefore, epigenetic changes, including DNA hypomethylation, that occur during normal aging are critical risk factors contributing to the development of chronic age-related pathological states. As long as global DNA demethylation with age is strong enough, it seems that it is mostly due to demethylation of highly repetitive DNA sequences that are highly methylated in young cells. This may compromise the integrity of genome via alteration of chromatin architecture and function. It has been suggested that the degree of global DNA methylation may serve as a sort of biological clock indicating the life span. The important unanswered questions are (a) how to slow this clock down or even, possibly, turn it back? and (b) how can we restore the DNA methylation profile of aging cells to that of young or mature cells and prevent significant age demethylation? These are very complicated tasks that rely on our ability to learn (a) how to modulate chromatin structure and make it available for respective DNA methylation and unavailable for its demethylation and (b) how to maintain the proper DNA methyltransferase and inhibit demethylase activities. However, considering the fact that the remarkable future of epigenetic abnormalities is their potential reversibility, it is clear that correction of epigenetic abnormalities, including hypomethylation of DNA, is a promising avenue to prevent, to some extent, the premature aging and the development of age-related diseases.

References

Agrawal, A., Murphy, R.F., and Agrawal, D.K. 2007. DNA methylation in breast and colorectal cancers. Mod. Pathol. 20:711–721.

Ahuja, N., and Issa, J.P. 2000. Aging, methylation and cancer. Histol. Histopathol. 15:835–842.

Allis, C.D., Jenuwein, T., and Reinberg, D. 2007. Epigenetics. Cold Spring Harbor: Cold Spring Harbor Laboratory Press.

Arita, K., Ariyoshi, M., Tochio, H., Nakamura, Y., and Shirakawa, M. 2008. Recognition of hemimethylated DNA by the SRA protein UHRF1 by a base flipping mechanism. Nature 455: 818–821.

Arnaud, M., Dante, R., and Niveleau, A. 1985. DNA methyltransferases in normal and avian sarcoma virus-transformed rat cells. Quantitation of 5-methyldeoxycytidine in DNA and enzyme kinetics study. Biochim. Biophys. Acta 826:108–112.

Avvakumov, G.V., Walker, J.R., Xue, S., Li, Y., Duan, S., Bronner, C., Arrowsmith, C.H., and Dhe-Paganon, S. 2008. Structural basis for recognition of hemimethylated DNA by the SRA domain of human UHRF1. Nature 455:822–825.

Bandyopadhyay, D., and Medrano, E.E. 2003. The emerging role of epigenetics in cellular and organismal aging. Exp. Gerontol. 38:1299–1307.

Barbot, W., Dupressoir, A., Lazar, V., and Heidmann, T. 2002. Epigenetic regulation of an IAP retrotransposon in the aging mouse: progressive demethylation and de-silencing of the element by its repetitive induction. Nucleic Acids Res. 30:2365–2373.

Bedford, M.T., and van Helden, P.D. 1987. Hypomethylation of DNA in pathological conditions of the human prostate. Cancer Res. 47:5274–5276.

Berdishev, G.D., Korotaev, G.K., Bojarskikh, G.V., and Vanyushin, B.F. 1967. Nucleotide composition of DNA and RNA from somatic tissues of humpback salmon and its changes during spawning. Biochemistry (Mosc.) 32:988–993.

Bertram, C., and Hass, R. 2008. Cellular responses to reactive oxygen species-induced DNA damage and aging. Biol. Chem. 389:211–220.

Bestor, T.H. 2000. The DNA methyltransferases of mammals. Hum. Mol. Genet. 9:2395–2402.

Casillas, M.A., Lopatina, N., Andrews, L.G., and Tollefsbol, T.O. 2003. Transcriptional control of the DNA methyltransferases is altered in aging and neoplastically-transformed human fibroblasts. Mol. Cell. Biochem. 252:33–43.

Castro, R., Rivera, I., Struys, E.A., Jansen, E.E.W., Ravasco, P., Camilo, M.E., Blom, H.J., Jakobs, C., and de Almeida, I.T. 2003. Increased homocysteine and S-adenosylhomocysteine concentrations and DNA hypomethylation in vascular disease. Clin. Chem. 49:1292–1296.

Cerda, S., and Weitzman, S.A. 1997. Influence of oxygen radical injury on DNA methylation. Mutat. Res. 386:141–152.

Chen, J.H., Hales, C.N., and Ozanne, S.E. 2007. DNA damage, cellular senescence and organismal ageing: causal or correlative? Nucleic Acids Res. 35:7417–7428.

Chen, R.Z., Pettersson, U., Beard, C., Jackson-Grusby, L., and Jaenisch, R. 1998. DNA hypomethylation leads to elevated mutation rates. Nature 395:89–93.

Chen, Z., Karaplis, A.C., Ackerman, S.L., Pogribny, I.P., Melnyk, S., Lussier-Cacan, S., Chen, M.F., Pai, A., John, S.W., Smith, R.S., Bottiglieri, T., Bagley, P., Selhub, J., Rudnicki, M.A., James, S.J., and Rozen, R.A. 2001. Mice deficient in methylenetetrahydrofolate reductase exhibit hyperhomocysteinemia and decreased methylation capacity, with neuropathology and aortic lipid deposition. Hum. Mol. Genet. 10:433–443.

Chiang, P.K., Gordon, R.K., Tal, J., Zeng, G.C., Doctor, B.P., Pardhasaradhi, K., and McCann, P.P. 1996. S-Adenosylmethionine and methylation. FASEB J. 10:471–480.

Coleman, W.B., and Tsongalis, G.J. 2006. Molecular mechanisms of human carcinogenesis. EXS. 96:321–349.

Creusot, F., Acs, G., and Christman, J.K. 1982. Inhibition of DNA methyltransferase and induction of Friend erythroleukemia cell differentiation by 5-azacytidine and 5-aza-2′-deoxycytidine. J. Biol. Chem. 257:2041–2048.

Das, H.K. 2008. Transcriptional regulation of the presenilin-1 gene: implication in Alzheimer's disease. Front. Biosci. 13:822–832.

Dennis, K., Fan, T., Geiman T., Yan, Q., and Muegge, K. 2001. Lsh, a member of the SNF2 family, is required for genome-wide methylation. Genes Dev. 15:2940–2944.

DePinho, R.A. 2000. The age of cancer. Nature 408:248–254.

Dong, C., Yoon, W., and Goldschmidt-Clermont, J. 2002. DNA methylation and atherosclerosis. J. Nutr. 132:2406S–2409S.

Drinkwater, R.D., Blake, T.J., Morley, A.A., and Turner, D.R. 1989. Human lymphocytes aged in vivo have reduced levels of methylation in transcriptionally active and inactive DNA. Mutat. Res. 219:29–37.

Dunn, B.K. 2003. Hypomethylation: one side of a larger picture. Ann. NY. Acad. Sci. 983:28–42.
Eden, A., Gaudet, F., Waghmare, A., and Jaenisch, R. 2003. Chromosomal instability and tumors promoted by DNA hypomethylation. Science 300:455.
Egger, G., Liang, G., Aparicio, A., and Jones, P.A. 2004. Epigenetics in human disease and prospects for epigenetic therapy. Nature 429:457–463.
Ehrlich, M. 2006. Cancer-linked DNA hypomethylation and its relationship to hypermethylation. Curr. Top. Microbiol. Immunol. 310:251–274.
Ehrlich, M., Woods, C.B., Yu, M.C., Dubeau, L., Yang, F., Campan, M., Weisenberger, D.J., Long, T., Youn, B., Fiala, E.S., and Laird, P.W. 2006. Quantitative analysis of associations between DNA hypermethylation, hypomethylation, and DNMT RNA levels in ovarian tumors. Oncogene 25:2636–2645.
Fairweather, D.S., Fox, M., and Margison, G.P. 1987. The in vitro lifespan of MRC-5 cells is shortened by 5-azacytidine-induced demethylation. Exp. Cell Res. 168:153–159.
Fallest-Strobl, P.C., Koch, D.D., Stein, J.H., and McBribe, P.E. 1997. Homocysteine: a new risk factor for atherosclerosis. Am. Fam. Physician. 56:1607–1612.
Feinberg, A.P. 2007. Phenotypic plasticity and the epigenetics of human disease. Nature 447: 433–440.
Feinberg, A.P., and Vogelstein, B. 1983. Hypomethylation distinguishes genes of some human cancers from their normal counterparts. Nature 301:89–92.
Feinberg, A.P., Gehrke, C.W., Kuo, K.C., and Ehrlich, M. 1988. Reduced genomic 5-methyl-cytosine in human colonic neoplasia. Cancer Res. 48:1159–1161.
Feinberg, A.P., and Tycko, B. 2004. The history of cancer epigenetics. Nat. Rev. Cancer 4:143–153.
Feinberg, A.P. 2008. Epigenetics at the epicenter of modern medicine. JAMA 2999:1345–1350.
Flatau, E., Bogenmann, E., and Jones, P.A. 1983. Variable 5-methylcytosine levels in human tumor cell lines and fresh pediatric tumor explants. Cancer Res. 43:4901–4905.
Fraga, M.F., Herranz, M., Espada, J., Ballestar, E., Paz, M.F., Ropero, S., Erkek, E., Bozdogan, O., Peinado, H., Niveleau, A., Mao, J.H., Balmain, A., Cano, A., and Esteller, M. 2004. A mouse skin multistage carcinogenesis model reflects the aberrant DNA methylation patterns of human cancers. Cancer Res. 64:5527–5534.
Fraga, M.A., and Esteller, M. 2007. Epigenetics and aging: the targets and the marks. Trends Genet. 23:413–418.
Fuso, A., Seminara, L., Cavallaro, R.A., D'Anselmi, F., and Scarpa, S. 2005. S-adenosylmethionine/homocysteine cycle alterations modify DNA methylation status with consequent deregulation of PS1 and BACE and beta-amyloid production. Mol. Cell. Neurosci. 28:195–204.
Fuso, A., Cavallaro, R.A., Zampelli, A., D'Anselmi, F., Piscopo, P., Confaloni, A., and Scarpa, S. 2007. gamma-Secretase is differentially modulated by alterations of homocysteine cycle in neuroblastoma and glioblastoma cells. J. Alzheimer's Dis. 11:275–290.
Gama-Sosa, M.A., Slagel, V.A., Trewyn, R.W., Oxenhandler, R., Kuo, K.C., Gehrke, C.M., and Erhlich, M. 1983a. The 5-methylcytosine content of DNA from human tumors. Nucleic Acids Res. 11:6883–6894.
Gama-Sosa, M.A., Wang, R.Y., Kuo, K.C., Gehrke, C.W., and Ehrlich, M. 1983b. The 5-methylcytosine content of highly repeated sequences in human DNA. Nucleic Acids Res. 11:3087–3095.
Gaudet, F., Hodgson, J.G., Eden, A., Jackson-Grusby, L., Dausman, J., Gray, J.W., Leonhardt, H., and Jaenisch, R. 2003. Induction of tumors in mice by genomic hypomethylation. Science 300:489–492.
Goll, M.G., and Bestor, T.H. 2005. Eukaryotic cytosine methyltransferases. Annu. Rev. Biochem. 74:481–514.
Gorbunova, V., Seluanov, A., Mao, Z., and Hine, C. 2007. Changes in DNA repair during aging. Nucleic Acids Res. 35:7466–7474.
Goronzy, J.J., and Weyand, C.M. 2003. Aging, autoimmunity and arthritis: T-cell senescence and contraction T-cell repertoire diversity – catalysts of autoimmunity and chronic inflammation. Arthritis Res. Ther. 5:225–234.

Gowher, H., Liebert, K., Hermann, A., Xu, G., and Jeltsch, A. 2005. Mechanism of stimulation of catalytic activity of Dnmt3A and Dnmt3B DNA-(cytosine-C5)-methyltransferases by Dnmt3L. J. Biol. Chem. 280:13341–13348.

Grunau, C., Brun, M.E., Rivals, I., Hindermann, W., Favre-Mercuret, M., Granier, G., and De Sario, A. 2008. BAGE hypomethylation, a new epigenetic biomarker for colon cancer detection. Cancer Epidemiol. Biomarkers Prev. 17:1374–1379.

Guerrero-Preston, R., Santella, R.M., Blanco, A., Desai, M., Berdasco, M., and Fraga, M. Global DNA hypomethylation in liver cancer cases and controls: a phase I preclinical biomarker development study. Epigenetics 2:223–226.

Hamm, S., Just, G., Lacoste, N., Moitessier, N., Szyf, M., and Mamer, O. 2008. On the mechanism of demethylation of 5-methylcytosine in DNA. Bioorg. Med. Chem. Lett. 18: 1046–1049.

Hardy, J., and Selkoe, D.J. 2002. The amyloid hypothesis of Alzheimer's disease: progress and problems on the road to therapeutics. Science 297:353–356.

Hashimoto, H., Horton, J.R., Zhang, X., Bostick, M., Jacobsen, S.E., and Cheng, X. 2008. The SRA domain of UHRF1 flips 5-methylcytosine out of the DNA helix. Nature 455:826–829.

Hiltunen, M.O., Turunen, M.P., Häkkinen, T.P., Rutanen, J., Hedman, M., Mäkinen, K., Turunen, A.M., Aalto-Setälä, K., and Ylä-Herttuala, S. 2002. DNA hypomethylation and methyltransferase expression in atherosclerotic lesions. Vasc. Med. 7:5–11.

Hiltunen, M.O., and Ylä-Herttuala, S. 2003. DNA methylation, smooth muscle cells, and atherogenesis. Arterioscler. Thromb. Vasc. Biol. 23:1750–1753.

Hoffmann, M.J., and Schulz, W.A. 2005. Causes and consequences of DNA hypomethylation in human cancer. Biochem. Cell. Biol. 83:296–321.

Holliday, R. 1985. The significance of DNA methylation in cellular aging. Basic Life Sci. 35: 269–283.

Holliday, R. 1986. Strong effects of 5-azacytidine on the in vitro lifespan of human diploid fibroblasts. Exp. Cell Res. 166:543–552.

Holliday, R. 1987. The inheritance of epigenetic defects. Science 238:163–170.

Hornsby, P.J., Yang, L., and Gunter, L.E. 1992. Demethylation of satellite I DNA during senescence of bovine adrenocortical cells in culture. Mutat. Res. 275:13–19.

Howard, G., Eiges, R., Gaudet, F., Jaenisch, R., and Eden, A. 2008. Activation and transposition of endogenous retroviral elements in hypomethylation induced tumors in mice. Oncogene 27: 404–408.

Issa, J.P. 2003. Age-related epigenetic changes and the immune system. Clin. Immunol. 109: 103–108.

Jager, R.D., Mieler W.F., and Miller, J.W. 2008. Age-related macular degeneration. N. Engl. J. Med. 358:2606–2617.

James, S.J., Melnyk, S., Pogribna, M., Pogribny, I.P., and Caudill, M.A. 2002. Elevation of S-adenosylhomocysteine and DNA hypomethylation: potential epigenetic mechanism for homocysteine-related pathology. J. Nutr. 132:2361S–2366S.

Jones, P.A., and Baylin, S.B. 2007. The epigenomics of cancer. Cell 128:683–692.

Kangaspeska, S., Stride, B., Métivier, R., Polycarpou-Schwarz, M., Ibberson, D., Carmouche, R.P., Benes, V., Gannon, F., and Reid, G. 2008. Transient cyclical methylation of promoter DNA. Nature 452:112–115.

Kaplan, M.J., Lu, Q., Wu, A., Attwood, J., and Richardson, B. 2004. Demethylation of promoter regulatory elements contributes to perforin overexpression in $CD4^+$ lupus T cells. J. Immunol. 172:3652–3661.

Karpf, A.R., and Matsui, S. 2005. Genetic disruption of cytosine DNA methyltransferase enzymes induces chromosomal instability in human cancer cells. Cancer Res. 65:8635–8639.

Kimura, H., and Shiota, K. 2003. Methyl-CpG-binding protein, MeCP2, is target molecule for maintenance DNA methyltransferase, Dnmt1. J. Biol. Chem. 278:4806–4812.

Kuo, H.K., Sorond, F.A., Chen, J.H., Hashmi, A., Milberg, W.P., and Lipsitz, L.A. 2005. The role of homocysteine in multisystem age-related problems: a systematic review. J. Gerontol. A Biol. Sci. Med. Sci. 60:1190–1201.

Laner, T., Schultz, W.A., Engers, R., Müller, M., and Florl, A.R. 2005. Hypomethylation of the XIST gene promoter in prostate cancer. Oncol. Res. 15:257–264.
Li, E., Bestor, T.H., and Jaenisch, R. 1992. Targeted mutation of the DNA methyltransferase gene results in embryonic lethality. Cell 69:915–926.
Li, G., Weyand, C., and Goronzy, J.J. 2008. Epigenetic mechanisms of age-dependent KIR2DL4 expression in T cells. J. Leukoc. Biol. 84:824–834.
Liang, G., Chan, M.F., Tomigahara, Y., Tsai, Y.C., Gonzales, F.A., Li, E., Laird, P.W., and Jones, P.A. 2002. Cooperativity between DNA methyltransferases in the maintenance methylation of repetitive elements. Mol. Cell. Biol. 22:480–491.
Liu, C., Xu, D., Sjöberg, J., Forsell, P., Björkholm, M., and Claesson, H.E. 2004. Transcriptional regulation of 15-lipoxygenase expression by promoter methylation. Exp. Cell Res. 297:61–67.
Lopatina, N., Haskell, J.F., Andrews, L.G., Poole, J.C., Saldanha, S., and Tollefsbol, T. 2002. Differential maintenance and de novo methylating activity by three DNA methyltransferases in aging and immortalized fibroblasts. J. Cell. Biochem. 84:324–334.
Loriot, A., De Plaen, E., Boon, T., and De Smet, C. 2006. Transient down-regulation of DNMT1 methyltransferase leads to activation and stable hypomethylation of *MAGE-A1* in melanoma cells. J. Biol. Chem. 281:10118–10126.
Lu, Q., Kaplan, M., Ray, D., Ray, D., Zacharek, S., Gutsch, D., and Richardson, B. 2002. Demethylation of *INGAL* (CD11a) regulatory sequences in systemic lupus erythematosus. Arthritis Rheum. 46:1282–1291.
Lund, G., Andersson, L., Lauria, M., Lindholm, M., Fraga, M.F., Villar-Garea, A., Ballestar, E., Esteller, M., and Zaina, S. 2004. DNA methylation polymorphisms precede any histological sign of atherosclerosis in mice lacking apolipoprotein E. J. Biol. Chem. 279: 29147–29154.
Mason, J.B. 2003. Biomarkers of nutrient exposure and status in one-carbon (methyl) metabolism. J. Nutr. 133:941S–947S.
Mays-Hoopes, L.L., Brown, A., and Huang, R.C.C. 1983. Methylation and rearrangement of mouse intracisternal A particle genes in development, adding, and myeloma. Mol. Cell. Biol. 3: 1371–1380.
Mensah, G.A., and Brown, D.W. 2007. An overview of cardiovascular disease burden in the United States. Health Aff. (Millwood). 26:38–48.
Métivier, R., Gallais, R., Tiffoche, C., Le Péron, C., Jurkowska, R.Z., Carmouche, R.P., Ibberson, D., Barath, P., Demay, F., Reid, G., Benes, V., Jeltsch, A., Gannon, F., and Salbert, G. 2008. Cyclical DNA methylation of a transcriptionally active promoter. Nature 452:45–50.
Moore, L.E., Pfeiffer, R.M., Poscablo, C., Real, F.X., Kogevinas, M., Silverman, D., García-Closas, R., Chanock, S., Tardón, A., Serra, C., Carrato, A., Dosemeci, M., García-Closas, M., Esteller, M., Fraga, M., Rothman, N., and Malats, N. 2008. Genomic DNA hypomethylation as a biomarker for bladder cancer susceptibility in the Spanish Bladder Cancer Study: a case-control study. Lancet Oncol. 9:359–366.
Morris, M.S. 2003. Homocysteine and Alzheimer's disease. Lancet Neurol. 2:425–428
Myant, K., and Stancheva, I. 2008. LSH cooperates with DNA methyltransferases to repress transcription. Mol. Cell. Biol. 28:215–226.
Nehler, M.R., Taylor, L.M., Jr., and Porter, J.M. 1997. Homocysteinemia as a risk factor for atherosclerosis: a review. Cardiovasc. Surg. 5:559–567.
Newman, P.E. 1999. Can reduced folic acid and vitamin B12 levels cause deficient DNA methylation producing mutations which initiate atherosclerosis? Med. Hypothesis 53:421–424.
Nguyen, S., Meletis, K., Fu, D., Jhaveri, S., and Jaenisch, R. 2007. Ablation of de novo DNA methyltransferase Dnmt3a in the nervous system leads to neuromuscular defects and shortened lifespan. Dev. Dyn. 236:1663–1676.
Niculescu, M.D., and Zeisel, S.H. 2002. Diet, methyl donors and DNA methylation: interactions between dietary folate, methionine and choline. J. Nutr. 132:2333S–2335S.
Nijhout, H.F., Reed, M.C., Anderson, D.F., Mattingly, J.C., James, S.J., and Ulrich, C.M. 2006. Long-range allosteric interactions between the folate and methionine cycles stabilize DNA methylation reaction rate. Epigenetics, 1:81–87.

Okano, M., Bell, D.W., Haber, D.A., and Li, E. 1999. DNA methyltransferases Dnmt3a and Dnmt3b are essential for de novo methylation and mammalian development. Cell 99: 247–257.
Ono, T., Shinya, K., Uehara, Y., and Okada, S. 1989. Endogenous virus genomes become hypomethylated tissue-specifically during aging process of C57BL mice. Mech. Ageing Dev. 50:27–36.
Ono, T., Takahashi, N., and Okada, S. 1989a. Age-associated changes in DNA methylation and mRNA level of the c-myc gene in spleen and liver of mice. Mutat. Res. 219:39–50.
Ono, T., Uehara, Y., Kurishita, A., Tawa, R., and Sakurai, H. 1993. Biological significance of DNA methylation in the ageing process. Age Ageing. 22:S34–S43.
Oomen, A.M., Griffin, J.B., Sarath, G., and Zempleni, J. 2005. Roles of nutrients in epigenetic events. J. Nutr. Biochem. 16:74–77.
Pastorino, L., and Lu, K.P. 2006. Pathogenic mechanisms in Alzheimer's disease. Eur. J. Pharmacol. 545:29–38.
Pogribny, I.P., Tryndyak, V.P., Boureiko, A., Melnyk, S., Bagnyukova, T.V., Montgomery, B., and Rusyn, I. 2008. Mechanisms of peroxisome proliferator-induced DNA hypomethylation in rat liver. Mut. Res. 644:17–23.
Prelog, M., 2006. Aging of the immune system: a risk factor for autoimmunity? Autoimmun. Rev. 5:136–139.
Raines, E.W., and Ross, R. 1995. Biology of atherosclerotic plaque formation: possible role of growth factor lesion development and the potential impact of soy. J. Nutr. 125: 624S–630S.
Ravindran, C.R., and Ticku, M.K. 2005. Methylation of NMDA receptor NR2B gene as a function of age in the mouse brain. Neurosci. Lett. 380:223–228.
Ray, D., Wu, A., Wilkinson, J.E., Murphy, H.S., Lu, Q., Kluve-Beckerman, B., Liepnieks, J.J., Benson, M., Yung, R., and Richardson, B. 2006. Aging in heterozygous Dnmt1-deficient mice: effects on survival, the DNA methylation genes, and the development of amyloidosis. J. Gerontol. A Biol. Sci. Med. Sci. 61:115–124.
Refsum, H., Nurk, E., Smith, A.D., Ueland, P.M., Gjesdal C.G., Bjelland, I., Tverdal, A., Tell, G.S., Nygård, O., and Vollset, S.E. 2006. The Hordaland Homocysteine Study: a community-based study of homocysteine, its determinants, and associations with disease. J. Nutr. 136: 1731S–1740S.
Richardson, B.C. 2002. Role of DNA methylation in the regulation of cell function: autoimmunity, aging and cancer. J. Nutr. 132:2401S–2405S.
Richardson, B. 2003a. DNA methylation and autoimmune disease. Clin. Immunol. 109:72–79.
Richardson, B. 2003b. Impact of aging on DNA methylation. Ageing Res. Rev. 2:245–261.
Rodriguez, J., Frigola, J., Vendrell, E., Riques, R.A., Fraga, M.F., Morales, C., Moreno, V., Esteller, M., Capellá, G., Ribas, M., and Peinado, M.A. 2006. Chromosomal instability correlates with genome-wide DNA demethylation in human primary colorectal cancers. Cancer Res. 66: 8462–8468.
Rollins, R.A., Haghighi, F., Edwards, J.R., Das, R., Zhang, M.Q., Ju, J., and Bestor, T.H. 2006. Large-scale structure of genomic methylation patterns. Genome Res. 16:157–163.
Romanov, G.A., and Vanyushin, B.F. 1981. Methylation of reiterated sequences in mammalian DNAs. Effects of the tissue type, age, malignancy and hormonal induction. Biochim. Biophys. Acta 653: 204–218.
Ross, R. 1999. Atherosclerosis – an inflammatory disease. N. Engl. J. Med. 340:115–126.
Rozen, R. 2000. Genetic modulation of homocysteinemia. Semin. Thromb. Hemost. 26:255–261.
Sato, N., Maitra, A., Fukushima, N., van Heek, N.T., Matsubayashi, H., Iacobuzio-Donahue, C.A., Rosty, C., and Coggins, M. 2003. Frequent hypomethylation of multiple genes overexpressed in pancreatic ductal adenocarcinoma. Cancer Res. 63:4158–4166.
Schulz, R.J. 2007. Homocysteine as a biomarker for cognitive dysfunction in the elderly. Curr. Opin. Clin. Nutr. Metab. Care. 10:718–723.
Selhub, J. 2002. Folate, vitamin B12 and vitamin B6 and one carbon metabolism. J. Nutr. Health Aging. 6:39–42.

Seivwright, C., Macnab, J.C., and Adams, R.L. 1993. S-adenosylmethionine metabolism in herpes simplex virus type 2-infected cells. J. Gen. Virol. 74:1405–1407.
Singhal, R.P., Mays-Hoopes, L.L., and Eichhorn, G.L. 1987. DNA methylation in aging of mice. Mech. Ageing Dev. 41:199–210.
Spry, M., Scott, T., Pierce, H., and D'Orazio, J.A. 2007. DNA repair pathways and hereditary cancer susceptibility syndromes. Front. Biosci. 12:4191–4207.
Sun, L.Q., Lee, D.W., Zhang, Q., Xiao, W., Raabe, E.H., Meeker, A., Miao, D., Huso, D.L., and Arceci, R.J. 2004. Growth retardation and premature aging phenotypes in mice with disruption of the SNF-2-like gene, PASG. Genes Dev. 18:1035–1046.
Sun, L.Q., and Arceci R.J. 2005. Altered epigenetic patterning leading to replicative senescence and reduced longevity. Cell Cycle 4:3–5.
Suuronen, T., Nuutinen, T., Ryhänen, T., Kaarniranta, K., and Salminen, A. 2007. Epigenetic regulation of clusterin/apolipoprotein J expression in retinal pigment epithelial cells. Biochem. Biophys. Res. Commun. 357:397–401.
Tohgi, H., Utsugisawa, K., Nagane, Y., Yoshimura, M., Genda, Y., and Ukitsu, M. 1999. Reduction with age in methylcytosine in the promoter region -224 approximately -101 of the amyloid precursor protein gene in autopsy human cortex. Brain Res. Mol. Brain Res. 70:288–292.
Tost, J. 2008. Epigenetics. Norfolk: Caister Academic Press.
Uhl, J., Klan, N., Rose, M., Entian, K.D., Werz, O., and Steinhilber, D. 2002. The 5-lipoxygenase promoter is regulated by DNA methylation. J. Biol. Chem. 277:4374–4379.
Ulrey, C.L., Liu, L., Andrews, L.G., and Tollefsbol, T.O. 2005. The impact of metabolism on DNA methylation. Hum. Mol. Genet. 14:R139–R147.
Valinluck, V., and Sowers, L.C. 2007. Endogenous cytosine damage products alter the site selectivity of human DNA maintenance methyltransferase DNMT1. Cancer Res. 67:946–950.
Vanyushin, B.F., Korotaev, G.K., Mazin, A.L., and Berdishev, G.D. 1969. Investigation of some characteristics of the primary and secondary structure of DNA from the liver of spawning humpback salmon. Biochemistry (Mosc.) 34:191–198.
Vanyushin, B.F., Tkacheva, S.G., and Belozersky, A.N. 1970. Rare bases in animal DNA. Nature 225:948–949.
Vanyushin, B.F., Nemirovsky, L.E., Klimenko, V.V., Vasiliev, V.K., and Belozersky, A.N. 1973. The 5-methylcytosine in DNA of rats. Tissue and age specificity and the changes induced by hydrocortisone and other agents. Gerontologia 19:138–152.
Verdile, G., Gandy, S.E., and Martins, R.N. 2007. The role of presenilin and its interacting proteins in the biogenesis of Alzheimer's beta amyloid. Neurochem. Res. 32:609–623.
Vijg, J. 2008. The role of DNA damage and repair in aging: New approaches to an old problem. Mech. Ageing Dev. 129:498–502.
West, R.L., Lee, J.M., and Maroun, L.E. 1995. Hypomethylation of the amyloid precursor protein gene in the brain of an Alzheimer's disease patient. J. Mol. Neurosci. 6:141–146.
Wilson, V.L., and Jones, P.A. 1983. DNA methylation decreases in aging but not in immortal cells. Science 220:1055–1057.
Wilson, V.L., Smith, R.A., Ma, S., and Cutler, R.G. 1987. Genomic 5-methyldeoxycytidine decreases with age. J. Biol. Chem. 262:9948–9951.
Wilson, A.S., Power, B.E., and Molloy, P.L. 2007. DNA hypomethylation and human diseases. Biochim. Biophys. Acta 1775:138–162.
Wolfe, M.S. 2007. When loss is gain: reduced presenilin proteolytic function leads to increased Aβ42/Aβ40. EMBO Reports 8:136–140.
Worthley, S.G., Osende, J.I., Helft, G., Badimon, J.J., and Fuster, V. 2001. Coronary artery disease: pathogenesis and acute coronary syndromes. Mt. Sinai J. Med. 68:167–181.
Wu, H., Chen, Y., Liang, J., Shi, B., Wu G., Zhang, Y., Wang, D., Li, R., Yi, X., Zhang, H., Sun, L., and Shang, Y. 2005. Hypomethylation-linked activation of PAX2 mediates tamoxifen-stimulated endometrial carcinogenesis. Nature 438:981–987.
Yenbutr, P., Hilakivi-Clarke, L., and Passaniti, A. 1998. Hypomethylation of an exon I estrogen receptor CpG island in spontaneous and carcinogen-induced mammary tumorigenesis in the rat. Mech. Ageing Dev. 106:93–102.

Yi, P., Melnyk, S., Pogribna, M., Pogribny, I.P., Hine, R.J., and James, S.J. 2000. Increase in plasma homocysteine associated with parallel increases in plasma S-adenosylhomocysteine and lymphocyte DNA hypomethylation. J. Biol. Chem. 275: 29318–29323.
Yideng, J., Jianzhong, Z., Ying, H., Juan, S., Jinge, Z., Shenglan, W., Xiaoqun, H., and Shuren, W. 2007. Homocysteine-mediated expression of SAHH, DMNTs, MBD2, and DNA hypomethylation potential pathogenic mechanism in VSMCs. DNA Cell. Biol. 26:603–611.
Yoder, J.A., Walsh, C.P., and Bestor, T.H. 1997. Cytosine methylation and the ecology of intragenomic parasites. Trends Genet. 13:335–340.
Yung, R.L., and Julius, A. 2008. Epigenetics, aging, and autoimmunity. Autoimmunity 41: 329–335.
Zaina, S., Lindholm, M.W., and Lund, G. 2005. Nutrition and aberrant DNA methylation patterns in atherosclerosis: more than just hyperhomocysteinemia? J. Nutr. 135:5–8.
Zhang, Z., Deng, C., Lu, Q., and Richardson, B. 2002. Age-dependent DNA methylation changes in the ITGAL (CD11a) promoter. Mech. Ageing Dev. 123:1257–1268.
Zhao, L., and Funk, C.D. 2004. Lipoxygenase pathways in atherosclerosis. Trends Cardiovasc. Med. 14:191–195.

Gene-Specific Hypermethylation in Aging

Adebayo D. Akintola and Alan R. Parrish

Abstract Aging is linked to a number of changes in gene expression that are associated with 'normal' or 'pathologic' aging (reviewed in Weindruch and Prolla, 2002; Toescu et al., 2004; Park and Prolla, 2005; Englander, 2005). As such, a number of studies have focused on the mechanism(s) underlying these changes in an attempt to identify conserved pathways altered during aging. Methylation of DNA is a critical epigenetic mechanism regulating gene expression that has received increased attention as a pathway affected by aging (reviewed in Issa, 2000; Attwood et al., 2002; Richardson, 2002, 2003). In this chapter, we will provide an overview of the regulation of gene expression by DNA methylation, the current methods used to investigate these changes, and specific genes that have been demonstrated to be regulated by methylation during aging.

Keywords Aging · Cancer · Methylation-specific PCR · CpG island

Hypermethylation of Promoter Elements: Regulation of Gene Expression

DNA methylation is an important epigenetic mechanism associated with the regulation of gene transcription. In mammalian cells, DNA methylation involves the postsynthetic transfer of a methyl group, donated by *S*-adenosyl-L-methioinine, to the 5′ position of the cytosine (C) present within cytosine–guanine dinucleotides (CpG). The reaction, catalyzed by DNA methyltransferases, results in the formation of 5-methylcytosine (m^5C). Importantly, DNA methylation does not occur in a random pattern in the mammalian genome (reviewed in Suzuki and Bird, 2008).

A.R. Parrish (✉)
Systems Biology and Translational Medicine, Texas A&M Health Science Center, College of Medicine, College Station, TX 77843-1114, USA
e-mail: parrish@medicine.tamhsc.edu

T.O. Tollefsbol (ed.), *Epigenetics of Aging*, DOI 10.1007/978-1-4419-0639-7_3,

About 60–90% of the C residues located within CpG dinucleotides of mammalian DNA are constitutively methylated, whereas most of the unmethylated cytosines are found within CpG islands – regions with clusters of CG sequences in higher proportion than would be expected (Bird, 1986) – located near the coding sequences and/or downstream of transcription initiation sites within the transcribed region (Ponger and Mouchiroud, 2002; Murakami et al., 2004; Yamashita et al., 2005). A CpG island is found at the 5′ end of nearly all housekeeping genes, and in many individual genes including murine MHC class I and II genes (Tykocinski and Max, 1984), β-globin gene (Waalwijk and Flavell, 1978), and N-cadherin (Akintola et al., 2008). In most cases, methylation of the individual genes occurs in a tissue-specific manner (Yamashita et al., 2005).

The CpG islands in the promoter regions of mammalian DNA contain multiple consensus sequences for binding of transcription factors (Murakami et al., 2004), thus making methylation modification of the CG dinucleotides in these regions critical for gene transcription. While DNA hypomethylation correlates with gene expression/activation, hypermethylation of the CpG island within the promoter region strongly correlates with transcriptional repression; in fact, DNA hypermethylation is a hallmark of gene silencing (Lu et al., 2006). There are at least three mechanisms by which hypermethylation of CG sequences in the promoter region leads to gene suppression. The most direct mechanism is to prevent the binding of basal transcriptional machinery or ubiquitous transcription factors by the presence of methylated cytosines at the recognition sites within the promoter regions or other *cis* regulatory elements (Bird and Wolffe, 1999). Such transcription factors may require contact with cytosine in the major groove of the double helix of promoter DNA for activity (Bird and Wolffe, 1999). For example, AP-2 binding to its recognition sequence on the promoter region of the pro-enkephalin gene is inhibited by methylation of a CCGG site within its recognition sequence, leading to suppression of expression of a pro-enkephalin–CAT fusion gene (Comb and Goodman, 1990). The transcription factor Sp1 has also been shown to be affected by hypermethylation of CG sequences (Clark et al., 1997); and more recently, Zhu et al. (2003) demonstrated the inhibition of Sp1/Sp3 binding and activity in the p21 promoter by the methylation of adjacent CG sites. Similarly, direct interference with transcription factor binding on the promoter region by methylated cytosine has been reported for the binding of ATF/CREB to the cyclic AMP (cAMP)-responsive element (Iguchi-Ariga and Schaffner, 1989) and for c-myc (Prendergast and Ziff, 1991).

A second mechanism of promoter hypermethylation-induced gene suppression is the exclusion of the transcriptional machinery from the promoter region by methylcytosine-binding proteins including MeCP2 and MBD-1, -2, -3, and -4 that are recruited to the methylated cytosine within or near the promoter region (Bird and Wolffe, 1999). Finally, the methylcytosine-binding proteins can in turn recruit large protein complexes containing corepressors and histone deacetylases to the CpG island on the promoter region (Lorincz et al., 2001). Both the MeCP2 and the MBD family of proteins contain a transcriptional repression-binding domain, and upon binding to methylcytosine-containing DNA facilitate the suppression of gene transcription through interaction with the corepressors and histone deacetylases that are chromatin inactivation complexes. The binding of these large complexes promotes a change in chromatin structure from an open active form accessible to transcriptional machinery, to a condensed, inactive form that is inaccessible to basal transcriptional machinery resulting in the repression of genes at a distant from the methylated region (Bird and Wolffe, 1999). The change in chromatin structure is accompanied by histone deacetylation, and both DNA methylation and histone deacetylation may act in a synergistic manner in the regulation of gene expression (reviewed in Attwood et al., 2002; Richardson, 2003). Demethylation of the methylcytosine can also occur either by passive demethylation during DNA replication or via the action of demethylases that occurs independent of DNA synthesis. In fact, it has been suggested that DNA methylation may be the default state and that either active or passive demethylation may be the key determinant of methylation patterns.

Methods to Detect Methylation Status

A wide range of techniques are available for detecting the methylation status of DNA, with each technique having its own advantages and limitations; the use of these methods to examine age-dependent methylation has previously been reviewed (Fraga and Esteller, 2002; Wojdacz and Hansen, 2006). The methylation techniques can be classified into two groups depending on the scope of investigation – genome-wide analysis of DNA methylation or the investigation of methylation of specific genes. Further, the available methods for detecting DNA methylation can also be grouped according to the type of techniques and the specific information revealed (Table 1).

Table 1 Common methylation assays

Method	Description
RLGS-M	Restriction landmark genomic scanning for methylation. Restriction digest and electrophoresis allows for genome-wide screening; can identify multiple targets in tissue
CpG microarray	Allows for high-throughput gene-specific analysis; can use limited starting material to identify multiple targets in tissue
MSP	Methylation-specific PRC. Rapid gene-specific analysis using limited starting material

Methods available for detecting genomic methylation include high-performance means liquid chromatography (HPLC) and capillary electrophoresis (HPCE), restriction landmark genomic scanning for methylation (RLGS-M), and CpG island microarray (Sulewska et al., 2007). The first two methods permit the quantification of overall methylcytosine in genomic DNA, the latter allow for genome-wide screening of methylation status. The problem with these methods is that the application requires access to sophisticated and expensive equipment. As such, the RLGS-M and microarray methods are increasingly utilized for studies. Other methods for studying global DNA methylation are in situ hybridization methods based on the accessibility of methylcytosine to specific antibodies (Adouard et al., 1985). The in situ method has the advantage that it can be carried out on a cell to cell basis rather than in a heterogeneous population (Fraga and Esteller, 2002).

The first available methods for the examination of single gene methylation involve digestion of the DNA samples with a pair of methylation-sensitive and insensitive restriction enzymes (Fraga and Esteller, 2002), such as *HpaII/MspI*. Following digestion with the enzyme pair, the methylation status of a particular gene is examined by southern blot hybridization or by PCR when the amount of tissue is limiting (Fraga and Esteller, 2002). Both methods rely on the length of the digested fragment, with the inability to digest methylated sequences resulting in longer fragments (Derks et al., 2004). The limitations of these methods include the fact that they identify methylation only in the sequences containing methylation restriction sensitive sites, and at times it may be difficult to differentiate between completely cut unmethylated DNA from a low quantity of methylated alleles (Sulewska et al., 2007). In addition, large amount of DNA is required for the southern hybridization method. However, this method has been used to examine methylation status during aging in multiple organs, including the brain, liver, and spleen (Ono et al., 1993).

An improved method for mapping methylcytosines in specific DNA sequences designed to overcome the limitations associated with the methylation restriction method is the bisulfite-based technique (Frommer et al., 1992; Clark et al., 1994). The method, which involves the sodium bisulfate modification of unmethylated cytosines to uracil is extensively used (Sasaki et al., 2003). Treatment with sodium bisulfite results in differences between unmethylated and methylated DNA, and the induced difference is accurately detected by a variety of methods including qualitative PCR (methylation-specific PCR, MSP), as well as several quantitative PCR methods; however, MSP is clearly the most commonly used method to investigate alterations in methylation in single genes. The quantitative PCR-based methods include PCR combined with MALDI-TOF MS, fluorescence-based real-time PCR (Methyl Light), and quantitative analysis of methylated alleles (QAMA) (Lehmann and Kreipe, 2004; Swift-Scanlan et al., 2005; Wong, 2006). Others are methylation-sensitive high-resolution melting (MS-HRM), methyl-binding (MB)-PCR, bisulfite sequencing of PCR products (BSP), analysis of deoxyribonucleoside monophosphates (dNMP), and methylation-dependent fragment separation (MDFS). The combined bisulfite restriction digestion analysis (COBRA) method is based on the creation or detection of a target for restriction endonuclease after bisulfite treatment, and the COMPARE-MS combines methylated-DNA precipitation and methylation-sensitive restriction enzymes techniques (Fraga and Esteller, 2002; Sulewska et al., 2007). These bisulfite-based methods are very sensitive and some are quick, such as MSP; however, the limitations include possible introduction of potential artifacts by incomplete denaturation before bisulfite modification (Fraga and Esteller, 2002).

Aging: Global Methylation

It has been established in number of studies that, in general, the levels of total deoxymethylcytosine decrease with aging as evidenced by a decrease in total 5-methylcytosine levels over passage number in fibroblasts (Wilson and Jones, 1983; Holliday, 1985). This finding also occurs in vivo in a number of vertebrate species in multiple organs (Vanyushin et al., 1970, 1973; Mays-Hoopes et al., 1986; Wilson et al., 1987; Singhal et al., 1987; Golbus et al., 1990). Interestingly, repetitive DNA sequences are the primary target of age-related hypomethylation (Romanov and Vanyushin, 1981; Mays-Hoopes et al., 1986). There are, however, notable exceptions to these findings – rat lung genomic DNA does not undergo significant demethylation during aging, while total methylcytosine content increases in the rat kidney during aging (Vanyushin et al., 1973). Total DNA methylation is also proposed to increase, based on the incorporation of labeled methylmethionine, in the aging rat brain and to a lesser extent the liver (Kanungo and Saran, 1991, 1992). While these studies were among the first to suggest that alterations in DNA methylation are a component of aging, they offer little mechanistic insight into the specific genes that underlie cell- and organ-specific changes associated with aging.

Aging: Gene-Specific Methylation

Over the past decade, identification of gene-specific alterations in methylation status during aging has become an increasing focus of research, primarily as a mechanism that may underlie age-dependent tumor formation. Although methylation of CpG islands in human and mice during aging is an infrequent event and is confined to small number of genes (Tra et al., 2002), a number of genes that are regulated by methylation during aging have now been identified (Table 2).

Table 2 Hypermethylated genes identified in multiple organs during aging in humans

Gene	Organ	Pathophysiology
ESR1	Colon, breast, heart, arterial SMC	Cancer, atherogenesis
GSTP1	Liver, prostate	Cancer
HIC-1	Colon, liver	Cancer
RASSF1A	Liver, breast, kidney, prostate	Cancer
SOCS1	Colon, liver	Cancer

Liver

Proto-oncogenes, including c-myc and c-fos, were first identified as methylation targets during aging. Using methylation-sensitive restriction enzyme sites, Ono and colleagues demonstrated hypermethylation of c-myc in the liver of aged mice (Ono et al., 1986). These changes were both gene and organ specific; methylation status of actin and dihydrofolate reductase genes was not significantly changed during age, and methylation status of c-myc was not affected by aging in the brain, but was hypomethylated in the spleen. Subsequent studies confirmed these results and demonstrated a decrease in c-myc mRNA levels in aging mice, although these methylation changes occur at distant locations from the promoter (Ono et al., 1989). Interestingly, hypermethylation of c-myc is attenuated by caloric restriction (Miyamura et al., 1993). The extent of DNA methylation of c-fos was also shown to increase with age in the mouse liver (Uehara et al., 1989). The finding that c-fos methylation increased in liver during aging in mice has been confirmed in humans, specifically that methylation from intron 1 to exon 4 increases with age (Choi et al., 1996). More recently, other genes that are affected by age-related methylation have been identified in the human liver. Using bisulfite modification, promoter methylation of hypermethylated in cancer 1 (HIC-1), caspase-8 (CASP8), glutathione *S*-transferase pi (GSTP1), suppressor of cytokine signaling 1 (SOCS1), RAS association family 1 (RASSF1A), p16, and adenomatosis polyposis coli (APC) are increased in aged patients (over 65 years) (Nishida et al., 2008).

Gastrointestinal Tract

A series of parallel studies by Issa and coworkers identified genes affected by methylation during aging that are also linked to carcinogenesis in the gastrointestinal tract. Using methylation-sensitive restriction digestion analysis, the estrogen receptor (ER) gene was methylated as a function of age in the human colon and paralleled the loss of expression (Issa et al., 1994). Interestingly, the methylation of the ER gene increased in a linear fashion over age (Issa, 2000). Since the initial finding, this group has identified a number of genes that are targets of methylation during aging, including insulin-like growth factor II (Issa et al., 1996), N33, and MYOD (Ahuja et al., 1998). These findings are gene specific as p16, thrombospondin 1 (THBS1), HIC-1, and calcitonin/calcitonin-related polypeptide (CALCA) are not hypermethylated with age (Ahuja et al., 1998). In colon, age-related hypermethylation involves 50% of genes hypermethylated in colon cancer; this may underlie the age-related risk of cancer development (Ahuja and Issa, 2000). More recent studies have used MSP analysis to demonstrate age-dependent methylation of a number of genes in the human duodenum, including helicase-like transcription factor (HLTF), suppression of tumorigenicity 14 (ST14), P-cadherin (CDH3), LIM homeobox protein 1 (LHX1), ubiquitin carboxy-terminal esterase L1 (UCHL1), serum deprivation response factor-related gene product that binds to c-kinase (SRBC), SOCS1, *O*-6-methylguanine-DNA methyltransferase (MGMT), and HPP1 (Matsubayashi et al., 2005). Again, these effects were gene specific, as the methylation of p16, E-cadherin, serine/threonine kinase 11 (STK11), VHL, and human mutL homolog 1 (hMLH1) was not affected by age. Cyclin D2 methylation is also increased in the duodenum as a function of age, as well as in the pancreas and gallbladder (Matsubayashi et al., 2003). In addition, stem cell crypts in human small intestine have a distinct age-related methylation pattern (Kim et al., 2005); while these studies did not focus on specific genes in relation to disease development, they are interesting given the role of stem cells in tumor development (reviewed in Gostjeva and Thilly, 2005; Pinto and Clevers, 2005).

Other Organs

Age-dependent methylation is also proposed to be important in other organs susceptible to tumor development in the elderly. Tumor suppressor gene (TSG) methylation increased in breast with age, specifically RASSF1A (Euhus et al., 2008). RARbeta2, RASSF1A, GSTP1, NKX-2 and -5, and estrogen receptor α (ESR1) methylation increased with age in human prostate cancer (Li et al., 2004; Kwabi-Addo et al., 2007). MSP revealed that promoter methylation of a number of genes (AP, DAP-kinase, E-cadherin, GSTP1, hMLH1, p16, RASSF1A, RUNX3) increased in a number of organs, including the liver, lung, and kidney (Waki et al., 2003). RASSF1A promoter methylation also increases with age in human kidney (Peters et al., 2007).

Age-dependent hypermethylation of genes may also be important in other age-related pathologies, although studies in this area are relatively sparse. Collagen α1(I) promoter is hypermethylated in periodontal ligament in humans and may be related to age-dependent degeneration of periodontal tissues (Takatsu et al., 1999). Similar to the studies cited above in human colon, the ERα promoter is methylated in the right atrium and arterial smooth muscle cells as a function of age (Post et al., 1999). Based on the finding that methylation is increased in coronary atherosclerotic plaques when compared to normal proximal aorta in the same study, it is suggested that inactivation of ERα by methylation is important in atherogenesis. In a study of peripheral blood harvested from Alzheimer's patients, telomerase reverse transcriptase (hTERT) methylation increased with age, but not SIRT3, SMARCA5, or E-cadherin (Silva et al., 2008). Our laboratory has shown that N-cadherin is lost in aging rat kidney due to promoter methylation, an effect that is attenuated by caloric restriction which suggests that this may be important in age-dependent alterations in renal structure and function (Akintola et al., 2008). Finally, an interesting study demonstrates that age-dependent methylation may differentially affect transcription factor-binding sites. In a study of human cerebral cortex, Tohgi et al. (1999) used MSP to demonstrate that AP-2-binding sites were not methylated, while methylcytosines in Sp1-binding sites increased with age; those in GCF-binding sites, a repressor of GC-rich promoters, decreased with age.

Conclusions

It is well established that aging is associated with global and gene-specific changes in DNA methylation. While an association between methylation-mediated silencing of tumor suppressor genes and the development of age-related tumors can be deduced, we still know very little about the cause-and-effect role of DNA methylation in age-dependent pathophysiologies. It is clear that these studies represent an exciting area of research and our knowledge will be facilitated by the continued development of techniques to assess gene-specific DNA methylation in a high-throughput manner. More important, however, will be the dissection of the mechanistic contribution of these changes to both normal and pathologic aging and to determine if this represents a point of therapeutic intervention in the latter.

References

Adouard V., Dante R., Niveleau A., Delain E., Revet B., Ehrlich M. 1985. The accessibility of 5-methylcytosine to specific antibodies in double-stranded DNA of Xanthomonas phage XP12. *Eur J Biochem* 152: 115–121.

Ahuja, N., Issa, J.P. 2000. Aging, methylation and cancer. *Histol Histopathol* 15: 835–842.

Ahuja, N., Li, Q., Mohan, A.L., Baylin, S.B., Issa, J.P. 1998. Aging and DNA methylation in colorectal mucosa and cancer. *Cancer Res* 58: 5489–5494.

Akintola, A.D., Crislip, Z.L., Catania, J.M., Chen, G., Zimmer, W.E., Burghard, R.C., Parrish, A.R. 2008. Promoter methylation is associated with the age-dependent loss of N-cadherin in the rat kidney. *Am J Physiol Renal Physiol* 294: F170–F176.

Attwood J.T., Yung R.L., Richardson B.C. 2002. DNA methylation and the regulation of gene transcription. *Cell Mol Life Sci* 59: 241–257.

Bird A.P. 1986. CpG-rich islands and the function of DNA methylation. *Nature* 321: 209–213.

Bird A.P., Wolffe A.P. 1999. Methylation-induced repression – belts, braces, and chromatin. *Cell* 99: 451–454.

Choi E.K., Uyeno S., Nishida N., Okumoto T., Fujimura S., Aoki Y., Nata M., Sagiska K., Fukuda Y., Nakao K., Yoshimoto T., Kim Y.S., Ono T. 1996. Alterations of c-fos gene methylation in the processes of aging and tumorigenesis in human liver. *Mutat Res* 354: 123–128.

Clark S.J., Harrison J., Molloy P.L. 1997. Sp1 binding is inhibited by (m)Cp(m)CpG methylation. *Gene* 195: 67–71.

Clark S.J., Harrison J., Paul C.L., Frommer M. 1994. High sensitivity mapping of methylated cytosines. *Nucleic Acids Res* 22: 2990–2997.

Comb M., Goodman H.M. 1990. CpG methylation inhibits proenkephalin gene expression and binding of the transcription factor AP-2. *Nucleic Acids Res* 18: 3975–3982.

Derks S., Lentjes M.H., Hellebrekers D.M., de Bruine A.P., Herman J.G., van Engeland M. 2004. Methylation-specific PCR unraveled. *Cell Oncol* 26: 291–299.

Englander E.W. 2005. Gene expression changes reveal patterns of aging in the rat digestive tract. *Ageing Res Rev* 4: 564–578.

Euhus, D.M., Bu D., Milchgrub S., Xie X.J., Bian A., Leitch A.M., Lewis, C.M. 2008. DNA methylation in benign breast epithelium in relation to age and breast cancer risk. *Cancer Epidemiol Biomarkers Prev* 17: 1051–1059.

Fraga M.F., Esteller M. 2002. DNA methylation: a profile of methods and applications. *Biotechniques* 33: 632, 634, 636–649.

Frommer M., McDonald L.E., Millar D.S., Collis C.M., Watt F., Grigg G.W., Molloy P.L., Paul C.L. 1992. A genomic sequencing protocol that yields a positive display of 5-methylcytosine residues in individual DNA strands. *Proc Natl Acad Sci USA* 89: 1827–1831.

Golbus J., Patella T.D., Richardson B.C. 1990. Quantitative changes in T cell DNA methylation occur during differentiation and ageing. *Eur J Immunol* 20: 1869–1872.

Gostjeva E.V., Thilly W.G. 2005. Stem cell stages and the origins of colon cancer: A multidisciplinary perspective. *Stem Cell Rev* 1: 243–251.

Holliday R. 1985. The significance of DNA methylation in cellular aging. *Basic Life Sci* 35: 269–283.
Iguchi-Ariga S.M., Schaffner W. 1989. CpG methylation of the cAMP-responsive enhancer/promoter sequence TGACGTCA abolishes specific factor binding as well as transcriptional activation. *Genes Dev* 3: 612–619.
Issa J.P., Ottaviano Y.L., Celano P., Hamilton S.R., Davidson N.E., Baylin S.B. 1994. Methylation of the oestrogen receptor CpG island links ageing and neoplasia in human colon. *Nat Genet* 7: 536–540.
Issa J-P. 2000. CpG-island methylation in aging and cancer. *Curr Top Microbiol Immunol* 249: 101–11.
Issa, J.P., Vertino P.M., Boehm C.D., Newsham I.F., Baylin S.B. 1996. Switch from monoallelic to biallelic human IGF2 promoter methylation during aging and carcinogenesis. *Proc Natl Acad Sci USA* 93: 11757–11762.
Kanungo M.S., Saran S. 1991. Methylation, acetylation and phosphorylation of the bases of DNA of young and old rats. *Indian J Biochem Biophys* 28: 96–99.
Kanungo M.S., Saran S. 1992. Methylation of DNA of the brain and liver of young and old rats. *Indian J Biochem Biophys* 29: 49–53.
Kim J.Y., Siegmund K.D., Tavare S., Shibata, D. 2005. Age-related human small intestine methylation: Evidence for stem cell niches. *BMC Med* 3: 10.
Kwabi-Addo B., Chung W., Shen L., Ittman M., Wheeler T., Jelinek J., Issa, J.P. 2007. Age-related DNA methylation changes in normal human prostate tissue. *Clin Cancer Res* 13: 3796–3802.
Lehmann U., Kreipe H. 2004. Real-time PCR-based assay for quantitative determination of methylation status. *Methods Mol Biol* 287: 207–218.
Li L.C., Shiina H., Deguchi M., Zhao H., Okino S.T., Kane C.J., Carroll P.R., Igawa, M., Dahiya, R. 2004. Age-dependent methylation of ESR1 gene in prostate cancer. *Biochem Biophys Res Commun* 321: 455–461.
Lorincz M.C., Schubeler D., Groudine M. 2001. Methylation-mediated proviral silencing is associated with MeCP2 recruitment and localized histone H3 deacetylation. *Mol Cell Biol* 21: 7913–7922.
Lu Q., Qiu X., Hu N., Wen H., Su Y., Richardson B.C. 2006. Epigenetics, disease, and therapeutic interventions. *Ageing Res Rev* 5: 449–467.
Matsubayashi H., Sato N., Fukushima N., Yeo C.J., Walter K.M., Brune K., Sahin, F., Hruban, R.H., Goggins M. 2003. Methylation of cyclin D2 is observed frequently in pancreatic cancer but is also an age-related phenomenon in gastrointestinal tissues. *Clin Cancer Res* 9: 1446–1452.
Matsubayashi, H., Sato N., Brune K., Blackford A.L., Hruban R.H., Canto M., Yeo C.J., Goggins, M. 2005. Age- and disease-related methylation of multiple genes in nonneoplastic duodenum and in duodenal juice. *Clin Cancer Res* 11: 573–583.
Mays-Hoopes L., Chao W., Butcher H.C., Huang R.C. 1986. Decreased methylation of the major mouse long interspersed repeated DNA during aging and in myeloma cells. *Dev Genet* 7: 65–73.
Miyamura Y., Tawa R., Koizumi A., Uehara Y., Kurishita A., Sakurai H., Kamiyama S., Ono T. 1993 Effects of energy restriction on age-associated changes of DNA methylation in mouse liver. *Mutat Res* 295: 63–69.
Murakami K., Kojima T., Sakaki Y. 2004. Assessment of clusters of transcription factor binding sites in relationship to human promoter, CpG islands and gene expression. *BMC Genomics* 5: 16.
Nishida N., Nagasaka T., Nishimura T., Ikai I., Boland C.R., Goel, A. 2008. Aberrant methylation of multiple tumor suppressor genes in aging liver, chronic hepatitis, and hepatocellular carcinoma. *Hepatology* 47: 908–918.
Ono T., Takahashi N., Okada S. 1989. Age-associated changes in DNA methylation and mRNA level of the c-myc gene in spleen and liver of mice. *Mutat Res* 219: 39–50.

Ono T., Uehara Y., Kurishita A., Tawa R., Sakurai H. 1993. Biological significance of DNA methylation in the ageing process. *Age Ageing* 22: S34–S43.

Ono, T., Tawa, R., Shinya, K., Hirose, S., Okada, S. 1986. Methylation of the c-myc gene changes during aging process of mice. *Biochem Biophys Res Commun* 139: 1299–1304.

Park S.K., Prolla T.A. 2005. Gene expression profiling studies of aging in cardiac and skeletal muscles. *Cardiovasc Res* 66: 205–212.

Peters I., Vaske B., Albrecht K., Kuczyk M.A., Jonas U., Serth J. 2007. Adiposity and age are statistically related to enhanced RASSF1A tumor suppressor gene promoter methylation in normal autopsy kidney tissue. *Cancer Epidemiol Biomarkers Prev* 16: 2526–2532.

Pinto D., Clevers H. 2005. Wnt, stem cells and cancer in the intestine. *Biol Cell* 97: 185–196.

Ponger L., Mouchiroud D. 2002. CpGProD: Identifying CpG islands associated with transcription factor start sites in large genomic mammalian species. *Bioinformatics* 18: 631–633.

Post W. S., Goldschmidt-Clermont P.J., Wilhide C.C., Heldman A.W., Sussman M.S., Ouyang P., Milliken E.E., Issa J.P. 1999. Methylation of the estrogen receptor gene is associated with aging and atherosclerosis in the cardiovascular system. *Cardiovasc Res* 43: 985–991.

Prendergast G.C., Ziff E.B. 1991. Methylation-sensitive sequence-specific DNA binding by the c-Myc basic region. *Science* 251: 186–189.

Richardson B. 2003. Impact of aging on DNA methylation. *Ageing Res Rev* 2: 245–261.

Richardson B.C. 2002. Role of DNA methylation in the regulation of cell function: Autoimmunity, aging and cancer. *J Nutr* 132: 2401S–2405S.

Romanov G.A., Vanyushin B.F. 1981. Methylation of reiterated sequences in mammalian DNAs. Effects of the tissue type, age, malignancy and hormonal induction. *Biochim Biophys Acta* 653: 204–218.

Sasaki M., Anast J., Bassett W., Kawakami T., Sakuragi N., Dahiya R. 2003. Bisulfite conversion-specific and methylation-specific PCR: A sensitive technique for accurate evaluation of CpG methylation. *Biochem Biophys Res Commun* 309: 305–309.

Silva P.N., Gigek C.O., Leal M.F., Bertolucci P.H., de Labio R.W., Payao S.L., Smith Mde A. 2008. Promoter methylation analysis of SIRT3, SMARCA5, HTERT, and CDH1 genes in aging and Alzheimer's disease. *J Alzheimers Dis* 13: 173–176.

Singhal R.P., Mays-Hoopes L.L., Eichhorn G.L. 1987. DNA methylation in aging of mice. *Mech Ageing Dev* 41: 199–210.

Sulewska A., Niklinska W., Kozlowski M., Minarowski L., Naumnik W., Niklinski J., Dabrowska K., Chyczewski L. 2007. Detection of DNA methylation in eucaryotic cells. *Folia Histochem Cytobiol* 45: 315–324.

Suzuki M.M., Bird A. 2008. DNA methylation landscapes: Provocative insights from epigenomics. *Nat Rev Genet* 9: 465–476.

Swift-Scanlan T., Blackford A., Argani P., Sukumar S., Fackler M.J. 2005. Two-color quantitative multiplex methylation-specific PCR. *Biotechniques* 40: 210–219.

Takatsu M., Uyeno S., Komura J., Watanabe M., Ono T. 1999. Age-dependent alterations in mRNA level and promoter methylation of collagen alpha1(I) gene in human periodontal ligament. *Mech Ageing Dev* 110: 37–48.

Toescu E.C., Verkhratsky A., Landfield P.W. 2004. Ca2+ regulation and gene expression in normal brain aging. *Trends Neurosci* 27: 614–620.

Tohgi H., Utsugisawa K., Nagane Y., Yoshimura M., Ukitsu M., Genda Y. 1999. The methylation status of cytosines in a tau gene promoter region alters with age to down regulate transcriptional activity in human cerebral cortex. *Neurosci Lett* 275: 89–92.

Tra J., Kondo T., Lu Q., Kuick R., Hanash S., Richardson B. 2002. Infrequent occurrence of age-dependent changes in CpG island methylation as detected by restriction landmark genome scanning. *Mech Ageing Dev* 123: 1487–1503.

Tykocinski M.L., Max E.E. 1984. CG dinucleotide clusters in MHC genes and in 5′ demethylated genes. *Nucleic Acids Res* 12: 4385–4396.

Uehara Y., Ono T., Kurishita A., Kokuryu H., Okada, S. 1989. Age-dependent and tissue-specific changes of DNA methylation within and around the c-fos gene in mice. *Oncogene* 4: 1023–1028.

Vanyushin B.F., Nemirovsky L.E., Klimenko V.V., Vasiliev V.K., Belozersky A.N. 1973. The 5-methylcytosine in DNA of rats. Tissue and age specificity and the changes induced by hydrocortisone and other agents. *Gerontologia* 19: 138–152.

Vanyushin B.F., Tkacheva S.G., Belozersky A.N. 1970. Rare bases in animal DNA. *Nature* 225: 948–949.

Waalwijk C., Flavell R.A. 1978. DNA methylation at a CCGG sequence in the large intron of the rabbit beta-globin gene: Tissue-specific variations. *Nucleic Acids Res* 5: 4631–4634.

Waki T., Tamura G., Sato M., Motoyama T. 2003. Age-related methylation of tumor suppressor and tumor-related genes: An analysis of autopsy samples. *Oncogene* 22: 4128–4133.

Weindruch R., Prolla T.A. 2002. Gene expression profile of the aging brain. *Arch Neurol* 59: 1712–1714.

Wilson V.L., Jones P.A. 1983. DNA methylation decreases in aging but not in immortal cells. *Science* 220: 1055–1057.

Wilson V.L., Smith R.A., Ma S., Cutler R.G. 1987. Genomic 5-methyldeoxycytidine decreases with age. *J Biol Chem* 262: 9948–9951.

Wojdacz T.K., Hansen L.L. 2006. Techniques used in studies of age-related DNA methylation changes. *Ann NY Acad Sci* 1067: 479–487.

Wong I.H. 2006. Qualitative and quantitative polymerase chain reaction-based methods for DNA methylation analyses. *Methods Mol Biol* 336: 33–43.

Yamashita R., Suzuki Y., Sugano S., Nakai K. 2005. Genome-wide analysis reveals strong correlation between CpG islands with nearly transcription factor start sites of genes and their tissue specificity. *Gene* 350: 129–136.

Zhu W.G., Srinivasan K., Dai Z, Duan W., Druhan L.J., Ding H., Yee L., Villalona-Calero M.A., Plass C., Otterson G.A. 2003. Methylation of adjacent CpG sites affects Sp1/Sp3 binding and activity in the p21(Cip1) promoter. *Mol Cell Biol* 23: 4056–4065.

Aging and Non-sirtuin Histone Modifications

Inga Kadish

Abstract Histones, once thought of as static structural elements, are now known to be dynamic and integral elements of the machinery responsible for regulating gene transcription. Modification of histones and/or DNA can alter the strength of their association and thus, together, modulate transcriptional activity. Acetylation of histones neutralizes their positively charged, lysine-rich amino-terminal tails, loosening the histone–DNA contacts, thus making DNA more accessible at these specific sites for transcription. It is widely accepted that there is a direct correlation between histone acetylation and transcriptional activity for a given segment of chromatin. Histone acetyltransferases (HATs) facilitate histone acetylation and are thus believed to act as transcriptional activators. In contrast, histone deacetylases (HDACs) remove acetyl groups from histones and thereby repress transcription by compacting DNA. The balance between the activity of HATs and HDACs regulates transcription, but with aging this balance is lost, leading to either gene overactivity (e.g., in cancer cells) or gene repression that can lead to neurodegeneration during the aging process.

Keywords Aging · Histone acetyltransferases · Histone deacetylases · Chromatin modification

In eukaryotic cells, DNA is associated with proteins (histones) that together form a complex known as chromatin. Together, this DNA–protein complex is packed compactly such that chromatin of all chromosomes is packed inside the nucleus of living cells. The first level of compaction is wrapping DNA around nucleosomes. Each nucleosome core consists of two molecules each of histones H2A, H2B, H3, and H4. A 146-bp segment of acidic DNA is wrapped around the outside of this core of basic histones, forming a 10-nm nucleosome fiber. The ability of transcription factors to access nucleosome-bound DNA at this level is largely controlled by the packaging of the chromatin and, thus, the DNA (Fig. 1). Chromatin

I. Kadish (✉)
Department of Cell Biology, University of Alabama, Birmingham AL USA; Center for Aging, University of Alabama, Birmingham, AL, USA
e-mail: ikadisha@uab.edu

T.O. Tollefsbol (ed.), *Epigenetics of Aging*, DOI 10.1007/978-1-4419-0639-7_4,

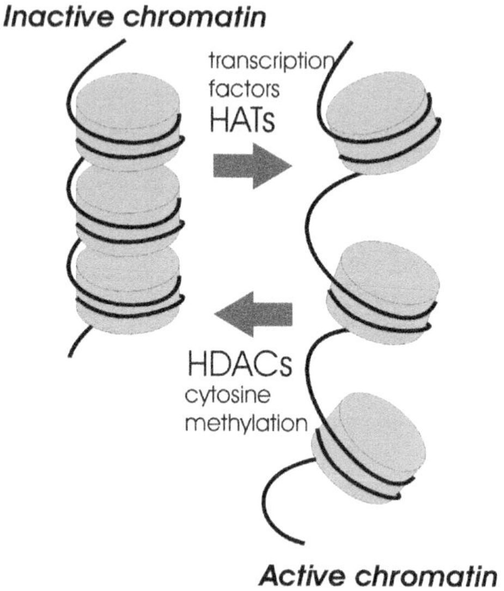

Fig. 1 Schematic illustrating the alteration in the nucleosome structure due to acetylation by histone acetyltransferases (HATs) and deacetylation by histone deacetylases (HDACs). The nucleosome is composed of eight core histones (i.e., two each of H2A, H2B, H3, and H4; cylinder) and DNA (line around the cylinders). In addition to acetylation, histones can be posttranslationally modified through methylation, phosphorylation, ubiquitination, and ADP-ribosylation; most of these modifications take place on their "tail" domains (not shown in Fig. 1.)

is further compacted into higher order structures known as 30-nm chromatin fibers, requiring the addition of histone H1 in many chromosomal regions (e.g., Marmorstein 2004).

Histones, once thought of as static structural elements, are now known to be dynamic and integral elements of the machinery responsible for regulating gene transcription, replication, repair, recombination, and chromosome segregation. Modification of histones and/or DNA can alter the strength of their association and can thus modulate transcriptional activity. Histones are basic proteins with a large proportion of positively charged amino acids, mainly arginine and lysine, and they can be posttranslationally modified through methylation, acetylation, phosphorylation, ubiquitination, and ADP-ribosylation. Most of these modifications take place on their "tail" domains (Bannister and Kouzarides 2005; Mai et al. 2005). The histone tails, which protrude from the surface of the chromatin polymer and are protease sensitive, compromise ~25–30% of the mass of individual histones (Wolffe and Hayes 1999), thus providing an exposed surface for potential interactions with other proteins. Modifications, performed by histone acetyltransferases (HATs), histone deacetylases (HDACs), histone methyltransferases (HMTs), and histone kinases (HKs), offer a mechanism through which upstream signaling pathways can converge on common targets to regulate gene expression.

Of the modifications listed above, histone acetylation has been the most studied. Acetylation of histones neutralizes their positively charged, lysine-rich amino-terminal tails, loosening the histone–DNA contacts, thus making DNA more accessible at these specific sites for transcription (e.g., Wolffe and Pruss 1996; Roth et al. 2001). It is widely accepted that there is a direct correlation between histone acetylation and the observed transcriptional activity for a given segment of chromatin. Histone acetyltransferases (HATs) facilitate histone acetylation and are thus believed to act as transcriptional activators. In contrast, histone deacetylases (HDACs) remove acetyl groups from histones and thereby repress transcription by compacting DNA (de Ruijter et al. 2003; Annunziato and Hansen 2000). Whereas this is a common pathway of gene regulation, in some cases, however, acetylation of histone tails can result in transcriptional repression and, similarly, decreased acetylation can lead to transcriptional activation (Deckert and Struhl, 2001). Thus, it is the interplay between HAT and HDAC activities that primarily governs local chromatin structure and, thus, gene expression. Likewise, the methylation status of histones can result in both transcriptional activation and repression of a particular genetic target depending on cellular conditions and signals (Lachner and Jenuwein 2002). Hyperacetylated histones are linked to transcriptionally active domains, whereas hypoacetylated histones are generally associated with transcriptionally silent loci (Strahl and Allis 2000) (Fig. 1). For example, many repression factors have been shown to operate by recruiting HDACs that alter the local chromatin structure to a more condensed organization (Moazed 2001). However, HDACs do not directly bind to the DNA sequence and require additional factors for target gene recognition (de Ruijter et al. 2003; Pazin and Kadonaga 1997). For instance, a linkage between some sequence-specific DNA-binding transcriptional repressors (e.g., Mad:Max, unliganded nuclear receptors, and Ume6) and the deacetylases is needed (de Ruijter et al. 2003). Factors that appear to establish the protein–protein link between the DNA-bound repressors and the deacetylases are SIN3 and NCoR-SMRT, which are related proteins that were originally identified as corepressors of unliganded nuclear receptors. Other proteins that are found to be associated with SIN3 proteins include SAP18 (SAP–mSIN3-associated polypeptide), SAP30, RbAp46, and RbAp48. Significantly, RbAp48 is associated with HDAC1 (Taunton et al. 1996) and it is also a subunit of chromatin assembly factor 1 (Roth and Allis 1996).

HDACs have additional activities that are not directed at histones: many HDACs are partially found in the cytoplasm (Fig. 2) and some have been shown to act on nonhistone substrates, such as the cytoskeletal protein tubulin and the transcription factors p53 and YY1 (Yao et al. 2001; Hubbert et al. 2002; Ito et al. 2002). Acetylation/deacetylation might thus be a widespread type of posttranscriptional modification, acting in a manner similar to phosphorylation/dephosphorylation in the regulation of protein activity (Kouzarides 2000).

Recent phylogenetic studies (Gregoretti et al. 2004) classify the non-sirtuin HDACs into three families: class 1 (including yeast RPD3, *Drosophila* RPD3, human HDACs 1, 2, 3, and 8), class 2 (including yeast HDA1, human HDACs 4–7, 9, and 10), and an additional class 4 defined by the recently identified human HDAC 11 (Gao et al. 2002) (Table 1). Class 1 HDACs, with the possible exception of HDAC3, are predominantly nuclear in localization (Taplick et al. 2001). Class 2 HDACs have a high degree of homology to the yeast HDAC Had-1, and

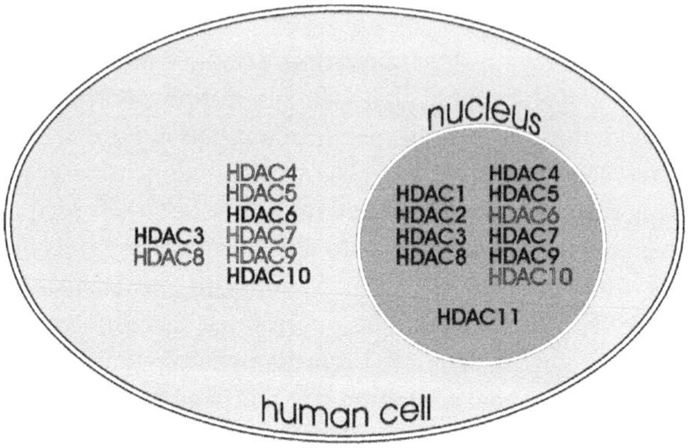

Fig. 2 Schematic illustrating the predominant localization of the non-sirtuin mammalian HDACs in the nucleus and the cytoplasm

they are larger in size compared to class 1 HDACs (Grozinger et al. 1999; Bertos et al. 2001) and are expressed in a tissue-specific manner. HDAC4 is expressed in brain, heart, and skeletal muscle tissues, HDAC5 expression partially overlaps with HDAC4, and HDAC6 has the highest expression levels in heart, liver, kidney, and pancreas. The differences in tissue expression may reflect the tissue-specific function of these enzymes. Class 2 HDACs exist in both the nucleus and the cytoplasm (Fig. 2), and the shuttling of class 2 HDACs 4, 5, 7, and 9 in and out of the nucleus is a major mechanism by which their activity is thought to be regulated (Bertos et al. 2001). These HDACs associate with 14-3-3 proteins, and this binding is absolutely dependent on phosphorylation of the conserved N-terminal serine residues of HDACs, and the association results in sequestration of HDACs to the cytoplasm. HDAC6, present predominantly in the cytoplasm, is capable of nucleocytoplasmic shuttling like most class 2 HDACs. Although HDAC6 does not bind to 14-3-3 proteins, the subcellular localization of HDAC6 also appears to be regulated, as cell-cycle arrest results in partial translocation of the protein into the nucleus (Verdel et al. 2000). In order to deacetylate histones and, thus, to repress transcription, HDACs must reside in the nucleus. Therefore, signals that enhance HDAC nuclear localization positively regulate HDAC activity. In contrast, signals that increase the cytoplasmic localization of HDACs negatively regulate their activities. Since HDAC1, 2, and 8 are predominantly nuclear proteins, it appears that the activity of these three class 1 HDACs is not regulated by subcellular localization. In contrast, HDAC3 can be found in both the nucleus and the cytoplasm, and the nuclear/cytoplasmic ratio depends on the different cell types. Class 1 and 2 HDACs share significant homology at the deacetylase domain but differ in their N-terminal sequence. Different HDACs recruit various types of cofactors and form distinct nuclear complexes that bring transcription factors to a local region in the chromatin to act as a transcriptional repressor (Khochbin et al. 2001). As a result, diverse gene expression patterns and physiological outcomes can be regulated by different HDACs and their binding partners. For example, class 1 HDACs such

Table 1 Names of the different class 1 and class 2 histone deacetylases in the three most commonly studied species

Species	Class 1	Class 2
Yeast	RPD3	HDA1
		HOS1
		HOS2
		HOS3
Drosophila	RPD3	dHDAC4
	dHDAC3	dHDAC6
Human	HDAC1	HDAC4
	HDAC2	HDAC5
	HDAC3	HDAC6
	HDAC8	HDAC7
		HDAC9
		HDAC10

as RPD3 and HDAC1 make complexes with SIN3 to alter the transcription of their target genes (Pile and Wassarman, 2000). Class 2 HDACs, instead of binding to SIN3, bind to MEF2 family of transcription factors through the MEF2-binding domain in the amino-terminal of the enzymes to inhibit MEF2-dependent transcription.

Epigenetic regulation of gene expression is recognized as an important mechanism during the developmental phase, and plays a role in some diseases, such as cancer. Correct development and cellular homeostasis are linked to the proper balance between histone acetylation (i.e., HAT activity) and histone deacetylation (i.e., HDAC activity) to regulate transcription (Fig. 1). In addition, DNA methylation and histone modifications are major players of epigenetic regulation in normal mammalian aging. For example, the status of histone acetylation plays an important role in senescence; studies using high-throughput screening of age-dependent chromatin remodeling, semirandom genome sampling, and chromatin immunoprecipitation revealed changes in histone H4 acetylation patterns spanning up to megabase distances when comparing young and old donor-derived fibroblasts (Russanova et al. 2004). Senescent cells display several characteristics, including histological changes, shortened telomeres, increased expression of p16^{INK4} and p21$^{Cip1/Waf1}$, and increased activity of senescence-associated β-galactosidase (Sasaki et al. 2008). Studies by Ocker and Schneider-Stock (2007) provide the evidence that the expression of p16^{INK4a} and cycline-dependent kinase inhibitor p21$^{Cip1/Waf1}$ is at least partially controlled through histone acetylation within promoter regions. For example, increased p21$^{Cip1/Waf1}$ transcription through HDAC inhibition is linked to increased H3 acetylation in that region.

Most of the current understanding of the role of HDACs in regulating aging comes from studies using the budding yeast as a model organism, and there SIR2, an NAD$^+$-dependent HDAC, is thought to regulate cellular senescence. However, the importance of HDACs in regulating longevity is also demonstrated by the role of RPD3, a class 1 HDAC. The deletion of yeast *rpd3* increases the budding yeast life span and leads to transcriptional silencing at several loci, including HM and

RDN1 (Kim et al. 1999). Deletion of another HDAC, HDA1, however, had no effect on longevity although an increase in subtelomeric silencing was observed. These results suggest that telomeric silencing alone is not sufficient to cause aging and that silencing at HM and RDN1 loci is important for regulating longevity. Thus, it has been proposed that different acetylation patterns of the core histones determine the intensity of heterochromatic silencing. Gene expression profile studies in yeast showed that *rpd3* deletion caused a twofold downregulation of 264 genes and upregulation of 170 transcripts (Bernstein et al. 2000). The *sir2* deletion, on the other hand, led to a twofold downregulation of 10 genes and upregulation of 57 genes. Bernstein and colleagues (2000) therefore propose that histone deacetylation by either RPD3 or SIR2 can alter the global gene expression pattern by both enhancing and repressing the transcription of numerous genes, not just genes in the telomeric, HM, and RDN1 loci. The gene expression studies also showed an interesting difference in the class of genes upregulated by *sir2* and *rpd3* deletions, suggesting that SIR2 and RPD3 normally affect genes of distinct physiological functions. A study on the effect of a general HDAC inhibitor, phenylbutyrate (PBA), on the extension of yeast life span also supports the idea that specific alteration of gene expression pattern is important for longevity.

So far, little is known about how aging changes epigenetic regulation of specific genes, especially histone acetylation. One problem in studying age-related epigenetic mechanisms is that there are, at this point in time, few well-established models of epigenetic changes in normal mammalian aging.

After the developmental phase, nearly all cells probably undergo chromatin remodeling that permanently prevents their reentry into the mitotic cycle. However, this stable chromatin repression has been suggested to be destabilized by age-related increases in HDAC activity. Thus, it has been postulated that senescence is mediated by a competition between histone acetyltransferases (HATs) and histone deacetylases (HDACs) to bind to promoters of cell-cycle genes, and thus increased HAT or HDAC levels can each trigger senescence. For instance, recently Bandyopadhyay et al. (2007) showed that the senescence response in cultured human melanocyte cells is likely due to the chromatin modifications because the retinoblastoma complexes from senescent melanocytes contained increased levels of HDAC activity and altered HDAC1. The authors demonstrated that HDAC1 was prominently detected in $p16^{INK4a}$-positive, senescent intradermal melanocytic nevi but not in the proliferating, recurrent nevus cells that localize to the epidermal–dermal junction. Based on the results, they propose a model by which upregulation of $p16^{INK4a}$ triggers retinoblastoma (RB) dephosphorylation and recruitment of increased levels of HDAC1. Changes in the stoichiometry and dynamics of RB macromolecular complexes, by upregulation of HDAC1 via increased transcription or by increased use of the cell's steady-state HDAC1 pool, in turn initiate a chain of events leading to chromatin remodeling and silencing of growth-promoting genes.

Further experimental evidence suggests that the HDAC1 deacetylase plays significant roles in aging and cancer. For example, HDAC1 stimulates angiogenesis both in vitro and in vivo; more specifically, it is hypoxia inducible and the HDAC inhibitor trichostatin A has a potent antiangiogenic effect, indicating that HDAC1 is closely involved in angiogenesis through suppression of hypoxia-responsive tumor suppressor genes (Kim et al., 2001).

The concept of a molecular "memory" regulating the pattern of gene expression has been very well characterized in the study of imprinting and inheritance of parental traits (Bantignies and Cavalli 2006). For instance, Shen et al. (2005) reported that for oligodendrocytes, "epigenetic memory" is established in two steps: the first step characterized by deacetylation of lysine residues in the tails of nucleosomal histones followed by a second step of more stable histone methylation. In a recently published study, Shen et al. (2008) have studied the age-related changes of the intrinsic properties of oligodendrocyte lineage cells, and they have identified the progressive loss of the "epigenetic memory" that is stored in the chromatin of oligodendrocytes (and that thus modulates gene expression). Namely, the authors show that the activity of histone methylation and histone deacetylation, which are required for the establishment and maintenance of the oligodendrocyte epigenetic memory, is defective in aged mice. They demonstrate that lack of histone methylation and increased acetylation in mature oligodendrocytes in the mouse corpus callosum are associated with global changes in gene expression, which include the reexpression of transcriptional inhibitors and persistent expression of precursor markers. The authors also demonstrate that reactivation of gene expression is not a random event, but rather it is selective for those genes whose levels progressively decrease during development, due to the establishment of the "epigenetic memory" in oligodendrocytes. The pattern of gene expression observed in oligodendrocytes in the aging brain, characterized by decreased *Olig2* expression concomitant with *Nkx2.2* upregulation, is reminiscent of the pattern detected in ventral glial progenitors of the zebra fish (Cunliffe and Cassacia-Bonnefil 2006), which lack a critical enzymatic activity (i.e., HDAC1) for the establishment of the epigenetic memory of gene expression in developing oligodendrocytes. Recently though, we made the unexpected finding that class 1 histone deacetylases show major downregulation in rat hippocampus and other brain regions with aging. This effect is especially strong in oligodendrocytes in white matter tracts such as the corpus callosum. Because HDACs are important molecules in epigenetic regulation (repressing transcription by deacetylating histone lysine residues), the study of the factors that regulate the activity of HDACs in oligodendrocytes in the aging brain appears to be a model system that is currently available for studying the role of epigenetics in normal aging.

In addition, HDACs have recently attracted considerable attention because chemical inhibitors of HDACs induce growth arrest, differentiation, and/or apoptosis of cancer cells and may thus represent a new class of antitumor agents (Marks et al. 2003). At present the application of HDAC inhibitors for the therapy of cancer (or other diseases) is mainly directed to class 1 and 2 HDACs (Table 1). HDAC inhibitors (HDACi) have been categorized in four different groups based on their chemical structures including hydroxamates (trichostatin A, suberoylanilide hydroxamic acid – SAHA), cyclic peptides (depsipeptide), aliphatic acids (sodium butyrate, valproic acid, phenylbutyrate), and benzamides. Of these, the most widely studied are sodium butyrate, phenylbutyrate, trichostatin A, and SAHA. These inhibitors of class 1 and 2 HDACs are in phase I/II trials for cancer therapy and potential cancer prevention. In the nervous system, the anticonvulsant and mood-stabilizing drug valproic acid has been identified as an inhibitor of HDAC1, thereby linking its antiepileptic effects to changes in histone acetylation. More recent work has revealed that inhibitors of class 1 and 2 HDACs also represent novel therapeutic approaches to treat neurodegenerative disorders (Gardian et al. 2005; Kontopoulos et al. 2006; Fischer et al. 2007), depression and anxiety (Schroeder et al., 2007), and possibly cognitive deficits that accompany many neurodevelopmental disorders. For example, Kontopoulos and colleagues (2006) using a *Drosophila* model of Parkinson's disease demonstrated that nuclear targeting of α-synuclein promotes its toxicity and that the sequestration of α-synuclein to the cytoplasm is protective. It was further shown that α-synuclein binds directly to histones, reduces levels of acetylated histone H3, and inhibits HAT-mediated acetyltransferase activity. Administration of HDACi in vivo or in vitro rescued α-synuclein-induced toxicity. A recent study by Faraco et al. (2006) demonstrated a role for HDACi in amelioration of neuronal death and cognitive deficits in postischemic neurons. The authors report that the potent HDACi SAHA administered intraperitoneally to mice at 0 and 6 hours after induction of stroke by middle cerebral artery occlusion prevented H3 deacetylation, promoted expression of neuroprotective proteins Bcl-2 and Hsp70, and reduced infarct volume, indicating a neuroprotective action of SAHA. Furthermore, although no direct demonstration of amelioration of synaptic plasticity or cognitive impairments has been documented for Alzheimer's disease, recent studies from Fischer and colleagues (2007) show that HDAC inhibitors restore histone acetylation status and learning and memory in a mouse model of neurodegeneration. This group conditionally and inducibly expressed p25, a cell-cycle protein implicated in neurodegenerative disease, under the control of the CaMKII promoter. The increased expression of p25 reduces learning and memory in this mouse line due to neuronal loss, but the treatment with an HDAC1 rescued these cognitive deficits. Furthermore, they demonstrated that environmental enrichment similarly improved these deficits and that this was correlated with chromatin modifications (i.e., increased histone-tail acetylation).

Since there are many different HDACs, it will be important to design therapies that can target individual enzymes and thus increase the precision of this approach. As discussed above, reduced histone acetylation is a final common endpoint in many neurodegenerative (including aging) and psychiatric disorders but may arise via diverse means including reduced HAT activity, increased HDAC activity, or increased DNA methylation. Further, combinations of drugs that inhibit DNMTs as well as HDACs may result in synergistic effects because DNA methylation and histone modifications can interact (Abel and Zukin 2008).

In conclusion, histone modifications have a defined profile during development, aging, and cell transformation. As organisms age, there is a general, marked decrease of histone acetylation, which in concert with altered methylation of DNA leads to changes in gene expression that together contribute to the progression of aging.

References

Abel, T., and Zukin, R. S. 2008. Epigenetic targets of HDAC inhibition in neurodegenerative and psychiatric disorders. Curr Opin Pharmacol 8:57–64.

Annunziato, A. T., and Allis, C. D. 2000. Role of histone acetylation in the assembly and modulation of chromatin structures. Gene Expr 9:37–61.

Bandyopadhyay, D., Curry, J. L., Lin, Q., Richards, H. W., Chen, D., Hornsby, P. J., Timchenko, N. A., and Medrano, E. E. 2007. Dynamic assembly of chromatin complexes during cellular senescence: implications for the growth arrest of human melanocytic nevi. Aging Cell 6: 577–591.

Bannister, A. J., and Kouzarides, T. 2005. Reversing histone methylation. Nature 436: 1103–1106.

Bantignies, F., and Cavalli, G. 2006. Cellular memory and dynamic regulation of polycomb group proteins. Curr Opin Cell Biol 18:275–283.

Bernstein, B. E., Tong, J. K., and Schreiber, S. T. 2000. Genomewide studies of histone deacetylase function in yeast. Proc Natl Acad Sci USA 97:13708–13713.

Bertos, N. R., Wang, A. J., and Yang, X. J. 2001. Class II histone deacetylases: structure, function and regulation. Biochem Cell Biol 79:243–252.

Cunliffe, V. T., Casaccia-Bonnefil, P. 2006. Histone deacetylase 1 is essential for oligodendrocyte specification in the zebrafish CNS. Mech Dev. 123:24–30.

Deckert, J., and Strhul, K. 2001. Histone acetylation at promoters is differently affected by specific activators and repressors. Mol Cell Biol 21:2726–2735.

de Ruijter, A. J., van Gennip, A. H., Caron, H. N., Kemp, S., and van Kuilenburg, A. B. 2003. Histone deacetylases (HDACs): Characterization of the classical HDAC family. Biochem J 370:737–749.

Faraco, G., Pancani, T., Formentini, L., Mascagni, P., Fossati, G., Leoni, F., Moroni, F., and Chiarugi, A. 2006. Pharmacological inhibition of histone deacetylases by suberoylanilide hydroxamic acid specifically alters gene expression and reduces ischemic injury in the mouse brain. Mol Pharmacol. 70:1876–1884.

Fischer, A., Sananbenesi, F., Wang, X., Dobbin, M., and Tsai, L. H. 2007. Recovery of learning and memory is associated with chromatin remodeling. Nature 447:178–182.

Gao, L., Cueto, M. A., Asselbergs, F., and Atadja, P. 2002. Cloning and functional characterization of HDAC 11, a novel member of the human histone deacetylase family. J Biol Chem 277:25748–25755.

Gardian, G., Browne, S. E., Choi, D. K., Klivenyi, P., Gregorio, J., Kubilus, J. K., Ryu, H., Langley, B., Ratan, R. R., Ferrante, R. J., and Beal, M. F. 2005. Neuroprotective effects of phenylbutyrate in the N171-82Q transgenic mouse model of Huntington's disease. J Biol Chem 7: 556–563.

Gariano, R. F., and Gardner, T.W. 2005. Retinal angiogenesis in development and disease. Nature 15:960–966.

Gregoretti, I. V., Lee, Y. M., Goodson, H. V. 2004. Molecular evolution of the histone deacetylase family: functional implications of phylogenetic analysis. J. Mol Biol. 338(1):17–31.

Grozinger, C. M., Hassig, C. A., Schreiber, S. L. 1999. Three proteins define a class of human histone deacetylases related to yeast Hda1p. Proc Natl Acad Sci USA 96:4868–4873.

Hubbert, C., Guardiola, A., Shao, R., Kawaguchi, Y., Ito, A., Nixon, A., Yoshida, M., Wang, X. F., and Yao, T. P. 2002. HDAC6 is a microtubule-associated deacetylase. Nature 417:455–458.

Ito, A., Kawaguchi, Y., Lai, C. H., Kovacs, J. J., Higashimoto, Y., Appella, E., and Yao, T. P. 2002. MDM2-HDAC1-mediated deacetylation of p53 is required for its degradation. EMBO J 21:6236–6245.

Khochbin, S., Verdel, A., Lemercier, C., and Seigneurin-Berny, D. 2001. Functional significance of histone deacetylase diversity. Curr Opin Genet Dev 11:162–166.

Kim, S., Benguria, A., Lai, C. Y., Jazwinski, S. M. 1999. Modulation of life-span by histone deacetylase genes in Saccharomyces cerevisiae. Mol Biol Cell. 10:3125–3136.

Kim, M. S., Kwon, H. J., Lee, Y. M., Baek, J. H., Jang, J. E., Lee, S. W., Moon, E. J., Kim, H. S., Lee, S. K., Chung, H. Y., Kim, C. W., and Kim, K. W. 2001. Histone deacetylases induce angiogenesis by negative regulation of tumor suppressor genes. Nat Med 7:437–443.

Kouzarides, T. 2000. Acetylation: a regulatory modification to rival phosphorylation? EMBO J 19:1176–1179.

Kontopoulos, E., Parvin, J. D., and Feany, M. B. 2006. Alpha-synuclein acts in the nucleus to inhibit histone acetylation and promote neurotoxicity. Hum Mol Genet 15:3012–3023.

Lachner, M., and Jenuwein, T. 2002. The many faces of histone lysine methylation. Curr Opin Cell Biol 14:286–298.

Mai, A., Massa, S., Rotili, D., Cerbara, I., Valente, S., Pezzi, R., Simeoni, S., and Ragno, R. 2005. Histone deacetylation in epigenetics: an attractive target for anticancer therapy. Med Res Rev 25:261–309.

Marks, P. A., Miller, T., and Richon, V. M. 2003. Histone deacetylases. Curr Opin Pharmacol 3:344–351.

Marmorstein, R. 2004. Structural and chemical basis of histone acetylation. Novartis Found Symp 259:78–98.

Moazed, D. 2001. Common themes in mechanisms of gene silencing. Mol Cell 8:489–498.

Ocker, M., and Schneider-Stock, R. 2007. Histone deacetylase inhibitors: signalling towards p21cip1/waf1. Int J Biochem Cell Biol 39:1367–1374.

Pazin, M. J., and Kadonaga, J. T. 1997. What's up and down with histone deacetylation and transcription? Cell 89:325–328.

Pile, L. A., and Wassarman, D. A. 2000. Chromosomal localization links the SIN3–RPD3 complex to regulation of chromatin condensation, histone acetylation and gene expression. EMBO J 19:6131–6140.

Roth, S. Y., and Allis, C. D. 1996. Histone acetylation and chromatin assembly: a single escort, multiple dances? Cell 4:5–8.

Roth, S. Y., Denu, J. M., and Allis, C. D. 2001. Histone acetyltransferases. Annu Rev Biochem 70:81–120.

Russanova, V. R., Hirai, T. H., Tchernov, A.V., and Howard, B. H. 2004. Mapping development-related and age-related chromatin remodeling by a high throughput ChIP-HPLC approach. J Geront Series A, Biol Sci Med Sci 59:1234–1243.

Sasaki, M., Ikeda, H., Yamaguchi, J., Nakada, S., and Nakanuma, Y. 2008. Telomere shortening in the damaged small bile ducts in primary biliary cirrhosis reflects ongoing cellular senescence. Hepatology 48:186–195.

Schroeder, F. A., Lin, C. L., Crusio, W. E., and Akbarian, S. 2007. Antidepressant-like effects of the histone deacetylase inhibitor, sodium butyrate, in the mouse. Biol Psychiatry 62:55–64.

Shen, S., Li, J., and Casaccia-Bonnefil, P. 2005. Histone modifications affect timing of oligodendrocyte progenitor differentiation in the developing rat brain. J Cell Biol 169:577–589.

Shen, S., Liu, A., Li, J., Wolubah, C., and Casaccia-Bonnefil, P. 2008. Neurobiol Aging 29: 452–463.

Strahl, B. D., and Allis, C. D. 2000. The language of covalent histone modification. Nature 403: 41–45.

Taplick, J., Kurtev, V., Kroboth, K., Posch, M., Lechner, T., and Seiser, C. 2001. Homo-oligomerisation and nuclear localisation of mouse histone deacetylase 1. J Mol Biol 308:27–38.

Taunton, J., Hassig, C.A., and Schreiber, S. L. 1996. A mammalian histone deacetylase related to the yeast transcriptional regulator Rpd3p. Science 19:408–411.

Verdel, A., Curtet, S., Brocard, M. P., Rousseaux, S., Lemercier, C., Yoshida, M., and Khochbin, S. 2000. Active maintenance of mHDA2/mHDAC6 histone-deacetylase in the cytoplasm. Curr Biol 10:747–749.

Wolffe, A.P., and Pruss D. 1996. Targeting chromatin disruption: transcription regulators that acetylate histones. Cell 84:817–819.

Wolffe, A. P., and Hayes, J. J. 1999. Chromatin disruption and modification. Nucleic Acids Res 27:711–720.

Yao, Y. L., Yang, W. M., and Seto, E. 2001. Regulation of transcription factor YYI by acetylation and deacetylation. Mol Cell Biol 21:5979–5991.

Sirtuins and Aging

Yuqing Dong and Sige Zou

Abstract Sirtuins are a family of highly conserved genes widely distributed in organisms ranging from bacteria to humans. Mounting evidence has revealed the important role of sirtuins in a variety of biological processes, including transcription regulation, apoptosis, DNA repair, metabolism, and more prominently, aging. Sirtuins regulate lifespan in evolutionarily diverse species partly through modulating calorie restriction pathways. Sirtuins link the nutritional status of the cell to transcription regulation through their nicotinamide adenine dinucleotide (NAD^+)-dependent deacetylase and/or ADP-ribosyltransferase. The unique features of sirtuins make them ideal targets for discovery of prolongevity compounds or aging interventions. This chapter will review the functions of sirtuins in aging and aging interventions related to sirtuins.

Keywords Sirtuins · Life span · Metabolism · (NAD^+)-dependent · Protein deacetylase · ADP-ribosyltransferase · Gene regulation

Introduction

It has been known for years that histone/protein deacetylases (HDACs) deacetylate lysine residues on histones and other proteins and epigenetically modify the activities and functions of their substrates (Marmorstein and Roth 2001; Roth et al. 2001). To date, four classes of HDACs have been defined in eukaryotes. The members in class I, II, and III are distinguished by their homology to yeast proteins, reduced potassium dependency 3 (Rpd3, class I), histone deacetylase 1 (Hda1, class II), and silent information regulator 2 (Sir2, class III) (Blander and Guarente 2004; Sengupta and Seto 2004). Human HDAC11 is the sole member of class IV, which carries traits

S. Zou (✉)
Functional Genomics Unit, Laboratory of Experimental Gerontology, National Institute on Aging, Baltimore MD 21224, USA
e-mail: zous@grc.nia.nih.gov

T.O. Tollefsbol (ed.), *Epigenetics of Aging*, DOI 10.1007/978-1-4419-0639-7_5,

of both class I and class II. HDAC11 protein is conserved only in fly, mouse, and higher organisms but not in yeast and nematode (Gao et al. 2002).

The Sir2 family consists of many evolutionarily conserved genes distributed in a broad range of organisms from yeast to mammals, collectively termed sirtuins. In comparison to other classes of HDACs, sirtuins are unique as their activity requires nicotinamide adenine dinucleotide (NAD^+) as a cofactor (Denu 2003; Blander and Guarente 2004). Sir2, the founding member of sirtuins, was originally found to regulate silencing at the silent mating-type loci in the budding yeast *Saccharomyces cerevisiae* (Klar et al. 1979). Sir2 has also been well documented to function in telomere silencing, ribosomal DNA (rDNA) recombination, and genome integrity maintenance (Gottschling et al. 1990; Bryk et al. 1997; Smith and Boeke 1997; Gasser and Cockell 2001; Rusche and Rine 2001). Notably, in addition to silencing, studies of aging by Guarente and colleagues a decade ago led to the discovery of Sir2 as a key regulator of replicative lifespan in *S. cerevisiae* (Kaeberlein et al. 1999). Since then, increasing interest has focused on the function of sirtuins in aging and aging-related biological processes in multiple organisms. This chapter will review these progresses and the effects of aging interventions linked to sirtuins.

Distribution of Sirtuins in Model Organisms

Sirtuins belong to a highly conserved gene family widely distributed in organisms from single-cell organisms to humans (Table 1) (Brachmann et al. 1995). Studies on the biochemistry of sirtuins indicate that they exhibit enzymatic activity as protein deacetylases and/or ADP-ribosyltransferases (ART) (Frye 1999; Liszt et al. 2005). Both of these activities are dependent on NAD^+(Fig. 1). Since NAD^+ is an indicator of metabolic status, these findings suggest that sirtuins are part of the nutritive sensing system in the cell.

Sir2, the founding member of sirtuins, was first discovered in *S. cerevisiae* when a spontaneous sterile mutation was found to cause silencing defects at the silent mating-type loci *HMRa* and *HML*α (Klar et al. 1979). To date, a total of five yeast sirtuins, Sir2 and four homologs of Sir2 (Hst1–Hst4), have been identified in *S. cerevisiae* (Derbyshire et al. 1996). All these sirtuins are involved in silencing at the silent mating loci, rDNA, and telomeric regions but in different ways. Overexpression of Hst1 can suppress transcription silencing defects caused by *sir2* deletion, while hst1 deletion appears to have no effect on rDNA recombination and silencing at the silent mating-type loci (Brachmann et al. 1995; Derbyshire et al. 1996). Hst2 overexpression disrupts silencing at the telomere but enhances silencing in rDNA (Perrod et al. 2001). Hst3 and Hst4 work together to participate in cell-cycle progression, genomic integrity, and telomeric silencing (Brachmann et al. 1995; Derbyshire et al. 1996; Smith et al. 1997). Although Sir2 functions as a transcription silencer through its protein deacetylase and/or ADP-ribosyltransferase activities, the predominant enzymatic activity of Sir2 is as a protein deacetylase. This notion is supported by findings that subtelomeric sequences (generally less

Table 1

Organism	Gene	Sub-cellular localization	Enzymatic activity
S. cerevisiae	*Sir2*	Nucleus	Deacetylase/ART
	Hst1	Nucleus	Deacetylase
	Hst2	Cytoplasm	Deacetylase
	Hst3	Nucleus	Unknown
	Hst4	Nucleus	Unknown
C. elegans	*Sir-2.1*	Nucleus	Deacetylase
	Sir-2.2	Unknown	Unknown
	Sir-2.3	Unknown	Unknown
	Sir-2.4	Unknown	Unknown
D. melanogaster	*dSir2*	Nucleus	Deacetylase
	dSirt2	Unknown	Deacetylase
	dSirt4	Unknown	Unknown
	dSirt6	Unknown	Unknown
	dSirt7	Unknown	Unknown
Mammals	*SIRT1*	Nucleus	Deacetylase
	SIRT2	Cytoplasm	Deacetylase/ART
	SIRT3	Mitochondria	Deacetylase/ART
	SIRT4	Mitochondria	ART
	SIRT5	Mitochondria	Unknown
	SIRT6	Nucleus	Deacetylase/ART
	SIRT7	Nucleus	Unknown

Note: ART, ADP-ribosyltransferase

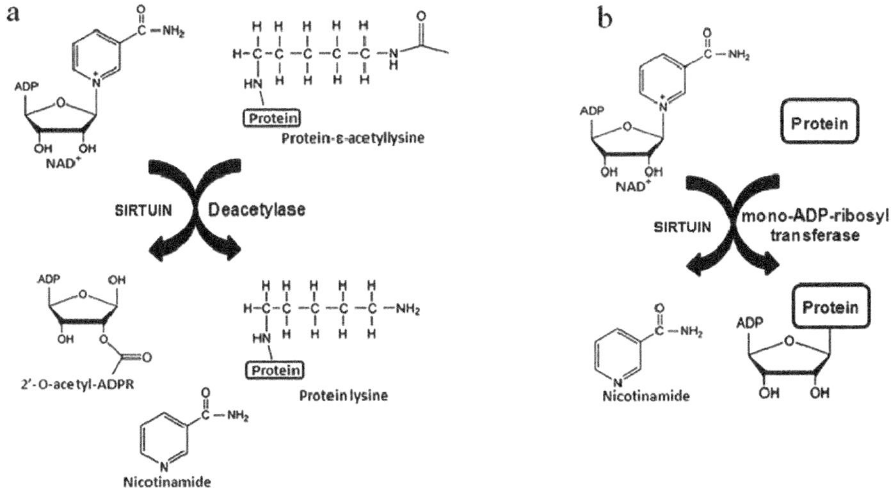

Fig. 1 Enzymatic activities of sirtuins. Sirtuins can function as protein deacetylase and/or ADP-ribosyltransferase to modulate activities of target proteins through post-translational and covalent but often reversible modifications

than 4 kb to the telomeric ends), the silent mating-type loci *HMRa* and *HMLα*, and the rDNA loci are hyperacetylated in the absence of *sir2* (Strahl-Bolsinger et al. 1997; Moazed 2001; Robyr et al. 2002). In the fission yeast *Schizosaccharomyces pombe,* there are three Sir2-like genes, Sir2, Hst2, and Hst4 (Ghidelli et al. 2001). Like those in the budding yeast, these sirtuins are involved in transcriptional silencing via NAD^+-dependent protein deacetylase at the silent mating loci, telomeres, centromeres, and the rDNA loci through different mechanisms (Shankaranarayana et al. 2003; Durand-Dubief et al. 2007).

Based on homology to yeast Sir2, four sirtuins, *sir-2.1* to *sir-2.4*, have been identified in the nematode *Caenorhabditis elegans*. SIR-2.1 appears to possess protein deacetylase activity based on an in vitro deacetylation assay (Wood et al. 2004). However, if and how the nematode sirtuins generally regulate gene expression is far less clear. SIR-2.2 appears to function in genomic stability and DNA repair since knockdown of *sir-2.2* by RNA-mediated interference (RNAi) results in increased spontaneous mutagenesis (Pothof et al. 2003). Among the *C. elegans* sirtuins, *sir-2.1* is the best studied and its biological function in longevity modulation has been well documented. Increasing SIR-2.1 activity with extra copies, but not *sir-2.2, sir-2.3*, or *sir-2.4*, extends lifespan in *C. elegans* (Tissenbaum and Guarente 2001).

In *Drosophila melanogaster,* five sirtuins, dSir2, dSirt2, dSirt4, dSirt6, and dSirt7, have been identified. Of these, dSir2 is most similar to the budding yeast Sir2. In vitro assays indicate that dSIR2 can act as a protein deacetylase to deacetylate acetylated histone peptides or full-length histone H4 protein (Barlow et al. 2001; Rosenberg and Parkhurst 2002). In vivo, dSir2 is implicated in regulating both euchromatic and heterochromatic silencing, which is known as position-effect variegation in *D. melanogaster* (Rosenberg and Parkhurst 2002). Consistent with other eukaryotic organisms, the *Drosophila* dSir2 gene is nonessential. Deletion of *dsir2* subtly affects position-effect variegation, but has no influences on viability and development (Astrom et al. 2003). Notably, however, *Drosophila dsir2* homozygous mutants have shortened lifespan (Astrom et al. 2003).

In mammals, the sirtuin family has seven members, Sirt1 to Sirt7. Each sirtuin is defined by a conserved catalytic core domain and its flanking sequence at the N- or C-termini. These flanking extensions of core domains are thought to promote target-binding specificity (Cuperus et al. 2000). Based on their sub-cellular localization, mammalian sirtuins can be classified into three groups. Group I consists of Sirt1, Sirt6, and Sirt7, which predominately localize in the nucleus. In the nucleus, Sirt1 is predominantly associated with euchromatin, Sirt6 mainly associates with heterochromatin and telomeres as a histone H3 lysine 9 deacetylase, while Sirt7 is predominantly found in the nucleolus (Michishita et al. 2005; Michishita et al. 2008). Group II has Sirt3, Sirt4, and Sirt5, which mainly reside in the mitochondria (Lombard et al. 2007; Ahuja et al. 2007; Haigis et al. 2006). Sirt2 is the sole member in group III and is located in the cytoplasm (Frye 1999; North et al. 2003).

Sirt1 has been found to deacetylate a growing number of protein targets, including tumor suppressor p53, forkhead transcription factor FOXO, and fat metabolism regulator PPARγ (Vaziri et al. 2001; Langley et al. 2002; Brunet et al. 2004; Rodgers et al. 2008). Sirt2 has been implicated in deacetylation of α-tubulin and FOXO3a (North et al. 2003; Wang et al. 2007). Sirt2 also shows mono-ADP-ribosyltransferase activities (Frye 1999; North et al. 2003). Sirt3 is another mammalian sirtuin with activities of both deacetylase and mono-ADP-ribosyltransferase (Frye 1999; Shi et al. 2005). As a protein deacetylase, Sirt3 is capable of deacetylating acetyl-coenzyme A in the mitochondria to regulate the tricarboxylic acid cycle (Hallows et al. 2006). Sirt4 seems not to be active as a histone deacetylase in vitro (North et al. 2003; Haigis et al. 2006). However, Sirt4 possesses activity of mono-ADP-ribosyltransferase to inhibit the mitochondrial glutamate dehydrogenase (GDH), which mediates amino acid-stimulated insulin secretion by regulating glutamine and glutamate oxidative metabolism (Haigis et al. 2006). Sirt5 is the most distant homolog of yeast Sir2, and little is known about its function (Frye 1999; North et al. 2003). Sirt6 has both deacetylase and mono-ADP-ribosyltransferase activities (Liszt et al. 2005; Mostoslavsky et al. 2006). As a protein deacetylase, Sirt6 deacetylates histones and DNA repair enzymes to modulate genomic DNA stability and DNA repair (Mostoslavsky et al. 2006; Michishita et al. 2008). Sirt7 resides in the nucleus and interacts with RNA Polymerase I (Pol I) and histones, as an activator of the RNA Pol I transcriptional machinery (Ford et al. 2006). Overexpression of Sirt7 increases Pol I-mediated rDNA transcription in the presence of NAD^+, whereas reduction of SIRT7 decreases the association of Pol I with rDNA (Ford et al. 2006). Depletion of Sirt7 inhibits cell proliferation and triggers apoptosis, for example, in cardiomyocytes due to hyperacetylation of p53, suggesting that SIRT7 regulates Pol I transcription and p53 and is required for cell viability in mammals (Ford et al. 2006; Vakhrusheva 2008).

Functions of Sirtuins

In the past few years, a growing number of studies have uncovered many important roles of sirtuins in a variety of biological processes, such as transcription, apoptosis, DNA repair, cellular response to stress, neurodegeneration, development and energetic metabolism. Notably, most of these biological processes are critical in aging and linked to regulation of lifespan in multiple model organisms.

Sirtuins and Lifespan

The function of sirtuins in regulating lifespan was first demonstrated in *S. cerevisiae* (Kaeberlein et al. 1999). Lifespan of *S. cerevisiae* can be measured by two methods (Jazwinski et al. 1989; Fabrizio and Longo 2007). One is the replicative lifespan measured by counting the total number a mother cell can divide asymmetrically to generate daughter cells. The other is the chronological lifespan, which is defined by counting the survival rate of yeast cells in the stationary phase over the time. In 1995, Guarente and colleagues conducted a genetic screen in an effort to identify longevity genes in *S. cerevisiae* and found that a mutation in Sir4 (*sir4-42*) significantly increased replicative lifespan (Kennedy et al. 1995). Sir4 is known to interact with Sir2 and Sir3 to form a silent chromatin complex at the telomeres and silent mating-type loci in *S. cerevisiae*. Homologs of Sir2, not Sir3 and Sir4, have been identified in evolutionarily divergent organisms (Guarente 2007). Subsequent studies have found that deletion of sir2 decreases, and an extra copy of Sir2 extends, replicative lifespan, which indicates that Sir2 is the key lifespan regulator (Kennedy et al. 1997; Kaeberlein et al. 1999). Given the role of Sir2 in rDNA recombination and based on a series of further genetic analyses, Guarente and colleagues have proposed that replicative lifespan in yeast is regulated by the accumulation of extrachromosomal rDNA circles (ERCs) within a mother cell (Sinclair and Guarente 1997). ERCs are formed during homologous recombination between rDNA repeats, and an extra copy of ERCs in daughter cells is sufficient to cause premature aging. Sir2 is capable of inhibiting rDNA recombination (Rusche et al. 2002). Deletion of sir2 increases the rate of ERC formation and, in turn, shortens lifespan, while Sir2 overexpression reduces ERCs and extends the replicative lifespan by up to 30% (Sinclair and Guarente 1997; Kaeberlein et al. 1999). This indicates that Sir2 activation is beneficial to replicative lifespan. The influence of Sir2 on chronological lifespan in yeast, however, appears to be different. Deletion of *sir2* extends chronological lifespan under an extreme calorie restriction growth condition in *S. cerevisiae* (Fabrizio et al. 2005). It appears that chronological lifespan is mediated by the yeast AKT and S6 kinase homolog, Sch9, and the Ras pathways (Longo 2008). The role of ERC formation in chronological lifespan is not clear. These results suggest that Sir2 is a double-edged sword in regulating lifespan.

Subsequently, sirtuins in metazoans, especially *sir-2.1* in worms and dSir2 in flies, have been investigated for their role in modulating lifespan (Tissenbaum and Guarente 2001; Rogina and Helfand 2004). In contrary to yeast, loss-of-function mutations of *sir-2.1* do not appear to shorten the lifespan (Wood et al. 2004). However, extra copies of *sir-2.1* induce a lifespan extension by up to 50% in *C. elegans* (Tissenbaum and Guarente 2001). This extension requires *daf-16*, the downstream target of the insulin/IGF (insulin-like growth factor)-1 signaling pathway. DAF-16 is the ortholog of mammalian forkhead transcription factors, FOXOs (Forkhead box type O), and regulates lifespan by modulating transcription of many target genes (Lin et al. 1997; Ogg et al. 1997; Lee et al. 2003). Moreover, Berdichevsky and colleagues showed that the 14-3-3 protein FTT-2 acts as a scaffold to direct the interaction of DAF-16 and SIR-2.1 under heat shock stress in *C. elegans* (Berdichevsky et al. 2006). Taken together, a potential mechanism of gene expression regulation by SIR-2.1 involves modification of the transcription factor, DAF-16. In *D. melanogaster*, ubiquitous overexpression of dSir2 increases lifespan in females and males by an average of 29 and 18%, respectively (Rogina and Helfand 2004). Similar to yeast, loss-of-function dsir2 mutations result in lifespan shortening in *D. melanogaster*. Although increasing Sir2 orthologs extend lifespan in worms and flies, there is no evidence linking ERCs to aging in metazoans. The mechanism by which Sir2 orthologs retard aging is largely unknown and merits further study.

Findings from lower organisms have generated considerable interest in the effects of mammalian sirtuins on lifespan regulation. To date, at least seven mammalian sirtuins Sirt1–Sirt7 have been identified (Blander and Guarente 2004). Unfortunately, we do not know yet whether these proteins modulate aging. A number of studies, however, have demonstrated that mammalian sirtuins are implicated in many aging-related diseases and cell defense mechanisms (de Nigris et al. 2002; Cheng et al. 2003; van der Horst et al. 2004; Kume et al. 2007). These findings have led to the speculation that sirtuins can modulate healthspan, if not longevity, in mammals.

Sirtuins and Calorie Restriction

A key to understanding lifespan regulation by sirtuins may center on their NAD^+-dependent protein deacetylase and/or ADP-ribosyltransferase activity (Guarente 2007; Jiang 2008). Sirtuins appear to be sensors of cellular energetic or nutritional status indicated by the level of NAD^+ or the ratio of NAD^+/NADH. Calorie restriction (CR) or dietary restriction (DR) in invertebrates is a well-known non-genetic and dietary intervention that has been shown to extend lifespan in almost all the species tested (Masoro 2003; Ingram et al. 2006). The known beneficial effects of CR are extensive. In mammals, CR increases insulin sensitivity and improves glucose tolerance, reducing the incidence of diabetes (Chen and Guarente 2007). CR significantly reduces cancer and delays the onset of age-related functional decline through modulating the expression of many genes (Weindruch et al. 2001; Masoro 2003). Intensive investigation has been directed at defining the mechanistic basis of CR effects. Perhaps the most pronounced cellular phenotype is that CR induces a dramatic shift in energetic status relative to organisms maintained under the ad libitum (AL) conditions (Masoro 2003). Compelling evidence has shown that sirtuins play an important role in CR response at least partly through sensing NAD^+ levels or the NAD^+/NADH ratio (Chen and Guarente 2007). In *S. cerevisiae*, moderate CR, implemented by reducing the glucose concentration in the culture medium from 2 to 0.5%, significantly increases lifespan, and this extension is abolished by *sir2* deletion (Lin et al. 2000; Lin et al. 2004). In addition, moderate CR fails to further extend lifespan in yeast constructed to overexpress Sir2. These observations, comprising the first to demonstrate the connection between CR and sirtuins, have been verified in higher organisms (Chen and Guarente 2007).

In *C. elegans*, DR can be modeled using genetic mutants with defects in feeding, including the widely studied *eat-2* mutant (Lakowski and Hekimi 1998). *eat-2* encodes a nicotinic acetylcholine receptor subunit that mediates pharyngeal muscle function (Raizen et al. 1995). These mutants display a significantly longer lifespan than wild-type worms, an effect that is attributable to a reduction of food intake, measured by the pumping rate of the pharynx (Lakowski and Hekimi 1998). Importantly, *sir-2.1* mutations block the lifespan extension observed in *eat-2* mutants under standard dietary conditions (Wang and Tissenbaum 2006). In *D. melanogaster*, a common means of implementing DR, by comparison, consists of diluting the concentrations of key food ingredients, mainly sugar and yeast extract (Tatar 2007). Under these conditions, most studies report that lifespan is increased by more than 30% (Magwere et al. 2004; Partridge et al. 2005). Mutation of dSir2 blocks these prolongevity effects, while overexpression of dSir2 does not further extend the lifespan of flies maintained under DR (Rogina and Helfand 2004). Thus, parallel results from multiple species converge on the conclusion that sirtuins are key mediators of the response to common dietary interventions.

Nonetheless, considerable debate surrounds the potential involvement of sirtuins in the effects of CR. It is clear, for example, that several CR paradigms effectively increase lifespan in multiple organisms but that sirtuins are not required in all of these conditions. In contrast to the results seen with moderate CR in some studies mentioned above, Sir2 does not suppress lifespan extension induced by greater CR (reducing glucose from 2 to 0.05%) and, in some other cases, even not by moderate CR in *S. cerevisiae* (Kaeberlein et al. 2005; Kaeberlein and Powers 2007). The Sch9 and TOR pathways have been implicated in mediating this effect (Kaeberlein et al. 2005; Longo 2008). Another extreme CR condition studied in *C. elegans*, termed dietary (DD) or food deprivation, extends lifespan by approximately 40% when applied to sterile adult worms (Kaeberlein et al. 2006; Lee et al. 2006), and this effect is also independent of Sir-2.1. The heat shock response pathway mediated by heat shock transcription factor 1 (Hsf1) appears to be critical for DD-induced lifespan extension in *C. elegans* (Steinkraus et al. 2008). Whether or not sirtuins are required for lifespan extension induced by CR in mammals remains to be determined. These results highlight that sirtuins are not the only critical factors mediating the response in CR, and the functions of sirtuins in CR depend on dietary conditions.

Sirtuins and Transcriptional Regulation

Aging is associated with alterations of transcript levels for many genes (Weindruch et al. 2001; McCarroll et al. 2004; Melov and Hubbard 2004; Zhang et al. 2008). In eukaryotes, genomic DNA is packaged into chromatin, which is comprised of repeating units of nucleosomes. The nucleosome core consists of approximately 147 base pairs of DNA wrapped around a histone octamer consisting of two copies each of the core histones H2A, H2B, H3, and H4 (Jenuwein and Allis 2001). Chromatin structures are altered by covalent but often reversible modifications of free histone tails through acetylation, phosphorylation, methylation, and ubiquitination. The acetylation status of histone proteins is critical in epigenetic regulation of gene transcription. Removal of acetyl groups from lysine residues of histones results in heterochromatin and thus, gene repression. In contrast, transcriptionally active euchromatin is typically hyperacetylated (Grant et al. 1999; Fouladi 2006). Therefore, sirtuins may mediate transcription regulation through their protein deacetylase and/or ADP-ribosyltransferase activities, exerting a repressive on either transcription activators or repressors involved in transcription (Chen and Guarente 2007; Jiang 2008; Longo 2008).

Sir2, the first member of sirtuins, was initially identified as a transcription regulator in the silencing chromatin (Klar et al. 1979). Sir2-mediated silencing involves two distinct protein complexes: SIR complex and RENT (regulator of nucleolar silencing and telophase) complex (Gottschling et al. 1990; Aparicio et al. 1991; Gasser and Cockell 2001). The core SIR complex consists of Sir2, Sir3, and Sir4, which recognizes yeast mating-type *loci* and telomeric repeats and thus, represses the gene expression in these regions. RENT is a Sir2-containing protein complex, primarily located at the rDNA locus and involved in rDNA silencing. Sir2-compacted chromatin is characterized by hypoacetylation of lysine residues on the N-terminal tails of histones H3 (K9 and K14) and H4 (K16) (Guarente 1999; Suka et al. 2001; Robyr et al. 2002). This partly explains why the sir4 mutant sir4-42 lives longer. The mutation in sir4-42 disrupts the SIR complex and results in release of Sir2 from the silent mating-type loci and telomeres, which may increase the dosage of Sir2 at rDNA locus and, in turn, reduce ERC formation by enhancing rDNA silencing and reducing rDNA recombination (Kennedy et al. 1997; Sinclair and Guarente 1997). Similar to yeast Sir2, mammalian Sirt1 has been found to facilitate heterochromatin formation by histone hypoacetylation, particularly deacetylation of lysine residues of histones at positions 9/14 of H3 and 16 of H4 (Imai et al. 2000). Little is known about the sirtuin-containing protein complexes and their functions in transcription regulation in mammals.

Although the members of HDAC III class (sirtuins) were originally characterized as histone deacetylases, more and more non-histone targets of sirtuins have been described. Among these, transcription factors/cofactors are prominent as a group of proteins modified by sirtuins. For instance, numerous studies have demonstrated that p53, the tumor suppressor protein, and the forkhead transcription factors FOXOs are directly deacetylated by sirtuins in various biological processes (Luo et al. 2001; Vaziri et al. 2001; Langley et al. 2002; Brunet et al. 2004; Daitoku et al. 2004; van der Horst et al. 2004; Wang et al. 2007). TAF(I)68, a TATA-box binding protein (TBP)-associated factor, is the second largest subunit of the transcription initiation factor IB/SL1, which promotes transcription controlled by RNA polymerase I (Pol I) (Muth et al. 2001). Deacetylation of TAF_I68 by SIRT1 decreases its binding to the rDNA promoter, thereby repressing RNA Pol I-mediated rDNA transcription in vitro (Muth et al. 2001).

Sirtuins and Apoptosis

Apoptosis is a well-characterized biological process of cell death and an intrinsic part of the aging process. Several key apoptotic genes have been identified as sirtuin substrates (Fig. 2) and some are implicated in aging (Langley et al. 2002). One such substrate, p53, is a central apoptotic gene and acts as a tumor suppressor in mammals. Following DNA damage, the functions of p53 are regulated in a very complex manner by reversible phosphorylation, ADP-ribosylation, and acetylation. Acetylation of p53 at multiple sites is thought to stabilize the molecule, preserve its active function, and thereby trigger apoptosis and cell-cycle arrest (Li et al. 2002; Gostissa et al. 2003; Li et al. 2003). Sirt1 has been found to physically interact

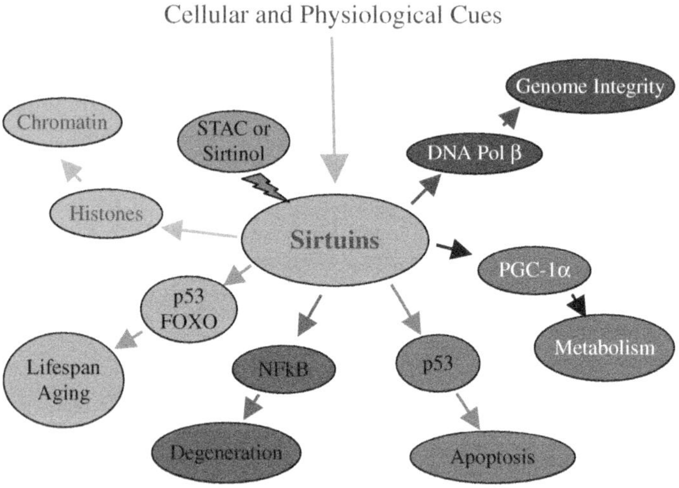

Fig. 2 Sirtuins modulate a number of biological processes. Sirtuins are highly conserved genes present from single cellular organisms to humans. Sirtuins modulate functions of many proteins through post-translational modifications and participate in various essential biological processes, such as aging, degeneration, metabolism, and apoptosis. Representative sirtuin targets are listed. Several activators (STACs) and inhibitors (Sirtinol) have been identified to at least partially affect the biological processes modulated by sirtuins

with p53 and repress DNA-damage-induced apoptosis (Luo et al. 2001; Vaziri et al. 2001). Deletion of Sirt1, or overexpression of truncated Sirt1 lacking part of the catalytic domain, results in hyperacetylation of p53 in embryo fibroblasts and a dramatic increase in p53-dependent apoptosis in thymocytes in mice (Cheng et al. 2003; Kamel et al. 2006; Solomon et al. 2006), suggesting that Sirt1 negatively regulates p53. Sirt1 also controls the p53-dependent apoptotic program through transcriptional regulation of the transcription factor, E2F1, which acts positively on p53 through the ARF/MDM2 pathway (Hershko and Ginsberg 2004; Chua et al. 2005). E2F1 is also capable of regulating Sirt1 expression, and interactions between E2F1 and Sirt1 may coordinate the apoptotic response to DNA damage (Wang et al. 2007). However, recent reports indicate that class I/II histone deacetylases, not just sirtuins (class III deacetylases), also attenuate p53 acetylation (Kamel et al. 2006; Solomon et al. 2006). These findings suggest that the regulation of p53 acetylation clearly extends beyond input from Sirt1.

Transcription factor FOXOs are also involved in apoptosis through regulating transcription of their target genes, such as the death receptor ligand FasL and proapoptotic BH3-only protein Bim3 (Gilley et al. 2003). The activities of FOXOs are regulated by phosphorylation and acetylation (Kenyon 2005). Growth factor-induced activation of phosphatidylinositol 3-kinase (PI3K) leads to an increase in the activity of the serine/threonine kinase AKT/protein kinase B (Anderson et al. 1998; Stephens et al. 1998), which in turn leads to phosphorylation and inactivation of FOXOs by preventing their translocation from the cytoplasm into the nucleus

(Cahill et al. 2001; Rena et al. 2001; Tzivion et al. 2001; Brunet et al. 2002). FOXOs function as transcription factors only after nuclear translocation. To a lesser extent than phosphorylation, the transcriptional activity of FOXOs is also regulated by acetylation/deacetylation. For instance, acetylation mediated by protein acetyltransferases PCAF and p300/CBP block the transcriptional function of FOXOs (Brunet et al. 2004), whereas deacetylation by Sirt1 and other class I and II HDACs restore this activity (Motta et al. 2004; van der Horst et al. 2004). Mammalian FOXOs include FOXO1 (FKHR), FOXO3 (FKHRL1), FOXO4 (AFX), and FOXO6. Among these, Sirt1 is known to deacetylate FOXO1, 3 and 4 (Brunet et al. 2004; Daitoku et al. 2004; van der Horst et al. 2004), and Sirt2 deacetylates FOXO3 (Wang et al. 2007). Thus, sirtuins are positioned to regulate apoptosis through the specific and bidirectional modulation of FOXOs.

Mitochondrial cytochrome *c* plays a major role in triggering apoptosis (Liu et al. 1996). Blc-2 and related proteins potently control the release of cytochrome *c* from mitochondria, thus regulating the apoptotic program (Wang 2001). Bax is a proapoptotic protein in this regulation complex. Ku70 is able to interact with the deacetylated form of Bax and sequester Bax from mitochondria. In this context it is noteworthy that Sirt1 deacetylates Bax, promoting the interaction of Bax and Ku70 to suppress Bax-mediated apoptosis (Cohen et al. 2004a).

Sirtuins and Metabolism

Growing evidence points to the importance of metabolism in the modulation of aging (Masoro 2003). As discussed in a previous section, sirtuins mediate the response of various organisms to at least a subset of CR paradigms that extend lifespan. To date, sirtuins have been shown to play key roles in metabolic control by deacetylating proteins involved in metabolism, such as PGC-1α and acetyl-CoA synthetase (AceCS) (Nemoto et al. 2005; Hallows et al. 2006; Schwer et al. 2006; Rodgers et al. 2008). PGC-1α is a member of a small family of transcriptional cofactors that possess a common function in mitochondrial biogenesis. As a transcriptional cofactor, PGC-1α initially was found to interact with peroxisome proliferator-activated receptor gamma (PPARγ). PGC-1α modulates target gene expression mainly through interaction with DNA-binding proteins and serves as a key regulator of adipogenesis and fat storage through controlling expression of many adipocyte-specific genes (Puigserver et al. 1998; Spiegelman et al. 2000; Picard and Auwerx 2002). Sirt1 physically binds to and deacetylates PGC-1α and thus maintains the interaction of PGC-1α and certain transcriptional factors (Nemoto et al. 2005; Rodgers et al. 2008). Notably, CR or fasting induces Sirt1 expression in liver, facilitating activation of PGC-1α to regulate genes involved in gluconeogenic and fatty acid oxidation and stimulating hepatic glucose output (Cohen et al. 2004b; Rodgers et al. 2008). AceCSs comprise another route of metabolic modulation regulated by sirtuins. Acetyl-CoA is a small molecule which enters fatty acid synthesis and tricarboxylic acid cycle. The synthesis of Acetyl-CoA in mammals is mediated by AceCS1 and AceCS2 predominantly in the cytoplasm and mitochondria, respectively. CR and fasting increase the activity of Sirt1 and Sirt3, which deacetylates, and in turn activates, AceCS1 and AceCS2 (Frye 1999; Tanny et al. 1999). Thus, elevated sirtuin levels induced by low blood glucose may regulate the rate of fatty acid synthesis and control carbon input for ATP production in the tricarboxylic acid cycle.

Sirtuins and Genomic Integrity

Human premature aging disorders, or progeroid syndromes, are associated with deficiency in DNA repair caused by mutations in a number of DNA damage response genes, such as RecQ helicases (Brosh and Bohr 2007). In *S. cerevisiae*, Sir2 along with other histone deacetylases are recruited to the double-strand DNA breaks (DSB) to mediate homologous recombination repair directly or indirectly by inducing deacetylation of histones (Lewis and Resnick 2000; Fernandez-Capetillo and Nussenzweig 2004; Jazayeri et al. 2004; Tamburini and Tyler 2005). In mammals, Sirt6 has been suggested to function in DNA repair as Sirt6-deficient cells exhibit elevated genomic instability and sensitivity to genotoxins. These deficiencies can be rescued by expression of DNA polymerase β (Pol β), a key enzyme involved in DNA repair (Mostoslavsky et al. 2006). However, the precise mechanism by which SIRT6 influences DNA repair is not clear yet. One possibility is that Sirt6 modulates the activity of DNA repair factors by deacetylation and/or ADP-ribosylation, consistent with the observation that Sirt6 weakly deacetylates Pol β in vitro (Mostoslavsky et al. 2006). Alternatively, Sirt6 may be indirectly involved by altering chromatin structure to allow access of repair factors to sites of DNA damage. Sirt6 has been shown to modulate telomeric chromatin by deacetylation of histone H3 lysine 9 (Michishita et al. 2008). The role of sirtuins in DSB repair also stems from the observation that DNA breaks activate poly-ADP ribose polymerase-1 (PARP-1), which competes with sirtuins for the cofactor NAD^+ and produces nicotinamide, a potent inhibitor of sirtuins (Anderson et al. 2003; Malanga and Althaus 2004). Activation of PARP-1 at the sites of DNA strand breaks may lead to local increases in nicotinamide concentration, inhibiting sirtuins precisely at the site of damage, enabling histone acetylation-mediated chromatin decondensation and recruitment of repair enzymes.

Sirtuins and Degenerative Diseases

Neurodegeneration and muscular degeneration are commonly associated with aging in many multiple-cellular organisms including humans. Emerging evident suggests that sirtuins play an important role in the progression of these disorders. CR has been shown to protect neurons against degeneration at least partly through modulation of Sirt1 as shown in transgenic mouse models of Alzheimer's (AD) and

Parkinson's disease (Luchsinger et al. 2002; Maswood et al. 2004; Patel et al. 2005; Wang et al. 2005). One of the pathological hallmarks of AD is the presence of amyloid plaques (Hardy 2006), which are composed of amyloid-β (Aβ) peptide derived from sequential process of the amyloid precursor protein by β-secretase and γ-secretase (Masters et al. 1985; Selkoe 2001; Haass 2004). Although the subject of continuing debate, the accumulation of aggregated Aβ is hypothesized to initiate the pathophysiological progression of AD (Hardy 2006). Aβ peptides induce NFκB activity and inflammation, while Sirt1 activation dramatically decreases NFκB signaling (Chen et al. 2005). These findings raise the possibility that Sirt1 activation may attenuate Aβ-stimulated neurotoxicity and AD-related inflammatory responses (Qin et al. 2006). Another neurodegenerative disease, Huntington's disease (HD), is accompanied by mitochondrial insufficiency. These mitochondrial impairments are coupled with dysregulation of PGC-1α by the mutant huntingtin (Cui et al. 2006; St-Pierre et al. 2006; Weydt et al. 2006). It has been reported that Sirt1 regulates mitochondrial metabolism by modulating PGC-1α (Lagouge et al. 2006; Rodgers et al. 2008). Taken together, available data suggest that activation of sirtuins may comprise a novel therapeutic approach for a variety of neurodegenerative diseases (Parker et al. 2005).

In the context of the present discussion it is important to recognize that suppression of sirtuin activity is not uniformly detrimental. Polyalanine expansion in poly (A)-binding protein, nuclear 1 (PABPN1), is associated with human oculopharyngeal muscular dystrophy (OPMD) (Fan et al. 2001; Abu-Baker and Rouleau 2007). An invertebrate model of OPMD has been established by introducing the mutant human PABPN1 in *C. elegans* (Catoire et al. 2008). In contrast to the outcome in AD and HD models, a sir-2.1 loss-of-function mutation delays, while extra copies of sir-2.1 exacerbate, muscle cell degeneration and abnormal motility in PABPN1 worms. This function of sir-2.1 also requires the coordinated functions of the transcription factor *daf-16/* FoxO and nutrient sensor *aak-2/*AMPK. It remains to be determined whether sirtuins can modify the pathogenesis of muscular dystrophy in mammals.

Sirtuin Activators and Suppressors

Considerable effort is being directed at the identification of compounds or drugs that can extend lifespan, increase healthspan, and mitigate aging-related functional decline. The unique features of sirtuins make them ideal as targets for such pro-longevity interventions (Chen and Guarente 2007). First, sirtuin activation can increase lifespan and delay the onset of age-related functional decline in organisms as diverse as yeast, worms, flies, and fish, as well as mice under a high-fat diet (Kaeberlein et al. 1999; Wood et al. 2004; Baur et al. 2006; Valenzano et al. 2006; Guarente 2007). In some cases, sirtuin inhibition also has beneficial effects on aging (Longo 2008). Second, sirtuins have been shown to have NAD^+-dependent protein deacetylase or ADP-ribosyltransferase activity (Chen and Guarente 2007). There are established and reliable methods of high throughput screening for activators and inhibitors of proteins or enzymes employing NAD^+ as a cofactor (Pallas et al. 2008). Therefore, active efforts to identify sirtuin-activating compounds (STACs) were launched not long after their discovery as NAD^+-dependent enzymes.

In 2003, Sinclair and colleagues published the first study to screen for activators of the protein deacetylase activity of Sirt1 using a number of small molecular libraries (Howitz et al. 2003). Although the reliability of the deacetylation assay has been questioned (Kaeberlein et al. 2005), this initial screen identified several structure-related polyphenolic compounds as STACs. Polyphenols comprise a large family of plant secondary metabolites including flavonoids, stilbene, and anthocyanidine, which mediate color diversity in fruits and plants (Bravo 1998). Resveratrol was among the most potent STACs to stimulate deacetylase activity of Sirt1 and Sir2. Resveratrol (3,5,4′-trihydroxystilbene) is present in many fruits and is especially enriched in grape skins. A growing list of health benefits of resveratrol has been reported including potent antioxidant activity, anti-cancer, and anti-inflammatory responses (Sun et al. 2002; Baur and Sinclair 2006). Significantly, resveratrol has been found to increase average replicative lifespan by 70% in *S. cerevisiae*. This effect is abolished by sir2 deletion, and resveratrol fails to extend lifespan further in yeast with extra copies of Sir2, suggesting that Sir2 is a target of resveratrol (Howitz et al. 2003).

The discovery of prolongevity function of resveratrol in yeast immediately prompted several studies to evaluate whether this longevity effect is evolutionarily conserved in higher organisms, with an eye toward potential intervention in human aging. Although some have challenged the finding on methodological grounds, Wood et al. found that resveratrol can stimulate protein deacetylase of *C. elegans* SIR-2.1 and *D. melanogaster* dSir2, indicating that resveratrol functions as a STAC in metazoans (Wood et al. 2004; Kaeberlein et al. 2005). Resveratrol extends lifespan of the nematode by up to 14%, the fly by up to 29%. Consistent with observations in yeast, these prolongevity effects required the worm and fly sirtuins, *sir-2.1* and dSir2, respectively. Together, these findings suggest that sirtuin is a prominent target of resveratrol. In addition, resveratrol fails to increase the lifespan extension normally seen in flies maintained under DR, i.e., an effect that requires dSir2. These results suggest that resveratrol extends lifespan in the metazoan partly through DR pathways mediated by sirtuins. As for higher organisms, Valenzano et al. demonstrated that resveratrol extends lifespan, delays the onset and age-related accumulation of damage, and improves cognitive function in a short-lived fish, *Nothobranchius furzeri* (Valenzano and Cellerino 2006; Valenzano et al. 2006). Extending this line of investigation to rodents, Baur et al. found that resveratrol can extend lifespan and improve motor function in middle-aged mice fed on a high-fat diet (Baur et al. 2006). In this study, resveratrol increased the deacetylase activity of Sirt1 and likely activated AMPK through increased phosphorylation level. Moreover, this investigation has shown that the genome-wide transcript changes induced by resveratrol share extensive overlap with those induced by a CR protocol known to extend lifespan in rodents (Masoro 2003; Baur et al. 2006). Consistent with these findings, Barger et al. used another standard CR protocol, providing mice 40% of daily ad libitum food and reported that supplementation with a low concentration of resveratrol induced a pattern of gene expression highly similar to the effects of CR (Barger et al. 2008). Thus, evidence from multiple species suggests that resveratrol is a potent prolongevity compound that functions as an effective CR mimetic. Although the prolongevity effect of resveratrol was originally discovered by screening for STACs, it targets a number of genes related to aging and age-related diseases, such as AMPK and cyclooxygenases (Harikumar and Aggarwal 2008; Pirola and Frojdo 2008). In addition, lifespan extension by CR is mediated by a number of diverse pathways, including sirtuin, AKT, and TOR pathways (Guarente 2007; Bartke et al. 2008; Cuervo 2008). On the basis of these findings, it is unlikely that the prolongevity effects of resveratrol are mediated exclusively by sirtuins.

Alongside this compelling evidence, several studies have failed to detect a prolongevity influence of resveratrol. Kaeberlein et al., for example, reported no effect of resveratrol on lifespan extension or Sir2 activation in three different *S. cerevisiae* strain backgrounds (Kaeberlein et al. 2005), and others obtained similarly negative results (Howitz et al. 2003). Testing a wide range of resveratrol concentrations, Bass et al. observed no consistent lifespan extension in two wild-type *D. melanogaster* strains (i.e., Dahomy and Canton S) or in the nematode N2 strain (Bass et al. 2007). Indeed the same authors found a slight lifespan extension in certain experiments using a worm *sir-2.1* mutant strain, in striking contrast to evidence that the effects of resveratrol are SIR-2.1 dependent (Wood et al. 2004; Bass et al. 2007). The basis of this apparent discrepancy remains to be determined and current speculation centers on possible species or strain specificity of sirtuins and resveratrol. Extending these negative results, Pearson et al. recently reported that resveratrol does not extend lifespan in mice fed a standard chow diet, but surprisingly that low-concentration resveratrol (approximately 8 mg/kg/day) led to modest lifespan extension in middle-aged mice maintained on an every-other-day feeding CR paradigm (Pearson et al. 2008). Together, these rodent studies point to potential synergistic, interactive effects of resveratrol and CR. Nonetheless, consistent with other studies, resveratrol supplementation has yielded a wide range of health benefits in animals under all diet conditions, including bone health, motor function, and cardiovascular function. Further investigations are needed to resolve these discrepancies and further evaluate the anti-aging effects of resveratrol. It remains to be determined, for example, whether resveratrol extends lifespan when treatment is initiated early, in relatively young animals. Nonetheless, current evidence calls into question the view that resveratrol will prove to be a magic anti-aging bullet.

Given the importance of sirtuin activation in aging and aging-related diseases, Milne et al. conducted another screen using a large collection of small molecules and identified additional STACs that are structurally distinct from resveratrol (Milne et al. 2007). Three such STACs, SRT1460, SRT 2183, and SRT1720, were found to be 1000-fold more potent than resveratrol and more selective for activation of SIRT1 versus its closest homologs, SIRT2 and SIRT3. One of these STACs, SRT1720, significantly improves glucose metabolism, increases insulin sensitivity and glucose tolerance in two rodent models of diabetes, the genetically obese mice ($Lep^{ob/ob}$) and diet-induced obesity (DIO). Unpublished results cited in that study indicate that analogues structurally related to SRT1460 can activate Sir2 and extend lifespan in yeast (Milne et al. 2007). Whether these compounds will prove to be effective prolongevity compounds in other species remains to be determined.

Although compelling evidence indicates that sirtuin activation provides health benefits, and sirtuin suppression can accelerate aging, data reviewed in previous sections also suggest that sirtuin suppression is associated with certain protective functions. Several Sirt1 inhibitors have been identified including nicotinamide and the Sir2 inhibitor napthol (sirtinol) (Grozinger et al. 2001; Anderson et al. 2003). Sirtinol delays muscular degeneration in a worm model of human muscular dystrophy OPMD, and this suppression requires Daf-16 (Catoire et al. 2008). In addition, sirtinol enhances the survival of mammalian cells expressing mutant PABPN1, a gene involved in human OPMD. In this degeneration model, the sirtuin activation activity of resveratrol accelerates muscular cell degeneration. This suggests that the beneficial effects of sirtuin activators and inhibitors are context dependent.

Conclusion and Future Directions

Sirtuins are evolutionarily conserved proteins and function in a wide range of biological processes (Guarente 2007). Many of these processes have direct and indirect effects on aging. Sirtuins function as protein deacetylase or ADP-ribosyltransferase to post-translationally modify many targets, which in turn regulates a broad constellation of biological processes through the epigenetic control of gene expression. Sirtuins can regulate lifespan and aging processes in evolutionarily diverse species partly through modulating CR pathways and linking the nutritional status of the cell to transcriptional regulation through its NAD^+-dependent deacetylase or ADP-ribosyltransferase.

There are still many questions and unsolved controversies in the sirtuin field. One of the most important is that the relationship between sirtuin and lifespan, especially CR-mediated lifespan, needs further clarification. Deletion of *sir2* decreases replicative lifespan but increases chronological lifespan depending on the dietary condition in *S. cerevisiae* (Kaeberlein et al. 1999; Lin et al. 2000; Lin et al. 2004; Fabrizio et al. 2005; Longo 2008). Loss-of-function of sirtuins does not shorten the lifespan in *C. elegans* but decreases lifespan in *D. melanogaster* (Tissenbaum and Guarente 2001, Wood et al. 2004; Rogina and Helfand 2004). In addition, it is not yet known whether sirtuin overexpression can extend lifespan in mammals, although Sirt1 overexpressing mice share many features with CR animals (Picard et al. 2004; Hasegawa et al. 2008; Pfluger et al. 2008). A second significant issue concerns the tissue-specific function of sirtuins in aging. In mammals, sirtuins are differentially expressed and regulated by various environmental and genetic factors, including CR, in different tissues (Chen et al. 2008; Pedersen et al. 2008). Sirt1 expression is induced in brain, liver, and muscle, but suppressed in adipose tissue under CR. It will be interesting to sort out the role of Sirt1 and other sirtuins in mediating prolongevity effects in different tissues. The third unsolved issue is the organelle-specific function of sirtuins in aging. Different members of the sirtuin family are targeted to different cellular compartments (Frye 1999; North et al. 2003; Michishita et al. 2005). For example, Sirt1 is primarily located in the nucleus, Sirt2 in the cytoplasm, and Sirt3 in the mitochondria (Blander and Guarente 2004). How these compartment-specific functions and their interaction are involved in aging will be an important focus for future investigation. Lastly, greater attention should be directed toward understanding the temporal dynamics of sirtuin function, particularly in an effort to develop STACs as potential drugs against aging-related disease and functional decline. Ideally, anti-aging drugs should be provided to the elderly instead of starting interventions at very young ages in humans. Resveratrol improves age-related functions of middle-aged mice (Baur et al. 2006; Pearson et al. 2008), suggesting that it retains efficacy when initiated relatively late in life. Nevertheless, as key epigenetic regulators, sirtuins play an important role in aging through mediating numerous biological pathways. Dissecting these pathways will no doubt provide insights into aging, and more importantly, lead to better interventions for promoting optimally healthy aging.

References

Abu-Baker, A., and Rouleau, G. A. 2007. Oculopharyngeal muscular dystrophy: recent advances in the understanding of the molecular pathogenic mechanisms and treatment strategies. Biochim Biophys Acta 1772: 173–185.

Ahuja N., Schwer B., Carobbio S., Waltregny D., North B. J., Castronovo V., Maechler P., Verdin E. 2007. Regulation of insulin secretion by SIRT4, a mitochondrial ADP-ribosyltransferase. J Biol Chem. 282:33583–92.

Anderson, K. E., Coadwell, J., Stephens, L. R., and Hawkins, P. T. 1998. Translocation of PDK-1 to the plasma membrane is important in allowing PDK-1 to activate protein kinase B. Curr Biol 8: 684–691.

Anderson, R. M., Bitterman, K. J., Wood, J. G., Medvedik, O., and Sinclair, D. A. 2003. Nicotinamide and PNC1 govern lifespan extension by calorie restriction in Saccharomyces cerevisiae. Nature 423: 181–185.

Aparicio, O. M., Billington, B. L., and Gottschling, D. E. 1991. Modifiers of position effect are shared between telomeric and silent mating-type loci in S. cerevisiae. Cell 66: 1279–1287.

Astrom, S. U., Cline, T. W., and Rine, J. 2003. The Drosophila melanogaster sir2+ gene is nonessential and has only minor effects on position-effect variegation. Genetics 163: 931–937.

Barger, J. L., Kayo, T., Vann, J. M., Arias, E. B., Wang, J., Hacker, T. A., Wang, Y., Raederstorff, D., Morrow, J. D., Leeuwenburgh, C., et al. 2008. A low dose of dietary resveratrol partially mimics caloric restriction and retards aging parameters in mice. PLoS ONE 3: e2264.

Barlow, A. L., van Drunen, C. M., Johnson, C. A., Tweedie, S., Bird, A., and Turner, B. M. 2001. dSIR2 and dHDAC6: two novel, inhibitor-resistant deacetylases in Drosophila melanogaster. Exp Cell Res 265: 90–103.

Bartke, A., Bonkowski, M., and Masternak, M. 2008. THow diet interacts with longevity genes. Hormones (Athens) 7: 17–23.

Bass, T. M., Weinkove, D., Houthoofd, K., Gems, D., and Partridge, L. 2007. Effects of resveratrol on lifespan in Drosophila melanogaster and Caenorhabditis elegans. Mech Ageing Dev 128: 546–552.

Baur, J. A., Pearson, K. J., Price, N. L., Jamieson, H. A., Lerin, C., Kalra, A., Prabhu, V. V., Allard, J. S., Lopez-Lluch, G., Lewis, K., et al. 2006. Resveratrol improves health and survival of mice on a high-calorie diet. Nature 444: 337–342.

Baur, J. A., and Sinclair, D. A. 2006. Therapeutic potential of resveratrol: the in vivo evidence. Nat Rev Drug Discov 5: 493–506.
Berdichevsky, A., Viswanathan, M., Horvitz, H. R., and Guarente, L. 2006. C. elegans SIR-2.1 interacts with 14-3-3 proteins to activate DAF-16 and extend life span. Cell 125: 1165–1177.
Blander, G., and Guarente, L. 2004. The Sir2 family of protein deacetylases. Annu Rev Biochem 73: 417–435.
Brachmann, C. B., Sherman, J. M., Devine, S. E., Cameron, E. E., Pillus, L., and Boeke, J. D. 1995. The SIR2 gene family, conserved from bacteria to humans, functions in silencing, cell cycle progression, and chromosome stability. Genes Dev 9: 2888–2902.
Bravo, L. 1998. Polyphenols: chemistry, dietary sources, metabolism, and nutritional significance. Nutr Rev 56: 317–333.
Brosh, R. M., Jr., and Bohr, V. A. 2007. Human premature aging, DNA repair and RecQ helicases. Nucleic Acids Res 35: 7527–7544.
Brunet, A., Kanai, F., Stehn, J., Xu, J., Sarbassova, D., Frangioni, J. V., Dalal, S. N., DeCaprio, J. A., Greenberg, M. E., and Yaffe, M. B. 2002. 14-3-3 transits to the nucleus and participates in dynamic nucleocytoplasmic transport. J Cell Biol 156: 817–828.
Brunet, A., Sweeney, L. B., Sturgill, J. F., Chua, K. F., Greer, P. L., Lin, Y., Tran, H., Ross, S. E., Mostoslavsky, R., Cohen, H. Y., et al. 2004. Stress-dependent regulation of FOXO transcription factors by the SIRT1 deacetylase. Science 303: 2011–2015.
Bryk, M., Banerjee, M., Murphy, M., Knudsen, K. E., Garfinkel, D. J., and Curcio, M. J. 1997. Transcriptional silencing of Ty1 elements in the RDN1 locus of yeast. Genes Dev 11: 255–269.
Cahill, C. M., Tzivion, G., Nasrin, N., Ogg, S., Dore, J., Ruvkun, G., and Alexander-Bridges, M. 2001. Phosphatidylinositol 3-kinase signaling inhibits DAF-16 DNA binding and function via 14-3-3-dependent and 14-3-3-independent pathways. J Biol Chem 276: 13402–13410.
Catoire, H., Pasco, M. Y., Abu-Baker, A., Holbert, S., Tourette, C., Brais, B., Rouleau, G. A., Parker, J. A., and Neri, C. 2008. Sirtuin inhibition protects from the polyalanine muscular dystrophy protein PABPN1. Hum Mol Genet 17: 2108–2117.
Chen, D., Bruno, J., Easlon, E., Lin, S. J., Cheng, H. L., Alt, F. W., and Guarente, L. 2008. Tissue-specific regulation of SIRT1 by calorie restriction. Genes Dev 22: 1753–1757.
Chen, D., and Guarente, L. 2007. SIR2: a potential target for calorie restriction mimetics. Trends Mol Med 13: 64–71.
Chen, J., Zhou, Y., Mueller-Steiner, S., Chen, L. F., Kwon, H., Yi, S., Mucke, L., and Gan, L. 2005. SIRT1 protects against microglia-dependent amyloid-beta toxicity through inhibiting NF-kappaB signaling. J Biol Chem 280: 40364–40374.
Cheng, H. L., Mostoslavsky, R., Saito, S., Manis, J. P., Gu, Y., Patel, P., Bronson, R., Appella, E., Alt, F. W., and Chua, K. F. 2003. Developmental defects and p53 hyperacetylation in Sir2 homolog (SIRT1)-deficient mice. Proc Natl Acad Sci USA 100: 10794–10799.
Chua, K. F., Mostoslavsky, R., Lombard, D. B., Pang, W. W., Saito, S., Franco, S., Kaushal, D., Cheng, H. L., Fischer, M. R., Stokes, N., et al. 2005. Mammalian SIRT1 limits replicative life span in response to chronic genotoxic stress. Cell Metab 2: 67–76.
Cohen, H. Y., Lavu, S., Bitterman, K. J., Hekking, B., Imahiyerobo, T. A., Miller, C., Frye, R., Ploegh, H., Kessler, B. M., and Sinclair, D. A. 2004a. Acetylation of the C terminus of Ku70 by CBP and PCAF controls Bax-mediated apoptosis. Mol Cell 13: 627–638.
Cohen, H. Y., Miller, C., Bitterman, K. J., Wall, N. R., Hekking, B., Kessler, B., Howitz, K. T., Gorospe, M., de Cabo, R., and Sinclair, D. A. 2004b. Calorie restriction promotes mammalian cell survival by inducing the SIRT1 deacetylase. Science 305: 390–392.
Cuervo, A. M. 2008. Calorie restriction and aging: the ultimate "cleansing diet". J Gerontol A Biol Sci Med Sci 63: 547–549.
Cui, L., Jeong, H., Borovecki, F., Parkhurst, C. N., Tanese, N., and Krainc, D. 2006. Transcriptional repression of PGC-1alpha by mutant huntingtin leads to mitochondrial dysfunction and neurodegeneration. Cell 127: 59–69.
Cuperus, G., Shafaatian, R., and Shore, D. 2000. Locus specificity determinants in the multifunctional yeast silencing protein Sir2. Embo J 19: 2641–2651.

Daitoku, H., Hatta, M., Matsuzaki, H., Aratani, S., Ohshima, T., Miyagishi, M., Nakajima, T., and Fukamizu, A. 2004. Silent information regulator 2 potentiates Foxo1-mediated transcription through its deacetylase activity. Proc Natl Acad Sci USA 101: 10042–10047.

de Nigris, F., Cerutti, J., Morelli, C., Califano, D., Chiariotti, L., Viglietto, G., Santelli, G., and Fusco, A. 2002. Isolation of a SIR-like gene, SIR-T8, that is overexpressed in thyroid carcinoma cell lines and tissues. Br J Cancer 86: 917–923.

Denu, J. M. 2003. Linking chromatin function with metabolic networks: Sir2 family of NAD(+)-dependent deacetylases. Trends Biochem Sci 28: 41–48.

Derbyshire, M. K., Weinstock, K. G., and Strathern, J. N. 1996. HST1, a new member of the SIR2 family of genes. Yeast 12: 631–640.

Durand-Dubief M., Sinha I., Fagerström-Billai F., Bonilla C., Wright A., Grunstein M., Ekwall K. 2007. Specific functions for the fission yeast Sirtuins Hst2 and Hst4 in gene regulation and retrotransposon silencing. EMBO J. 26(10):2477–88.

Fabrizio, P., Gattazzo, C., Battistella, L., Wei, M., Cheng, C., McGrew, K., and Longo, V. D. 2005. Sir2 blocks extreme life-span extension. Cell 123: 655–667.

Fabrizio, P., and Longo, V. D. 2007. The chronological life span of Saccharomyces cerevisiae. Methods Mol Biol 371: 89–95.

Fan, X., Dion, P., Laganiere, J., Brais, B., and Rouleau, G. A. 2001. Oligomerization of polyalanine expanded PABPN1 facilitates nuclear protein aggregation that is associated with cell death. Hum Mol Genet 10: 2341–2351.

Fernandez-Capetillo, O., and Nussenzweig, A. 2004. Linking histone deacetylation with the repair of DNA breaks. Proc Natl Acad Sci USA 101: 1427–1428.

Ford, E., Voit, R., Liszt, G., Magin, C., Grummt, I., and Guarente, L. 2006. Mammalian Sir2 homolog SIRT7 is an activator of RNA polymerase I transcription. Genes Dev 20: 1075–1080.

Fouladi, M. 2006. Histone deacetylase inhibitors in cancer therapy. Cancer Invest 24: 521–527.

Frye, R. A. 1999. Characterization of five human cDNAs with homology to the yeast SIR2 gene: Sir2-like proteins (sirtuins) metabolize NAD and may have protein ADP-ribosyltransferase activity. Biochem Biophys Res Commun 260: 273–279.

Gao, L., Cueto, M. A., Asselbergs, F., and Atadja, P. 2002. Cloning and functional characterization of HDAC11, a novel member of the human histone deacetylase family. J Biol Chem 277: 25748–25755.

Gasser, S. M., and Cockell, M. M. 2001. The molecular biology of the SIR proteins. Gene 279: 1–16.

Ghidelli, S., Donze, D., Dhillon, N., and Kamakaka, R. T. 2001. Sir2p exists in two nucleosome-binding complexes with distinct deacetylase activities. Embo J 20: 4522–4535.

Gilley, J., Coffer, P. J., and Ham, J. 2003. FOXO transcription factors directly activate bim gene expression and promote apoptosis in sympathetic neurons. J Cell Biol 162: 613–622.

Gostissa, M., Hofmann, T. G., Will, H., and Del Sal, G. 2003. Regulation of p53 functions: let's meet at the nuclear bodies. Curr Opin Cell Biol 15: 351–357.

Gottschling, D. E., Aparicio, O. M., Billington, B. L., and Zakian, V. A. 1990. Position effect at S. cerevisiae telomeres: reversible repression of Pol II transcription. Cell 63: 751–762.

Grant, P. A., Berger, S. L., and Workman, J. L. 1999. Identification and analysis of native nucleosomal histone acetyltransferase complexes. Methods Mol Biol 119: 311–317.

Grozinger, C. M., Chao, E. D., Blackwell, H. E., Moazed, D., and Schreiber, S. L. 2001. Identification of a class of small molecule inhibitors of the sirtuin family of NAD-dependent deacetylases by phenotypic screening. J Biol Chem 276: 38837–38843.

Guarente, L. 1999. Diverse and dynamic functions of the Sir silencing complex. Nat Genet 23: 281–285.

Guarente, L. 2007. Sirtuins in aging and disease. Cold Spring Harb Symp Quant Biol 72: 483–488.

Haass, C. 2004. Take five – BACE and the gamma-secretase quartet conduct Alzheimer's amyloid beta-peptide generation. Embo J 23: 483–488.

Haigis, M. C., Mostoslavsky, R., Haigis, K. M., Fahie, K., Christodoulou, D. C., Murphy, A. J., Valenzuela, D. M., Yancopoulos, G. D., Karow, M., Blander, G., et al. 2006. SIRT4 inhibits

glutamate dehydrogenase and opposes the effects of calorie restriction in pancreatic beta cells. Cell 126: 941–954.
Hallows, W. C., Lee, S., and Denu, J. M. 2006. Sirtuins deacetylate and activate mammalian acetyl-CoA synthetases. Proc Natl Acad Sci USA 103: 10230–10235.
Hardy, J. 2006. Alzheimer's disease: the amyloid cascade hypothesis: an update and reappraisal. J Alzheimers Dis 9: 151–153.
Harikumar, K. B., and Aggarwal, B. B. 2008. Resveratrol: a multitargeted agent for age-associated chronic diseases. Cell Cycle 7: 1020–1035.
Hasegawa, K., Wakino, S., Yoshioka, K., Tatematsu, S., Hara, Y., Minakuchi, H., Washida, N., Tokuyama, H., Hayashi, K., and Itoh, H. 2008. Sirt1 protects against oxidative stress-induced renal tubular cell apoptosis by the bidirectional regulation of catalase expression. Biochem Biophys Res Commun 372: 51–56.
Hershko, T., and Ginsberg, D. 2004. Up-regulation of Bcl-2 homology 3 (BH3)-only proteins by E2F1 mediates apoptosis. J Biol Chem 279: 8627–8634.
Howitz, K. T., Bitterman, K. J., Cohen, H. Y., Lamming, D. W., Lavu, S., Wood, J. G., Zipkin, R. E., Chung, P., Kisielewski, A., Zhang, L. L., et al. 2003. Small molecule activators of sirtuins extend Saccharomyces cerevisiae lifespan. Nature 425: 191–196.
Imai, S., Armstrong, C. M., Kaeberlein, M., and Guarente, L. 2000. Transcriptional silencing and longevity protein Sir2 is an NAD-dependent histone deacetylase. Nature 403: 795–800.
Ingram, D. K., Zhu, M., Mamczarz, J., Zou, S., Lane, M. A., Roth, G. S., and deCabo, R. 2006. Calorie restriction mimetics: an emerging research field. Aging Cell 5: 97–108.
Jazayeri, A., McAinsh, A. D., and Jackson, S. P. 2004. Saccharomyces cerevisiae Sin3p facilitates DNA double-strand break repair. Proc Natl Acad Sci USA 101: 1644–1649.
Jazwinski, S. M., Egilmez, N. K., and Chen, J. B. 1989. Replication control and cellular life span. Exp Gerontol 24: 423–436.
Jenuwein, T., and Allis, C. D. 2001. Translating the histone code. Science 293: 1074–1080.
Jiang, W. J. 2008. Sirtuins: novel targets for metabolic disease in drug development. Biochem Biophys Res Commun 373: 341–344.
Kaeberlein, M., McDonagh, T., Heltweg, B., Hixon, J., Westman, E. A., Caldwell, S. D., Napper, A., Curtis, R., DiStefano, P. S., Fields, S., et al. 2005. Substrate-specific activation of sirtuins by resveratrol. J Biol Chem 280: 17038–17045.
Kaeberlein, M., McVey, M., and Guarente, L. 1999. The SIR2/3/4 complex and SIR2 alone promote longevity in Saccharomyces cerevisiae by two different mechanisms. Genes Dev 13: 2570–2580.
Kaeberlein, M., and Powers, R. W., 3rd 2007. Sir2 and calorie restriction in yeast: a skeptical perspective. Ageing Res Rev 6: 128–140.
Kaeberlein, M., Powers, R. W., 3rd, Steffen, K. K., Westman, E. A., Hu, D., Dang, N., Kerr, E. O., Kirkland, K. T., Fields, S., and Kennedy, B. K. 2005. Regulation of yeast replicative life span by TOR and Sch9 in response to nutrients. Science 310: 1193–1196.
Kaeberlein, T. L., Smith, E. D., Tsuchiya, M., Welton, K. L., Thomas, J. H., Fields, S., Kennedy, B. K., and Kaeberlein, M. 2006. Lifespan extension in Caenorhabditis elegans by complete removal of food. Aging Cell 5: 487–494.
Kamel, C., Abrol, M., Jardine, K., He, X., and McBurney, M. W. 2006. SirT1 fails to affect p53-mediated biological functions. Aging Cell 5: 81–88.
Kennedy, B. K., Austriaco, N. R., Jr., Zhang, J., and Guarente, L. 1995. Mutation in the silencing gene SIR4 can delay aging in S. cerevisiae. Cell 80: 485–496.
Kennedy, B. K., Gotta, M., Sinclair, D. A., Mills, K., McNabb, D. S., Murthy, M., Pak, S. M., Laroche, T., Gasser, S. M., and Guarente, L. 1997. Redistribution of silencing proteins from telomeres to the nucleolus is associated with extension of life span in S. cerevisiae. Cell 89: 381–391.
Kenyon, C. 2005. The plasticity of aging: insights from long-lived mutants. Cell 120: 449–460.
Klar, A. J., Fogel, S., and Macleod, K. 1979. MAR1-a Regulator of the HMa and HMalpha Loci in Saccharomyces Cerevisiae. Genetics 93: 37–50.

Kume, S., Haneda, M., Kanasaki, K., Sugimoto, T., Araki, S., Isshiki, K., Isono, M., Uzu, T., Guarente, L., Kashiwagi, A., and Koya, D. 2007. SIRT1 inhibits transforming growth factor beta-induced apoptosis in glomerular mesangial cells via Smad7 deacetylation. J Biol Chem 282: 151–158.

Lagouge, M., Argmann, C., Gerhart-Hines, Z., Meziane, H., Lerin, C., Daussin, F., Messadeq, N., Milne, J., Lambert, P., Elliott, P., et al. 2006. Resveratrol improves mitochondrial function and protects against metabolic disease by activating SIRT1 and PGC-1alpha. Cell 127: 1109–1122.

Lakowski, B., and Hekimi, S. 1998. The genetics of caloric restriction in Caenorhabditis elegans. Proc Natl Acad Sci USA 95: 13091–13096.

Langley, E., Pearson, M., Faretta, M., Bauer, U. M., Frye, R. A., Minucci, S., Pelicci, P. G., and Kouzarides, T. 2002. Human SIR2 deacetylates p53 and antagonizes PML/p53-induced cellular senescence. Embo J 21: 2383–2396.

Lee, G. D., Wilson, M. A., Zhu, M., Wolkow, C. A., de Cabo, R., Ingram, D. K., and Zou, S. 2006. Dietary deprivation extends lifespan in Caenorhabditis elegans. Aging Cell 5: 515–524.

Lee, S. S., Kennedy, S., Tolonen, A. C., and Ruvkun, G. 2003. DAF-16 target genes that control C. elegans life-span and metabolism. Science 300: 644–647.

Lewis, L. K., and Resnick, M. A. 2000. Tying up loose ends: nonhomologous end-joining in Saccharomyces cerevisiae. Mutat Res 451: 71–89.

Li, M., Brooks, C. L., Wu-Baer, F., Chen, D., Baer, R., and Gu, W. 2003. Mono-versus polyubiquitination: differential control of p53 fate by Mdm2. Science 302: 1972–1975.

Li, M., Luo, J., Brooks, C. L., and Gu, W. 2002. Acetylation of p53 inhibits its ubiquitination by Mdm2. J Biol Chem 277: 50607–50611.

Lin, K., Dorman, J. B., Rodan, A., and Kenyon, C. 1997. daf-16: An HNF-3/forkhead family member that can function to double the life-span of Caenorhabditis elegans. Science 278: 1319–1322.

Lin, S. J., Defossez, P. A., and Guarente, L. 2000. Requirement of NAD and SIR2 for life-span extension by calorie restriction in Saccharomyces cerevisiae. Science 289: 2126–2128.

Lin, S. J., Ford, E., Haigis, M., Liszt, G., and Guarente, L. 2004. Calorie restriction extends yeast life span by lowering the level of NADH. Genes Dev 18: 12–16.

Liszt, G., Ford, E., Kurtev, M., and Guarente, L. 2005. Mouse Sir2 homolog SIRT6 is a nuclear ADP-ribosyltransferase. J Biol Chem 280: 21313–21320.

Liu, X., Kim, C. N., Yang, J., Jemmerson, R., and Wang, X. 1996. Induction of apoptotic program in cell-free extracts: requirement for dATP and cytochrome c. Cell 86: 147–157.

Lombard, D. B., Alt, F. W., Cheng, H. L., Bunkenborg, J., Streeper, R. S., Mostoslavsky, R., Kim, J., Yancopoulos, G., Valenzuela, D., Murphy, A., Yang, Y., Chen, Y., Hirschey, M. D., Bronson, R. T., Haigis, M., Guarente, L. P., Farese, R. V., Jr, Weissman, S., Verdin, E., Schwer, B. 2007. Mammalian Sir2 homolog SIRT3 regulates global mitochondrial lysine acetylation. Mol Cell Biol. Dec;27(24): 8807–14. Epub 2007 Oct 8.

Longo, V. D. 2008. Linking sirtuins, IGF-I signaling, and starvation. Exp Gerontol. 2009 Jan–Feb; 44(1–2): 70–4. Epub 2008 Jun 24.

Luchsinger, J. A., Tang, M. X., Shea, S., and Mayeux, R. 2002. Caloric intake and the risk of Alzheimer disease. Arch Neurol 59: 1258–1263.

Luo, J., Nikolaev, A. Y., Imai, S., Chen, D., Su, F., Shiloh, A., Guarente, L., and Gu, W. 2001. Negative control of p53 by Sir2alpha promotes cell survival under stress. Cell 107: 137–148.

Magwere, T., Chapman, T., and Partridge, L. 2004. Sex differences in the effect of dietary restriction on life span and mortality rates in female and male Drosophila melanogaster. J Gerontol A Biol Sci Med Sci 59: 3–9.

Malanga, M., and Althaus, F. R. 2004. Poly(ADP-ribose) reactivates stalled DNA topoisomerase I and Induces DNA strand break resealing. J Biol Chem 279: 5244–5248.

Marmorstein, R., and Roth, S. Y. 2001. Histone acetyltransferases: function, structure, and catalysis. Curr Opin Genet Dev 11: 155–161.

Masoro, E. J. 2003. Subfield history: caloric restriction, slowing aging, and extending life. Sci Aging Knowledge Environ 2003: RE2.

Masters, C. L., Simms, G., Weinman, N. A., Multhaup, G., McDonald, B. L., and Beyreuther, K. 1985. Amyloid plaque core protein in Alzheimer disease and Down syndrome. Proc Natl Acad Sci USA 82: 4245–4249.

Maswood, N., Young, J., Tilmont, E., Zhang, Z., Gash, D. M., Gerhardt, G. A., Grondin, R., Roth, G. S., Mattison, J., Lane, M. A., et al. 2004. Caloric restriction increases neurotrophic factor levels and attenuates neurochemical and behavioral deficits in a primate model of Parkinson's disease. Proc Natl Acad Sci USA 101: 18171–18176.

McCarroll, S. A., Murphy, C. T., Zou, S., Pletcher, S. D., Chin, C. S., Jan, Y. N., Kenyon, C., Bargmann, C. I., and Li, H. 2004. Comparing genomic expression patterns across species identifies shared transcriptional profile in aging. Nat Genet 36: 197–204.

Melov, S., and Hubbard, A. 2004. Microarrays as a tool to investigate the biology of aging: a retrospective and a look to the future. Sci Aging Knowledge Environ 2004: re7.

Michishita, E., McCord, R. A., Berber, E., Kioi, M., Padilla-Nash, H., Damian, M., Cheung, P., Kusumoto, R., Kawahara, T. L., Barrett, J. C., et al. 2008. SIRT6 is a histone H3 lysine 9 deacetylase that modulates telomeric chromatin. Nature 452: 492–496.

Michishita, E., Park, J. Y., Burneskis, J. M., Barrett, J. C., and Horikawa, I. 2005. Evolutionarily conserved and nonconserved cellular localizations and functions of human SIRT proteins. Mol Biol Cell 16: 4623–4635.

Milne, J. C., Lambert, P. D., Schenk, S., Carney, D. P., Smith, J. J., Gagne, D. J., Jin, L., Boss, O., Perni, R. B., Vu, C. B., et al. 2007. Small molecule activators of SIRT1 as therapeutics for the treatment of type 2 diabetes. Nature 450: 712–716.

Moazed, D. 2001. Enzymatic activities of Sir2 and chromatin silencing. Curr Opin Cell Biol 13: 232–238.

Mostoslavsky, R., Chua, K. F., Lombard, D. B., Pang, W. W., Fischer, M. R., Gellon, L., Liu, P., Mostoslavsky, G., Franco, S., Murphy, M. M., et al. 2006. Genomic instability and aging-like phenotype in the absence of mammalian SIRT6. Cell 124: 315–329.

Motta, M. C., Divecha, N., Lemieux, M., Kamel, C., Chen, D., Gu, W., Bultsma, Y., McBurney, M., and Guarente, L. 2004. Mammalian SIRT1 represses forkhead transcription factors. Cell 116: 551–563.

Muth, V., Nadaud, S., Grummt, I., and Voit, R. 2001. Acetylation of TAF(I)68, a subunit of TIF-IB/SL1, activates RNA polymerase I transcription. Embo J 20: 1353–1362.

Nemoto, S., Fergusson, M. M., and Finkel, T. 2005. SIRT1 functionally interacts with the metabolic regulator and transcriptional coactivator PGC-1{alpha}. J Biol Chem 280: 16456–16460.

North, B. J., Marshall, B. L., Borra, M. T., Denu, J. M., and Verdin, E. 2003. The human Sir2 ortholog, SIRT2, is an NAD+-dependent tubulin deacetylase. Mol Cell 11: 437–444.

Ogg, S., Paradis, S., Gottlieb, S., Patterson, G. I., Lee, L., Tissenbaum, H. A., and Ruvkun, G. 1997. The Fork head transcription factor DAF-16 transduces insulin-like metabolic and longevity signals in C. elegans. Nature 389: 994–999.

Pallas, M., Verdaguer, E., Tajes, M., Gutierrez-Cuesta, J., and Camins, A. 2008. Modulation of sirtuins: new targets for antiageing. Recent Patents CNS Drug Discov 3: 61–69.

Parker, J. A., Arango, M., Abderrahmane, S., Lambert, E., Tourette, C., Catoire, H., and Neri, C. 2005. Resveratrol rescues mutant polyglutamine cytotoxicity in nematode and mammalian neurons. Nat Genet 37: 349–350.

Partridge, L., Piper, M. D., and Mair, W. 2005. Dietary restriction in Drosophila. Mech Ageing Dev 126: 938–950.

Patel, N. V., Gordon, M. N., Connor, K. E., Good, R. A., Engelman, R. W., Mason, J., Morgan, D. G., Morgan, T. E., and Finch, C. E. 2005. Caloric restriction attenuates Abeta-deposition in Alzheimer transgenic models. Neurobiol Aging 26: 995–1000.

Pearson, K. J., Baur, J. A., Lewis, K. N., Peshkin, L., Price, N. L., Labinskyy, N., Swindell, W. R., Kamara, D., Minor, R. K., Perez, E., et al. 2008. Resveratrol delays age-related deterioration

and mimics transcriptional aspects of dietary restriction without extending life span. Cell Metab 8: 157–168.
Pedersen, S. B., Olholm, J., Paulsen, S. K., Bennetzen, M. F., and Richelsen, B. 2008. Low Sirt1 expression, which is upregulated by fasting, in human adipose tissue from obese women. Int J Obes (Lond) 32: 1250–1255.
Perrod S., Cockell M. M., Laroche T., Renauld H., Ducrest A. L., Bonnard C., Gasser S. M. 2001. A cytosolic NAD-dependent deacetylase, Hst2p, can modulate nucleolar and telomeric silencing in yeast. EMBO J. 20:197–209.
Pfluger, P. T., Herranz, D., Velasco-Miguel, S., Serrano, M., and Tschop, M. H. 2008. Sirt1 protects against high-fat diet-induced metabolic damage. Proc Natl Acad Sci USA 105: 9793–9798.
Picard, F., and Auwerx, J. 2002. PPAR(gamma) and glucose homeostasis. Annu Rev Nutr 22: 167–197.
Picard, F., Kurtev, M., Chung, N., Topark-Ngarm, A., Senawong, T., Machado De Oliveira, R., Leid, M., McBurney, M. W., and Guarente, L. 2004. Sirt1 promotes fat mobilization in white adipocytes by repressing PPAR-gamma. Nature 429: 771–776.
Pirola, L., and Frojdo, S. 2008. Resveratrol: one molecule, many targets. IUBMB Life 60: 323–332.
Pothof, J., van Haaften, G., Thijssen, K., Kamath, R. S., Fraser, A. G., Ahringer, J., Plasterk, R. H., and Tijsterman, M. 2003. Identification of genes that protect the C. elegans genome against mutations by genome-wide RNAi. Genes Dev 17: 443–448.
Puigserver, P., Wu, Z., Park, C. W., Graves, R., Wright, M., and Spiegelman, B. M. 1998. A cold-inducible coactivator of nuclear receptors linked to adaptive thermogenesis. Cell 92: 829–839.
Qin, W., Yang, T., Ho, L., Zhao, Z., Wang, J., Chen, L., Zhao, W., Thiyagarajan, M., MacGrogan, D., Rodgers, J. T., et al. 2006. Neuronal SIRT1 activation as a novel mechanism underlying the prevention of Alzheimer disease amyloid neuropathology by calorie restriction. J Biol Chem 281: 21745–21754.
Raizen, D. M., Lee, R. Y., and Avery, L. 1995. Interacting genes required for pharyngeal excitation by motor neuron MC in Caenorhabditis elegans. Genetics 141: 1365–1382.
Rena, G., Prescott, A. R., Guo, S., Cohen, P., and Unterman, T. G. 2001. Roles of the forkhead in rhabdomyosarcoma (FKHR) phosphorylation sites in regulating 14-3-3 binding, transactivation and nuclear targetting. Biochem J 354: 605–612.
Robyr, D., Suka, Y., Xenarios, I., Kurdistani, S. K., Wang, A., Suka, N., and Grunstein, M. 2002. Microarray deacetylation maps determine genome-wide functions for yeast histone deacetylases. Cell 109: 437–446.
Rodgers, J. T., Lerin, C., Gerhart-Hines, Z., and Puigserver, P. 2008. Metabolic adaptations through the PGC-1 alpha and SIRT1 pathways. FEBS Lett 582: 46–53.
Rogina, B., and Helfand, S. L. 2004. Sir2 mediates longevity in the fly through a pathway related to calorie restriction. Proc Natl Acad Sci USA 101: 15998–16003.
Rosenberg, M. I., and Parkhurst, S. M. 2002. Drosophila Sir2 is required for heterochromatic silencing and by euchromatic Hairy/E(Spl) bHLH repressors in segmentation and sex determination. Cell 109: 447–458.
Roth, S. Y., Denu, J. M., and Allis, C. D. 2001. Histone acetyltransferases. Annu Rev Biochem 70: 81–120.
Rusche, L. N., Kirchmaier, A. L., and Rine, J. 2002. Ordered nucleation and spreading of silenced chromatin in Saccharomyces cerevisiae. Mol Biol Cell 13: 2207–2222.
Rusche, L. N., and Rine, J. 2001. Conversion of a gene-specific repressor to a regional silencer. Genes Dev 15: 955–967.
Schwer, B., Bunkenborg, J., Verdin, R. O., Andersen, J. S., and Verdin, E. 2006. Reversible lysine acetylation controls the activity of the mitochondrial enzyme acetyl-CoA synthetase 2. Proc Natl Acad Sci USA 103: 10224–10229.
Selkoe, D. J. 2001. Presenilin, Notch, and the genesis and treatment of Alzheimer's disease. Proc Natl Acad Sci USA 98: 11039–11041.

Sengupta, N., and Seto, E. 2004. Regulation of histone deacetylase activities. J Cell Biochem 93: 57–67.

Shankaranarayana, G. D., Motamedi, M. R., Moazed, D., and Grewal, S. I. 2003. Sir2 regulates histone H3 lysine 9 methylation and heterochromatin assembly in fission yeast. Curr Biol 13: 1240–1246.

Shi, T., Wang, F., Stieren, E., and Tong, Q. 2005. SIRT3, a mitochondrial sirtuin deacetylase, regulates mitochondrial function and thermogenesis in brown adipocytes. J Biol Chem 280: 13560–13567.

Sinclair, D. A., and Guarente, L. 1997. Extrachromosomal rDNA circles – a cause of aging in yeast. Cell 91: 1033–1042.

Smith, J. S., and Boeke, J. D. 1997. An unusual form of transcriptional silencing in yeast ribosomal DNA. Genes Dev 11: 241–254.

Solomon, J. M., Pasupuleti, R., Xu, L., McDonagh, T., Curtis, R., DiStefano, P. S., and Huber, L. J. 2006. Inhibition of SIRT1 catalytic activity increases p53 acetylation but does not alter cell survival following DNA damage. Mol Cell Biol 26: 28–38.

Spiegelman, B. M., Puigserver, P., and Wu, Z. 2000. Regulation of adipogenesis and energy balance by PPARgamma and PGC-1. Int J Obes Relat Metab Disord 24 Suppl 4: S8–10.

St-Pierre, J., Drori, S., Uldry, M., Silvaggi, J. M., Rhee, J., Jager, S., Handschin, C., Zheng, K., Lin, J., Yang, W., et al. 2006. Suppression of reactive oxygen species and neurodegeneration by the PGC-1 transcriptional coactivators. Cell 127: 397–408.

Steinkraus, K. A., Smith, E. D., Davis, C., Carr, D., Pendergrass, W. R., Sutphin, G. L., Kennedy, B. K., and Kaeberlein, M. 2008. Dietary restriction suppresses proteotoxicity and enhances longevity by an hsf-1-dependent mechanism in Caenorhabditis elegans. Aging Cell 7: 394–404.

Stephens, L., Anderson, K., Stokoe, D., Erdjument-Bromage, H., Painter, G. F., Holmes, A. B., Gaffney, P. R., Reese, C. B., McCormick, F., Tempst, P., et al. 1998. Protein kinase B kinases that mediate phosphatidylinositol 3,4,5-trisphosphate-dependent activation of protein kinase B. Science 279: 710–714.

Strahl-Bolsinger, S., Hecht, A., Luo, K., and Grunstein, M. 1997. SIR2 and SIR4 interactions differ in core and extended telomeric heterochromatin in yeast. Genes Dev 11: 83–93.

Suka, N., Suka, Y., Carmen, A. A., Wu, J., and Grunstein, M. 2001. Highly specific antibodies determine histone acetylation site usage in yeast heterochromatin and euchromatin. Mol Cell 8: 473–479.

Sun, A. Y., Simonyi, A., and Sun, G. Y. 2002. The "French Paradox" and beyond: neuroprotective effects of polyphenols. Free Radic Biol Med 32: 314–318.

Tamburini, B. A., and Tyler, J. K. 2005. Localized histone acetylation and deacetylation triggered by the homologous recombination pathway of double-strand DNA repair. Mol Cell Biol 25: 4903–4913.

Tanny, J. C., Dowd, G. J., Huang, J., Hilz, H., and Moazed, D. 1999. An enzymatic activity in the yeast Sir2 protein that is essential for gene silencing. Cell 99: 735–745.

Tatar, M. 2007. Diet restriction in Drosophila melanogaster. Design and analysis. Interdiscip Top Gerontol 35: 115–136.

Tissenbaum, H. A., and Guarente, L. 2001. Increased dosage of a sir-2 gene extends lifespan in Caenorhabditis elegans. Nature 410: 227–230.

Tzivion, G., Shen, Y. H., and Zhu, J. 2001. 14-3-3 proteins; bringing new definitions to scaffolding. Oncogene 20: 6331–6338.

Vakhrusheva, O., Smolka, C., Gajawada, P., Kostin, S., Boettger, T., Kubin, T., Braun, T., and Bober, E. 2008. Sirt7 increases stress resistance of cardiomyocytes and prevents apoptosis and inflammatory cardiomyopathy in mice. Circ Res. 102:703–10

Valenzano, D. R., and Cellerino, A. 2006. Resveratrol and the pharmacology of aging: a new vertebrate model to validate an old molecule. Cell Cycle 5: 1027–1032.

Valenzano, D. R., Terzibasi, E., Genade, T., Cattaneo, A., Domenici, L., and Cellerino, A. 2006. Resveratrol prolongs lifespan and retards the onset of age-related markers in a short-lived vertebrate. Curr Biol 16: 296–300.

van der Horst, A., Tertoolen, L. G., de Vries-Smits, L. M., Frye, R. A., Medema, R. H., and Burgering, B. M. 2004. FOXO4 is acetylated upon peroxide stress and deacetylated by the longevity protein hSir2(SIRT1). J Biol Chem 279: 28873–28879.

Vaziri, H., Dessain, S. K., Ng Eaton, E., Imai, S. I., Frye, R. A., Pandita, T. K., Guarente, L., and Weinberg, R. A. 2001. hSIR2(SIRT1) functions as an NAD-dependent p53 deacetylase. Cell 107: 149–159.

Wang, C., Ko, H. S., Thomas, B., Tsang, F., Chew, K. C., Tay, S. P., Ho, M. W., Lim, T. M., Soong, T. W., Pletnikova, O., et al. 2005. Stress-induced alterations in parkin solubility promote parkin aggregation and compromise parkin's protective function. Hum Mol Genet 14: 3885–3897.

Wang, F., Nguyen, M., Qin, F. X., and Tong, Q. 2007. SIRT2 deacetylates FOXO3a in response to oxidative stress and caloric restriction. Aging Cell 6: 505–514.

Wang, X. 2001. The expanding role of mitochondria in apoptosis. Genes Dev 15: 2922–2933.

Wang, Y., and Tissenbaum, H. A. 2006. Overlapping and distinct functions for a Caenorhabditis elegans SIR2 and DAF-16/FOXO. Mech Ageing Dev 127: 48–56.

Weindruch, R., Kayo, T., Lee, C. K., and Prolla, T. A. 2001. Microarray profiling of gene expression in aging and its alteration by caloric restriction in mice. J Nutr 131: 918S–923S.

Weydt, P., Pineda, V. V., Torrence, A. E., Libby, R. T., Satterfield, T. F., Lazarowski, E. R., Gilbert, M. L., Morton, G. J., Bammler, T. K., Strand, A. D., et al. 2006. Thermoregulatory and metabolic defects in Huntington's disease transgenic mice implicate PGC-1alpha in Huntington's disease neurodegeneration. Cell Metab 4: 349–362.

Wood, J. G., Rogina, B., Lavu, S., Howitz, K., Helfand, S. L., Tatar, M., and Sinclair, D. 2004. Sirtuin activators mimic caloric restriction and delay ageing in metazoans. Nature 430: 686–689.

Zhang, W., Zou, S., and Song, J. 2008. Term-tissue specific models for prediction of gene ontology biological processes using transcriptional profiles of aging in drosophila melanogaster. BMC Bioinformatics 9: 129.

Chromatin in Senescent Cells: A Conduit for the Anti-Aging Effects of Wnt Signaling?

Gowrishankar Banumathy and Peter D. Adams

Abstract Cellular senescence is thought to contribute to tissue aging. The role of Wnt signaling in aging is little studied and poorly understood, but there is evidence that Wnt signaling can be both pro- and anti-aging. Here we discuss the idea that, in at least some cell and tissue contexts, Wnt signaling might antagonize aging through its ability to regulate chromatin structure in senescent cells.

Keywords Senescence · Chromatin · Senescence-associated heterochromatin foci · Wnt signaling · Aging

An Overview of Cellular Senescence

Senescence, Aging, and Tumor Suppression

Cellular senescence is characterized by an irreversible arrest of cell proliferation (Campisi and d'Adda di Fagagna, 2007; Hayflick, 1965). Senescence is caused by shortened telomeres that result from repeated rounds of cell division, inadequate in vitro growth conditions, and other cellular stresses (Campisi, 2005; Herbig and Sedivy, 2006; Ramirez et al., 2001; Wright and Shay, 2002). Because of senescence, most primary human cells have a finite proliferative life span, and evidence has been presented that senescence contributes to tissue aging in vivo, in part by limiting the proper self-renewal of tissues due to exhaustion and/or functional impairment of renewable tissue stem cell populations (Campisi, 2005; Collado et al., 2007; Finkel et al., 2007; Herbig and Sedivy, 2006; Serrano and Blasco, 2007; Wright and Shay, 2002). Specifically, senescent cells and/or molecular markers of the

---

P.D. Adams (✉)
University of Glasgow, Cancer Research UK Beatson Labs,
Glasgow, G61 1BD, United Kingdom
e-mail: p.adams@beatson.gla.ac.uk

T.O. Tollefsbol (ed.), *Epigenetics of Aging*, DOI 10.1007/978-1-4419-0639-7_6,

senescent phenotype have been reported to increase in some aging tissues (Dimri et al., 1995; Francis et al., 2004; Herbig et al., 2006; Janzen et al., 2006; Jeyapalan et al., 2007; Krishnamurthy et al., 2006; Krishnamurthy et al., 2004; Molofsky et al., 2006) and are linked to some age-associated tissue pathologies, such as osteoarthritis, atherosclerosis, and liver cirrhosis (Minamino and Komuro, 2007; Price et al., 2002; Wiemann et al., 2002). Moreover, manipulation of the cell signals that initiate senescence can modulate some aspects of organismal aging (Baker et al., 2008; Blasco et al., 1997; Du et al., 2004; Garcia-Cao et al., 2006; Gonzalez-Suarez et al., 2005; Gonzalez-Suarez et al., 2001; Herrera et al., 1999; Janzen et al., 2006; Krishnamurthy et al., 2006; Lee et al., 1998; Molofsky et al., 2006; Rudolph et al., 1999; Wiemann et al., 2002).

Although not the main focus of this review, senescence in vivo is also thought to be an important tumor suppression process (Prieur and Peeper, 2008). Cells with cellular and molecular characteristics of senescence have been observed in several benign pre-cancerous neoplasms, both in humans and in mouse models (Braig et al., 2005; Chen et al., 2005b; Collado et al., 2005; Courtois-Cox et al., 2006; Dai et al., 2000; Dankort et al., 2007; Gray-Schopfer et al., 2006; Ha et al., 2007; Michaloglou et al., 2005; Sarkisian et al., 2007; Sun et al., 2007). Senescence caused by shortened telomeres limits cell proliferation capacity and so suppresses tumor formation (Cosme-Blanco et al., 2007; Feldser and Greider, 2007; Rudolph et al., 1999). Activated oncogenes also trigger cell senescence, and this is thought to block progression to a transformed cell phenotype (Prieur and Peeper, 2008; Serrano et al., 1997). Importantly, in murine breast epithelium, the proliferation arrest induced by high-level oncogenic activation appears to be irreversible, in at least a proportion of the cells (Sarkisian et al., 2007). In mouse models, inactivation of the senescence program frequently allows oncogenes to drive formation of cancerous neoplasms, instead of benign ones (Dankort et al., 2007; Ha et al., 2007; Sarkisian et al., 2007; Sun et al., 2007). Further underscoring the importance of senescence as a tumor suppression mechanism as well as its therapeutic potential, recent studies have shown that reactivation of cell senescence in murine tumors causes tumor regression (Ventura et al., 2007; Xue et al., 2007).

In keeping with the theme of this book, this review will focus mostly on senescence and associated epigenetic changes as they pertain to tissue aging, rather than tumor suppression. However, many of the molecular mechanisms underlying senescence in the two contexts are similar. Therefore, we will frequently discuss results from studies of oncogene-induced senescence, because of the light that they shed on senescence likely to be associated with aging.

Molecular Features of Senescence

Senescence has been most widely studied in fibroblasts in vitro, but is also well defined in melanocytes and epithelial cells (Denoyelle et al., 2006; Michaloglou et al., 2005; Wright and Shay, 2001). Other cell types suggested to undergo senescence include hematopoietic and neural progenitor cells (Geiger and Van Zant, 2002; Palmer et al., 2001). In many human cells, cellular senescence is characterized by several molecular and cytological markers, such as a large flat morphology; expression of a *S* enescence-*A* ssociated *β-gal* actosidase activity (SA β-gal); formation of intracellular vacuoles; resistance of proliferation-promoting genes to mitogenic stimulation; persistent activation of DNA damage signaling pathways at specific subnuclear sites, called *S* enescence-Associated DNA *D* amage *F* oci (SDFs) (Campisi, 2005; d'Adda di Fagagna, 2008; Herbig and Sedivy, 2006; Wright and Shay, 2002); and formation of punctate highly condensed domains of facultative heterochromatin, called *S* enescence-*A* ssociated *H* eterochromatin *F* oci (SAHF) (Narita et al., 2003).

However, specific senescence markers vary in magnitude depending on the species, cell type, and the trigger of senescence. For example, although some mouse cells exhibit a general increase in the amount of nuclear heterochromatin as judged by histone modifications (Braig et al., 2005), senescent mouse cells do not accumulate domains of facultative heterochromatin as pronounced as the punctate SAHF observed in human cells. Moreover, in mouse cells, SAHF should not be confused with the highly condensed domains of constitutive pericentromeric heterochromatin that are present even in growing mouse cells (Guenatri et al., 2004). Of human cells, WI38 and IMR90 fibroblasts and primary human melanocytes form pronounced SAHF, whereas BJ fibroblasts do not, except in response to an activated oncogene (Denoyelle et al., 2006; Narita et al., 2003). In response to an activated Ras oncogene, primary human melanocytes express SA β-gal and assume the classical large, flat, extensively vacuolarized morphology. In response to an activated BRAF oncogene, the same cells express SA β-gal, but the large flat morphology and intracellular vacuoles are much less apparent (Denoyelle et al., 2006).

The physiological significance of these variations in the senescence phenotype is unknown. To understand their significance it is necessary to define the impact of the molecular phenotype, e.g., SA β-gal and SAHF formation, on potential physiological endpoints of senescence, such as tissue aging and tumor suppression. SAHF formation and how it impacts aging, the major topic of this review, is considered at the end of the chapter.

Senescence-Inducing Pathways

The pRB and p53 tumor suppressor pathways are master regulators of senescence. Inactivation of these two pathways typically abolishes senescence in mouse and human cells, regardless of the initial senescence trigger (Campisi, 2005; Herbig and Sedivy, 2006; Wright and Shay, 2002). Although human cells lacking pRB and p53 circumvent senescence, most such cells ultimately still cease proliferation through "crisis" due to erosion of telomeres to a critically short length (Counter et al., 1992). The p53 pathway is comprised of at least three proteins whose activity is altered in human cancer – p53, p19ARF, and hdm2 (Sherr and McCormick,

2002). This pathway exerts its effects through activation of downstream target genes, including the cell cycle inhibitor p21CIP1, whose expression is increased in senescent cells. The pRB pathway is comprised of at least four proteins whose activity is frequently perturbed by genetic mutations or altered level of expression in human cancers – p16INK4a, cyclin D1, cdk4, and pRB (Nevins, 2001). By inhibiting cyclin D/cdk4 kinases, p16INK4a activates pRB. The pRB pathway inhibits cell proliferation through numerous downstream effectors. For example, pRB inhibits the E2F family of transcription factors, whose target genes are necessary for progression through S-phase (Nevins, 2001).

The mechanisms by which activated oncogenes, short telomeres, and cellular stresses drive senescence through activation of the pRB and p53 pathways have been extensively reviewed elsewhere (Ben-Porath and Weinberg, 2005; Campisi, 2005; Campisi and d'Adda di Fagagna, 2007; d'Adda di Fagagna, 2008; Herbig and Sedivy, 2006) and will only be briefly summarized here to emphasize the major points, recent findings, and some outstanding questions. The p53 pathway is activated by DNA damage, in response to either short telomeres or activated oncogenes. The free DNA ends of short telomeres are sensed by the cell as a form of DNA damage and activate p53 via the ATM and Chk2 DNA damage signaling cascade (d'Adda di Fagagna et al., 2003; Herbig et al., 2004). Activated oncogenes cause a proliferative burst and rounds of error-prone DNA synthesis. The DNA damage that accumulates during these rounds of unscheduled DNA synthesis also activates p53 via the ATM and ATR damage signaling pathways (Bartkova et al., 2006; Di Micco et al., 2006; Mallette et al., 2007). In mouse cells, activated oncogenes also activate p53 by upregulation of p14ARF. However, this pathway is apparently not conserved in human cells (Brookes et al., 2002; Ferbeyre et al., 2000; Wei et al., 2001). Activation of the p53 pathway contributes to activation of the pRB pathway. The p53 target gene, p21CIP1, inhibits cyclin/cdk2 complexes, thereby activating pRB.

The pRB pathway is also activated by upregulation of p16INK4a. The mechanism by which expression of p16INK4a is increased is poorly understood, although several mechanisms have been implicated (Collado et al., 2007). A likely contributor to aging is reactive oxygen species (ROS) and the cellular damage that is caused by them (Chen et al., 2007). ROS activate the stress-responsive p38 MAP kinases, and several studies have established that these kinases contribute to senescence-associated upregulation of p16INK4a (Bulavin et al., 2004; Deng et al., 2004; Ito et al., 2006; Iwasa et al., 2003). However, the molecular mechanism linking p38 MAP kinase to p16INK4a has been unclear. Interestingly, one target of p38 MAP kinases is another kinase, MAPKAPK3, which is activated by p38 MAP kinase (Gaestel, 2006; Ludwig et al., 1996). In turn, MAPKAP3 has been reported to phosphorylate members of the multi-subunit polycomb (PcG) complex of transcriptional repressors that repress expression of p16INK4a (Bernard et al., 2005; Gil et al., 2004; Itahana et al., 2003; Voncken et al., 2005). Phosphorylation of PcG proteins causes their dissociation from chromatin and derepression of the CDKN2A/INK4a locus that codes for p16INK4a (Voncken et al., 2005). Senescence-associated upregulation of p16INK4a is, in part, a consequence of inactivation of PcG proteins (Bracken et al., 2007). Some studies have found that aging is associated with increased p38 MAP-kinase activity (Abidi et al., 2008; Hsieh and Papaconstantinou, 2006), suggesting that aging upregulates p16INK4a through a ROS/stress-p38 MAPK-MAPKAPK3-PcG-p16INK4a pathway. Helin and coworkers also showed that senescence-associated decreased expression of one PcG protein, EZH2, contributes to upregulation of p16INK4a (Bracken et al., 2007). Significantly, expression of EZH2 has also been reported to be repressed in the aging hematopoietic system (Rossi et al., 2007). Thus, age-associated upregulation of p16INK4A might be a consequence of age-associated repression of polycomb activity, by both transcriptional and post-transcriptional mechanisms. The mechanisms controlling p16INK4a expression are still areas of active investigation.

Chromatin Remodeling in Senescent Cells

Description of SAHF

Researchers have frequently considered the possibility that chromatin structure has a major impact on cell senescence and tissue aging (Han et al., 2006; Howard, 1996; Imai and Kitano, 1998; Oberdoerffer and Sinclair, 2007; Rogakou and Sekeri-Pataryas, 1999; Sedivy et al., 2008; Villeponteau, 1997; Zhang and Adams, 2007). However, SAHF were first explicitly described by Scott Lowe and coworkers (Narita et al., 2003). When stained with 4′-6-diamidino-2-phenylindole (DAPI), normal human cells exhibit a relatively even, diffuse distribution of DNA through the cell nucleus. However, in DAPI-stained senescent human cells, SAHF appear as approximately 30–50 bright, punctate DNA foci. The chromatin in these foci appears much more compact than the chromatin in normal interphase growing cells. Indeed, chromatin from cells with SAHF is more resistant to nuclease digestion than chromatin from growing cells (Narita et al., 2003). Inclusion of proliferation-promoting genes, such as cyclin A, into these compact chromatin foci is thought to silence expression of those genes, thereby contributing to senescence-associated cell cycle arrest.

Remarkably, each SAHF focus in a senescent cell results from condensation of an individual chromosome (Funayama et al., 2006; Zhang et al., 2007a). SAHF contain several common markers of heterochromatin, including histones that are hypoacetylated, methylation of lysine 9 of histone H3 (H3K9Me), and bound *H*eterochromatin *P*rotein 1 (HP1) proteins. However, SAHF do not contain some other markers of condensed chromatin in mitotic and apoptotic cells, such as phosphoserine 10 of histone H3 (H3S10P), H2BS14P, and H3S28P (Funayama et al., 2006; Peterson and Laniel, 2004). SAHF are also characterized by their depletion of linker histone H1 and enrichment in at least two other proteins, namely the histone variant macroH2A and HMGA proteins (Funayama et al., 2006; Narita et al., 2006; Zhang et al., 2005).

MacroH2A is actually a family of three variants, macroH2A1.1, 1.2, and 2 (where 1.1 and 1.2 are splice variants). MacroH2As contain an N-terminal histone H2A-like domain and a C-terminal "macro domain" of more than 200 residues that is unrelated to other histones. MacroH2A clearly contributes to gene silencing, since it is depleted from active genes (Changolkar and Pehrson, 2006), inserted into the inactive X chromosome (Chadwick and Willard, 2002; Changolkar and Pehrson, 2006; Costanzi and Pehrson, 1998, 2001; Hernandez-Munoz et al., 2005), and macroH2A-containing chromatin is resistant to ATP-dependent remodeling proteins and binding of transcription factors in vitro and in vivo (Agelopoulos and Thanos, 2006; Angelov et al., 2003; Doyen et al., 2006). Moreover, genetic inactivation of macroH2A1.1 and 1.2 in mice caused increased expression of some genes that are normally silenced and enriched in macroH2A1.1 and 1.2 (Changolkar et al., 2007) and derepressed expression of endogenous murine leukemia viruses (Changolkar et al., 2008).

The HMGA1 and HMGA2 proteins (previously called HMGIY and HMGIC, respectively) are abundant non-histone chromatin proteins (Reeves, 2001; Sgarra et al., 2004). Paradoxically, the presence of HMGA proteins in chromatin is normally associated with gene activation, cell proliferation, and cell transformation (Reeves, 2001; Sgarra et al., 2004). Both proteins are expressed in embryos, repressed during cell differentiation, and their expression is stimulated by mitogens. Mice expressing an HMGA transgene develop tumors (Fedele et al., 2002; Xu et al., 2004). In human tumors, HMGA proteins are sometimes overexpressed and the genes coding for these proteins are targets of amplification and translocations (Reeves, 2001; Sgarra et al., 2004). In light of these prior observations, it was a surprise when SAHF were shown to be enriched in HMGA proteins and that these proteins contribute to senescence-associated proliferation arrest and transformation suppression in fibroblasts (Funayama et al., 2006; Narita et al., 2006). However, others have also confirmed that in some contexts HMGA proteins play a tumor suppression role (Fedele et al., 2006). In sum, HMGA proteins contribute to senescence-associated SAHF formation and cell cycle exit in fibroblasts, and the extent to which this role is conserved in other cell types remains to be determined.

Remarkably, although SAHF appear to result from the condensation of almost entire chromosomes, DNA sequences that are typically contained in constitutive heterochromatin, such as pericentromeres and telomeres, actually appear to be excluded from the bulk of the condensed chromosome (Funayama et al., 2006; Narita et al., 2003; Ye et al., 2007b; Zhang and Adams, 2007; Zhang et al., 2007a) (Fig. 1A). This suggests that these normally constitutively heterochromatic regions are perhaps deheterochromatinized in senescent cells. Consistent with this idea at least for telomeres, Maria Blasco and coworkers have shown that the shortened telomeres in mice lacking telomerase have reduced heterochromatin compared to telomeres from normal cells (Benetti et al., 2007). Extending this observation, inactivation of SIRT6, a histone deacetylase which acts at telomeres, causes telomere deheterochromatinization, end-to-end fusions, and senescence (Michishita et al., 2008). This raises the possibility that deheterochromatinization of telomeres is actually a trigger for senescence and not just

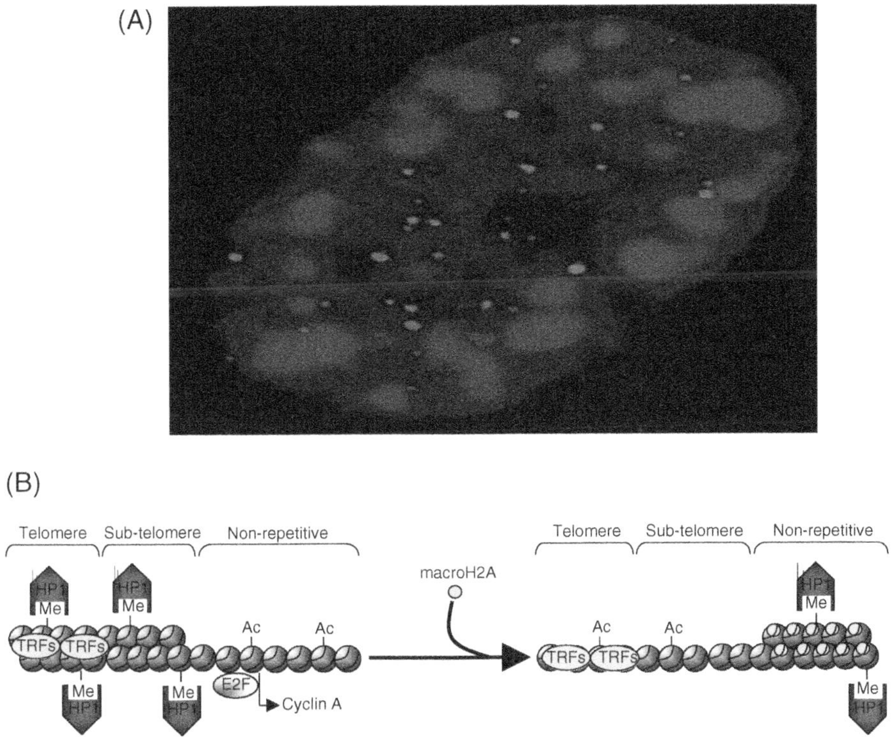

Fig. 1 Formation of SAHF in senescent human cells is linked to apparent heterochromatin redistribution. (**A**) Senescent human WI38 cells were stained with DAPI to visualize DNA (*blue*) and a telomeric FISH probe (*red*) to visualize telomeres. Note the predominant localization of telomeres to less DNA-stained inter-chromatin spaces, consistent with a less compact chromatin structure for these regions. (**B**) A schematic to illustrate the concept of heterochromatin redistribution in senescent human cells. *Left*, chromosome structure in a proliferating cell; *Right*, chromosome structure in a senescent cell. In proliferating cells, telomeric and subtelomeric regions are contained in compact heterochromatin, with HP1 proteins bound to methylated lysine 9 of histone H3 (Me). TRF1 and TRF2 proteins (TRFs) are bound to telomeres. In non-repetitive DNA sequences, in less compact acetylated (Ac) euchromatin, expression of proliferation-promoting genes, such as cyclin A, are driven by transcription factors, such as E2F. In senescent cells, telomeric and subtelomeric regions are less condensed and less heterochromatic. TRFs remain bound. On the other hand, non-repetitive DNA sequences become more compact or more heterochromatic. This equates to a direct or indirect redistribution of heterochromatin from telomeres and subtelomeres to non-repetitive DNA to form SAHF

a consequence of telomere shortening or senescence. Regardless of the extent to which telomere deheterochromatinization is a physiological trigger for senescence, cellular senescence appears to be accompanied by a redistribution of heterochromatin from constitutive heterochromatin to other normally euchromatic sites, specifically to specialized domains of facultative heterochromatin, called SAHF (Fig. 1B).

Formation of SAHF Is a Multi-step Process

Two lines of evidence indicate that SAHF form through a cascade of temporally and mechanistically separable events. First, a kinetic analysis of cells forming SAHF showed that condensed chromosomes, in the form of DAPI foci, are detectable before their enrichment with H3K9Me, HP1 proteins, and macroH2A (Zhang et al., 2005). Second, a dominant-negative HP1 mutant that removes 50–80% of all three HP1 isoforms from chromatin in primary human cells has no effect on chromosome condensation or incorporation of macroH2A into condensed chromosomes (Zhang et al., 2007a). Together, these results indicate that formation of SAHF is a multi-step process. The earliest detectable event to date is chromosome condensation to form a SAHF focus that is detectable by DAPI staining of DNA, followed by methylation of lysine 9 of histone H3 to create H3K9Me, binding of HP1 proteins, and incorporation of macroH2A (Fig. 2).

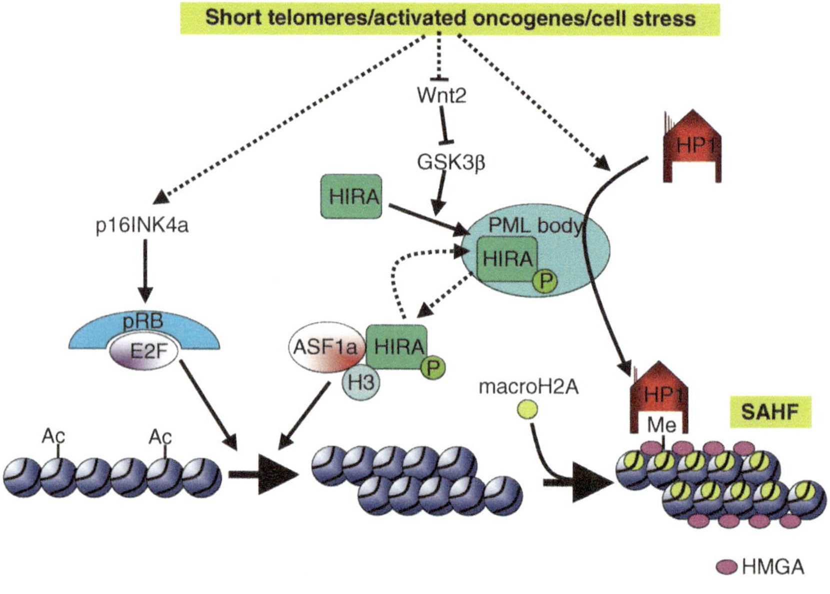

Fig. 2 A model for formation of SAHF in senescent human cells. Senescence is triggered by short telomeres, activated oncogenes, and other cell stresses. The HIRA/ASF1a pathway cooperates with the p16INK4a/pRB pathway to drive chromosome condensation. After chromosome condensation, HP1 proteins and histone variant macroH2A are incorporated into SAHF. Recruitment of HP1γ to SAHF depends on HP1γ phosphorylation. Dashed lines indicate steps that are poorly defined at present. See text for further details

Remarkably, these studies also indicated that loading of abundant HP1 proteins onto chromatin is not required for two additional hallmarks of the senescent phenotype, expression of SA β-gal and senescence-associated cell cycle exit (Zhang et al., 2007a). Conceivably, the residual chromatin-bound HP1 proteins in cells expressing the dominant-negative HP1 mutant are sufficient to mediate HP1 functions that are required for these senescence phenotypes. However, these results raise the possibility that HP1 proteins do not contribute to acute onset of the senescent phenotype. Instead, HP1 proteins might be required for long-term maintenance of SAHF and the senescent state. Alternatively, HP1 proteins might secure the senescent state in the face of genetic alterations or cellular perturbations that compromise other aspects of the senescence program. These ideas remain to be tested.

Chromosome Condensation Is Driven by Histone Chaperones HIRA and ASF1a

Based on the studies described above, senescence-associated cell cycle exit appears to be linked to the process of chromosome condensation through SAHF formation. Two chromatin regulators, HIRA and ASF1a, drive chromosome condensation during SAHF assembly in human cells (Zhang et al., 2005) (Fig. 2). HIRA and ASF1a are the human orthologs of proteins known to create transcriptionally silent heterochromatin in yeast, flies, and plants (Goodfellow et al., 2007; Kaufman et al., 1998; Moshkin et al., 2002; Phelps-Durr et al., 2005; Rocha and Verreault, 2008; Sharp et al., 2002; Singer et al., 1998). Yeast Asf1p is required for heterochromatin-mediated silencing of telomeres and mating loci and has histone deposition activity in vitro (Krawitz et al., 2002; Sharp et al., 2001; Singer et al., 1998; Tyler et al., 1999). Yeast Asf1p is a multi-functional protein that is also required for histone eviction and subsequent replacement at transcribed genes (Adkins et al., 2004; Kim et al., 2007; Korber et al., 2006; Rufiange et al., 2007; Schwabish and Struhl, 2006) and serves as a platform for post-translational modification of histones (Adkins et al., 2007; Driscoll et al., 2007; Recht et al., 2006; Tsubota et al., 2007). In yeast and other species, Asf1 proteins also play roles in DNA replication and repair-coupled chromatin assembly (Franco et al., 2005; Grigsby and Finger, 2008; Groth et al., 2007; Groth et al., 2005; Myung et al., 2003; Schulz and Tyler, 2006; Tyler et al., 1999).

Yeast Hir1p and Hir2p share several biological and biochemical properties with Asf1p (Rocha and Verreault, 2008). Like Asf1p, they are required for heterochromatin-mediated silencing of telomeres and mating loci and are also required for formation of proper pericentromeric chromatin (but in an Asf1p-independent manner) (Kaufman et al., 1998; Krawitz et al., 2002; Sharp et al., 2001; Sharp et al., 2002). The Hir proteins also contribute to suppression of spurious transcripts during transcription elongation, likely by nucleosome re-assembly in the wake of RNA polymerase transcription (Nourani et al., 2006; Prather et al., 2005). Consistent with their partially overlapping functions, Asf1 and Hir proteins physically interact and in yeast this interaction is necessary for telomeric silencing (Daganzo et al., 2003; Green et al., 2005; Sharp et al., 2001).

The role of Asf1 and Hir proteins in formation of heterochromatin is conserved in mammalian cells. Mouse ES cells lacking HIRA have a larger pool of loosely bound histones than wild-type cells, consistent with a role for HIRA in generation of compact, nucleosome-dense, transcriptionally silent heterochromatin (Meshorer et al., 2006). Recently, both HIRA and ASF1a emerged from unbiased whole-genome screens for genes required for gene silencing and/or nuclear heterochromatinization in response to an activated oncogene (Gazin et al., 2007; Wajapeyee et al., 2008). More specifically, in human cells, HIRA and ASF1a play a key role in formation of SAHF. Ectopic expression of HIRA or ASF1a in primary human cells accelerates formation of SAHF (Zhang et al., 2005). This activity requires binding of HIRA and ASF1a to each other, and shRNA-mediated knockdown of ASF1a blocks formation of SAHF triggered by an activated Ras-oncogene (Zhang et al., 2005).

Consistent with the idea that formation of SAHF depends on histone chaperone activity of the HIRA/ASF1a complex, formation of SAHF also requires an interaction between ASF1a and histone H3 (Zhang et al., 2007a). Significantly, many previous reports indicate that transcriptionally active chromatin is depleted of nucleosomes, at both a genome-wide and local chromatin level (Adkins et al., 2004; Agalioti et al., 2000; Angermayr and Bandlow, 2003; Angermayr et al., 2002; Bernstein et al., 2004; Chen et al., 2005a; Lee et al., 2004; Mito et al., 2005; Zhao et al., 2005). Moreover, a previous study reported that the facultative heterochromatin of the inactive X chromosome has higher nucleosome density than most other regions of the nucleus (Perche et al., 2000). Together, these results suggest that chromosome condensation associated with SAHF formation may depend, in part, on increased nucleosome density due to HIRA/ASF1a-mediated nucleosome deposition.

Recently, several groups have described molecular structures of Asf1 proteins, either as free proteins or bound to histones, HIRA, or fragments of either (Agez et al., 2007; Antczak et al., 2006; Daganzo et al., 2003; English et al., 2006; Malay et al., 2008; Mousson et al., 2005; Natsume et al., 2007; Tang et al., 2006). Regardless of the species, the Asf1 protein forms an elongated immunoglobulin-like β-sandwich fold, with three α-helices in the loops between the β-strands. Together, these studies indicate that HIRA and the histone H3/H4 heterodimer bind to distinct faces of the Asf1 polypeptide (Malay et al., 2008; Tang et al., 2006). HIRA binds to a shallow hydrophobic groove on ASF1a, perpendicular to the strands of the β-sandwich, and is anchored at one end of the groove by a cluster of salt bridge interactions. The histone H3/H4 heterodimer binds largely to the opposite face of Asf1 (Agez et al., 2007; Antczak et al., 2006; English et al., 2006; Mousson et al., 2005; Natsume et al., 2007). Interestingly, in the Asf1/H3/H4 trimeric complex, Asf1 binds to the C-terminus of histone H4 that normally interacts with histone H2A in the nucleosome. This suggests that release of histone H3/H4 from Asf1 will facilitate nucleosome assembly by exposing the histone H4 tail to histone H2A (English et al., 2006).

The specific histone substrate utilized by HIRA/ASF1a to make SAHF is not well defined. Like ASF1a, HIRA also binds to histones. HIRA-containing chaperone complexes preferentially bind to and deposit the histone variant histone H3.3, over the canonical histone H3.1, in a DNA replication-independent manner (Green et al., 2005; Loppin et al., 2005; Nakayama et al., 2007; Prochasson et al., 2005; Ray-Gallet et al., 2002; Tagami et al., 2004; van der Heijden et al., 2007). Significantly, histone H3.3 accumulates in fibroblasts approaching senescence and in non-dividing differentiated cells, in some cases to about 90% of the total histone H3, with presumably the majority being in largely inactive chromatin (Bosch and Suau, 1995; Brown et al., 1985; Grove and Zweidler, 1984; Krimer et al., 1993; Pantazis and Bonner, 1984; Pina and Suau, 1987; Rogakou and Sekeri-Pataryas, 1999; Urban and Zweidler, 1983; Wunsch and Lough, 1987). Unfortunately, because human histone H3.3 and canonical H3.1 only differ by 5 amino acids, differentiating between them immunologically is challenging, making it difficult to ask whether endogenous histone H3.3 is specifically enriched in SAHF. The idea that SAHF contains histone H3.3 may initially seem unlikely, because deposition of histone H3.3 is typically linked to transcription activation (Ahmad and Henikoff, 2002; McKittrick et al., 2004; Mito et al., 2005; Schwartz and Ahmad, 2005; Wirbelauer et al., 2005), whereas SAHF is thought to be predominantly transcriptionally silent (Narita et al., 2006; Narita et al., 2003; Zhang et al., 2005). However, although histone H3.3-containing nucleosomes are less stable to high salt and immunoprecipitation in vitro (Jin and Felsenfeld, 2007), when ectopically expressed in cells histone H3.3 activates expression of some genes that incorporate histone H3.3, but represses others (Jin and Felsenfeld, 2006). Although histone H3.3 is enriched in modifications that correlate with active transcription, a proportion of histone H3.3 does carry post-translational marks characteristic of transcriptionally silent chromatin (Hake et al., 2006; Loyola et al., 2006; McKittrick et al., 2004). Therefore, histone H3.3 is unlikely to be exclusively linked to transcription activation. Instead, deposition of histone H3.3 may be associated with any major remodeling of chromatin, perhaps as a way to "re-set" histone modifications. In principle, this can be associated with gene activation, repression, or neither. Concordant with this proposal, after egg fertilization in flies, dHIRA activity is required for replacement of protamines by histone H3.3-containing nucleosomes in decondensing sperm chromatin (Bonnefoy et al., 2007; Loppin et al., 2005). Also, after treatment of primary human cells with histone deacetylase inhibitors that disrupt heterochromatin structure, HIRA and histone H3.3 are required for a chromatin "repair" process that recruits HP1 proteins to pericentromeres, thereby maintaining structure and function of the adjacent chromosome kinetochores (Zhang et al., 2007b). Recently, van der Heijden and coworkers showed that histone H3.3 is incorporated into the X and Y chromosomes during formation of the transcriptionally silent sex body by meiotic sex chromosome inactivation (van der Heijden et al., 2007). Replacement of canonical histone H3.1 by variant histone H3.3 is linked to HIRA's localization to the developing sex body. Thus, the HIRA/ASF1a complex might drive formation of SAHF by deposition of histone H3.3-containing nucleosomes.

Alternatively, under some conditions, for example, during SAHF formation, HIRA/ASF1a might utilize histone H3.1 as a substrate. Consistent with this idea, ASF1a interacts with histone H3.1 and histone H3.3 and inactivation of HIRA in mouse ES cells affects the nuclear mobility of both histone H3.3 and H3.1 (Meshorer et al., 2006; Mousson et al., 2005; Tagami et al., 2004). Although HIRA is thought to act in a DNA replication-independent manner and histone H3.1 is incorporated predominantly in S-phase of the cell cycle, recent studies have shown that histone H3.1 can be deposited outside of S-phase (Polo et al., 2006). In sum, the histone H3 variant utilized by HIRA/ASF1a to make SAHF and the key events in chromosome condensation are still not fully defined.

Activation of the HIRA/ASF1a Pathway by Wnt Signaling

In proliferating primary human cells, HIRA is evenly dispersed throughout the cell nucleus (Zhang et al., 2005). However, in cells approaching senescence, regardless of whether the trigger is an activated oncogene, short telomeres, or cell stress, HIRA is translocated into a specific subnuclear organelle, the PML nuclear body. Most human cells contain 20–30 PML nuclear bodies, which are typically 0.1–1 μM in diameter and enriched in the protein PML, as well as many other nuclear regulatory proteins (Borden, 2002; Salomoni and Pandolfi, 2002). PML bodies have been previously implicated in various cellular processes, including tumor suppression and cellular senescence (de Stanchina et al., 2004; Ferbeyre et al., 2000; Pearson et al., 2000). Significantly, HIRA is translocated to PML bodies prior to formation of SAHF and prior to exit of the cells from the cell cycle (Zhang et al., 2005) (Fig. 2). Two lines of evidence indicate that HIRA's translocation to PML bodies is essential for formation of SAHF. First, a dominant-negative HIRA mutant, which is targeted to PML bodies but does not bind to ASF1a and which blocks localization of endogenous HIRA to PML bodies, also blocks formation of SAHF (Ye et al., 2007b). Second, expression of the PML-RARα fusion protein, that is known to inhibit the function of PML bodies (Salomoni and Pandolfi, 2002), also blocks formation of SAHF. At a molecular level, PML bodies have been proposed to serve as sites of assembly of macromolecular regulatory complexes (Fogal et al., 2000; Guo et al., 2000; Pearson et al., 2000). Therefore, it seems likely that PML bodies serve as a molecular "staging ground" for assembly or modification and activation of HIRA-containing complexes, prior to export of these complexes to sites of nascent SAHF.

If recruitment of HIRA to PML bodies is a key step in the SAHF assembly process, then how is relocalization triggered in pre-senescent cells? Obvious candidates to control this event are the pRB and p53 tumor suppressor pathways, both master regulators of the senescence program. Indeed, Lowe and coworkers originally showed that formation of SAHF depends on an active pRB pathway (Narita et al., 2003), and others have confirmed this and demonstrated a requirement for p53 for formation of SAHF (Chan et al., 2005; Ye et al., 2007b). However, our studies also showed that translocation of HIRA to PML bodies is independent of pRB and p53 activity (Ye et al., 2007b), suggesting that pRB and p53 act downstream or in parallel to HIRA. Significantly, formation of SAHF driven by ectopically expressed p16INK4a, an activator of the pRB pathway, also requires ASF1a, suggesting that the p16INK4a/pRB pathway is not downstream of HIRA. Therefore, these observations suggest that the HIRA/ASF1a and pRB pathways act in parallel to form SAHF (Fig. 2).

Consistent with this idea, the pRB tumor suppressor protein and the HIRA and ASF1a histone chaperones are all well-established regulators of chromatin structure and function (Blackwell et al., 2004; Braig et al., 2005; Gonzalo et al., 2005; Greenall et al., 2006; Kaufman et al., 1998; Krawitz et al., 2002; Phelps-Durr et al., 2005; Sharp et al., 2001; Zhang et al., 2005; Zhu, 2005). Therefore, one simple model to describe the cooperation of the pRB and HIRA/ASF1a pathways is that pRB initiates heterochromatin formation at the promoters of E2F target genes, and this heterochromatin acts as a nucleation site for HIRA/ASF1a-mediated large-scale chromosome condensation. Consistent with this, E2F target genes, such as cyclin A, are incorporated into SAHF, and pRB has been reported to partially colocalize with SAHF (Narita et al., 2003). This model is also consistent with the observation that some human cell types, such as BJ cells, and mouse cells do not form SAHF efficiently (Braig et al., 2005; Narita et al., 2003; Zhang et al., 2005). Senescence in these cell types is less dependent on the p16INK4a/pRB pathway (Beausejour et al., 2003; Campisi, 2005; Itahana et al., 2003).

A likely role for the pRB pathway at the level of chromatin structure is quite satisfying, but it does not answer the original question: what is responsible for activation of HIRA/ASF1a? Recently, we defined a role for Wnt signaling in this event in the senescence program (Ye et al., 2007a) (Fig. 2). Canonical Wnt signaling is known to maintain proliferation of a diverse range of tissue progenitor cells by stimulation of cell division and inhibition of differentiation and apoptosis, for example, in the intestinal epithelium (Pinto et al., 2003), melanoblasts, and neural and embryonal stem cells (Cai et al., 2007; Chenn and Walsh, 2003; Cole et al., 2008; Dunn et al., 2005; Dunn et al., 2000; Ikeya et al., 1997; Kielman et al., 2002; Ogawa et al., 2006; Sato et al., 2004; Singla et al., 2006). Extracellular Wnt proteins bind to their cognate transmembrane receptors, members of the Frizzled family (Logan and Nusse, 2004). This ultimately results in inhibition of the serine/threonine kinase, GSK3, via a signaling cascade that depends on the disheveled protein and is antagonized by two negative regulators of Wnt signaling, axin and the *A* denomatous *P* olyposis *C* oli (APC) protein. GSK3 phosphorylates the transcription factor β-catenin, causing its proteolytic destruction. Consequently, Wnt signaling acts to stabilize soluble β-catenin, which then acts in the nucleus, together with the Lef/TCF family of transcription factors, to drive expression of proliferative genes, such as c-myc and cyclin D1. Underscoring its role as a promoter of cell proliferation, the Wnt-signaling pathway is frequently activated in human cancers, by mutations in the proto-oncogene, β-catenin, or the tumor suppressor genes, APC and axin (Kinzler and Vogelstein, 1996; Reya and Clevers, 2005; Zurawel et al., 1998).

Recently, we found that as primary human fibroblasts approach senescence, triggered by activated oncogenes or extended growth in culture, expression of a specific Wnt ligand, Wnt2, is repressed in a pRB- and p53-independent manner (Ye et al., 2007a). This is accompanied by decreased soluble β-catenin and increased GSK3β kinase activity, indicative of repressed canonical Wnt signaling in senescent cells. GSK3β phosphorylates HIRA on a specific serine residue, serine 697, which causes its translocation to PML nuclear bodies. As discussed above, HIRA's localization to PML bodies is required for activation of the HIRA/ASF1a SAHF assembly pathway (Ye et al., 2007b; Zhang et al., 2005). Consistent with these observations, activation of canonical Wnt signaling by addition of the canonical Wnt ligand, Wnt3a, to growth medium delayed SAHF formation and oncogene-induced senescence, whereas premature repression of Wnt2 expression by shRNA knockdown accelerates SAHF formation and senescence (Ye et al., 2007a). In sum, these results identify repression of Wnt signaling as a novel early step in the onset of cell senescence, which specifically initiates HIRA's recruitment to PML bodies and formation of SAHF.

Wnt Signaling, SAHF, and Tissue Aging

As discussed above, accumulating evidence implicates cellular senescence as a contributor to tissue aging. Therefore, our demonstration that Wnt signaling can antagonize senescence in vitro implies that elevated Wnt signaling might suppress aspects of aging in vivo (Fig. 3). Moreover, formation of SAHF in cells might contribute to tissue aging. Remarkably, there is evidence to support both of these ideas, although the emerging story is not yet clear.

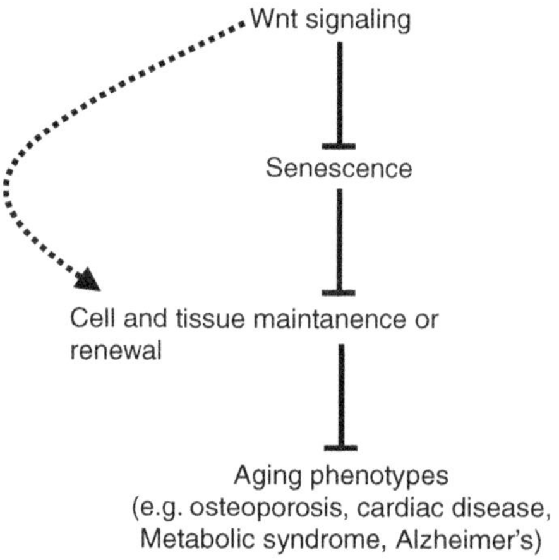

Fig. 3 Wnt signaling might be anti-aging. At least under certain conditions, elevated Wnt signaling can inhibit cell senescence. Cell senescence is thought to contribute to aging by impairing tissue function and renewal. Thus, Wnt signaling might suppress aspects of aging. Specifically, low Wnt signaling has been linked to osteoporosis, coronary disease, metabolic syndrome, and Alzheimer's disease

Mutations or polymorphic variants that lead to low Wnt-signaling activity are linked to premature onset of some age-associated phenotypes. One hallmark of the aging process, and a major public health problem, is declining bone mass, which in the most extreme cases manifest as osteoporosis. Wnt signaling promotes bone formation and osteogenesis. Loss-of-function mutations in the human gene LRP5 cause the low bone mass and fragile skeleton syndrome, osteoporosis-pseudoglioma (OPPG) (Gong et al., 2001; van Meurs et al., 2008). Conversely, mutations in human LRP5 (e.g., G171V) that reduce its affinity for the Wnt-signaling inhibitor, Dkk1, cause high bone mass (Ai et al., 2005; Boyden et al., 2002). The same is true in mice.

Lrp5–/– mice have low bone mass (Kato et al., 2002), whereas mice overexpressing LRP5G171V in osteoblasts have high bone mass (Babij et al., 2003). Activated Wnt signaling promotes bone formation via several molecular mechanisms, acting at several steps during osteogenic differentiation. First, early during differentiation, Wnt signaling drives multipotent osteochondro-progenitors into the osteogenic lineage (Day et al., 2005; Hill et al., 2005; Hu et al., 2005; Rodda and McMahon, 2006). Second, Wnt signaling appears to drive proliferation of pre-osteoblasts or osteoblasts themselves, which is linked to bone formation (Kato et al., 2002). Third, Wnt signaling has been reported to drive expression of some osteogenic genes, such as the transcription factor RUNX2 (Bennett et al., 2005; Gaur et al., 2005). Fourth, Wnt signaling regulates activity of bone-resorbing osteoclasts. For example, Wnt signaling in osteoblasts drives expression of osteoprotegerin (Glass et al., 2005), an inhibitor of osteoclast bone-resorbing activity. In addition to being linked to osteoporosis, low Wnt signaling is also linked to other age-associated phenotypes, such as coronary disease (Mani et al., 2007), age-associated metabolic syndrome, and Alzheimer's disease (De Ferrari et al., 2007). In sum, low levels of Wnt signaling are linked to various age-associated pathologies, consistent with the idea that suppression of Wnt signaling might contribute to aging and elevated Wnt signaling might delay aging. Contrary to this idea, two recent reports linked aging in mice to high levels of Wnt signaling (Brack et al., 2007; Liu et al., 2007). Specifically, Brack et al. showed that aging of mice is associated with increased systemic Wnt-signaling activity and this promotes muscle fibrosis. In this study, the pro-aging effects of Wnt signaling were due to a muscle to fibrotic cell fate switch, rather than an effect of Wnt signaling on senescence (Brack et al., 2007). Liu et al. presented evidence that elevated systemic Wnt-signaling activity contributes to premature aging of the Klotho mouse. Although it was not shown that inactivation of Wnt signaling rescues the premature aging in the Klotho mouse, these authors did show that Wnt signals can promote cell senescence in vitro (Liu et al., 2007). The specific reconciliation of this work with the work of Ye et al. (2007a) is not yet clear. On balance, it seems likely that the impact of Wnt signaling on aging, like its effects on development, is likely to be highly complex and tissue and context dependent.

As discussed above, formation of SAHF appears to depend on redistribution of heterochromatin throughout the cell nucleus (Fig. 1). Remarkably, this also seems to occur in normal tissue aging. In mammals, there is an age-associated decline in total genomic DNA methylation (Romanov and Vanyushin, 1981; Singhal et al., 1987; Wilson et al., 1987). This occurs mostly at repetitive DNA sequences, and so probably occurs predominantly in domains of constitutive heterochromatin. Since DNA methylation promotes formation of transcriptionally silent heterochromatin (Kouzarides, 2007), this change will facilitate deheterochromatinization of these regions. However, although genome-wide levels of methylation decrease with age in mammals, at specific sites there is a tendency for DNA methylation to increase (Ahuja et al., 1998; Issa et al., 2001; Issa et al., 1994; Issa et al., 1996; Kim et al., 2005a, b; So et al., 2006; Waki et al., 2003; Yatabe et al., 2001). This can occur at CpG islands, CG-rich sequences that are typically unmethylated. Many CpG islands are in the promoter regions of genes, and their methylation often silences gene expression. In addition, the total abundance of histone H4 methylated on lysine 20 (H4K20Me) has also been reported to increase with age in rat liver and kidney (Sarg et al., 2002). Like DNA methylation, H4K20Me is linked to transcriptional repression (Berger, 2007), supporting the notion that heterochromatin accumulates with tissue aging, at least at some sites. Together, these observations suggest that mammalian aging is also associated with remodeling of chromatin structure. In particular, analysis of DNA methylation patterns suggests that mammalian aging is associated with an overall decrease in heterochromatin, but an increase at specific sites in the genome. Conceivably, there is a "redistribution" of transcription-silencing heterochromatin from repetitive DNA, which is normally packaged into constitutive heterochromatin, to regions of the genome that are normally transcribed (Imai and Kitano, 1998; Villeponteau, 1997; Zhang and Adams, 2007). Significantly, the histone chaperone HIRA has been shown to increase in expression or undergo some level of regulation in aging baboon skin (Herbig et al., 2006; Jeyapalan et al., 2007), suggesting that the redistribution of heterochromatin in cells of aging tissue might be mechanistically linked to formation of SAHF in senescent cells in vitro. Thus, although the story is very preliminary, there is evidence to indicate that both chromatin structure and Wnt signaling impact aging of tissues in vivo similar to the ways that they impact cell senescence in vitro. Even more speculatively, Wnt signaling and chromatin structure impact aging, in part, through a single linear pathway (Fig. 3).

Summary

In this review, we have described the chromatin changes that occur in senescent cells and what is known of the molecular mechanisms underlying them. We have described a role for repression of Wnt signaling in initiation of these changes. We have also compared the chromatin changes in senescent cells to chromatin changes that have been described in association with normal tissue aging. Extending the comparison, we have suggested that elevated Wnt signaling can suppress some aspects of tissue aging. Synthesizing these findings, we have hypothesized the existence of a pathway through which Wnt signaling can affect aging, in part, via chromatin structure (Fig. 3). Whether this pathway is a reality remains to be determined. In particular, it is important to determine whether the chromatin changes that have been reported during tissue aging are mechanistically related to SAHF. We also need to know whether the phenotypes linked to hypomorphic Wnt signaling are just phenocopies of aging or really are premature aging. Finally, we need to know the molecular mechanism by which low Wnt signaling drives age-like phenotypes.

Acknowledgments PDA is funded by the NIH by grants RO1 GM062281, R01 CA129334-01, and P01 AG031862. PDA is a Leukemia and Lymphoma Society scholar and also thanks the Mary Kay Ash Charitable Foundation for support.

References

Abidi, P., Leers-Sucheta, S., Cortez, Y., Han, J., and Azhar, S. (2008). Evidence that age-related changes in p38 MAP kinase contribute to the decreased steroid production by the adrenocortical cells from old rats. Aging Cell *7*, 168–178.

Adkins, M.W., Carson, J.J., English, C.M., Ramey, C.J., and Tyler, J.K. (2007). The histone chaperone anti-silencing function 1 stimulates the acetylation of newly synthesized histone H3 in S-phase. J Biol Chem *282*, 1334–1340.

Adkins, M.W., Howar, S.R., and Tyler, J.K. (2004). Chromatin disassembly mediated by the histone chaperone Asf1 is essential for transcriptional activation of the yeast PHO5 and PHO8 genes. Mol Cell *14*, 657–666.

Agalioti, T., Lomvardas, S., Parekh, B., Yie, J., Maniatis, T., and Thanos, D. (2000). Ordered recruitment of chromatin modifying and general transcription factors to the IFN-beta promoter. Cell *103*, 667–678.

Agelopoulos, M., and Thanos, D. (2006). Epigenetic determination of a cell-specific gene expression program by ATF-2 and the histone variant macroH2A. Embo J *25*, 4843–4853.

Agez, M., Chen, J., Guerois, R., van Heijenoort, C., Thuret, J.Y., Mann, C., and Ochsenbein, F. (2007). Structure of the histone chaperone ASF1 bound to the histone H3 C-terminal helix and functional insights. Structure *15*, 191–199.

Ahmad, K., and Henikoff, S. (2002). The histone variant h3.3 marks active chromatin by replication-independent nucleosome assembly. Mol Cell *9*, 1191–1200.

Ahuja, N., Li, Q., Mohan, A.L., Baylin, S.B., and Issa, J.P. (1998). Aging and DNA methylation in colorectal mucosa and cancer. Cancer Res *58*, 5489–5494.

Ai, M., Holmen, S.L., Van Hul, W., Williams, B.O., and Warman, M.L. (2005). Reduced affinity to and inhibition by DKK1 form a common mechanism by which high bone mass-associated missense mutations in LRP5 affect canonical Wnt signaling. Mol Cell Biol *25*, 4946–4955.

Angelov, D., Molla, A., Perche, P.Y., Hans, F., Cote, J., Khochbin, S., Bouvet, P., and Dimitrov, S. (2003). The histone variant macroH2A interferes with transcription factor binding and SWI/SNF nucleosome remodeling. Mol Cell *11*, 1033–1041.

Angermayr, M., and Bandlow, W. (2003). Permanent nucleosome exclusion from the Gal4p-inducible yeast GCY1 promoter. J Biol Chem *278*, 11026–11031.

Angermayr, M., Oechsner, U., Gregor, K., Schroth, G.P., and Bandlow, W. (2002). Transcription initiation in vivo without classical transactivators: DNA kinks flanking the core promoter of the housekeeping yeast adenylate kinase gene, AKY2, position nucleosomes and constitutively activate transcription. Nucleic Acids Res *30*, 4199–4207.

Antczak, A.J., Tsubota, T., Kaufman, P.D., and Berger, J.M. (2006). Structure of the yeast histone H3-ASF1 interaction: implications for chaperone mechanism, species-specific interactions, and epigenetics. BMC Struct Biol *6*, 26.

Babij, P., Zhao, W., Small, C., Kharode, Y., Yaworsky, P.J., Bouxsein, M.L., Reddy, P.S., Bodine, P.V., Robinson, J.A., Bhat, B., et al. (2003). High bone mass in mice expressing a mutant LRP5 gene. J Bone Miner Res *18*, 960–974.

Baker, D.J., Perez-Terzic, C., Jin, F., Pitel, K., Niederlander, N.J., Jeganathan, K., Yamada, S., Reyes, S., Rowe, L., Hiddinga, H.J., et al . (2008). Opposing roles for p16Ink4a and p19Arf in senescence and ageing caused by BubR1 insufficiency. Nat Cell Biol *10*, 825–836.

Bartkova, J., Rezaei, N., Liontos, M., Karakaidos, P., Kletsas, D., Issaeva, N., Vassiliou, L.V., Kolettas, E., Niforou, K., Zoumpourlis, V.C., et al. (2006). Oncogene-induced senescence is part of the tumorigenesis barrier imposed by DNA damage checkpoints. Nature *444*, 633–637.

Beausejour, C.M., Krtolica, A., Galimi, F., Narita, M., Lowe, S.W., Yaswen, P., and Campisi, J. (2003). Reversal of human cellular senescence: roles of the p53 and p16 pathways. Embo J *22*, 4212–4222.

Ben-Porath, I., and Weinberg, R.A. (2005). The signals and pathways activating cellular senescence. Int J Biochem Cell Biol *37*, 961–976.

Benetti, R., Garcia-Cao, M., and Blasco, M.A. (2007). Telomere length regulates the epigenetic status of mammalian telomeres and subtelomeres. Nat Genet *39*, 243–250.

Bennett, C.N., Longo, K.A., Wright, W.S., Suva, L.J., Lane, T.F., Hankenson, K.D., and MacDougald, O.A. (2005). Regulation of osteoblastogenesis and bone mass by Wnt10b. Proc Natl Acad Sci U S A *102*, 3324–3329.

Berger, S.L. (2007). The complex language of chromatin regulation during transcription. Nature *447*, 407–412.

Bernard, D., Martinez-Leal, J.F., Rizzo, S., Martinez, D., Hudson, D., Visakorpi, T., Peters, G., Carnero, A., Beach, D., and Gil, J. (2005). CBX7 controls the growth of normal and tumor-derived prostate cells by repressing the Ink4a/Arf locus. Oncogene *24*, 5543–5551.

Bernstein, B.E., Liu, C.L., Humphrey, E.L., Perlstein, E.O., and Schreiber, S.L. (2004). Global nucleosome occupancy in yeast. Genome Biol *5*, R62.

Blackwell, C., Martin, K.A., Greenall, A., Pidoux, A., Allshire, R.C., and Whitehall, S.K. (2004). The Schizosaccharomyces pombe HIRA-like protein Hip1 is required for the periodic expression of histone genes and contributes to the function of complex centromeres. Mol Cell Biol *24*, 4309–4320.

Blasco, M.A., Lee, H.W., Hande, M.P., Samper, E., Lansdorp, P.M., DePinho, R.A., and Greider, C.W. (1997). Telomere shortening and tumor formation by mouse cells lacking telomerase RNA. Cell *91*, 25–34.

Bonnefoy, E., Orsi, G.A., Couble, P., and Loppin, B. (2007). The essential role of Drosophila HIRA for de novo assembly of paternal chromatin at fertilization. PLoS Genet *3*, 1991–2006.

Borden, K.L. (2002). Pondering the promyelocytic leukemia protein (PML) puzzle: possible functions for PML nuclear bodies. Mol Cell Biol *22*, 5259–5269.

Bosch, A., and Suau, P. (1995). Changes in core histone variant composition in differentiating neurons: the roles of differential turnover and synthesis rates. Eur J Cell Biol *68*, 220–225.

Boyden, L.M., Mao, J., Belsky, J., Mitzner, L., Farhi, A., Mitnick, M.A., Wu, D., Insogna, K., and Lifton, R.P. (2002). High bone density due to a mutation in LDL-receptor-related protein 5. N Engl J Med *346*, 1513–1521.

Brack, A.S., Conboy, M.J., Roy, S., Lee, M., Kuo, C.J., Keller, C., and Rando, T.A. (2007). Increased Wnt signaling during aging alters muscle stem cell fate and increases fibrosis. Science *317*, 807–810.

Bracken, A.P., Kleine-Kohlbrecher, D., Dietrich, N., Pasini, D., Gargiulo, G., Beekman, C., Theilgaard-Monch, K., Minucci, S., Porse, B.T., Marine, J.C. (2007). The Polycomb group proteins bind throughout the INK4A-ARF locus and are disassociated in senescent cells. Genes Dev *21*, 525–530.

Braig, M., Lee, S., Loddenkemper, C., Rudolph, C., Peters, A.H., Schlegelberger, B., Stein, H., Dorken, B., Jenuwein, T., and Schmitt, C.A. (2005). Oncogene-induced senescence as an initial barrier in lymphoma development. Nature *436*, 660–665.

Brookes, S., Rowe, J., Ruas, M., Llanos, S., Clark, P.A., Lomax, M., James, M.C., Vatcheva, R., Bates, S., Vousden, K.H., et al. (2002). INK4a-deficient human diploid fibroblasts are resistant to RAS-induced senescence. Embo J *21*, 2936–2945.

Brown, D.T., Wellman, S.E., and Sittman, D.B. (1985). Changes in the levels of three different classes of histone mRNA during murine erythroleukemia cell differentiation. Mol Cell Biol *5*, 2879–2886.

Bulavin, D.V., Phillips, C., Nannenga, B., Timofeev, O., Donehower, L.A., Anderson, C.W., Appella, E., and Fornace, A.J., Jr. (2004). Inactivation of the Wip1 phosphatase inhibits mammary tumorigenesis through p38 MAPK-mediated activation of the p16(Ink4a)-p19(Arf) pathway. Nat Genet *36*, 343–350.

Cai, L., Ye, Z., Zhou, B.Y., Mali, P., Zhou, C., and Cheng, L. (2007). Promoting human embryonic stem cell renewal or differentiation by modulating Wnt signal and culture conditions. Cell Res *17*, 62–72.

Campisi, J. (2005). Senescent cells, tumor suppression, and organismal aging: good citizens, bad neighbors. Cell *120*, 513–522.

Campisi, J., and d'Adda di Fagagna, F. (2007). Cellular senescence: when bad things happen to good cells. Nat Rev Mol Cell Biol *8*, 729–740.
Chadwick, B.P., and Willard, H.F. (2002). Cell cycle-dependent localization of macroH2A in chromatin of the inactive X chromosome. J Cell Biol *157*, 1113–1123.
Chan, H.M., Narita, M., Lowe, S.W., and Livingston, D.M. (2005). The p400 E1A-associated protein is a novel component of the p53 –> p21 senescence pathway. Genes Dev *19*, 196–201.
Changolkar, L.N., Costanzi, C., Leu, N.A., Chen, D., McLaughlin, K.J., and Pehrson, J.R. (2007). Developmental changes in histone macroH2A1 mediated gene regulation. Mol Cell Biol.
Changolkar, L.N., and Pehrson, J.R. (2006). macroH2A1 histone variants are depleted on active genes but concentrated on the inactive X chromosome. Mol Cell Biol *26*, 4410–4420.
Changolkar, L.N., Singh, G., and Pehrson, J.R. (2008). macroH2A1-dependent silencing of endogenous murine leukemia viruses. Mol Cell Biol *28*, 2059–2065.
Chen, J.H., Hales, C.N., and Ozanne, S.E. (2007). DNA damage, cellular senescence and organismal ageing: causal or correlative? Nucleic Acids Res *35*, 7417–7428.
Chen, X., Wang, J., Woltring, D., Gerondakis, S., and Shannon, M.F. (2005a). Histone dynamics on the interleukin-2 gene in response to T-cell activation. Mol Cell Biol *25*, 3209–3219.
Chen, Z., Trotman, L.C., Shaffer, D., Lin, H.K., Dotan, Z.A., Niki, M., Koutcher, J.A., Scher, H.I., Ludwig, T., Gerald, W. (2005b). Crucial role of p53-dependent cellular senescence in suppression of Pten-deficient tumorigenesis. Nature *436*, 725–730.
Chenn, A., and Walsh, C.A. (2003). Increased neuronal production, enlarged forebrains and cytoarchitectural distortions in beta-catenin overexpressing transgenic mice. Cereb Cortex *13*, 599–606.
Cole, M.F., Johnstone, S.E., Newman, J.J., Kagey, M.H., and Young, R.A. (2008). Tcf3 is an integral component of the core regulatory circuitry of embryonic stem cells. Genes Dev *22*, 746–755.
Collado, M., Blasco, M.A., and Serrano, M. (2007). Cellular senescence in cancer and aging. Cell *130*, 223–233.
Collado, M., Gil, J., Efeyan, A., Guerra, C., Schuhmacher, A.J., Barradas, M., Benguria, A., Zaballos, A., Flores, J.M., Barbacid, M. (2005). Tumour biology: senescence in premalignant tumours. Nature *436*, 642.
Cosme-Blanco, W., Shen, M.F., Lazar, A.J., Pathak, S., Lozano, G., Multani, A.S., and Chang, S. (2007). Telomere dysfunction suppresses spontaneous tumorigenesis in vivo by initiating p53-dependent cellular senescence. EMBO Rep *8*, 497–503.
Costanzi, C., and Pehrson, J.R. (1998). Histone macroH2A1 is concentrated in the inactive X chromosome of female mammals. Nature *393*, 599–601.
Costanzi, C., and Pehrson, J.R. (2001). MACROH2A2, a new member of the MARCOH2A core histone family. J Biol Chem *276*, 21776–21784.
Counter, C.M., Avilion, A.A., LeFeuvre, C.E., Stewart, N.G., Greider, C.W., Harley, C.B., and Bacchetti, S. (1992). Telomere shortening associated with chromosome instability is arrested in immortal cells which express telomerase activity. Embo J *11*, 1921–1929.
Courtois-Cox, S., Genther Williams, S.M., Reczek, E.E., Johnson, B.W., McGillicuddy, L.T., Johannessen, C.M., Hollstein, P.E., MacCollin, M., and Cichowski, K. (2006). A negative feedback signaling network underlies oncogene-induced senescence. Cancer Cell *10*, 459–472.
d'Adda di Fagagna, F. (2008). Living on a break: cellular senescence as a DNA-damage response. Nat Rev Cancer *8*, 512–522.
d'Adda di Fagagna, F., Reaper, P.M., Clay-Farrace, L., Fiegler, H., Carr, P., Von Zglinicki, T., Saretzki, G., Carter, N.P., and Jackson, S.P. (2003). A DNA damage checkpoint response in telomere-initiated senescence. Nature *426*, 194–198.
Daganzo, S.M., Erzberger, J.P., Lam, W.M., Skordalakes, E., Zhang, R., Franco, A.A., Brill, S.J., Adams, P.D., Berger, J.M., and Kaufman, P.D. (2003). Structure and function of the conserved core of histone deposition protein Asf1. Curr Biol *13*, 2148–2158.

Dai, C.Y., Furth, E.E., Mick, R., Koh, J., Takayama, T., Niitsu, Y., and Enders, G.H. (2000). p16(INK4a) expression begins early in human colon neoplasia and correlates inversely with markers of cell proliferation. Gastroenterology *119*, 929–942.

Dankort, D., Filenova, E., Collado, M., Serrano, M., Jones, K., and McMahon, M. (2007). A new mouse model to explore the initiation, progression, and therapy of BRAFV600E-induced lung tumors. Genes Dev *21*, 379–384.

Day, T.F., Guo, X., Garrett-Beal, L., and Yang, Y. (2005). Wnt/beta-catenin signaling in mesenchymal progenitors controls osteoblast and chondrocyte differentiation during vertebrate skeletogenesis. Dev Cell *8*, 739–750.

De Ferrari, G.V., Papassotiropoulos, A., Biechele, T., Wavrant De-Vrieze, F., Avila, M.E., Major, M.B., Myers, A., Saez, K., Henriquez, J.P., Zhao, A. (2007). Common genetic variation within the low-density lipoprotein receptor-related protein 6 and late-onset Alzheimer's disease. Proc Natl Acad Sci USA *104*, 9434–9439.

de Stanchina, E., Querido, E., Narita, M., Davuluri, R.V., Pandolfi, P.P., Ferbeyre, G., and Lowe, S.W. (2004). PML is a direct p53 target that modulates p53 effector functions. Mol Cell *13*, 523–535.

Deng, Q., Liao, R., Wu, B.L., and Sun, P. (2004). High intensity ras signaling induces premature senescence by activating p38 pathway in primary human fibroblasts. J Biol Chem *279*, 1050–1059.

Denoyelle, C., Abou-Rjaily, G., Bezrookove, V., Verhaegen, M., Johnson, T.M., Fullen, D.R., Pointer, J.N., Gruber, S.B., Su, L.D., Nikiforov, M.A., et al. (2006). Anti-oncogenic role of the endoplasmic reticulum differentially activated by mutations in the MAPK pathway. Nat Cell Biol *8*, 1053–1063.

Di Micco, R., Fumagalli, M., Cicalese, A., Piccinin, S., Gasparini, P., Luise, C., Schurra, C., Garre, M., Nuciforo, P.G., Bensimon, A., et al. (2006). Oncogene-induced senescence is a DNA damage response triggered by DNA hyper-replication. Nature *444*, 638–642.

Dimri, G.P., Lee, X., Basile, G., Acosta, M., Scott, G., Roskelley, C., Medrano, E.E., Linskens, M., Rubelj, I., Pereira-Smith, O., et al. (1995). A biomarker that identifies senescent human cells in culture and in aging skin in vivo. Proc Natl Acad Sci USA *92*, 9363–9367.

Doyen, C.M., An, W., Angelov, D., Bondarenko, V., Mietton, F., Studitsky, V.M., Hamiche, A., Roeder, R.G., Bouvet, P., and Dimitrov, S. (2006). Mechanism of polymerase II transcription repression by the histone variant macroH2A. Mol Cell Biol *26*, 1156–1164.

Driscoll, R., Hudson, A., and Jackson, S.P. (2007). Yeast Rtt109 promotes genome stability by acetylating histone H3 on lysine 56. Science *315*, 649–652.

Du, X., Shen, J., Kugan, N., Furth, E.E., Lombard, D.B., Cheung, C., Pak, S., Luo, G., Pignolo, R.J., DePinho, R.A., et al. (2004). Telomere shortening exposes functions for the mouse Werner and Bloom syndrome genes. Mol Cell Biol *24*, 8437–8446.

Dunn, K.J., Brady, M., Ochsenbauer-Jambor, C., Snyder, S., Incao, A., and Pavan, W.J. (2005). WNT1 and WNT3a promote expansion of melanocytes through distinct modes of action. Pigment Cell Res *18*, 167–180.

Dunn, K.J., Williams, B.O., Li, Y., and Pavan, W.J. (2000). Neural crest-directed gene transfer demonstrates Wnt1 role in melanocyte expansion and differentiation during mouse development. Proc Natl Acad Sci USA *97*, 10050–10055.

English, C.M., Adkins, M.W., Carson, J.J., Churchill, M.E., and Tyler, J.K. (2006). Structural basis for the histone chaperone activity of Asf1. Cell *127*, 495–508.

Fedele, M., Battista, S., Kenyon, L., Baldassarre, G., Fidanza, V., Klein-Szanto, A.J., Parlow, A.F., Visone, R., Pierantoni, G.M., Outwater, E., et al. (2002). Overexpression of the HMGA2 gene in transgenic mice leads to the onset of pituitary adenomas. Oncogene *21*, 3190–3198.

Fedele, M., Fidanza, V., Battista, S., Pentimalli, F., Klein-Szanto, A.J., Visone, R., De Martino, I., Curcio, A., Morisco, C., Del Vecchio, L. (2006). Haploinsufficiency of the Hmga1 gene causes cardiac hypertrophy and myelo-lymphoproliferative disorders in mice. Cancer Res *66*, 2536–2543.

Feldser, D.M., and Greider, C.W. (2007). Short telomeres limit tumor progression in vivo by inducing senescence. Cancer Cell *11*, 461–469.
Ferbeyre, G., de Stanchina, E., Querido, E., Baptiste, N., Prives, C., and Lowe, S.W. (2000). PML is induced by oncogenic ras and promotes premature senescence. Genes Dev *14*, 2015–2027.
Finkel, T., Serrano, M., and Blasco, M.A. (2007). The common biology of cancer and ageing. Nature *448*, 767–774.
Fogal, V., Gostissa, M., Sandy, P., Zacchi, P., Sternsdorf, T., Jensen, K., Pandolfi, P.P., Will, H., Schneider, C., and Del Sal, G. (2000). Regulation of p53 activity in nuclear bodies by a specific PML isoform. Embo J *19*, 6185–6195.
Francis, M.K., Appel, S., Meyer, C., Balin, S.J., Balin, A.K., and Cristofalo, V.J. (2004). Loss of EPC-1/PEDF expression during skin aging in vivo. J Invest Dermatol *122*, 1096–1105.
Franco, A.A., Lam, W.M., Burgers, P.M., and Kaufman, P.D. (2005). Histone deposition protein Asf1 maintains DNA replisome integrity and interacts with replication factor C. Genes Dev *19*, 1365–1375.
Funayama, R., Saito, M., Tanobe, H., and Ishikawa, F. (2006). Loss of linker histone H1 in cellular senescence. J Cell Biol *175*, 869–880.
Gaestel, M. (2006). MAPKAP kinases – MKs – two's company, three's a crowd. Nat Rev Mol Cell Biol *7*, 120–130.
Garcia-Cao, I., Garcia-Cao, M., Tomas-Loba, A., Martin-Caballero, J., Flores, J.M., Klatt, P., Blasco, M.A., and Serrano, M. (2006). Increased p53 activity does not accelerate telomere-driven ageing. EMBO Rep *7*, 546–552.
Gaur, T., Lengner, C.J., Hovhannisyan, H., Bhat, R.A., Bodine, P.V., Komm, B.S., Javed, A., van Wijnen, A.J., Stein, J.L., Stein, G.S., et al. (2005). Canonical WNT signaling promotes osteogenesis by directly stimulating Runx2 gene expression. J Biol Chem *280*, 33132–33140.
Gazin, C., Wajapeyee, N., Gobeil, S., Virbasius, C.M., and Green, M.R. (2007). An elaborate pathway required for Ras-mediated epigenetic silencing. Nature *449*, 1073–1077.
Geiger, H., and Van Zant, G. (2002). The aging of lympho-hematopoietic stem cells. Nat Immunol *3*, 329–333.
Gil, J., Bernard, D., Martinez, D., and Beach, D. (2004). Polycomb CBX7 has a unifying role in cellular lifespan. Nat Cell Biol *6*, 67–72.
Glass, D.A., 2nd, Bialek, P., Ahn, J.D., Starbuck, M., Patel, M.S., Clevers, H., Taketo, M.M., Long, F., McMahon, A.P., Lang, R.A., et al. (2005). Canonical Wnt signaling in differentiated osteoblasts controls osteoclast differentiation. Dev Cell *8*, 751–764.
Gong, Y., Slee, R.B., Fukai, N., Rawadi, G., Roman-Roman, S., Reginato, A.M., Wang, H., Cundy, T., Glorieux, F.H., Lev, D., et al. (2001). LDL receptor-related protein 5 (LRP5) affects bone accrual and eye development. Cell *107*, 513–523.
Gonzalez-Suarez, E., Geserick, C., Flores, J.M., and Blasco, M.A. (2005). Antagonistic effects of telomerase on cancer and aging in K5-mTert transgenic mice. Oncogene *24*, 2256–2270.
Gonzalez-Suarez, E., Samper, E., Ramirez, A., Flores, J.M., Martin-Caballero, J., Jorcano, J.L., and Blasco, M.A. (2001). Increased epidermal tumors and increased skin wound healing in transgenic mice overexpressing the catalytic subunit of telomerase, mTERT, in basal keratinocytes. Embo J *20*, 2619–2630.
Gonzalo, S., Garcia-Cao, M., Fraga, M.F., Schotta, G., Peters, A.H., Cotter, S.E., Eguia, R., Dean, D.C., Esteller, M., Jenuwein, T., et al. (2005). Role of the RB1 family in stabilizing histone methylation at constitutive heterochromatin. Nat Cell Biol *7*, 420–428.
Goodfellow, H., Krejci, A., Moshkin, Y., Verrijzer, C.P., Karch, F., and Bray, S.J. (2007). Gene-specific targeting of the histone chaperone asf1 to mediate silencing. Dev Cell *13*, 593–600.
Gray-Schopfer, V.C., Cheong, S.C., Chong, H., Chow, J., Moss, T., Abdel-Malek, Z.A., Marais, R., Wynford-Thomas, D., and Bennett, D.C. (2006). Cellular senescence in naevi and immortalisation in melanoma: a role for p16? Br J Cancer *95*, 496–505.
Green, E.M., Antczak, A.J., Bailey, A.O., Franco, A.A., Wu, K.J., Yates, J.R., 3rd, and Kaufman, P.D. (2005). Replication-independent histone deposition by the HIR complex and Asf1. Curr Biol *15*, 2044–2049.

Greenall, A., Williams, E.S., Martin, K.A., Palmer, J.M., Gray, J., Liu, C., and Whitehall, S.K. (2006). Hip3 interacts with the HIRA proteins Hip1 and Slm9 and is required for transcriptional silencing and accurate chromosome segregation. J Biol Chem *281*, 8732–8739.

Grigsby, I.F., and Finger, F.P. (2008). UNC-85, a C. elegans homolog of the histone chaperone Asf1, functions in post-embryonic neuroblast replication. Dev Biol *319*, 100–109.

Groth, A., Corpet, A., Cook, A.J., Roche, D., Bartek, J., Lukas, J., and Almouzni, G. (2007). Regulation of replication fork progression through histone supply and demand. Science *318*, 1928–1931.

Groth, A., Ray-Gallet, D., Quivy, J.P., Lukas, J., Bartek, J., and Almouzni, G. (2005). Human Asf1 regulates the flow of S phase histones during replicational stress. Mol Cell *17*, 301–311.

Grove, G.W., and Zweidler, A. (1984). Regulation of nucleosomal core histone variant levels in differentiating murine erythroleukemia cells. Biochemistry *23*, 4436–4443.

Guenatri, M., Bailly, D., Maison, C., and Almouzni, G. (2004). Mouse centric and pericentric satellite repeats form distinct functional heterochromatin. J Cell Biol *166*, 493–505.

Guo, A., Salomoni, P., Luo, J., Shih, A., Zhong, S., Gu, W., and Paolo Pandolfi, P. (2000). The function of PML in p53-dependent apoptosis. Nat Cell Biol *2*, 730–736.

Ha, L., Ichikawa, T., Anver, M., Dickins, R., Lowe, S., Sharpless, N.E., Krimpenfort, P., Depinho, R.A., Bennett, D.C., Sviderskaya, E.V., et al. (2007). ARF functions as a melanoma tumor suppressor by inducing p53-independent senescence. Proc Natl Acad Sci USA *104*, 10968–10973.

Hake, S.B., Garcia, B.A., Duncan, E.M., Kauer, M., Dellaire, G., Shabanowitz, J., Bazett-Jones, D.P., Allis, C.D., and Hunt, D.F. (2006). Expression patterns and post-translational modifications associated with mammalian histone H3 variants. J Biol Chem *281*, 559–568.

Han, X., Berardi, P., and Riabowol, K. (2006). Chromatin modification and senescence: linkage by tumor suppressors? Rejuvenation Res *9*, 69–76.

Hayflick, L. (1965). The limited in vitro lifetime of human diploid cell strains. Exp Cell Res *37*, 614–636.

Herbig, U., Ferreira, M., Condel, L., Carey, D., and Sedivy, J.M. (2006). Cellular senescence in aging primates. Science *311*, 1257.

Herbig, U., Jobling, W.A., Chen, B.P., Chen, D.J., and Sedivy, J.M. (2004). Telomere shortening triggers senescence of human cells through a pathway involving ATM, p53, and p21(CIP1), but not p16(INK4a). Mol Cell *14*, 501–513.

Herbig, U., and Sedivy, J.M. (2006). Regulation of growth arrest in senescence: telomere damage is not the end of the story. Mech Ageing Dev *127*, 16–24.

Hernandez-Munoz, I., Lund, A.H., van der Stoop, P., Boutsma, E., Muijrers, I., Verhoeven, E., Nusinow, D.A., Panning, B., Marahrens, Y., and van Lohuizen, M. (2005). Stable X chromosome inactivation involves the PRC1 Polycomb complex and requires histone MACROH2A1 and the CULLIN3/SPOP ubiquitin E3 ligase. Proc Natl Acad Sci USA *102*, 7635–7640.

Herrera, E., Samper, E., Martin-Caballero, J., Flores, J.M., Lee, H.W., and Blasco, M.A. (1999). Disease states associated with telomerase deficiency appear earlier in mice with short telomeres. Embo J *18*, 2950–2960.

Hill, T.P., Spater, D., Taketo, M.M., Birchmeier, W., and Hartmann, C. (2005). Canonical Wnt/beta-catenin signaling prevents osteoblasts from differentiating into chondrocytes. Dev Cell *8*, 727–738.

Howard, B.H. (1996). Replicative senescence: considerations relating to the stability of heterochromatin domains. Exp Gerontol *31*, 281–293.

Hsieh, C.C., and Papaconstantinou, J. (2006). Thioredoxin-ASK1 complex levels regulate ROS-mediated p38 MAPK pathway activity in livers of aged and long-lived Snell dwarf mice. Faseb J *20*, 259–268.

Hu, H., Hilton, M.J., Tu, X., Yu, K., Ornitz, D.M., and Long, F. (2005). Sequential roles of Hedgehog and Wnt signaling in osteoblast development. Development *132*, 49–60.

Ikeya, M., Lee, S.M., Johnson, J.E., McMahon, A.P., and Takada, S. (1997). Wnt signalling required for expansion of neural crest and CNS progenitors. Nature *389*, 966–970.

Imai, S., and Kitano, H. (1998). Heterochromatin islands and their dynamic reorganization: a hypothesis for three distinctive features of cellular aging. Exp Gerontol *33*, 555–570.

Issa, J.P., Ahuja, N., Toyota, M., Bronner, M.P., and Brentnall, T.A. (2001). Accelerated age-related CpG island methylation in ulcerative colitis. Cancer Res *61*, 3573–3577.

Issa, J.P., Ottaviano, Y.L., Celano, P., Hamilton, S.R., Davidson, N.E., and Baylin, S.B. (1994). Methylation of the oestrogen receptor CpG island links ageing and neoplasia in human colon. Nat Genet *7*, 536–540.

Issa, J.P., Vertino, P.M., Boehm, C.D., Newsham, I.F., and Baylin, S.B. (1996). Switch from monoallelic to biallelic human IGF2 promoter methylation during aging and carcinogenesis. Proc Natl Acad Sci USA *93*, 11757–11762.

Itahana, K., Zou, Y., Itahana, Y., Martinez, J.L., Beausejour, C., Jacobs, J.J., Van Lohuizen, M., Band, V., Campisi, J., and Dimri, G.P. (2003). Control of the replicative life span of human fibroblasts by p16 and the polycomb protein Bmi-1. Mol Cell Biol *23*, 389–401.

Ito, K., Hirao, A., Arai, F., Takubo, K., Matsuoka, S., Miyamoto, K., Ohmura, M., Naka, K., Hosokawa, K., Ikeda, Y., et al. (2006). Reactive oxygen species act through p38 MAPK to limit the lifespan of hematopoietic stem cells. Nat Med *12*, 446–451.

Iwasa, H., Han, J., and Ishikawa, F. (2003). Mitogen-activated protein kinase p38 defines the common senescence-signalling pathway. Genes Cells *8*, 131–144.

Janzen, V., Forkert, R., Fleming, H.E., Saito, Y., Waring, M.T., Dombkowski, D.M., Cheng, T., Depinho, R.A., Sharpless, N.E., and Scadden, D.T. (2006). Stem-cell ageing modified by the cyclin-dependent kinase inhibitor p16(INK4a). Nature *443*, 421–426.

Jeyapalan, J.C., Ferreira, M., Sedivy, J.M., and Herbig, U. (2007). Accumulation of senescent cells in mitotic tissue of aging primates. Mech Ageing Dev *128*, 36–44.

Jin, C., and Felsenfeld, G. (2006). Distribution of histone H3.3 in hematopoietic cell lineages. Proc Natl Acad Sci USA *103*, 574–579.

Jin, C., and Felsenfeld, G. (2007). Nucleosome stability mediated by histone variants H3.3 and H2A.Z. Genes Dev *21*, 1519–1529.

Kato, M., Patel, M.S., Levasseur, R., Lobov, I., Chang, B.H., Glass, D.A., 2nd, Hartmann, C., Li, L., Hwang, T.H., Brayton, C.F., et al. (2002). Cbfa1-independent decrease in osteoblast proliferation, osteopenia, and persistent embryonic eye vascularization in mice deficient in Lrp5, a Wnt coreceptor. J Cell Biol *157*, 303–314.

Kaufman, P.D., Cohen, J.L., and Osley, M.A. (1998). Hir proteins are required for position-dependent gene silencing in Saccharomyces cerevisiae in the absence of chromatin assembly factor I. Mol Cell Biol *18*, 4793–4806.

Kielman, M.F., Rindapaa, M., Gaspar, C., van Poppel, N., Breukel, C., van Leeuwen, S., Taketo, M.M., Roberts, S., Smits, R., and Fodde, R. (2002). Apc modulates embryonic stem-cell differentiation by controlling the dosage of beta-catenin signaling. Nat Genet *32*, 594–605.

Kim, H.J., Seol, J.H., Han, J.W., Youn, H.D., and Cho, E.J. (2007). Histone chaperones regulate histone exchange during transcription. Embo J *26*, 4467–4474.

Kim, J.Y., Siegmund, K.D., Tavare, S., and Shibata, D. (2005a). Age-related human small intestine methylation: evidence for stem cell niches. BMC Med *3*, 10.

Kim, J.Y., Tavare, S., and Shibata, D. (2005b). Counting human somatic cell replications: methylation mirrors endometrial stem cell divisions. Proc Natl Acad Sci USA *102*, 17739–17744.

Kinzler, K.W., and Vogelstein, B. (1996). Lessons from hereditary colorectal cancer. Cell *87*, 159–170.

Korber, P., Barbaric, S., Luckenbach, T., Schmid, A., Schermer, U.J., Blaschke, D., and Horz, W. (2006). The histone chaperone Asf1 increases the rate of histone eviction at the yeast PHO5 and PHO8 promoters. J Biol Chem *281*, 5539–5545.

Kouzarides, T. (2007). Chromatin modifications and their function. Cell *128*, 693–705.

Krawitz, D.C., Kama, T., and Kaufman, P.D. (2002). Chromatin assembly factor i mutants defective for PCNA binding require Asf1/Hir proteins for silencing. Mol Cell Biol *22*, 614–625.

Krimer, D.B., Cheng, G., and Skoultchi, A.I. (1993). Induction of H3.3 replacement histone mRNAs during the precommitment period of murine erythroleukemia cell differentiation. Nucleic Acids Res *21*, 2873–2879.
Krishnamurthy, J., Ramsey, M.R., Ligon, K.L., Torrice, C., Koh, A., Bonner-Weir, S., and Sharpless, N.E. (2006). p16(INK4a) induces an age-dependent decline in islet regenerative potential. Nature *443*, 453–457.
Krishnamurthy, J., Torrice, C., Ramsey, M.R., Kovalev, G.I., Al-Regaiey, K., Su, L., and Sharpless, N.E. (2004). Ink4a/Arf expression is a biomarker of aging. J Clin Invest *114*, 1299–1307.
Lee, C.K., Shibata, Y., Rao, B., Strahl, B.D., and Lieb, J.D. (2004). Evidence for nucleosome depletion at active regulatory regions genome-wide. Nat Genet *36*, 900–905.
Lee, H.W., Blasco, M.A., Gottlieb, G.J., Horner, J.W., 2nd, Greider, C.W., and DePinho, R.A. (1998). Essential role of mouse telomerase in highly proliferative organs. Nature *392*, 569–574.
Liu, H., Fergusson, M.M., Castilho, R.M., Liu, J., Cao, L., Chen, J., Malide, D., Rovira, II., Schimel, D., Kuo, C.J., et al. (2007). Augmented Wnt signaling in a mammalian model of accelerated aging. Science *317*, 803–806.
Logan, C.Y., and Nusse, R. (2004). The Wnt signaling pathway in development and disease. Annu Rev Cell Dev Biol *20*, 781–810.
Loppin, B., Bonnefoy, E., Anselme, C., Laurencon, A., Karr, T.L., and Couble, P. (2005). The histone H3.3 chaperone HIRA is essential for chromatin assembly in the male pronucleus. Nature *437*, 1386–1390.
Loyola, A., Bonaldi, T., Roche, D., Imhof, A., and Almouzni, G. (2006). PTMs on H3 variants before chromatin assembly potentiate their final epigenetic state. Mol Cell *24*, 309–316.
Ludwig, S., Engel, K., Hoffmeyer, A., Sithanandam, G., Neufeld, B., Palm, D., Gaestel, M., and Rapp, U.R. (1996). 3pK, a novel mitogen-activated protein (MAP) kinase-activated protein kinase, is targeted by three MAP kinase pathways. Mol Cell Biol *16*, 6687–6697.
Malay, A.D., Umehara, T., Matsubara-Malay, K., Padmanabhan, B., and Yokoyama, S. (2008). Crystal structures of fission yeast histone chaperone Asf1 complexed with the Hip1 B-domain or the Cac2 C terminus. J Biol Chem *283*, 14022–14031.
Mallette, F.A., Gaumont-Leclerc, M.F., and Ferbeyre, G. (2007). The DNA damage signaling pathway is a critical mediator of oncogene-induced senescence. Genes Dev *21*, 43–48.
Mani, A., Radhakrishnan, J., Wang, H., Mani, A., Mani, M.A., Nelson-Williams, C., Carew, K.S., Mane, S., Najmabadi, H., Wu, D., et al. (2007). LRP6 mutation in a family with early coronary disease and metabolic risk factors. Science *315*, 1278–1282.
McKittrick, E., Gafken, P.R., Ahmad, K., and Henikoff, S. (2004). Histone H3.3 is enriched in covalent modifications associated with active chromatin. Proc Natl Acad Sci USA *101*, 1525–1530.
Meshorer, E., Yellajoshula, D., George, E., Scambler, P.J., Brown, D.T., and Misteli, T. (2006). Hyperdynamic plasticity of chromatin proteins in pluripotent embryonic stem cells. Dev Cell *10*, 105–116.
Michaloglou, C., Vredeveld, L.C., Soengas, M.S., Denoyelle, C., Kuilman, T., van der Horst, C.M., Majoor, D.M., Shay, J.W., Mooi, W.J., and Peeper, D.S. (2005). BRAFE600-associated senescence-like cell cycle arrest of human naevi. Nature *436*, 720–724.
Michishita, E., McCord, R.A., Berber, E., Kioi, M., Padilla-Nash, H., Damian, M., Cheung, P., Kusumoto, R., Kawahara, T.L., Barrett, J.C., et al. (2008). SIRT6 is a histone H3 lysine 9 deacetylase that modulates telomeric chromatin. Nature *452*, 492–496.
Minamino, T., and Komuro, I. (2007). Vascular cell senescence: contribution to atherosclerosis. Circ Res *100*, 15–26.
Mito, Y., Henikoff, J.G., and Henikoff, S. (2005). Genome-scale profiling of histone H3.3 replacement patterns. Nat Genet *37*, 1090–1097.
Molofsky, A.V., Slutsky, S.G., Joseph, N.M., He, S., Pardal, R., Krishnamurthy, J., Sharpless, N.E., and Morrison, S.J. (2006). Increasing p16(INK4a) expression decreases forebrain progenitors and neurogenesis during ageing. Nature *443*, 448–452.

Moshkin, Y.M., Armstrong, J.A., Maeda, R.K., Tamkun, J.W., Verrijzer, P., Kennison, J.A., and Karch, F. (2002). Histone chaperone ASF1 cooperates with the Brahma chromatin-remodelling machinery. Genes Dev *16*, 2621–2626.
Mousson, F., Lautrette, A., Thuret, J.Y., Agez, M., Courbeyrette, R., Amigues, B., Becker, E., Neumann, J.M., Guerois, R., Mann, C., et al. (2005). Structural basis for the interaction of Asf1 with histone H3 and its functional implications. Proc Natl Acad Sci USA *102*, 5975–5980.
Myung, K., Pennaneach, V., Kats, E.S., and Kolodner, R.D. (2003). Saccharomyces cerevisiae chromatin-assembly factors that act during DNA replication function in the maintenance of genome stability. Proc Natl Acad Sci USA *100*, 6640–6645.
Nakayama, T., Nishioka, K., Dong, Y.X., Shimojima, T., and Hirose, S. (2007). Drosophila GAGA factor directs histone H3.3 replacement that prevents the heterochromatin spreading. Genes Dev *21*, 552–561.
Narita, M., Narita, M., Krizhanovsky, V., Nunez, S., Chicas, A., Hearn, S.A., Myers, M.P., and Lowe, S.W. (2006). A novel role for high-mobility group a proteins in cellular senescence and heterochromatin formation. Cell *126*, 503–514.
Narita, M., Nunez, S., Heard, E., Lin, A.W., Hearn, S.A., Spector, D.L., Hannon, G.J., and Lowe, S.W. (2003). Rb-mediated heterochromatin formation and silencing of E2F target genes during cellular senescence. Cell *113*, 703–716.
Natsume, R., Eitoku, M., Akai, Y., Sano, N., Horikoshi, M., and Senda, T. (2007). Structure and function of the histone chaperone CIA/ASF1 complexed with histones H3 and H4. Nature *446*, 338–341.
Nevins, J.R. (2001). The Rb/E2F pathway and cancer. Hum Mol Genet *10*, 699–703.
Nourani, A., Robert, F., and Winston, F. (2006). Evidence that Spt2/Sin1, an HMG-like factor, plays roles in transcription elongation, chromatin structure, and genome stability in Saccharomyces cerevisiae. Mol Cell Biol *26*, 1496–1509.
Oberdoerffer, P., and Sinclair, D.A. (2007). The role of nuclear architecture in genomic instability and ageing. Nat Rev Mol Cell Biol *8*, 692–702.
Ogawa, K., Nishinakamura, R., Iwamatsu, Y., Shimosato, D., and Niwa, H. (2006). Synergistic action of Wnt and LIF in maintaining pluripotency of mouse ES cells. Biochem Biophys Res Commun *343*, 159–166.
Palmer, T.D., Schwartz, P.H., Taupin, P., Kaspar, B., Stein, S.A., and Gage, F.H. (2001). Cell culture. Progenitor cells from human brain after death. Nature *411*, 42–43.
Pantazis, P., and Bonner, W.M. (1984). Specific alterations in the pattern of histone-3 synthesis during conversion of human leukemic cells to terminally differentiated cells in culture. Differentiation *28*, 186–190.
Pearson, M., Carbone, R., Sebastiani, C., Cioce, M., Fagioli, M., Saito, S., Higashimoto, Y., Appella, E., Minucci, S., Pandolfi, P.P., et al. (2000). PML regulates p53 acetylation and premature senescence induced by oncogenic Ras. Nature *406*, 207–210.
Perche, P.Y., Vourc'h, C., Konecny, L., Souchier, C., Robert-Nicoud, M., Dimitrov, S., and Khochbin, S. (2000). Higher concentrations of histone macroH2A in the Barr body are correlated with higher nucleosome density. Curr Biol *10*, 1531–1534.
Peterson, C.L., and Laniel, M.A. (2004). Histones and histone modifications. Curr Biol *14*, R546–551.
Phelps-Durr, T.L., Thomas, J., Vahab, P., and Timmermans, M.C. (2005). Maize rough sheath2 and its Arabidopsis orthologue ASYMMETRIC LEAVES1 interact with HIRA, a predicted histone chaperone, to maintain knox gene silencing and determinacy during organogenesis. Plant Cell *17*, 2886–2898.
Pina, B., and Suau, P. (1987). Changes in histones H2A and H3 variant composition in differentiating and mature rat brain cortical neurons. Dev Biol *123*, 51–58.
Pinto, D., Gregorieff, A., Begthel, H., and Clevers, H. (2003). Canonical Wnt signals are essential for homeostasis of the intestinal epithelium. Genes Dev *17*, 1709–1713.
Polo, S.E., Roche, D., and Almouzni, G. (2006). New histone incorporation marks sites of UV repair in human cells. Cell *127*, 481–493.

Prather, D., Krogan, N.J., Emili, A., Greenblatt, J.F., and Winston, F. (2005). Identification and characterization of Elf1, a conserved transcription elongation factor in Saccharomyces cerevisiae. Mol Cell Biol *25*, 10122–10135.

Price, J.S., Waters, J.G., Darrah, C., Pennington, C., Edwards, D.R., Donell, S.T., and Clark, I.M. (2002). The role of chondrocyte senescence in osteoarthritis. Aging Cell *1*, 57–65.

Prieur, A., and Peeper, D.S. (2008). Cellular senescence in vivo: a barrier to tumorigenesis. Curr Opin Cell Biol *20*, 150–155.

Prochasson, P., Florens, L., Swanson, S.K., Washburn, M.P., and Workman, J.L. (2005). The HIR corepressor complex binds to nucleosomes generating a distinct protein/DNA complex resistant to remodeling by SWI/SNF. Genes Dev *19*, 2534–2539.

Ramirez, R.D., Morales, C.P., Herbert, B.S., Rohde, J.M., Passons, C., Shay, J.W., and Wright, W.E. (2001). Putative telomere-independent mechanisms of replicative aging reflect inadequate growth conditions. Genes Dev *15*, 398–403.

Ray-Gallet, D., Quivy, J.P., Scamps, C., Martini, E.M., Lipinski, M., and Almouzni, G. (2002). HIRA is critical for a nucleosome assembly pathway independent of DNA synthesis. Mol Cell *9*, 1091–1100.

Recht, J., Tsubota, T., Tanny, J.C., Diaz, R.L., Berger, J.M., Zhang, X., Garcia, B.A., Shabanowitz, J., Burlingame, A.L., Hunt, D.F., et al. (2006). Histone chaperone Asf1 is required for histone H3 lysine 56 acetylation, a modification associated with S phase in mitosis and meiosis. Proc Natl Acad Sci USA *103*, 6988–6993.

Reeves, R. (2001). Molecular biology of HMGA proteins: hubs of nuclear function. Gene *277*, 63–81.

Reya, T., and Clevers, H. (2005). Wnt signalling in stem cells and cancer. Nature *434*, 843–850.

Rocha, W., and Verreault, A. (2008). Clothing up DNA for all seasons: Histone chaperones and nucleosome assembly pathways. FEBS Lett *582*, 1938–1949.

Rodda, S.J., and McMahon, A.P. (2006). Distinct roles for Hedgehog and canonical Wnt signaling in specification, differentiation and maintenance of osteoblast progenitors. Development *133*, 3231–3244.

Rogakou, E.P., and Sekeri-Pataryas, K.E. (1999). Histone variants of H2A and H3 families are regulated during in vitro aging in the same manner as during differentiation. Exp Gerontol *34*, 741–754.

Romanov, G.A., and Vanyushin, B.F. (1981). Methylation of reiterated sequences in mammalian DNAs. Effects of the tissue type, age, malignancy and hormonal induction. Biochim Biophys Acta *653*, 204–218.

Rossi, D.J., Bryder, D., and Weissman, I.L. (2007). Hematopoietic stem cell aging: mechanism and consequence. Exp Gerontol *42*, 385–390.

Rudolph, K.L., Chang, S., Lee, H.W., Blasco, M., Gottlieb, G.J., Greider, C., and DePinho, R.A. (1999). Longevity, stress response, and cancer in aging telomerase-deficient mice. Cell *96*, 701–712.

Rufiange, A., Jacques, P.E., Bhat, W., Robert, F., and Nourani, A. (2007). Genome-wide replication-independent histone H3 exchange occurs predominantly at promoters and implicates H3 K56 acetylation and Asf1. Mol Cell *27*, 393–405.

Salomoni, P., and Pandolfi, P.P. (2002). The role of PML in tumor suppression. Cell *108*, 165–170.

Sarg, B., Koutzamani, E., Helliger, W., Rundquist, I., and Lindner, H.H. (2002). Postsynthetic trimethylation of histone H4 at lysine 20 in mammalian tissues is associated with aging. J Biol Chem *277*, 39195–39201.

Sarkisian, C.J., Keister, B.A., Stairs, D.B., Boxer, R.B., Moody, S.E., and Chodosh, L.A. (2007). Dose-dependent oncogene-induced senescence in vivo and its evasion during mammary tumorigenesis. Nat Cell Biol 9, 493–505.

Sato, N., Meijer, L., Skaltsounis, L., Greengard, P., and Brivanlou, A.H. (2004). Maintenance of pluripotency in human and mouse embryonic stem cells through activation of Wnt signaling by a pharmacological GSK-3-specific inhibitor. Nat Med *10*, 55–63.

Schulz, L.L., and Tyler, J.K. (2006). The histone chaperone ASF1 localizes to active DNA replication forks to mediate efficient DNA replication. Faseb J *20*, 488–490.
Schwabish, M.A., and Struhl, K. (2006). Asf1 Mediates Histone Eviction and Deposition during Elongation by RNA Polymerase II. Mol Cell *22*, 415–422.
Schwartz, B.E., and Ahmad, K. (2005). Transcriptional activation triggers deposition and removal of the histone variant H3.3. Genes Dev *19*, 804–814.
Sedivy, J.M., Banumathy, G., and Adams, P.D. (2008). Aging by epigenetics – a consequence of chromatin damage? Exp Cell Res *314*, 1909–1917.
Serrano, M., and Blasco, M.A. (2007). Cancer and ageing: convergent and divergent mechanisms. Nat Rev Mol Cell Biol *8*, 715–722.
Serrano, M., Lin, A.W., McCurrach, M.E., Beach, D., and Lowe, S.W. (1997). Oncogenic ras provokes premature cell senescence associated with accumulation of p53 and p16INK4a. Cell *88*, 593–602.
Sgarra, R., Rustighi, A., Tessari, M.A., Di Bernardo, J., Altamura, S., Fusco, A., Manfioletti, G., and Giancotti, V. (2004). Nuclear phosphoproteins HMGA and their relationship with chromatin structure and cancer. FEBS Lett *574*, 1–8.
Sharp, J.A., Fouts, E.T., Krawitz, D.C., and Kaufman, P.D. (2001). Yeast histone deposition protein Asf1p requires Hir proteins and PCNA for heterochromatic silencing. Curr Biol *11*, 463–473.
Sharp, J.A., Franco, A.A., Osley, M.A., Kaufman, P.D., Krawitz, D.C., Kama, T., Fouts, E.T., and Cohen, J.L. (2002). Chromatin assembly factor I and Hir proteins contribute to building functional kinetochores in S. cerevisiae. Genes Dev *16*, 85–100.
Sherr, C.J., and McCormick, F. (2002). The RB and p53 pathways in cancer. Cancer Cell *2*, 103–112.
Singer, M.S., Kahana, A., Wolf, A.J., Meisinger, L.L., Peterson, S.E., Goggin, C., Mahowald, M., and Gottschling, D.E. (1998). Identification of high-copy disruptors of telomeric silencing in Saccharomyces cerevisiae. Genetics *150*, 613–632.
Singhal, R.P., Mays-Hoopes, L.L., and Eichhorn, G.L. (1987). DNA methylation in aging of mice. Mech Ageing Dev *41*, 199–210.
Singla, D.K., Schneider, D.J., LeWinter, M.M., and Sobel, B.E. (2006). wnt3a but not wnt11 supports self-renewal of embryonic stem cells. Biochem Biophys Res Commun *345*, 789–795.
So, K., Tamura, G., Honda, T., Homma, N., Waki, T., Togawa, N., Nishizuka, S., and Motoyama, T. (2006). Multiple tumor suppressor genes are increasingly methylated with age in non-neoplastic gastric epithelia. Cancer Sci *97*, 1155–1158.
Sun, P., Yoshizuka, N., New, L., Moser, B.A., Li, Y., Liao, R., Xie, C., Chen, J., Deng, Q., Yamout, M., et al. (2007). PRAK is essential for ras-induced senescence and tumor suppression. Cell *128*, 295–308.
Tagami, H., Ray-Gallet, D., Almouzni, G., and Nakatani, Y. (2004). Histone H3.1 and H3.3 complexes mediate nucleosome assembly pathways dependent or independent of DNA synthesis. Cell *116*, 51–61.
Tang, Y., Poustovoitov, M.V., Zhao, K., Garfinkel, M., Canutescu, A., Dunbrack, R., Adams, P.D., and Marmorstein, R. (2006). Structure of a human ASF1a-HIRA complex and insights into specificity of histone chaperone complex assembly. Nat Struct Mol Biol *13*, 921–929.
Tsubota, T., Berndsen, C.E., Erkmann, J.A., Smith, C.L., Yang, L., Freitas, M.A., Denu, J.M., and Kaufman, P.D. (2007). Histone H3-K56 acetylation is catalyzed by histone chaperone-dependent complexes. Mol Cell *25*, 703–712.
Tyler, J.K., Adams, C.R., Chen, S.R., Kobayashi, R., Kamakaka, R.T., and Kadonaga, J.T. (1999). The RCAF complex mediates chromatin assembly during DNA replication and repair. Nature *402*, 555–560.
Urban, M.K., and Zweidler, A. (1983). Changes in nucleosomal core histone variants during chicken development and maturation. Dev Biol *95*, 421–428.
van der Heijden, G.W., Derijck, A.A., Posfai, E., Giele, M., Pelczar, P., Ramos, L., Wansink, D.G., van der Vlag, J., Peters, A.H., and de Boer, P. (2007). Chromosome-wide nucleosome

replacement and H3.3 incorporation during mammalian meiotic sex chromosome inactivation. Nat Genet *39*, 251–258.

van Meurs, J.B., Trikalinos, T.A., Ralston, S.H., Balcells, S., Brandi, M.L., Brixen, K., Kiel, D.P., Langdahl, B.L., Lips, P., Ljunggren, O., et al. (2008). Large-scale analysis of association between LRP5 and LRP6 variants and osteoporosis. Jama *299*, 1277–1290.

Ventura, A., Kirsch, D.G., McLaughlin, M.E., Tuveson, D.A., Grimm, J., Lintault, L., Newman, J., Reczek, E.E., Weissleder, R., and Jacks, T. (2007). Restoration of p53 function leads to tumour regression in vivo. Nature *445*, 661–665.

Villeponteau, B. (1997). The heterochromatin loss model of aging. Exp Gerontol *32*, 383–394.

Voncken, J.W., Niessen, H., Neufeld, B., Rennefahrt, U., Dahlmans, V., Kubben, N., Holzer, B., Ludwig, S., and Rapp, U.R. (2005). MAPKAP kinase 3pK phosphorylates and regulates chromatin association of the polycomb group protein Bmi1. J Biol Chem *280*, 5178–5187.

Wajapeyee, N., Serra, R.W., Zhu, X., Mahalingam, M., and Green, M.R. (2008). Oncogenic BRAF Induces Senescence and Apoptosis through Pathways Mediated by the Secreted Protein IGFBP7. Cell *132*, 363–374.

Waki, T., Tamura, G., Sato, M., and Motoyama, T. (2003). Age-related methylation of tumor suppressor and tumor-related genes: an analysis of autopsy samples. Oncogene *22*, 4128–4133.

Wei, W., Hemmer, R.M., and Sedivy, J.M. (2001). Role of p14(ARF) in replicative and induced senescence of human fibroblasts. Mol Cell Biol *21*, 6748–6757.

Wiemann, S.U., Satyanarayana, A., Tsahuridu, M., Tillmann, H.L., Zender, L., Klempnauer, J., Flemming, P., Franco, S., Blasco, M.A., Manns, M.P., et al. (2002). Hepatocyte telomere shortening and senescence are general markers of human liver cirrhosis. Faseb J *16*, 935–942.

Wilson, V.L., Smith, R.A., Ma, S., and Cutler, R.G. (1987). Genomic 5-methyldeoxycytidine decreases with age. J Biol Chem *262*, 9948–9951.

Wirbelauer, C., Bell, O., and Schubeler, D. (2005). Variant histone H3.3 is deposited at sites of nucleosomal displacement throughout transcribed genes while active histone modifications show a promoter-proximal bias. Genes Dev *19*, 1761–1766.

Wright, W.E., and Shay, J.W. (2001). Cellular senescence as a tumor-protection mechanism: the essential role of counting. Curr Opin Genet Dev *11*, 98–103.

Wright, W.E., and Shay, J.W. (2002). Historical claims and current interpretations of replicative aging. Nat Biotechnol *20*, 682–688.

Wunsch, A.M., and Lough, J. (1987). Modulation of histone H3 variant synthesis during the myoblast-myotube transition of chicken myogenesis. Dev Biol *119*, 94–99.

Xu, Y., Sumter, T.F., Bhattacharya, R., Tesfaye, A., Fuchs, E.J., Wood, L.J., Huso, D.L., and Resar, L.M. (2004). The HMG-I oncogene causes highly penetrant, aggressive lymphoid malignancy in transgenic mice and is overexpressed in human leukemia. Cancer Res *64*, 3371–3375.

Xue, W., Zender, L., Miething, C., Dickins, R.A., Hernando, E., Krizhanovsky, V., Cordon-Cardo, C., and Lowe, S.W. (2007). Senescence and tumour clearance is triggered by p53 restoration in murine liver carcinomas. Nature *445*, 656–660.

Yatabe, Y., Tavare, S., and Shibata, D. (2001). Investigating stem cells in human colon by using methylation patterns. Proc Natl Acad Sci USA *98*, 10839–10844.

Ye, X., Zerlanko, B., Kennedy, A., Banumathy, G., Zhang, R., and Adams, P.D. (2007a). Downregulation of Wnt signaling is a trigger for formation of facultative heterochromatin and onset of cell senescence in primary human cells. Mol Cell *27*, 183–196.

Ye, X., Zerlanko, B., Zhang, R., Somaiah, N., Lipinski, M., Salomoni, P., and Adams, P.D. (2007b). Definition of pRB- and p53-dependent and -independent steps in HIRA/ASF1a-mediated formation of senescence-associated heterochromatin foci. Mol Cell Biol *27*, 2452–2465.

Zhang, R., and Adams, P.D. (2007). Heterochromatin and its relationship to cell senescence and cancer therapy. Cell Cycle *6*, 784–789.

Zhang, R., Chen, W., and Adams, P.D. (2007a). Molecular dissection of formation of senescent associated heterochromatin foci. Mol Cell Biol *27*, 2343–2358.

Zhang, R., Liu, S.-T., Chen, W., Bonner, B., Pehrson, J., Yen, T.J., and Adams, P.D. (2007b). HP1 proteins are essential for a dynamic nuclear response that rescues the function of perturbed heterochromatin in primary human cells. Mol Cell Biol *27*, 949–962.

Zhang, R., Poustovoitov, M.V., Ye, X., Santos, H.A., Chen, W., Daganzo, S.M., Erzberger, J.P., Serebriiskii, I.G., Canutescu, A.A., Dunbrack, R.L., et al. (2005). Formation of MacroH2A-containing senescence-associated heterochromatin foci and senescence driven by ASF1a and HIRA. Dev Cell *8*, 19–30.

Zhao, J., Herrera-Diaz, J., and Gross, D.S. (2005). Domain-wide displacement of histones by activated heat shock factor occurs independently of Swi/Snf and is not correlated with RNA polymerase II density. Mol Cell Biol *25*, 8985–8999.

Zhu, L. (2005). Tumour suppressor retinoblastoma protein Rb: a transcriptional regulator. Eur J Cancer *41*, 2415–2427.

Zurawel, R.H., Chiappa, S.A., Allen, C., and Raffel, C. (1998). Sporadic medulloblastomas contain oncogenic beta-catenin mutations. Cancer Res *58*, 896–899.

S-Adenosylmethionine: Simple Agent of Methylation and Secret to Aging and Metabolism?

Wil A.M. Loenen

Abstract *S*-adenosylmethionine (SAM) is a sulfur-containing molecule at the heart of metabolism of all organisms. It is well known as the methyl donor for the majority of methyltransferases that modify DNA, RNA, histones, and many other proteins such as Tp53, lipids, and a variety of other small molecules including toxic compounds, such as arsenic, whose actions affect replication, transcription, and translation, DNA repair and chromatin modeling, epigenetic modifications, and imprinting. Transmethylation by SAM generates *S*-adenosylhomocysteine, which is converted back to SAM via the methionine cycle or to the antioxidant glutathione via the transsulfuration pathway. So far 15 superfamilies of SAM-binding proteins have been identified with additional vital roles in polyamine synthesis and in the generation of radicals for difficult chemical reactions such as the synthesis of biotin. Surprisingly, SAM also serves as an essential cofactor in specific recognition and cutting of DNA by nucleases, such as EcoKI, and in FeS cluster-containing proteins such as the transcription elongator Elp3. Finally, on a completely different track, SAM can bind certain RNA structures called riboswitches that control transcription and/or translation. In this way, gene expression can be regulated in a SAM-dependent manner, a recent finding that opens up new avenues into gene control by alternative RNA secondary structure formation. Deregulation of SAM through folate or vitamin shortage and/or radical surplus via dietary insufficiency, alcohol abuse, arsenic poisoning, irradiation, and/or other environmental or hereditary factors leads to a wide variety of human diseases, e.g., autoimmune disease, cancer, depression, and other neurological illnesses, and is implicated in longevity and aging. This review gives an overview of the roles of this small metabolite and discusses the implications of deregulation of SAM to longevity and aging.

Keywords Methylation · Epigenetics · Aging · Metabolism · Folate · Human disease

W.A.M. Loenen (✉)
Department of Toxicogenetics, Division 5, Leiden University Medical Centre, Einthovenweg 20, Po Box 9600, 2300RC Leiden, The Netherlands
e-mail: w.a.m.loenen@lumc.nl

T.O. Tollefsbol (ed.), *Epigenetics of Aging*, DOI 10.1007/978-1-4419-0639-7_7,

Introduction

S-adenosylmethionine (SAM, also called AdoMet, Fig. 1) was discovered 55 years ago by Giulio Cantoni as an important molecule in methylation reactions (reviewed by Kresge et al. 2005). It is a conjugate of methionine at the sulfur atom with adenosine (derived from ATP), a reaction catalyzed by methionine adenosyltransferase (MAT or SAM synthetase). In the 1960s it became clear that the role of SAM was not confined to methylation when, unexpectedly, the first nuclease was described that required SAM for activity, the *Escherichia coli* K-12 restriction enzyme (Meselson and Yuan 1968; see Murray 2000; Loenen 2003; Loenen, 2006 for reviews). This intriguing enzyme, EcoKI, is a large pentameric complex belonging to the type IA family of ATP-dependent helicase cum nuclease cum methylase proteins. The enzyme binds SAM, which event alters the DNA contacts of the methylase (MTase or methyltransferase) moiety, and thus distinguishes unmethylated DNA from hemi-methylated DNA. In the case of the latter, EcoKI acts as an MTase and modifies the second strand. However, if the DNA is unmethylated, EcoKI acts as an endonuclease in an unusual way. It uses helicase domains to translocate the DNA past itself and cuts up to 5 kb away from the recognition site. Some other restriction systems also require SAM as allosteric effector and for specific cleavage of DNA (Sistla and Rao 2008; http://rebase.neb.com). Interestingly, the presence of SAM limits cleavage of DNA by EcoP15I to supercoiled DNA, and linear DNA cannot be cut under these circumstances (Raghavendra and Rao 2005). SAM-dependent nucleases are not confined to bacteria, as one has recently been identified in wheat (Fedoreyeva et al. 2007; Vanyushin 2007). It is plausible that such unforeseen SAM-dependent mechanisms for specific DNA recognition, translocation, and/or nuclease activity dependent on higher-ordered structure of the DNA are more widespread, though so far no data on this interesting topic have emerged with respect to evolutionarily related mammalian enzymes.

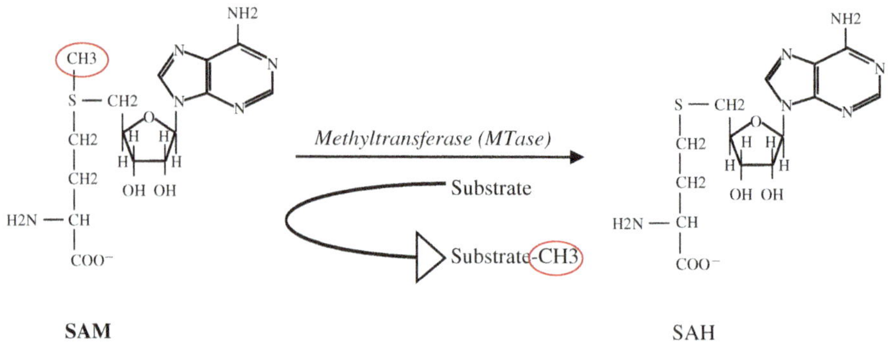

Fig. 1 *S*-adenosylmethionine (SAM) is the universal methyl donor. SAM is the major methyl donor used for transfer by methylases (methyltransferases or MTases) to DNA, RNA, protein, lipids, small molecules, arsenic, etc., the methyl group is indicated with a circle

In this review an overview of the types of SAM-dependent enzymes is given, followed by a description of important metabolic pathways that involve SAM and that are all under tight multilayered control to avoid a biochemical tug of war between methylation and the other jobs of SAM. This is followed by a discussion of the implications of the central position of SAM to life and proven or potential risks of alterations in SAM levels to human health, disease, and aging. The chromosomal changes that modify gene expression via methylation of CpG dinucleotides and post-translational modifications of histone tails that use SAM as methyl donor dictate tissue-specific expression, imprinting, and inactivation of the X chromosome, which are amply described elsewhere in this book and will not be discussed here further. Due to lack of space each section lists only few original papers and the reader is mainly referred to reviews for further reading.

SAM-Dependent Enzymes

Over the past 50 years, it has become obvious that SAM serves not only as a universal methyl donor as well as allosteric effector but also as other biochemical transfer reactions, making it perhaps the most frequently used substrate after ATP. A selection of these reactions in mammalian cells is shown in Table 1. Additional bacterial inventions exist that are currently unknown in mammals (Martin and McMillan 2002; Schubert et al. 2003; Fontecave et al. 2004; Kozbial and Muzhegian 2005; Grillo and Colombatto 2007).

Fifteen SAM superfamilies have been identified with rather different structural domains without obvious homology and folds that perform these chemical reactions. It has been claimed that SAM already played these diverse roles in the last universal common ancestor (LUCA) of bacteria, archaea, and eukaryotes (Kozbial and Muzhegian 2005). LUCA could probably synthesize SAM de novo, methylate RNA and proteins, decarboxylate SAM as source for the polyamine pathway, and generate SAM radicals in order to catalyze difficult chemical reactions in the cell. Currently, it is estimated that 95% of all SAM is used for methylation of a wide variety of molecules and 3–5% for the generation of decarboxylated SAM (dcSAM) (Merali and Clarkson 2004). In humans, 85% of all of these methylation reactions and 50% of all methionine metabolism takes place in a single organ, the liver (Mato et al. 2002). A small proportion of SAM is used by perhaps as many as 1000 different proteins for formation of 5′-deoxyadenosyl radicals that perform vital reactions in the cell (Buckel and Golding 2006).

The SAM Domain/Rossmann Fold

The majority of SAM-dependent MTases share a common structure, the Rossmann fold (see http://rebase.neb.com for details and crystal structures), which is conserved in evolution, though the residues that contact SAM are not (see Martin and

Table 1 Selection of chemical reactions occurring in mammalian cells using different parts of SAM

Transfer reaction*	Enzymes	Target or pathway	Conserved structure
Methyl	Majority of MTases **	DNA, RNA, protein, lipid, small molecules (e.g., arsenic)	Rossmann fold, MS fold, corrinoid-like MTases, SPOUT domain, SET domain
Methylene	CFA synthase	Unsaturated fatty acids in phospholipids	Rossmann fold
Amino	DAPA synthase	Biotin synthesis	SAM-dependent type I fold of aminotransferases
Ribosyl	Acp^3U synthase	Uridine in tRNA	Related to TIM barrel
Decarboxylation	SAMDC	Decarboxylation of SAM as substrate for polyamine pathway	Two half barrels
Aminopropyl	Spermidine synthase	Substrate dcSAM, conversion putrescine to spermidine (polyamine)	Rossmann fold
Aminopropyl	Spermine synthase	Substrate dcSAM, conversion spermidine to spermine (polyamine)	Rossmann fold
Methionine + ATP	MAT	Synthesis methionine	Unique fold
5′-deoxyadenosine radicals	Radical SAM enzymes with SAM-dependent [4Fe-4S] cluster	Generation free radicals in controlled way for, e.g., DNA precursor, vitamin, cofactor synthesis; biodegradation, tRNA modification, and DNA repair	TIM barrel
Regulatory binding of SAM	CBS domain	Cystathionine-beta-synthase domain in transsulfuration pathway	Small molecule binding domain
Regulatory binding of SAM	Transcription factor mtTFB	MTase family member that has lost catalytic activity	Rossmann fold

*Sources: Martin and McMillan 2002; Schubert et al. 2003; Fontecave, Atta and Mulliez 2004; Kozbial and Muzhegian 2005; Krauetler 2005; Toohey 2006; Grillo and Colombatto 2007. See text, and these reviews for figures and further details on other organisms.

**Abbreviations: methyltransferase (MTase), *m*ethionine *s*ynthase (MS), corrinoid-like enzymes bind metal ions (such as cobalt, iron, magnesium, and nickel), common fold between *Spo*U and *T*rmD RNA MTases (SPOUT), *s*uppressor of variegation 3–9 [Su(var)3–9], *e*nhancer of zeste and *t*rithorax (SET), cyclopropane fatty acid (CFA) synthase, 7,8-diaminopelargonic acid (DAPA), 3-(3-amino-3-carboxypropyl)uridine (Acp^3U), SAM decarboxylase (SAMDC), decarboxylated SAM (dcSAM), SAM synthetase or methionine adenosyltransferase (MAT).

McMillan 2002; Schubert et al. 2003; Fontecave et al. 2004; Kozbial and Muzhegian 2005 for reviews). Classification of this large superfamily is based on substrate specificity (e.g., DNA, RNA, protein, lipid, and small molecules such as arsenic) and on the atom targeted for methylation (e.g., N, O, C, or S). Several other proteins contain this fold too, despite being inactive as MTases per se, e.g., DNA MTase 2 (DNMT2) and DNA MTase 3-like (DNMT3L). Another important example is spermidine synthase, which is active in the polyamine pathway (see below). The polyamines (putrescine, spermidine, and spermine) are small positively charged molecules in the cell that bind tightly to DNA, RNA, proteins, phospholipids, and many other negatively charged molecules. In this way, they can affect DNA bending and transition from B to Z DNA, cause frame shifts and other infidelities at the RNA level, and modulate signal transduction (see Pegg et al. 2003; Thomas and Thomas 2003; Wallace et al. 2003; Wallace and Fraser 2004; Ivanov and Atkins 2007; Pegg and Feith 2007 for reviews and references therein). Spermidine synthase shares 70% identity to putrescine *N*-methyltransferase, but fuses the methionine backbone of dcSAM to putrescine (Martin and McMillan 2002). Overexpression of enzymes in the polyamine pathway is heavily implicated in cancer, while homozygous knockout of the enzymes in mice proves lethal (Pegg et al. 2003).

Other SAM-Binding Domains

A so-called TIM barrel, also known as triose phosphate isomerase-like domain, is present in SAM radical enzymes, which use SAM to generate methionine and a 5′-deoxyadenosyl radical that can be used to generate further radicals on the same protein or on a coupled enzyme (Kozbial and Mushegian 2005). One important example of this class is SAM decarboxylase (SAMDC or AMD), the enzyme that provides the precursor dcSAM for the synthesis of spermidine and spermine. The manifold control of SAMDC synthesis and activity reflects the importance of careful regulation of polyamine levels in the cell (see for reviews on this topic Wallace et al. 2003; Pegg et al. 2003; Thomas and Thomas 2003; Wallace and Fraser 2004; Kozbial and Mushegian 2005). A third class of important SAM-dependent enzymes contains the SET domain. This domain was discovered as a conserved domain shared by the chromatin remodeling protein suppressor of variegation 3–9 (*S*u(var)3–9), *e*nhancer of zest and *t*rithorax. These enzymes affect chromatin function and transcription by methylating lysines in, e.g., histones and p53, the importance of which is obvious (Kozbial and Mushegian 2005).

The SPOUT fold was originally identified as a domain shared by the *SpoU* and *T*rmD MTases and is present in a superfamily of enzymes that methylate tRNA and rRNA (Tkaczuk et al. 2007). Some enzymes in the methionine cycle have unusual or even unique folds (Kozbial and Mushegian 2005). Different MAT proteins, the all-important enzymes of de novo SAM biosynthesis of ancient LUCA origin, are still closely related at the sequence level in all kingdoms, with a unique wedge-shaped structure. Interestingly, a rare fold is present in methionine synthetase (MS), which

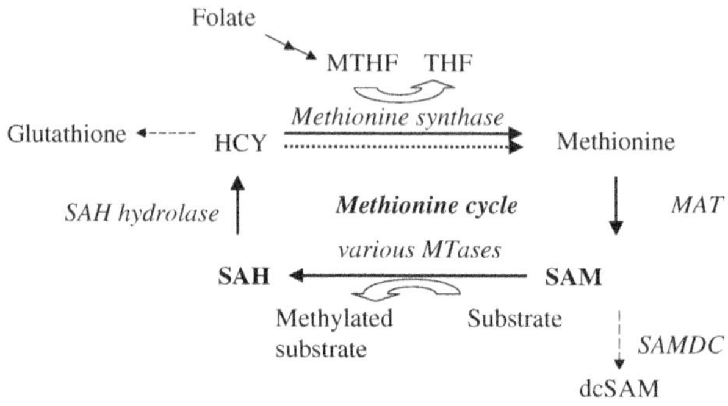

Fig. 2 Simplified diagram of SAM as methyl donor in the methionine cycle. Also called SAM cycle or one-carbon cycle. The methyl group is derived from food and/or from recycling in various other ways (see text for details). In the methionine cycle SAM is generated from methionine by methionine adenosyltransferase (MAT or SAM synthetase). SAM is converted to *S*-adenosylhomocysteine (SAH) by many MTases, in which process it donates the methyl group. SAH is hydrolyzed to homocysteine (HCY). The latter is the substrate for methionine synthase, which uses a derivative of folate, MTHF, as a methyl donor to regenerate methionine, producing THF (tetrahydrofolate). HCY is also substrate for the route to glutathione (Fig. 3). About 3–5% of SAM is diverted to the polyamine pathway after decarboxylation by SAM decarboxylase (SAMDC) to dcSAM (Fig. 4)

regenerates methionine and derives the methyl group from methylated tetrahydrofolate (MTHF) (Fig. 2), while the repressor of the methionine operon MetJ, which uses SAM as a co-repressor, is the only known SAM-binding protein with a RHH (ribbon–helix–helix) domain, again an evolutionarily ancient class of DNA-binding proteins (Kozbial and Mushegian 2005). It is tempting to speculate that such crucial enzymes with their rare folds may get priority access to SAM when methionine/SAM levels in the cell become dangerously low.

SAM and Radical Formation: The Iron–Sulfur Cluster

As mentioned above, organisms use SAM to generate radicals for difficult steps in biochemistry. These enzymes contain FeS clusters, which were originally discovered in electron-transport proteins, but are also present in enzymes involved in DNA and RNA metabolism, including some base excision repair (BER) enzymes (Johnson et al. 2005; Lill and Muehlenhoff 2005; Lukianova and David 2005). An interesting model has been proposed, in which base lesions might be detected speedily using DNA-mediated charge transfer between BER enzymes located on different parts of DNA molecules (Lukianova and David 2005; Merino et al. 2008). More recently, four helicases were identified with FeS clusters that are involved in genome integrity and transcription, i.e., Xeroderma pigmentosum D (XPD) protein with a role in nucleotide excision repair (NER), transcription, and aging (discussed below), the Fanconi J protein (FancJ, which interacts with BRCA1), Chl1 (also called chromosome transmission fidelity 1 (Ctf1), involved in sister-chromatid cohesion in budding yeast), and Rtel1 with a role in telomere maintenance (Rudolf et al. 2006). Disruption of the FeS cluster led to clinically relevant mutations, which have been confirmed in yeast (Rudolf et al. 2006). It will be interesting to see whether any of these helicases with a FeS cluster will prove capable of binding SAM. This is obviously currently a matter of pure speculation, but such a notion may be fostered by recent studies on Elp3, a protein in the elongator complex and highly conserved from archaea to humans (Paraskevopoulou et al. 2006). An archaeal Elp3 was shown to have a (4Fe-4S) cluster that binds SAM. This is an exciting finding that warrants further investigation into therole of SAM in human helicases, especially in light of the story of the SAM-mediated distinction between supercoiled and linear DNA by nucleases (see Introduction).

Many other FeS-containing enzymes exist, and two cases are worth mentioning here: the bacterial oxygen-sensitive cluster (discussed below) and the two FeS clusters of biotin synthase (BS), which show the extraordinary versatility of SAM-dependent FeS clusters. The (4Fe-4S) cluster of BS binds three cysteines at the first three positions, and SAM at the fourth. The reduced cluster donates one electron to SAM producing a 5′-deoxyadenosine radical, after which a second half cluster (2Fe-2S) is involved in the insertion of a sulfur atom into the biotin precursor (Lotierzo et al. 2005; Brosnan and Brosnan 2006).

SAM and RNA Riboswitches

In addition to proteins, on a completely different track it has been observed that SAM is capable of binding RNA (see Corbino et al. 2005; McDaniel et al. 2005; Winkler and Breaker 2005; Batey 2006; Stoddard and Batey 2006; Sudarsan et al, 2006; Fuchs et al. 2006; Cochrane and Strobel 2008; Montange and Batey 2008; Wang and Breaker 2008; Weinberg et al. 2008 for reviews). The RNA regions involved are 5′ leader (*un*translated *r*egion or UTR) sequences and the principle rests on the formation of stem loops that either open up or close the ribosome binding site, or alternatively a transcription termination signal, thus affecting translation and/or transcription. The latter mechanism is strongly reminiscent of classic examples of gene control such as attenuation at the tryptophan operon (Oxender et al. 1979). These metabolite-dependent structures have been named riboswitches and were first discovered in *Bacillus subtilis* in certain metabolite-sensing mRNAs, e.g., for thiamine, purine, glycine, SAM, and vitamin B12. So far four SAM riboswitches have been identified in bacteria and are involved in methionine synthesis, as expected, but also sulfur metabolism and others (Shivji et al. 2005). Interestingly, riboswitches may also cooperate with each other, e.g., the SAM-I riboswitch can act in tandem with a coenzyme B12 riboswitch in the *B. subtilis metE* mRNA (Stoddard and Batey 2006; Sudarsan et al., 2006). The riboswitch principle has entered the laboratory to control gene expression (http://aptamer.icmb.utexas.edu), e.g., a tetracycline-dependent riboswitch allows expression of a reporter gene in yeast in a reversible manner (Suess 2005).

Though most data on riboswitches concern bacteria, thiamine riboswitches have recently been described in plants and fungi (Winkler and Breaker 2005, McDaniel et al. 2005), suggesting that SAM riboswitches may be just around the corner from our own human RNA world. With respect to the latter, Batey (2006) compared the riboswitch with the internal ribosome entry site (IRES), another highly structured mRNA element present in a number of important mammalian genes involved in control of life and death (e.g., insulin-like growth factor 2, c-myc, and AML1/Runx1, but also ODC, the rate-limiting enzyme of the polyamine pathway [see below] and some viruses [e.g., hepatitis C virus] allowing cap-independent mRNA translation (Holcik et al. 2000; Vagner et al. 2001; http://www.iresite.org). The question raised by the story of the riboswitches is inevitable: Will RNA structures that bind SAM and thus control gene expression appear on the human horizon and dictate the activity of non-coding and microRNAs?

The Central Role of SAM in Metabolism

The Methionine Cycle

As mentioned earlier, about 95% of SAM is usurped in methylation reactions. Demethylation of SAM yields *S*-adenosylhomocysteine (SAH) and SAM is regenerated in the methionine cycle (also called one-carbon cycle or SAM cycle [Fig. 2]). This involves a homocysteine (HCY) intermediate, in turn the substrate for methionine synthase (MS) with methylated tetrahydrofolate (MTHF) as methyl donor, and the cycle is complete after fusion of methionine and the adenosine part of ATP by methionine adenosyltransferase (MAT, also called *S*-adenosylmethionine synthetase). The name 'folate,' the precursor for MTHF, describes a family of related molecules that are capable of one-carbon transfer. Folate-derived tetrahydrofolate (THF or vitamin B9) can be synthesized by plants in mitochondria, and by microorganisms, but must be ingested with food by animals, including humans (Sahr et al. 2005).

Methionine can also be regenerated in other ways and one route is worth mentioning: the conversion of HCY by betaine-homocysteine MTase 1 (BHMT1), indicated as a dotted arrow in Fig. 2. This is an important enzyme in the cell, because, in contrast to most MTases, BHMT1 is insensitive to feedback inhibition by SAH, thus preventing increased levels of HCY in, e.g., plasma, which is linked to human cardiovascular disease and diabetes (Wijekoon et al. 2007). The literature on SAM levels in these and other diseases warrants an extensive discussion beyond the scope of this review, but, briefly, low levels of folate in serum and/or a polymorphism in the MTHF receptor (MTHFR C677T) have been implicated in SAM depletion and a wide range of diseases in humans (Coppen and Bolander-Gouaille 2005; Purohit et al. 2005; Pogribny et al. 2005; Davis and Uthus 2004; Lamprecht and Lipkin

2003a; McCabe and Caudill 2005; Ulrey et al. 2005). SAM supplements may be helpful in at least some cases, e.g., SAM may protect against deleterious effects of TNF (tumor necrosis factor) alpha in liver disease (Veal et al. 2005). A folate-deficient diet leads to decreased amounts of SAM, with the danger of hypomethylation of promoters of oncogenes and activation of silent transposons, resulting in chromosome instability and cancer, but also a decrease in folate-derived factors with concomitant incorporation of uracil in DNA instead of thymine, as well as futile cycles of DNA repair and chromosome breaks (Ames and Wakimoto 2002; Lamprecht and Witkin 2003a; Lamprecht and Witkin 2003b).

Due to its central role in methylation, it is understandable that deregulation of SAM has major developmental and neurological implications, resulting among others in various well-known imprinting diseases. Interestingly, the autoimmune disease systemic lupus erythematosus (SLE) has been linked to overexpression of the ligand of CD27 (CD27L, CD70 or TNFSF7) due to decreased methylation of the promoter (Lu et al. 2005). CD27 is an important lymphocyte-specific member of the tumor necrosis factor receptor (TNFR) superfamily (Loenen 1998). The enhanced CD70 expression is linked to diminished activity of methylation and extracellular signal-regulated kinase (Sawalha et al. 2008). Such reports make it likely that the list of human diseases linked to faulty (de)methylation will keep growing.

SAM has been shown to be unstable within the cell and generates a variety of degradation products (Vinci and Clarke 2007). After synthesis, SAM is present in an *S,S* configuration, which is the biologically active form. However, the sulfur is unstable and forms an *R,S* form over time, which is inactive. This form may usurp precious SAM, but there is concern that it is also toxic. This is of importance, as synthetic SAM sold as dietary supplement may contain up to 20–40% *R,S* SAM. Recently, two HCY MTases have been identified in yeast that are capable of recognizing this compound and convert it back to methionine (Vinci and Clarke 2007).

Finally, SAM is also involved in detoxification in tissues exposed to metalloids such as arsenic (Qin et al. 2006; Reguera et al. 2007; Coppin et al. 2008). Apparently, some bacteria excrete arsenic as soluble and gaseous methylated species, which leads to global cycling of arsenic. This poison and human carcinogen targets tissues such as skin, lung, and bladder. The body defends itself by methylating inorganic arsenic using SAM as methyl donor and excretes dimethylarsenic via the urine. High levels of arsenic thus deplete the SAM pool, which can be partly rescued by stimulating the methionine cycle with a folate-rich diet. However, long-term exposure to low levels of arsenic results in hypomethylation, enhanced glutathione production and increased HCY levels, and a progressive reduction in SAM cycling, with deleterious results.

SAM and the Transsulfuration Pathway

In the methionine cycle HCY is back-converted to methionine by MS, which uses vitamin B12 as cofactor, but another enzyme, cystathionine beta-synthase

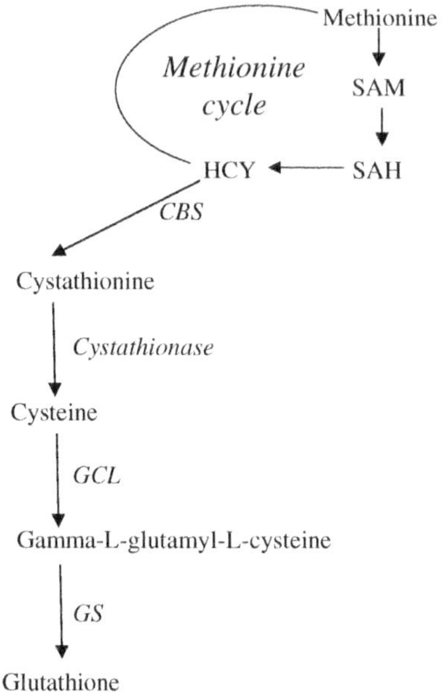

Fig. 3 Simplified diagram of the transsulfuration route from the methionine cycle to glutathione. HCY is the substrate of cystathionine beta-synthase (CBS or cystathionine synthase), which is converted by cystathionase (or CTH) to cysteine. This amino acid is used to generate the tripeptide glutathione, an important antioxidant and detoxifier in the cell via glutamate cysteine ligase (GCL). Other conversion routes are not shown for clarity (see text for further details)

(cystathionine synthase or CBS), generates glutathionine (GSH) via an alternative route, the transsulfuration pathway with vitamin B6 as cofactor (Prudova et al. 2006; Reguera et al. 2007; Lu 2008) (Fig. 3). GSH is an ubiquitous tripeptide with antioxidant and detoxification properties that is involved in cell proliferation. The balance between the two routes is dependent on the level of methionine on the one hand, but also controlled by allosteric activation of CBS by SAM. When methionine is abundant, the transsulfuration route generates cysteine and GSH for diverse cellular functions, while low levels of methionine favor transmethylation. In the latter case, decreased binding of SAM to CBS destabilizes the protein, and thus CBS affects viability under conditions of oxidative stress, e.g., in human liver cancer. The SAM-binding domain of CBS (the CBS domain) is conserved in evolution and can also bind the adenosine moieties of ATP and AMP, which led to the hypothesis that CBS domains function as sensors of intracellular metabolites. In line with such an important role for CBS-containing proteins, mutations in the CBS domain are linked to a number of human diseases, and manipulation of GSH synthesis may relieve symptoms (Ignoul and Eggermont 2005; Lu 2008). The methionine flux to transsulfuration is involved in longevity and is enhanced in Ames dwarf mice (Uthus and Brown-Borg 2006), see below.

SAM and the Polyamine Pathway

While the majority of SAM is used for methylation purposes, an estimated 3–5% is used by SAMDC to generated dcSAM, the substrate for another vital biochemical pathway that generates the polyamines spermidine and spermine from putrescine (Fig. 4) (see, e.g., Coffino 2001; Wallace et al. 2003; Thomas and Thomas 2003; Gerner and Meyskens 2004; Ikeguchi et al. 2006; Pegg, 2006; Gilmour 2007 for reviews of the vast literature on these topics). Already observed by Van Leeuwenhoeck back in 1678 as crystals in semen, the polyamines are organic cations that can bind to most if not virtually all negatively charged molecules in the cell, including DNA, RNA, and protein. Polyamines are necessary for growth and development of many tissues and in tissue repair. The rate-limiting enzyme in this pathway is ornithine decarboxylase (ODC), which converts ornithine (derived from arginine in the urea cycle) to putrescine. While the enzymes of the urea cycle are expressed primarily in the liver and intestine, the polyamines are generated in all tissues and are also present in diets containing cheese and red meat. The next step in polyamine synthesis involves two molecules of dcSAM, used by spermidine synthase to convert putrescine to spermidine and by spermine synthase to generate spermine. Donation of the methionine backbone by dcSAM to putrescine leads to the generation of methylthioadenosine (MTA), which can also be recycled to methionine (Avila et al. 2004; Merali and Clarkson 2004).

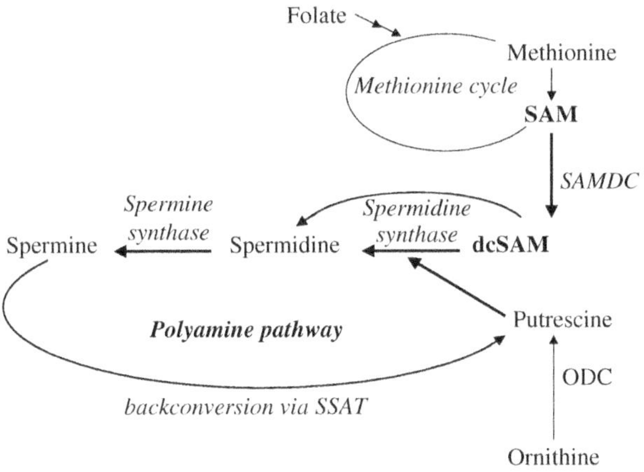

Fig. 4 Simplified diagram of SAM as methionine donor in the polyamine pathway. SAM is first decarboxylated by SAMDC. The methionine backbone of dcSAM is used by spermidine synthetase to convert putrescine into spermidine. A second molecule of dcSAM is used by spermine synthase to convert spermidine into spermine (back-conversion routes to putrescine involve spermidine/spermine-N1-acetyltransferase SSAT and other enzymes (not shown). De novo putrescine production requires the activity of ornithine decarboxylase, the rate-limiting step in this pathway

The polyamine levels are very tightly regulated at the transcriptional, translational, and post-translational levels of the ODC and SAMDC genes, as well as that of spermidine/spermine-*N*1-acetyltransferase (SSAT), which is involved in the back-conversion of spermidine and spermine to putrescine. As appears the case with methylation, polyamine levels may decline during adulthood and in senescent tissues. Deregulation of the activity of any of these genes is poorly tolerated and many clinical trials target this route in the battle against cancer (see also Keren-Paz et al. 2007; Ivanov and Atkins 2007; Pegg 2008).

Though both putrescine and agmatine are present in food, de novo synthesis of putrescine appears to be limited to the ODC route, and the regulation of enzymatic activity of ODC is a beautiful example of tight control at many levels, involving, e.g., polyamine-dependent programmed frame shifting, proteasome-independent degradation, control by transcription factors such as c-myc and NQO1, the tumor suppressor adenomatous polyposis coli (APC), and an internal ribosome entry site (IRES), to name just a few (Gerner and Meyskens 2004; Pegg 2006; Zoumas-Morse et al. 2007; http://recode.genetics.utah.edu). Interestingly, a link was made in our laboratory by Van der Eb and colleagues between ODC and cancer in patients with different mutations in Xeroderma pigmentosum (XP) genes: while the majority of these individuals are highly sensitive to UV-induced skin cancer, some patients are not cancer prone, and this trait is linked to lack of ODC activity after UV irradiation (Terleth et al. 1997). Preliminary analysis of expression profiling data in our laboratory comparing some of these patients reveals alterations in a number of genes that are linked to SAM in one way or another (unpublished observations).

Polyamine levels are not only implicated in neoplastic growth but also capable of stimulating differentiation, e.g., reversing the phenotype of F9 teratocarcinoma stem cells (Frostesjo et al. 1997). Depletion of spermidine and putrescine leads to differentiation into embryonic endoderm-like cells, which is associated with alterations in DNA methylation, and, concomitantly, a dramatic decrease in the levels of dcSAM. Inhibition of SAMDC blocks this differentiation and simultaneously halts the gradual demethylation observed, evidence for the causal relationship between SAM concentrations, DNA methylation, and differentiation.

SAM and Aging

The central role of SAM in methylation and the methionine cycle, the transsulfuration and polyamine pathways, as well as radical formation for difficult chemistry, puts this small molecule at the heart of metabolic processes in the cell, which obviously will have its impact on aging. In this final section, an overview of several theories of aging will be given in relationship to SAM, followed by some examples from the literature on aging individuals that emphasize the need for a proper balance between the methionine cycle for methylation purposes on

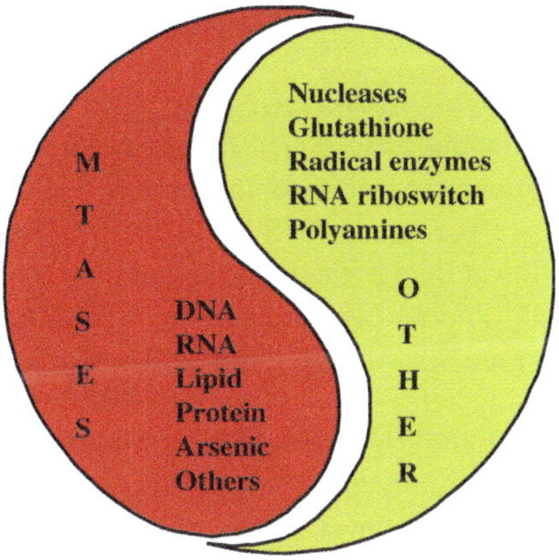

Fig. 5 Yin and Yang of SAM. About 95% of SAM is used in methylation reactions by the majority of methylases (MTases), while the remainder is used by a wide variety of other enzymes. A proper balance is crucial for a long and healthy life

the one hand, and the availability of SAM to polyamine and transsulfuration pathways, and radical enzymes on the other hand in order to live to a healthy old age (Fig. 5).

Aging appears to be evolutionary conserved, as even bacteria are capable of asymmetric cell division, in which old material is preferentially located in one of the two new cells: this results in a damage-enriched and a low-damage population, the former with reduced division potential (Ackermann et al. 2007; Nystrom 2007). While aging is a natural time-dependent deterioration, it varies greatly from days to >100 years and appears to be both genetically and environmentally determined. The evidence for the necessity for a restricted and balanced diet to enhance longevity and slow down aging is accumulating (see elsewhere in this book). Interventions in mice that extend life span and/or delay aging often result in smaller animals and thus may be the product of fewer cell divisions (de Magalhaes and Faragher 2008). At least in *E. coli* K-12 SAM plays an important role in cell division, as the assembly of the septal ring is prevented by SAM depletion leading to long filaments (Wang et al. 2005). Whether SAM plays a direct role in the aforementioned asymmetry itself and/or in higher organisms remains to be investigated.

According to Libert and Pletcher (2007) the ability to sense the total environment using eyes, ears, nose, tongue, and touch or other sensory systems would determine longevity and would trigger well-known pathways, e.g., those involving insulin and Daf-16/FOXO in *Caenorhabditis elegans*. In bacteria, SAM is implicated in environmental sensing via a process called quorum sensing, which involves small diffusible molecules like autoinducer-2 (AI-2) (de Keersmaecker et al. 2006; Zhu et al. 2008). AI-2 synthase proved to be identical to LuxS and *S*-ribosylhomocysteine cleavage enzyme, and thus is comparable to SAH hydrolase in the methionine cycle (which converts SAH into HCY, Fig. 2). This directly links SAM as precursor to extracellular communication and longevity (Libert and Pletcher 2007).

SAM, Mitochondria, and Life Span

Several theories have been put forward to explain life span, including the mitochondrial and free radical theories of development and aging, which involve the harmful effects of free radicals such as reactive oxygen species like superoxide and hydrogen peroxide (ROS). It is well known that overall DNA repair in mitochondria is lower than in the nucleus and mitochondrial DNA is not so well protected as nuclear DNA. The question here is, would mitochondria be more sensitive to oxidation and aging, and does this involve SAM? First, SAM cannot be made inside these important energy-generating organelles, which require rapid exchange of many molecules with the cytosol. Among a whole series of mitochondrial carriers that transport different molecules, several transporters relate to SAM metabolism, including SAM itself, ornithine, folate, and ATP/ADP exchange (Del Arco and Satrustegui 2005). Therefore, mitochondria might be particularly vulnerable to SAM depletion, which would also affect the production of antioxidants like GSH, and disruption of transport of SAM into the mitochondria would be detrimental.

Second, mitochondria might be more sensitive to DNA damage, as mitochondrial polymerase (Pol gamma) and the absence of NER would increase the mutation rate in the genome of this organelle considerably, thus increasing mitochondrial instability and limiting mammalian life span (Del Arco and Satrustegui 2005; de Souza-Pinto et al. 2008; Vermulst et al. 2008).

Third, impaired mitochondrial activity has been linked to the methionine cycle in the case of human and mouse diabetes type 2, which is characterized by insulin resistance and high glucose production in the liver. Overexpression of a protein with an important role in energy metabolism, angiopoietin-like protein 4 (ANGPTL4), in diabetic (*db*/*db*) mice affected expression of key enzymes in the methionine cycle, resulting in restoration of the SAM/SAH ratio to normal as well as the activity of proteins of the mitochondrial respiratory chain (Santamaria et al. 2003; Wang et al. 2007). The SAM/SAH ratio is also important for the flux into the transsulfuration pathway. In this respect Moosmann and Behl (2008) make a strong argument for a role of oxygen in longevity due to deleterious effects on mitochondrial proteins. They hypothesized that proteins of the mitochondrial respiratory chain would be prime targets for damage by ROS resulting in cross-linking of cysteine residues, which would either (temporarily or permanently) inactivate the enzymes and/or target them for degradation. In line with this assumption, they report a strikingly low level of cysteine residues in, e.g., NADH dehydrogenase (NADHDH) in many aerobic species, but not in those of anaerobic organisms. Such redox-dependent cross-links in proteins would reduce the number of free thiol groups, a known feature of aging. The greatly decreased number of cysteines in NADHDH enzymes of aerobic species would reduce the risk of such oxidations and slow down cellular senescence and organ aging. In vivo support for this theory comes from long-lived mouse strains, which tend to have higher GSH levels in the mitochondria, thus allowing rescue of oxidized cysteines in proteins, as further discussed below.

In a paper related to this topic called 'An epigenetic perspective on the free radical theory of development' Hitchler and Domann (2007) take the role of oxygen in development one step further. They hypothesized that oxygen shapes development via ROS, which would stimulate transsulfuration and GSH production, thus limiting SAM availability for methylation of DNA and histones. The discovery of histone demethylases such as LSD1 that require oxygen as a cofactor would directly link epigenetic processes to oxygen gradients during development, which are dependent on the distance to blood vessels. According to this theory the redox potential would become more oxidized during development, which was confirmed in vivo. Overexpression in *Drosophila* of subunit C of glutamate cysteine ligase (GCLC), the rate-limiting enzyme in the transsulfuration pathway (Fig. 3), results in overproduction of GSH and delays aging, while overproduction of both subunits (GCLCM) blocks the development of larvae to adult flies (see Hitchler and Domann 2007). Finally, on a completely different but important track, the redox state also influences SAM directly via the activity of two of the three MAT enzymes in an interesting way, as oxidation of a particular cysteine residue (Cys150) diminishes the activity of MAT (mentioned in Hitchler and Domann 2007), the importance of which observation with respect to the regulation of enzymatic activity will be obvious. Taken together such theories make a strong case for a link between SAM and longevity due to altered flux through the transsulfuration route and the generation of radicals.

Interestingly, another link between accumulation of reactive oxygen species in mitochondria, life span, and SAM was discovered 8 years ago in *Podospora anserina*, when a protein called PaMTH1 that accumulated during senescence of cultures of this fungus showed sequence homology to SAM-dependent MTases (Averbeck et al. 2000). The authors' hypothesis that post-translationally modified proteins might play a role in protecting the fungus against oxidative stress during aging was recently confirmed, when PaMTH1 was shown to be a bona fide MTase capable of methylating flavonoids (Kunstmann and Osiewacz 2008). The earlier observed accumulation of PaMTH1 protein took place in the mitochondrial matrix of senescent cells, and constitutive overexpression of the enzyme did indeed increase the life span of the fungus (Kunstmann and Osiewacz 2008). Such post-translational modifications are not confined to this fungus, as, e.g., mitochondrial ATP synthase was differentially affected in both *P. anserina* and in the brains of young versus old rats, as well as in human cells (Groebe et al. 2007; Soskic et al. 2008).

Finally, life span will be affected not only by damage to mitochondria but also by uncorrected accumulation of damage to cellular DNA, RNA, and protein. As discussed above, low levels of SAM lead to misincorporation of uracil in DNA and disrupt maintenance methylation, which may cause deregulated expression of silenced transposons and other genes, and, if this concerns activation of oncogenes, this could affect life span. On the other hand, deregulation of genes and pathways that affect genome maintenance and decrease fidelity at the DNA, RNA, or post-translational level are also considered aging factors (and linked to cancer predisposition), while the ability to faithfully conserve and/or repair damage and/or misincorporations would be involved in longevity. Thus, an inverse correlation would exist between longevity and mutations in genes that repair DNA damage due to exogenous or endogenous agents, depending on the type of damage. The damage to DNA may be confined to single bases and strand breaks or, worse, cross-linking or double-strand breaks. If the damage is too severe or mutations in repair genes prevent removal, the cells become committed to programmed cell death (suicide or apoptosis) or become permanently stuck in the cell cycle (senescence), which would both contribute to aging. In this respect Hasty (2008) considers, e.g., the proteins involved in non-homologous end joining (NHEJ) of double-strand breaks aging suppressors based on mouse models that show premature aging.

The Yin and Yang of SAM in Real Life

SAM and Neurodegenerative Disease: Alzheimer and Cognitive Dysfunction

In the much-feared dementia of Alzheimer's disease, increased HCY in patients' blood concomitant with a decrease in folate and vitamin B12 suggested altered methionine metabolism and led Scarpa and colleagues to investigate the loss of epigenetic control of the gene encoding the presenilin 1 (PS1) protein (Scarpa et al. 2003). PS1 is a component of the proteolytic gamma–secretase complex that generates amyloid-beta, accumulation of which is characteristic for this disease. Addition of SAM to neuroblastoma cell cultures downregulated PS1 and reduced beta-amyloid production in vitro, while in mice studies B-vitamin deprivation led to an imbalance of the SAM/SAH ratio and concomitant amyloid-beta deposition (Scarpa et al. 2003; Scarpa et al. 2006, Fuso et al. 2005; Fuso et al. 2008). In line with these findings, in these mice the vitamin B deprivation induced hyperhomocysteinemia and brain SAH, as well as upregulation of PS1 (as expected). The authors also noted a decreased performance in a water maze, a tool to measure cognitive functioning in rodent model systems (see further Van Groen on Alzheimer in the chapter 'DNA Methylation and Alzheimer's Disease').

In similar folate-deprivation studies, Shea and colleagues reported cognitive impairment, but in addition enhanced aggression (Chan and Shea 2007, 2008; Tchantchou et al. 2006a, b; Tchantchou et al. 2008). These authors investigated mice with genetic mutations in genes encoding MTHFR, ApoE (risk factor for Alzheimer), or transgenics with various human ApoE variants, and these folate deprivation-induced symptoms could be alleviated by supplementation of the diet with SAM. A crucial observation was the restoration of acetylcholine levels to normal as a result of the availability of choline as an alternative methyl donor (see Chan et al. (2008) for further details). The same laboratory also reported oxidative damage and cognitive impairment of *apoE*$^{-/-}$ mice after folate deprivation due to low levels of SAM, which prohibits activation of glutathione *S*-transferase (Tchantchou et al. 2006b; Tchantchou et al. 2008). Also in this case, SAM supplements ameliorated the effects of folate deprivation. Finally, in *apoE*$^{-/-}$ mice, lack of folate led to upregulation of the PS1 gene and amyloid-beta deposition in the brain (similar to the above studies with B vitamin deprivation in Alzheimer's disease), whose symptoms were reversible with SAM supplements (Chan and Shea 2007).

The above studies clearly showed a causal relationship between disturbances in the methionine cycle and its related routes (folate, glutathione), SAM methylation potential, and impaired brain function. Work by Troen, Shukitt-Hale, and coworkers provided yet another dimension to this story, as they investigated the association between observed mildly elevated plasma levels of HCY in elderly people and increased risk of Alzheimer's disease and neurodegeneration, but also cerebrovascular disease (Troen et al. 2008). The tentative conclusion from B vitamin deprivation studies with mice was that the induced higher HCY levels were linked to changes in the microvasculature of the hippocampus, resulting in cognitive impairment in the absence (or preceding symptoms) of neurodegeneration per se. This was in line with their earlier studies with aged transgenic mice overexpressing the amyloid-beta precursor protein, or with a disruption of the *apoE* gene (Troen et al. 2006; Bernardo et al. 2007). In the former animals increased HCY levels were observed with or without a diet designed to specifically elevate HCY levels, and their results suggested that the disturbed HCY metabolism caused memory deficits and neurodegeneration in the transgenics. In young $apoE^{-/-}$ mice B vitamin deprivation impaired spatial memory and learning (using the maze test), but had no effect on other behavior. While histology appeared normal, brain methylation potential appeared to be moderately down. As will be discussed elsewhere in this book, reports from the same laboratory discussed the beneficial effects of fruit on the aging brain (see, e.g., Shukitt-Hale et al. 2008).

SAM and Dopamine Metabolism in Parkinson's Disease

Finally, neurodegeneration in Parkinson's disease, a motor disorder due to loss of dopamine-producing cells in the brain, is also linked to the methionine cycle, in this case via a SAM-dependent MTase, catechol-*O*-methyltransferase (COMT), which plays a role in dopamine metabolism (Martignoni et al. 2007). *O*-methylation of dopamine by COMT leads to inactivation and degradation of this important neurotransmitter, but COMT also methylates various catechol drugs used for, e.g., the treatment of Parkinson patients. The authors argue that deficiencies in folate, vitamin B6, and B12, or genetic mutations that affect catabolism of HCY and increase levels of HCY in plasma, would not be sufficient to cause the clinical neurological symptoms of these patients per se, but altered methylation and degradation of dopamine (e.g., due to low enzyme activity of COMT) would affect HCY metabolism and predispose to and/or enhance neurotoxicity in combination with other factors (Martignoni et al. 2007, see OMIM at http://www.ncbi.nlm.nih.gov/sites/entrez for further details).

SAM and Down's Syndrome

An enhancement of the transsulfuration route in favor of the methionine cycle has been described in patients with Down's syndrome. Due to trisomy 21 they suffer from mental retardation and heart problems, as well as accelerated aging and Alzheimer's disease. This is in part due to overexpression of superoxide dismutase (SOD1), which leads to high amounts of peroxide and is normally controlled by the activity of catalase and GSH peroxidase. The premature cellular aging and chronic oxidative stress have been linked to damage of asparagine residues in red blood cells of these patients, which can be repaired by a specific SAM-dependent MTase (Galletti et al. 2007). Though the MTase itself is functional, overexpression of cystathionine beta-synthase (CBS or cystathionine synthase) fuels the transsulfuration pathway, leading to a low level of both HCY and methionine (though not SAM itself) and increased protein instability due to damaged asparagines residues. Hence in this case it is radical damage to proteins that involves SAM-dependent MTases for repair.

SAM and the Ames Dwarf Mouse

Imbalance between the methionine cycle and the other SAM-dependent routes also lies at the bottom of the symptoms of the Ames dwarf mouse, a long-lived mouse with an increase in life span of more than 50% compared to normal littermates. Due to a pituitary gland problem this mouse is unable to make, among others, growth hormone (GH), which plays a role in the regulation of the levels of SAM and SAH in tissues (Uthus and Brown-Borg 2006). Using radiolabeled methionine, these authors report an enhanced methionine flux toward the transsulfuration pathway in the liver, brain, and kidney of these mice (in which the transsulfuration route is active), with increased activity in the liver of MAT1a, SAHH, CBS, cystathionase (Fig. 3), as well as another enzyme, glycine-*N*-methyltransferase (GNMT), which controls tissue levels of SAM. The latter is an important enzyme controlled by SAM and folate: when SAM is low in the diet, GNMT is inhibited, making more SAM available for methylation reactions. When SAM is high, GNMT demethylates SAM, thus reducing SAM concentrations and fuelling the transsulfuration route. GNMT activity is under control of GH, as administration of this hormone to the dwarf mice nearly halves liver GNMT activity. This and other data support the notion that the longevity of these mice is linked to enhancement of the oxidative defense system through altered GSH metabolism and an increased GSH pool.

SAM and Mutations in DNA Repair Genes

Several animal models have been developed over the past decades that try to address this issue of aging and longevity with respect to mutations in genes that cause defects in repair of double-strand breaks (Li et al. 2008), and in NER, such as various XP genes, which show tissue-specific aging and neurological disease (Niedernhofer 2008a, b).

In one of these mouse models for premature aging, that of XPD/TTD (XPD mutation linked to the disease trichothiodystrophy), an imbalance was reported between apoptosis and cell renewal in the liver of the mice (Park et al. 2008). Microarray analyses of young and old mice versus age-matched controls revealed a decrease in metabolism and IGF-1 signaling, predictors of life span. The authors link this premature aging to the DNA repair deficiency in the XPD/TTD mice. But it should be noted that in their supplemental table 2 (old mice at 20 months) several genes are significantly altered that are linked to SAM via the transsulfuration, methionine, or polyamine routes. These include cystathionase and the catalytic subunit of GCL (Fig. 3), as well as GSH peroxidase 6, GST zeta1 and alpha3, mitochondrial ornithine transcarbamylase, and BHMT2. This may indicate disturbances in the transsulfuration and polyamine pathways, which will warrant further investigation at the protein level.

Finally, in another mouse model, that of $csb^{m/m}/xpa^{-/-}$ (double mutant of the Cockayne syndrome B and Xeroderma pigmentosum A genes), impaired genome maintenance suppresses the GH/IGF1 axis, which results in a phenotype resembling that of human Cockayne patients (Van der Pluijm et al. 2007). These mice have significantly altered expression levels of a number of liver genes involved in the methionine cycle and/or the transsulfuration and polyamine routes, as compared to wild-type littermates (their supplemental table 2). In addition, these authors compared livers of older mice (16, 96, and 130 weeks) with those of young ones (8 weeks). The latter comparisons indicate significant differences of all three age groups compared with the young one with respect to genes in the methionine cycle and polyamine and transsulfuration routes (their supplemental tables 3, 4, and 5), though not the same genes in each age group. Though data on enzymatic activity of these enzymes and knowledge on the flux through these routes are lacking, these results fit the pattern of imbalance between the methylation and other jobs of SAM in the cell being the root of evil.

Conclusion

SAM is one of the most versatile compounds in life, second perhaps only to ATP in a myriad of biochemical processes, many of which date back to the early origins of life. In this way, SAM can be considered a five-pointed star on the biochemical firmament, which extends from methylation to polyamine and glutathione synthesis, and last but not least radical enzymes and RNA riboswitches.

References

Ackermann, M., A. Schauerte, S.C. Stearns, U. Jenal. Experimental evolution of aging in a bacterium, BMC. Evol. Biol., 7, (2007)126.

Ames, B.N., P. Wakimoto. Are vitamin and mineral deficiencies a major cancer risk? Nat. Rev. Cancer, 2, (2002) 694–704.

Arco, A.D., J. Satrustegui. New mitochondrial carriers: an overview, Cell Mol. Life Sci., 62, (2005) 2204–2227.
Averbeck, N.B., O.N. Jensen, M. Mann, H. Schagger, H.D. Osiewacz. Identification and characterization of PaMTH1, a putative O-methyltransferase accumulating during senescence of Podospora anserina cultures, Curr. Genet., 37, (2000) 200–208
Avila, M.A., E.R. Garcia-Trevijano, S.C. Lu, F.J. Corrales, J.M. Mato. Methylthioadenosine, Int. J. Biochem. Cell Biol., 36, (2004) 2125–2130.
Batey, R.T. Structures of regulatory elements in mRNAs, Curr. Opin. Struct. Biol., 16, (2006) 299–306.
Bernardo, A., M. McCord, A.M. Troen, J.D. Allison, M.P. McDonald. Impaired spatial memory in APP-overexpressing mice on a homocysteinemia-inducing diet, Neurobiol. Aging, 28, (2007) 1195–1205.
Brosnan, J.T., M.E. Brosnan. The sulfur-containing amino acids: an overview, J. Nutr., 136, (2006) 1636S–1640S.
Buckel, W., B.T. Golding. Radical enzymes in anaerobes, Annu. Rev. Microbiol., 60, (2006) 27–49.
Chan, A., T.B. Shea. Folate deprivation increases presenilin expression, gamma-secretase activity, and Abeta levels in murine brain: potentiation by ApoE deficiency and alleviation by dietary S-adenosyl methionine, J. Neurochem., 102, (2007) 753–760.
Chan, A., F. Tchantchou, V. Graves, R. Rozen, T.B. Shea. Dietary and genetic compromise in folate availability reduces acetylcholine, cognitive performance and increases aggression: critical role of S-adenosyl methionine, J. Nutr. Health Aging, 12, (2008) 252–261.
Cochrane, J.C., S.A. Strobel. Riboswitch effectors as protein enzyme cofactors, RNA., 14, (2008) 993–1002.
Coffino, P. Regulation of cellular polyamines by antizyme, Nat. Rev. Mol. Cell Biol., 2, (2001) 188–194.
Coppen, A., C. Bolander-Gouaille. Treatment of depression: time to consider folic acid and vitamin B12, J. Psychopharmacol., 19, (2005) 59–65.
Coppin, J.F., W. Qu, M.P. Waalkes. Interplay between cellular methyl metabolism and adaptive efflux during oncogenic transformation from chronic arsenic exposure in human cells, J. Biol. Chem., (2008).
Corbino, K.A., J.E. Barrick, J. Lim, R. Welz, B.J. Tucker, I. Puskarz, M. Mandal, N.D. Rudnick, R.R. Breaker. Evidence for a second class of S-adenosylmethionine riboswitches and other regulatory RNA motifs in alpha-proteobacteria, Genome Biol., 6, (2005) R70.
Davis, C.D., E.O. Uthus. DNA methylation, cancer susceptibility, and nutrient interactions, Exp. Biol. Med. (Maywood.), 229, (2004) 988–995.
Fedoreyeva, L.I., D.E. Sobolev, B.F. Vanyushin. Wheat endonuclease WEN1 dependent on S-adenosyl-L-methionine and sensitive to DNA methylation status, Epigenetics., 2, (2007) 50–53.
Fontecave, M., M. Atta, E. Mulliez. S-adenosylmethionine: nothing goes to waste, Trends Biochem. Sci., 29, (2004) 243–249.
Frostesjo, L., I. Holm, B. Grahn, A.W. Page, T.H. Bestor, O. Heby. Interference with DNA methyltransferase activity and genome methylation during F9 teratocarcinoma stem cell differentiation induced by polyamine depletion, J. Biol. Chem., 272, (1997) 4359–4366.
Fuchs, R.T., F.J. Grundy, T.M. Henkin. The S(MK) box is a new SAM-binding RNA for translational regulation of SAM synthetase, Nat. Struct. Mol. Biol., 13, (2006) 226–233.
Fuso, A., L. Seminara, R.A. Cavallaro, F. D'Anselmi, S. Scarpa. S-adenosylmethionine/ homocysteine cycle alterations modify DNA methylation status with consequent deregulation of PS1 and BACE and beta-amyloid production, Mol. Cell Neurosci., 28, (2005) 195–204.
Fuso, A., V. Nicolia, R.A. Cavallaro, L. Ricceri, F. D'Anselmi, P. Coluccia, G. Calamandrei, S. Scarpa. B-vitamin deprivation induces hyperhomocysteinemia and brain S-adenosylhomocysteine, depletes brain S-adenosylmethionine, and enhances PS1 and BACE expression and amyloid-beta deposition in mice, Mol. Cell Neurosci., 37, (2008) 731–746.

Galletti, P., M.L. De Bonis, A. Sorrentino, M. Raimo, S. D'Angelo, I. Scala, G. Andria, A. D'Aniello, D. Ingrosso, V. Zappia. Accumulation of altered aspartyl residues in erythrocyte proteins from patients with Down's syndrome, FEBS J., 274, (2007) 5263–5277.

Gerner, E.W., F.L. Meyskens, Jr. Polyamines and cancer: old molecules, new understanding, Nat. Rev. Cancer, 4, (2004) 781–792.

Gilmour, S.K. Polyamines and nonmelanoma skin cancer, Toxicol. Appl. Pharmacol., 224, (2007) 249–256.

Grillo, M.A., S. Colombatto. S-adenosylmethionine and radical-based catalysis, Amino. Acids, 32, (2007) 197–202.

Groebe, K., F. Krause, B. Kunstmann, H. Unterluggauer, N.H. Reifschneider, C.Q. Scheckhuber, C. Sastri, W. Stegmann, W. Wozny, G.P. Schwall, S. Poznanovic, N.A. Dencher, P. Jansen-Durr, H.D. Osiewacz, A. Schrattenholz. Differential proteomic profiling of mitochondria from Podospora anserina, rat and human reveals distinct patterns of age-related oxidative changes, Exp. Gerontol., 42, (2007) 887–898.

Hasty, P. Is NHEJ a tumor suppressor or an aging suppressor?, Cell Cycle, 7, (2008) 1139–1145.

Hitchler, M.J., F.E. Domann. An epigenetic perspective on the free radical theory of development, Free Radic. Biol. Med., 43, (2007) 1023–1036.

Holcik, M., N. Sonenberg, R.G. Korneluk. Internal ribosome initiation of translation and the control of cell death, Trends Genet., 16, (2000) 469–473.

Ignoul, S., J. Eggermont. CBS domains: structure, function, and pathology in human proteins, Am. J. Physiol Cell Physiol, 289, (2005) C1369–C1378.

Ikeguchi, Y., M.C. Bewley, A.E. Pegg. Aminopropyltransferases: function, structure and genetics, J. Biochem., 139, (2006) 1–9.

Ivanov, I.P., J.F. Atkins. Ribosomal frameshifting in decoding antizyme mRNAs from yeast and protists to humans: close to 300 cases reveal remarkable diversity despite underlying conservation, Nucleic Acids Res., 35, (2007) 1842–1858.

Johnson, D.C., D.R. Dean, A.D. Smith, M.K. Johnson. Structure, function, and formation of biological iron-sulfur clusters, Annu. Rev. Biochem., 74, (2005) 247–281

de Keersmaecker, S.C., K. Sonck, J. Vanderleyden. Let LuxS speak up in AI-2 signaling, Trends Microbiol., 14, (2006) 114–119.

Keren-Paz, A., Z. Bercovich, C. Kahana. Antizyme inhibitor: a defective ornithine decarboxylase or a physiological regulator of polyamine biosynthesis and cellular proliferation, Biochem. Soc. Trans., 35, (2007) 311–313.

Kozbial, P.Z., A.R. Mushegian. Natural history of S-adenosylmethionine-binding proteins, BMC. Struct. Biol., 5, (2005) 19.

Kresge, N., H. Tabor, R.D. Simoni, R.L. Hill. An escape from Italy, the discovery of S-adenosylmethionine, and the biosynthesis of creatine by Giulio L. Cantoni. 1953, J. Biol. Chem., 280, (2005) e35.

Kunstmann, B., H.D. Osiewacz. Over-expression of an S-adenosylmethionine-dependent methyltransferase leads to an extended lifespan of Podospora anserina without impairments in vital functions, Aging Cell, 7, (2008) 651–662.

Lamprecht, S.A., M. Lipkin. Chemoprevention of colon cancer by calcium, vitamin D and folate: molecular mechanisms, Nat. Rev. Cancer, 3, (2003a) 601–614.

Lamprecht, S.A., M. Lipkin. Mouse models of gastrointestinal tumorigenesis for dietary cancer prevention studies, Nutr. Rev. 61, (2003b) 255–258.

Li, H., J.R. Mitchell, P. Hasty. DNA double-strand breaks: A potential causative factor for mammalian aging?, Mech. Ageing Dev., 129, (2008) 416–424.

Libert, S., S.D. Pletcher. Modulation of longevity by environmental sensing, Cell, 131, (2007) 1231–1234.

Lill, R., U. Muhlenhoff. Iron-sulfur-protein biogenesis in eukaryotes, Trends Biochem. Sci., 30, (2005) 133–141.

Loenen, W.A.M. CD27 and (TNFR) relatives in the immune system: their role in health and disease, Semin. Immunol., 10, (1998) 417–422.

Loenen, W.A.M. Tracking EcoKI and DNA fifty years on: a golden story full of surprises, Nucleic Acids Res., 31, (2003) 7059–7069.
Loenen, W.A.M. S-adenosylmethionine: jack of all trades and master of everything?, Biochem. Soc. Trans., 34, (2006) 330–333.
Lotierzo, M., B.B. Tse Sum, D. Florentin, F. Escalettes, A. Marquet. Biotin synthase mechanism: an overview, Biochem. Soc. Trans., 33, (2005) 820–823.
Lu, S.C. Regulation of glutathione synthesis, Mol. Aspects Med. 30, (2008) 42–59.
Lu, Q., A. Wu, B.C. Richardson. Demethylation of the same promoter sequence increases CD70 expression in lupus T cells and T cells treated with lupus-inducing drugs, J. Immunol. 174, (2005) 6212–6219.
Lukianova, O.A., S.S. David. A role for iron-sulfur clusters in DNA repair, Curr. Opin. Chem. Biol., 9, (2005) 145–151.
de Magalhaes, J.P., R.G. Faragher. Cell divisions and mammalian aging: integrative biology insights from genes that regulate longevity, Bioessays, 30, (2008) 567–578.
Martignoni, E., C. Tassorelli, G. Nappi, R. Zangaglia, C. Pacchetti and F.J. Blandini. Homocysteine and Parkinson's disease: a dangerous liaison? J. Neurol. Sci. 257, (2007) 31–7.
Martin, J.L., F.M. McMillan. SAM (dependent) I AM: the S-adenosylmethionine-dependent methyltransferase fold, Curr. Opin. Struct. Biol., 12, (2002) 783–793.
Mato, J.M., F.J. Corrales, S.C. Lu, M.A. Avila. S-Adenosylmethionine: a control switch that regulates liver function, FASEB J., 16, (2002) 15–26.
McCabe, D.C., M.A. Caudill. DNA methylation, genomic silencing, and links to nutrition and cancer, Nutr. Rev., 63, (2005) 183–195.
McDaniel, B.A., F.J. Grundy, T.M. Henkin. A tertiary structural element in S box leader RNAs is required for S-adenosylmethionine-directed transcription termination, Mol. Microbiol., 57, (2005) 1008–1021.
Merali, S., A.B. Clarkson, Jr. S-adenosylmethionine and Pneumocystis, FEMS Microbiol. Lett., 237, (2004) 179–186.
Merino, E.J., A.K. Boal, J.K. Barton. Biological contexts for DNA charge transport chemistry, Curr. Opin. Chem. Biol., 12, (2008) 229–237.
Meselson, M., R. Yuan. DNA restriction enzyme from E. coli, Nature, 217, (1968) 1110–1114.
Montange, R.K., R.T. Batey. Riboswitches: emerging themes in RNA structure and function, Annu. Rev. Biophys., 37, (2008) 117–133.
Moosmann, B., C. Behl. Mitochondrially encoded cysteine predicts animal lifespan, Aging Cell, 7, (2008) 32–46.
Murray, N.E. Type I restriction systems: sophisticated molecular machines (a legacy of Bertani and Weigle), Microbiol. Mol. Biol. Rev., 64, (2000) 412–434.
Niedernhofer, L.J. Tissue-specific accelerated aging in nucleotide excision repair deficiency, Mech. Ageing Dev., (2008a).
Niedernhofer L.J. Nucleotide excision repair deficient mouse models and neurological disease, DNA Repair (Amst), 7, (2008b) 1180–1189.
Nystrom T. A bacterial kind of ageing, PLoS. Genet., 3, (2007) e224.
Oxender, D.L., G. Zurawski, C. Yanofsky. Attenuation in the Escherichia coli tryptophan operon: role of RNA secondary structure involving the tryptophan codon region, Proc. Natl. Acad. Sci. USA, 76, (1979) 5524–5528.
Paraskevopoulou, C., S.A. Fairhurst, D.J. Lowe, P. Brick, S. Onesti. The Elongator subunit Elp3 contains a Fe4S4 cluster and binds S-adenosylmethionine, Mol. Microbiol., 59, (2006) 795–806.
Park, J.Y., M.O. Cho, S. Leonard, B. Calder, I.S. Mian, W.H. Kim, S. Wijnhoven, H.van Steeg, J. Mitchell, G.T. van der Horst, J. Hoeijmakers, P. Cohen, J. Vijg, Y. Suh. Homeostatic imbalance between apoptosis and cell renewal in the liver of premature aging Xpd mice, PLoS. ONE., 3, (2008) e2346.

Pegg, A.E., D.J. Feith, L.Y. Fong, C.S. Coleman, T.G. O'Brien, L.M. Shantz. Transgenic mouse models for studies of the role of polyamines in normal, hypertrophic and neoplastic growth, Biochem. Soc. Trans., 31, (2003) 356–360.

Pegg, A.E. Regulation of ornithine decarboxylase, J. Biol. Chem., 281, (2006) 14529–14532.

Pegg, A.E., D.J. Feith. Polyamines and neoplastic growth, Biochem. Soc. Trans., 35, (2007) 295–299.

Pegg, A.E. Spermidine/spermine-N1-acetyltransferase: a key metabolic regulator, Am. J. Physiol Endocrinol. Metab, 294, (2008) E995–E1010.

Pluijm, I. vd, G.A. Garinis, R.M. Brandt, T.G. Gorgels, S.W. Wijnhoven, K.E. Diderich, J.de Wit, J.R. Mitchell, C. van Oostrom, R. Beems, L.J. Niedernhofer, S. Velasco, E.C. Friedberg, K. Tanaka, H. van Steeg, J.H. Hoeijmakers, G.T. van der Horst. Impaired genome maintenance suppresses the growth hormone – insulin-like growth factor 1 axis in mice with Cockayne syndrome, PLoS. Biol., 5, (2007) e2.

Pogribny, I., I. Koturbash, V. Tryndyak, D. Hudson, S.M. Stevenson, O. Sedelnikova, W. Bonner, O. Kovalchuk. Fractionated low-dose radiation exposure leads to accumulation of DNA damage and profound alterations in DNA and histone methylation in the murine thymus, Mol. Cancer Res., 3, (2005) 553–561.

Prudova, A., Z. Bauman, A. Braun, V. Vitvitsky, S.C. Lu, R. Banerjee. S-adenosylmethionine stabilizes cystathionine beta-synthase and modulates redox capacity, Proc. Natl. Acad. Sci. USA, 103, (2006) 6489–6494

Purohit, V., J. Khalsa, J. Serrano. Mechanisms of alcohol-associated cancers: introduction and summary of the symposium, Alcohol, 35, (2005) 155–160.

Qin, J., B.P. Rosen, Y. Zhang, G. Wang, S. Franke, C. Rensing. Arsenic detoxification and evolution of trimethylarsine gas by a microbial arsenite S-adenosylmethionine methyltransferase, Proc. Natl. Acad. Sci. USA, 103, (2006) 2075–2080.

Raghavendra, N.K., D.N. Rao. Exogenous AdoMet and its analogue sinefungin differentially influence DNA cleavage by R. EcoP15I – usefulness in SAGE, Biochem. Biophys. Res. Commun., 334, (2005) 803–811.

Reguera, R.M., C.M. Redondo, Y. Perez-Pertejo, R. Balana-Fouce. S-Adenosylmethionine in protozoan parasites: functions, synthesis and regulation, Mol. Biochem. Parasitol., 152, (2007) 1–10.

Rudolf, J., V. Makrantoni, W.J. Ingledew, M.J. Stark, M.F. White. The DNA repair helicases XPD and FancJ have essential iron-sulfur domains, Mol. Cell, 23, (2006) 801–808.

Sahr, T., S. Ravanel, F. Rebeille. Tetrahydrofolate biosynthesis and distribution in higher plants, Biochem. Soc. Trans., 33, (2005) 758–762.

Santamaria, E., M.A. Avila, M.U. Latasa, A. Rubio, A. Martin-Duce, S.C. Lu, J.M. Mato, F.J. Corrales. Functional proteomics of nonalcoholic steatohepatitis: mitochondrial proteins as targets of S-adenosylmethionine, Proc. Natl. Acad. Sci. USA, 100, (2003) 3065–3070.

Sawalha, A.H., M. Jeffries, R. Webb, Q. Lu, G. Gorelik, D. Ray, J. Osban, N. Knowlton, K. Johnson, B. Richardson. Defective T-cell ERK signaling induces interferon-regulated gene expression and overexpression of methylation-sensitive genes similar to lupus patients, Genes Immun., 9, (2008) 368–378.

Scarpa, S., A. Fuso, F. D'Anselmi, R.A. Cavallaro. Presenilin 1 gene silencing by S-adenosylmethionine: a treatment for Alzheimer disease?, FEBS Lett., 541, (2003) 145–148.

Scarpa, S., R.A. Cavallaro, F. D'Anselmi, A. Fuso. Gene silencing through methylation: an epigenetic intervention on Alzheimer disease, J. Alzheimers. Dis., 9, (2006) 407-414.

Schubert, H.L., R.M. Blumenthal, X. Cheng. Many paths to methyltransfer: a chronicle of convergence, Trends Biochem. Sci., 28, (2003) 329–335.

Shivji, M., S. Burger, C.A. Moncada, A.B. Clarkson, Jr., S. Merali. Effect of nicotine on lung S-adenosylmethionine and development of Pneumocystis pneumonia, J. Biol. Chem., 280, (2005) 15219–15228.

Shukitt-Hale, B., F.C. Lau, J.A. Joseph. Berry fruit supplementation and the aging brain. J. Agric. Food Chem., 56, (2008) 636–641

Sistla, S., D.N. Rao. S-Adenosyl-L-methionine-dependent restriction enzymes, Crit Rev. Biochem. Mol. Biol., 39, (2004) 1–19.
Soskic, V., K. Groebe, A. Schrattenholz. Nonenzymatic posttranslational protein modifications in ageing, Exp. Gerontol., 43, (2008) 247–257.
Souza-Pinto, N.C., D.M. Wilson, III, T.V. Stevnsner, V.A. Bohr. Mitochondrial DNA, base excision repair and neurodegeneration, DNA Repair (Amst), 7, (2008) 1098–1109.
Stoddard, C.D., R.T. Batey. Mix-and-match riboswitches, ACS Chem. Biol., 1, (2006) 751–754.
Sudarsan, N., M.C. Hammond, K.F. Block, R. Welz, J.E. Barrick, A. Roth, R.R. Breaker. Tandem riboswitch architectures exhibit complex gene control functions, Science, 314, (2006) 300–304.
Suess, B. Engineered riboswitches control gene expression by small molecules, Biochem. Soc. Trans., 33, (2005) 474–476.
Tchantchou, F., M. Graves, D. Ortiz, A. Chan, E. Rogers, T.B. Shea. S-adenosyl methionine: A connection between nutritional and genetic risk factors for neurodegeneration in Alzheimer's disease, J. Nutr. Health Aging, 10, (2006a) 541–544.
Tchantchou, F., M. Graves, T.B. Shea. Expression and activity of methionine cycle genes are altered following folate and vitamin E deficiency under oxidative challenge: modulation by apolipoprotein E-deficiency, Nutr. Neurosci., 9, (2006b) 17–24.
Tchantchou, F., M. Graves, D. Falcone, T.B. Shea. S-adenosylmethionine mediates glutathione efficacy by increasing glutathione S-transferase activity: implications for S-adenosyl methionine as a neuroprotective dietary supplement, J. Alzheimers. Dis., 14, (2008) 323–328.
Terleth, C., T.van Laar, R. Schouten, H. van Steeg, H. Hodemaekers, T. Wormhoudt, P.D. Cornelissen-Steijger, P.J. Abrahams, A.J. van der Eb. A lack of radiation-induced ornithine decarboxylase activity prevents enhanced reactivation of herpes simplex virus and is linked to non-cancer proneness in xeroderma pigmentosum patients, Cancer Res., 57, (1997) 4384–4392.
Thomas, T., T.J. Thomas. Polyamine metabolism and cancer, J. Cell Mol. Med., 7, (2003) 113–126.
Tkaczuk, K.L., S. Dunin-Horkawicz, E. Purta, J.M. Bujnicki. Structural and evolutionary bioinformatics of the SPOUT superfamily of methyltransferases, BMC. Bioinformatics., 8, (2007) 73.
Troen, A.M., B. Shukitt-Hale, W.H. Chao, B. Albuquerque, D.E. Smith, J. Selhub, J. Rosenberg. The cognitive impact of nutritional homocysteinemia in apolipoprotein-E deficient mice, J. Alzheimers. Dis., 9, (2006) 381–392.
Troen, A.M., M. Shea-Budgell, B. Shukitt-Hale, D.E. Smith, J. Selhub, I.H. Rosenberg. B-vitamin deficiency causes hyperhomocysteinemia and vascular cognitive impairment in mice, Proc. Natl. Acad. Sci. USA, 105, (2008) 12474–12479.
Ulrey, C.L., L. Liu, L.G. Andrews, T.O. Tollefsbol. The impact of metabolism on DNA methylation, Hum. Mol. Genet., 14 Spec No 1, (2005) R139–R147.
Uthus, E.O., H.M. Brown-Borg. Methionine flux to transsulfuration is enhanced in the long living Ames dwarf mouse, Mech. Ageing Dev., 127, (2006) 444–450.
Vanyushin, B.F. A view of an elemental naturalist at the DNA world (base composition, sequences, methylation), Biochemistry (Mosc.), 72, (2007) 1289–1298.
Veal, V., C.L. Hsieh, S. Xiong, J.M. Mato, S. Lu, H. Tsukamoto. Inhibition of lipopolysaccharide-stimulated TNF-alpha promoter activity by S-adenosylmethionine and 5′-methylthioadenosine, Am. J. Physiol Gastrointest. Liver Physiol, 287, (2004) G352–G362.
Vermulst, M., J. Wanagat, G.C. Kujoth, J.H. Bielas, P.S. Rabinovitch, T.A. Prolla, L.A. Loeb. DNA deletions and clonal mutations drive premature aging in mitochondrial mutator mice, Nat. Genet., 40, (2008) 392–394.
Vinci, C.R., S.G. Clarke. Recognition of age-damaged (R,S)-adenosyl-L-methionine by two methyltransferases in the yeast Saccharomyces cerevisiae, J Biol Chem., 282, (2007) 8604–8612.
Wallace, H.M., A.V. Fraser, A. Hughes. A perspective of polyamine metabolism, Biochem. J., 376, (2003) 1–14.
Wallace, H.M., A.V. Fraser. Inhibitors of polyamine metabolism: review article, Amino. Acids, 26, (2004) 353–365.

Vagner, S., B. Galy, S. Pyronnet. Irresistible IRES. Attracting the translation machinery to internal ribosome entry sites, EMBO Rep., 2, (2001) 893–898.
Wang, S., S.J. Arends, D.S. Weiss, E.B. Newman. A deficiency in S-adenosylmethionine synthetase interrupts assembly of the septal ring in Escherichia coli K-12, Mol. Microbiol., 58, (2005) 791–799.
Wang, Y., K.S. Lam, J.B. Lam, M.C. Lam, P.T. Leung, M. Zhou, A. Xu. Overexpression of angiopoietin-like protein 4 alters mitochondria activities and modulates methionine metabolic cycle in the liver tissues of db/db diabetic mice, Mol. Endocrinol., 21, (2007) 972–986.
Wang, J.X., R.R. Breaker. Riboswitches that sense S-adenosylmethionine and S-adenosylhomocysteine, Biochem. Cell Biol., 86, (2008) 157–168.
Weinberg, Z., J.E. Barrick, Z. Yao, A. Roth, J.N. Kim, J. Gore, J.X. Wang, E.R. Lee, K.F. Block, N. Sudarsan, S. Neph, M. Tompa, W.L. Ruzzo, R.R. Breaker. Identification of 22 candidate structured RNAs in bacteria using the CMfinder comparative genomics pipeline, Nucleic Acids Res., 35, (2007) 4809–4819.
Wijekoon, E.P., M.E. Brosnan, J.T. Brosnan. Homocysteine metabolism in diabetes, Biochem. Soc. Trans., 35, (2007) 1175–1179.
Winkler, W.C., R.R. Breaker. Regulation of bacterial gene expression by riboswitches, Annu. Rev. Microbiol., 59, (2005) 487–517.
Zhu, H., Y.L. Shen, D.Z. Wei, J.W. Zhu. Cloning and characterizations of the Serratia marcescens metK and pfs genes involved in AI-2-dependent quorum-sensing system, Mol. Cell Biochem., 315, 2(2008) 25–30.
Zoumas-Morse, C., C.L. Rock, E.L. Quintana, M.L. Neuhouser, E.W. Gerner, F.L. Meyskens, Jr. Development of a polyamine database for assessing dietary intake, J. Am. Diet. Assoc., 107, (2007) 1024–1027.

Part II
Other Epigenetic Processes and Aging

Polycomb Group of Genes and the Epigenetics of Aging

Krishnaveni Mishra and Rakesh K. Mishra

Abstract Polycomb group (PcG) members are the key components of the cellular machinery that maintains the expression state of a large number of genes right from the early development through the entire life span of an individual. This evolutionarily well–conserved group of proteins create and maintain epigenetic chromatin marks that are translated into means of maintaining a temporally and spatially regulated gene expression status. Among the many different targets that PcG regulate are transcripts from *Ink4a* locus; the gene products of these transcripts being key regulators of cell division and cellular senescence. Additionally, PcG mutations affect self renewal capacity of several stem cells and also interact with another epigenetic chromatin modifier, Sir2, a key negative regulator of aging in yeast and mammals. In this article we bring together the various observations associated with PcG function in the context of cellular senescence and organismal aging.

Keywords Aging · Polycomb · Epigenetics · Chromatin · Gene regulation

Aging of an organism is characterized by the progressive impairment of function of cells with increased susceptibility to disease and death. Even though we observe and register this process in our daily lives, our understanding of aging and its molecular basis is very primitive. However, in the recent past, several studies using diverse model organisms have given us a glimpse of the molecular events associated with aging. These studies converge upon a theme that there is increased accumulation of cellular damage as organisms age. There are multiple mechanisms that lead to this accumulation of damage and several pathways to counter it. However, aging tilts the balance toward reduced ability to repair this damage. The molecular basis of this inability to effectively repair damage is manifold and one of the emerging candidates is epigenetic changes in chromatin (Kaeberlein, 2007).

R.K. Mishra (✉)
Center for Cellular and Molecular Biology, Uppal Road, Hyderabad 500007, India
e-mail: mishra@ccmb.res.in

T.O. Tollefsbol (ed.), *Epigenetics of Aging*, DOI 10.1007/978-1-4419-0639-7_8,

Epigenetic Cellular Memory System

Variety of cellular processes in complex organisms is established by differential expression states of genes in different cell types during development. These expression states are then maintained by epigenetic cellular memory system involving the *Polycomb* group (PcG) of genes throughout the life of the individual. PcG genes have been isolated as mutations that mis-regulate homeotic genes in *Drosophila* (Kennison, 1995). Several members of PcG were identified by genetic approaches and were found to be responsible for the maintenance of repressed state of homeotic genes (Francis and Kingston, 2001; Ringrose and Paro, 2004; Brock and Fisher, 2005). Another group of genes called the *trithorax* group (trxG) of genes counteract the effect of PcG mutations and are positive regulators of homeotic genes (Kennison, 1995; Gould, 1997; Ringrose and Paro, 2004). Following these key initial findings it was later shown that both these groups of genes are conserved during evolution, Table 1, and they regulate a large number of loci in addition to the homeotic genes (Gould, 1997; Levine et al., 2002; Schuettengruber et al., 2007; Vasanthi and Mishra, 2008).

Both PcG and trxG genes work at the level of chromatin structure and interact in an opposing manner to maintain the expression state of genes that is set early during development. These groups of proteins remodel chromatin by modifying histones and assembling active or repressive state of chromatin at a number of loci (Shao et al., 1999; Muller et al., 2002; Diop et al., 2008). Nature of these proteins as indicated by the kind of functional motifs present in them indicates their ability to interact with many other factors and generate versatile regulatory features, Table 1. Studies in *Drosophila* have shown that a complex of proteins called PRC2 (*Polycomb* repressive complex 2) consisting of ESC, E(Z), Su(Z)12, and NURF-55 is first recruited to the target loci and the histone methyl transferase activity of the complex adds trimethyl group to the lysine 27 of histone H3 in the nucleosome (H3K27) (Otte and Kwaks, 2003). This modification is recognized by the chromodomain motif of PC (Min et al., 2003), which comes in as a component of the complex called PRC1 that consists of PH, PC, SCE, and PSC as the core components. Dynamics in the composition – constituents and their proportion – in the PcG complexes is the source of specificity and variety that is expected for this mechanism (Pirrotta, 1997a, b; Levine et al., 2004). Large number of loci are regulated by these proteins and several studies indicate that in addition to the core components of the complex there are several other members that are dynamic and may generate variety and confer specificity of function to different kinds of such complexes (Mishra et al., 2003; Cao and Zhang, 2004; Chopra and Mishra, 2005).

Another aspect of PcG complexes that is less well understood but is extremely important for understanding their function is the recruitment of these complexes to different loci. How the expression state is interpreted by different kinds of complexes is not known at all, although several studies have revealed sequence requirements for the site in genomes to recruit such complexes. These sequences are referred to as *Polycomb* response elements (PREs). Several DNA binding proteins have now been found to be members of PcG of genes. Although the DNA motifs

Table 1 Conservation of PcG/trxG genes and their functional domains

Drosophila melanogaster	Vertebrates	Functional domains
PcG		
Polycomb, Pc	*M33/Cbx*	Chromodomain, C-box, AT hook
Posterior sex comb, Psc	*Bmi-1/Mel18*	RING finger
extra sex combs, esc	*Eed/eed*	WD40 repeat
Sex combs extra, Sce/Ring	*Ring1a/Rnf2*	RING finger
Enhancer of zeste, E(z)	*Enx1/EZH1, Ezh1*	SET domain
pleiohomeotic, pho	*YY1*	Zn finger
pleiohomeotic like, phol		
polyhomeotic, ph	*Rae28*	Zn finger, RNA binding
Sex comb on midleg, Scm	*Scmh1*	SPM/SAM, Zn finger, MBT repeat
Dsfmbt	*Sfmbt*	MBT repeat, SAM
Suppressor of zeste12, Su(z)12	*SU(Z)12*	VEFS box, Zn finger
Additional sex comb, Asx	*ASXL1, 2, 3*	PHD finger, ASXH/N/M, NR binding
Polycomb like, Pcl	*HPcl1*	PHD finger
Enhancer of Polycomb, E(Pc)	*Epc1, 2*	EpcA, EpcB, and Epc C domains
Sex comb reduced, Scr		
cramped, crm		Myb DNA binding, PEST
Chromosome condensation factor, croto		Chromodomain
lola like, lolal/batman, ban		BTB/POZ
Drosophila Mi2, dMi2		PHD finger, Chromodomain, Helicase/ ATPase, DNA binding
trxG		
trithorax, trx	ALR/ MLL	AT hook, PHD finger, SET, Bromodomain
trithorax related, trr		SET domain
Trithorax like, Trl		BTB/POZ, Zn finger
Absent small or homeotic1, ash1	HuASH1	SET, PHD finger, AT hook
ash2		2PHD Zn fingers, SPRY
bramha, brm	hBRM	Bromodomain, BRK, ATPase
moira, mor	SMARCC2/BAF155	Chromodomain, SWIRM, SANT
osa/eye lid, osa	HELD1	ARID, EHD1, EHD2
Little imaginal discs 1, Lid	RBP2	ARID, PHD finger, Leucine zipper
zeste, z		Leucine zipper, DNA binding
modifier of mdg4, mod(mdg4)		BTB
kismet, kis		BRK
kohtalo, kto		
toutatis		TAM, PHD finger, Bromodomain

bound by these proteins are very small – 4–6 base pairs – a cluster of such motifs is needed for a functional PRE. It has also emerged from recent studies that a cooperative interaction of a set of DNA binding members of PcG proteins in the context of their interaction with DNA is critical for creating the appropriate platform for recruiting appropriate kind of PcG complexes (Mulholland et al., 2003; Wang et al., 2004; Blastyak et al., 2006).

Similarly, trxG of genes function as complexes, the Brm complex (consisting of trxG members BRAHMA, MOIRA, OSA, and SNR1) and Trx complex (consisting of TRX, dCBPSBF1), and use 'histone codes' to remodel the chromatin and maintain the active expression state (Diop et al., 2008). Sites where these complexes are recruited are called *trithorax* response elements (TRE) (Rozovskaia et al., 1999; Tillib et al., 1999). Whether a completely different set of DNA binding proteins are needed for recruitment of trxG complexes or they share the same set of proteins that recruit PcG complexes is not clear. Since studies in *Drosophila* indicate that PREs and TREs often overlap, the possibility that they share similar set of cues and proteins but in altered combinations to recruit the respective complexes is highly likely. The two groups of proteins have antagonistic functions, and mutations in the member of one group suppress the phenotype of mutations in other group members, suggesting that PcG and trG signal converges to common genomic sites.

Although most of the studies on PcG/trxG proteins have been done in the context of their role in the regulation of homeotic genes, it has begun to emerge that these proteins are involved in several epigenetic regulatory events (Francis and Kingston, 2001; Brock and Fisher, 2005). In general PcG/trxG genes have been found to maintain the expression state of a large number of genes throughout the life span of the cell. It is only expected that any mis-regulation of this maintenance may lead to disease conditions. Indeed, PcG/trxG genes have been found to be directly linked to diseases like cancer.

Maintenance of the expression state is also critical for the appropriate functioning of the cell during the life of an organism. A weakening of this epigenetic memory may lead to improper functioning of a large number of genes – a situation similar to aging where large degree of mis-regulation of genes takes place. So does this class of proteins contribute to aging? Several recent studies suggest an emphatic yes. During the lifetime of a cell or organism damage by several means keeps accumulating that at some point starts to weaken the repair or maintenance system. This can happen, for example, when the efficiency of PcG/trxG system is overpowered by the degree of damage due to extrinsic agents or intrinsic processes. Changes in the epigenetic state of genome packaging can also lead to a similar situation altering the activity/expression state of PcG/trxG proteins and influencing the process of aging.

Before we delve into this question, we need to understand the link between cellular senescence, tumor suppression, and organismal aging. Cellular senescence was first used to describe the inability of in vitro cultured fibroblasts to divide and propagate themselves unlimitedly. These cells tend to divide a limited number of times and then stop dividing and are not responsive to mitogenic stimuli. But they are metabolically active for several weeks and then enter into a period of senescence. Senescence is characterized by the flattened cell shape, expression of senescence-associated β-gal activity, and appearance of a large number of vacuoles. They also show extensive nuclear reorganization like the assembly of senescence-associated heterochromatic foci (Narita et al., 2003; Narita et al., 2006). Recently, senescent cells have been identified in tissues in vivo (Collado et al., 2005; Lazzerini Denchi et al., 2005; Michaloglou et al., 2005). What is the link between organismal aging and cellular senescence? First, fibroblasts taken from older animals go through fewer cell division cycles before entering senescence than those from younger animals. Second, some mutations like p53 and Wrn helicase influence both organismal aging and cellular senescence. These observations suggest that both processes are associated with at least a few common molecular changes.

In higher eukaryotes most of the adult cells are post-mitotic, i.e., they do not divide. But a few cells are capable of dividing and tissues that contain these cells (called progenitor cells) have the capacity to renew and repair themselves when damaged. This confers increased life span on these tissues. The other side of this coin is that these tissues are more prone to cancer. Their susceptibility comes from increased risk of somatic mutations being introduced during DNA replication leading to cancer. In order to balance this increased risk, tumor suppressors have coevolved with the cell proliferation genes. These genes induce cell-cycle arrest or in some cases cell death to prevent uncontrolled proliferation, a hallmark of cancer cells. This pathway is also activated when cells experience prolonged stress or damage. Thus senescence, in vivo, is a mechanism that protects tissues from cancer. Aging reduces the capacity of tissues to renew and regenerate. One of the leading current hypotheses is that senescence of progenitor cells is the cause for reduced renewal in older tissue leading to decline in tissue regeneration and finally aging. A clear role for *Polycomb* proteins in aging comes from the study of tumor suppression and senescence in stem cells, especially hematopoietic stem cells.

How Do Cells Senesce?

A diverse set of stimuli induce cellular senescence. In cultured human cells the primary reason is the non-availability of telomerase to replicate the chromosome ends. In the absence of telomerase, each cell division reduces the length of the telomeres by a few base pairs until the telomeres become critically short and signal a cell-cycle arrest. In the absence of this signaling, cells continue to divide and unprotected, short telomeres increase recombination and genome instability eventually leading to cancer. However, mouse cells which have long telomeres and may even express telomerase also senesce and this is believed to be induced by stress due to culture conditions. Additionally, senescence can be induced, in both mouse and human cells, by irreparable damage to DNA, for example, during radiation assault or exposure to chemical mutagens. Sometimes this may be triggered by lack of DNA repair enzymes (due to a genetic defect) to mend the damage. In sum, in multicellular organisms, senescence is a mechanism of protection from tumorigenesis, being induced by both irreparable DNA damage and uncontrolled external stimulation for cell division.

The induction and maintenance of senescence is governed by two key proteins, p53 and Rb. p53 is a key mediator of cellular senescence induced by telomere shortening and DNA damage. p53 is a transcriptional regulator and has multiple roles as the guardian of the genome, for example, promoting DNA repair and cell-cycle arrest. However, its key role in inducing senescence is the transcriptional upregulation of p21, a cell-cycle inhibitor, leading to cell-cycle arrest. While this cell-cycle arrest is reversible, the Rb-activated senescence is irreversible. Rb is activated by p16, a tumor suppressor that works as a cell-cycle inhibitor, upon stimulation by stress. Rb represses the expression of E2F target genes possibly through induction of heterochromatinization (Narita et al., 2003). The two pathways are not completely independent: p21 is a more general inhibitor and inhibits phosphorylation of Rb as well and Rb can also bind to MDM2 and influence MDM2 action on p53. In mouse embryo fibroblasts and most human fibroblasts the p53 pathway is predominant; however, many human cells also invoke the Rb pathway via p16 (Itahana et al., 2004).

Role of Polycomb Group Genes in Rb-Induced Senescence

A PcG gene *Bmi-1* (also called PCGF4; homologue of *Drosophila, Psc*) was first identified as an oncogene required for c-myc-induced lymphoma in mice (Haupt et al., 1993). *Bmi1*$^{-/-}$ mice show a whole spectrum of phenotypes like posterior transformations, a hallmark of PcG mutants, neurological defects, skeletal abnormalities, severe defects in hematopoiesis (van der Lugt et al., 1994). This suggests that, as expected, *Bmi1* has multiple targets in the genome and hence the knockout has pleiotropic effects. However, in the context of cell division and senescence, it has been unequivocally demonstrated to be a key requirement for self-renewing cell divisions of both adult hematopoietic stem cells (HSC) and neural stem cells. Interestingly, in both cases, even though *Bmi1* was required for self-renewing divisions it was not essential for producing differentiated progenitors (Park et al., 2003). This was elegantly demonstrated by transplanting *Bmi1*$^{-/-}$ fetal liver cells in mice. These cells continued hematopoiesis for only 4–8 weeks suggesting that although they could differentiate correctly, they did not self-renew. Similarly, it was shown that self-renewal of neural stem cells was *Bmi1* dependent but progenitor generation and proliferation were not. Recently, Bmi1 has also been shown to be important for the proliferation of mammary and intestinal stem cells (Liu et al., 2006; Reinisch et al., 2006). These studies reinforce Bmi1 requirement to maintain the stem cell properties of a wide variety of stem cells. These studies further indicate that maintenance of a particular state of differentiation (or lack of differentiation) is an important event in self-renewing and loss of this property leads to lack of self-renewal and therefore inability to regenerate tissue and begin the march toward aging.

Several lines of evidence point to a clear role for *Bmi1* in preventing senescence. Human fetal lung fibroblasts downregulate BMI1 during senescence but not quiescence (Vonlanthen et al., 2001). *Bmi1* overexpression increases life span in fibroblasts of mice and man (Jacobs et al., 1999). Similarly, normal mouse embryo fibroblasts undergo senescence after 7–8 passages in culture, but those from *Bmi1*$^{-/-}$ undergo only three passages before senescence. Also re-expression of *Bmi1* reversed the premature senescence phenotype and overexpression of *Bmi1* increased life span. How does Bmi1 bring about these effects?

One key target of *Bmi1* is the *Ink4a* locus that encodes p16 (also called CDKN2A) and p19^{ARF} (p14^{ARF} in humans). The two proteins are generated from the same locus using two different promoters and although they have two common exons, due to translation in different frames, encode structurally different proteins. This intriguing genome organization is rather unique in mammals. p16INK4A is an inhibitor of the cell cycle; while p14ARF regulates p53 stability. P16INK4 belongs to a class of cyclin-dependent kinase inhibitors and mice lacking p16Ink4a show increased tumorigenesis (Krimpenfort et al., 2001). The p16INK4 protein binds to Cdk4 and Cdk6 and this binding prevents the interaction of these Cdks with D-type cyclins, which is required for catalytic activity of Cdks (Russo et al., 1998). Rb is an important target of the Cdks and Cdk inhibition leads to hypophosphorylation of Rb, which in turn leads to E2F repression and G1 arrest (Shapiro et al., 1995; Pomerantz et al., 1998; Zhang et al., 1998). p19^{ARF} also plays an important role in senescence. ARF binds to and inhibits MDM2 (Kamijo et al., 1998; Pomerantz et al., 1998; Zhang et al., 1998). MDM2 is an E3 ubiquitin ligase that targets p53 for proteasomal degradation. Therefore, by inhibiting MDM2, ARF stabilizes p53 leading to growth arrest or apoptosis depending on cellular context. Thus Bmi1-regulated expression of this locus influences both the Rb and p53 branches of senescence pathways. That this locus plays a very important role in controlling cell proliferation is emphasized by the fact that it is deleted in many cancers.

Importantly, *Ink4a/Arf* locus (both p16 and p19) upregulation is not only seen in senescencing cultured cell lines, but has also been shown in tissues of aging animals (Krishnamurthy et al., 2004). Its expression correlated with SA-β-galactosidase. Additionally, the expression of both INK4/ARF locus and SA-β-galactosidase is also reduced in calorie-restricted animals, calorie restriction being an important means of retarding aging. Another intriguing observation is that the transcriptional activator of *Ink4a* locus, Ets-1, was also expressed at the same time but its levels were not increased in proportion. So it has been speculated that perhaps *Ink4a/Arf* locus becomes more susceptible to transcriptional activation with age. What could be the mechanism of this susceptibility? A strong candidate is of course Bmi1 and other PcG regulators because these proteins function upstream of transcription factors by remodeling the genome to make it inaccessible for transcription; in their absence genes become more susceptible to activation. This is corroborated by the observation that Bmi1 levels are reduced in senescent diploid fibroblasts (Itahana et al., 2003). In contrast, Dietrich et al., (2007) see no change in levels of Bmi1 or Cbx8 and propose instead that a modification like phosphorylation of these factors may reduce their activity at these loci. Currently we have no evidence for either of the possibilities and future research will address these issues.

Even though Bmi1 is the most studied PcG protein that regulates the *Ink4a* locus, it has recently emerged that other PcG members including CBX7, CBX8,

Mel18, and M33 also regulate this locus. Cbx8 was shown to bind physically to the *Ink4a/Arf* locus and repress p16Ink4a expression. Moreover, BMI1 and CBX8 were both required for maintaining the repression, and loss of one protein affected the binding of the other by 50% (Dietrich et al., 2007). CBX8 does not seem to be

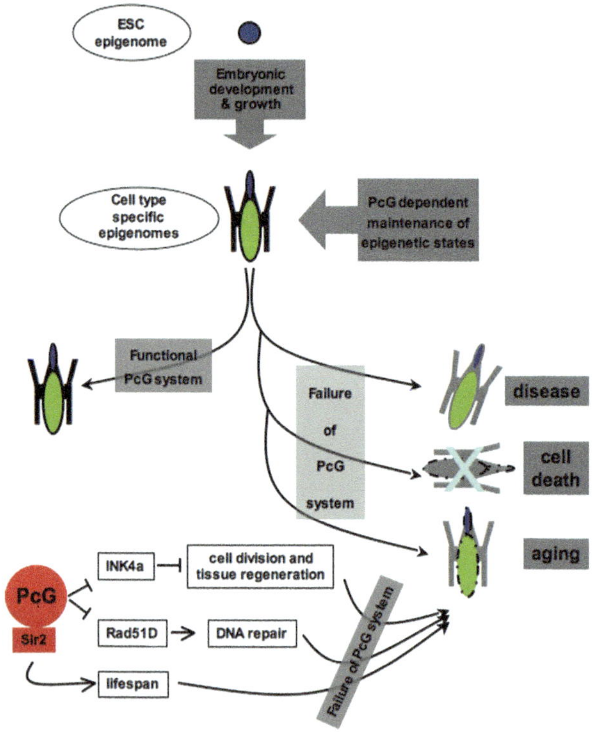

Fig. 1 PcG-mediated epigenetic regulation and its link to aging. *Polycomb* group of proteins play major roles in variety of nuclear events that regulate gene expression. Every cell in the human body contains identical DNA sequence. The packaging of this genome is, however, different in each of the >200 cell types in our body that leads to specific and precise expression pattern of genes. This distinct functional form of the genome in each cell type is referred to as the epigenome. It is suggested that the same genome, by means of unique packaging and epigenetic modifications, becomes a functionally distinct epigenome that determines the cell type. Embryonic stem cell (ESC) epigenome is differentiated into cell type-specific epigenomes of an organism and PcG system maintains this epigenome state during the growth and development for the entire life span. The aging phenotype can be due partly to the breakdown or weakening of the PcG-mediated memory system. This failure of PcG system can lead to, depending on the extent of failure, cell death, disease, or aging state. For example, even minor relaxations in the PcG-mediated repressed state can lead to ectopic expression of several genes, even though at a very low level, since PcG genes have a large number of targets. Such a misexpression event can lead to gradual emergence of aging phenotype. As shown in the lower part of the figure, the targets of the PcG genes include INK4A and Rad51D that are involved in cell division, tissue regeneration, and DNA repair. Nature of these targets indicates that failure of PcG system and thereby mis-regulation of the proteins involved in tissue regeneration and DNA repair will lead to aging phenotype. Direct involvement of PcG interacting protein Sir2 in life span determination and aging has already been demonstrated

involved in regulating ARF, which correlates with its peak binding to p16INK4a exon. CBX7 also represses the *Ink4a/Arf* locus and affects both the p16INK4a and ARF levels. Although CBX7 regulates the expression, it does not physically interact with Bmi1 and their binding to the locus is independent of each other (Gil et al., 2004). This suggests that distinct complexes of PcG proteins are assembled and involved in fine-tuning gene regulation (Otte and Kwaks, 2003). Similarly mouse embryo fibroblasts derived from PcG-deficient mice, including knockouts of Phc1, Phc2, M33, show premature senescence and this is due to the altered regulation of INK4 locus (Core et al., 1997, 2004; Isono et al., 2005). Role of Mel18, another PcG protein, seems a little more complicated. One set of evidences points to Mel18 directly binding to and reducing ARF expression (Miki et al., 2007). On the other hand, it has also been shown to be overexpressed in senescent cells in contrast to Bmi1 and overexpression of Mel18 promotes senescence in cultured cells. These authors also show that Mel18 directly binds to c-Myc promoter and represses c-Myc and since Bmi1 is a direct target of c-Myc, it reduces Bmi1 expression. Therefore, in this case the two PcG proteins have antagonistic functions (Guo et al., 2007). Is it possible that both ARF and c-Myc are targets of Mel18? This would at the outset seem futile because they trigger opposing effects, but it could be possible that they only bind to these promoters under distinct conditions and not simultaneously.

The emerging theme from all the studies discussed above is that PcG genes work in concert to control expression of *Ink4a/Arf* locus. They cooperate to repress transcription of genes in the *Ink4a/Arf* locus. In their absence, transcription is activated from this locus. *Ink4a/Arf* is one of the few established markers for senescence that is seen in both cultured cells and older animals. Expression of the genes from *Ink4a/Arf* locus leads to arrest of cell cycle and triggering of the senescence program. A gradual loss of function of Bmi1 and other PcG proteins is observed as an organism ages (Gil et al., 2004; Gil and Peters, 2006; Sasaki et al., 2006) and this could lead to upregulation of genes that promote senescence, Fig. 1. If this is so, could shoring up their function improve life span? Unfortunately, the answer is unlikely to be that simple because *Ink4a/Arf* locus expression is also controlled by other transcription factors like FOXO, SIRT1, etc. (Gil and Peters, 2006), and there are likely to be other effectors of aging that are yet to be identified.

Polycomb and Sir2 Connection

Sir2 (silent information regulator 2) was first identified in budding yeast as a protein required to establish and maintain heterochromatin at the silent mating-type loci (Loo and Rine, 1995). It is the most evolutionarily conserved histone deacetylase with homologues from bacteria to humans (Brachmann et al., 1995). Sir2 class of protein deacetylases are unique in that they use NAD+ as cofactor for their activity (Imai et al., 2000; Tanner et al., 2000). This feature provides a link between the energy metabolism in the cell that generates NAD+ and Sir2 activity. Calorie restriction is one of the earliest reported means of extension of life span in rodents (Masoro, 2003) and the observation has been expanded to yeast to metazoans like *Drosophila* and *Caenorhabditis elegans* (discussed in other chapters). Sir2 has been clearly demonstrated to be the means through which calorie restriction extends life span: Overexpression of Sir2 in flies expands life span and so does calorie restriction but (a) calorie restriction does not further increase life span in Sir2 overexpressing cells and (b) calorie restriction leads to overexpression of Sir2 (Rogina and Helfand, 2004; Kusama et al., 2006).

Interestingly, recent observations link PcG-mediated gene silencing to Sir2-dependent gene silencing (Furuyama et al., 2004). In flies, it has been shown that Sir2 is physically associated with the E(Z)-containing PRC complex and co-localizes with the E(Z)-containing chromosomal sites. It is also functionally involved in Polycomb-mediated gene repression as Sir2 mutations enhance the phenotypes of PcG mutants. Therefore, given that Sir2 overexpression increases life span and that it is a part of PRC, it can be extrapolated that Polycomb group genes are involved in life span extension, Fig. 1. However, this needs to be directly demonstrated because even though both proteins physically interact and co-localize on several sites on the chromosomes, there are also sites where the two proteins do not co-localize. Also, experiments testing directly the effect of overexpression of E(Z) on life span or on calorie restriction have not been reported.

However, in mammals it has been shown that SirT1 (Sir2 homologue) is found in distinct complexes containing EZH2 and EED2 (Kuzmichev et al., 2005). This complex, termed PRC4, shows a unique methylating activity: methylation of histone H1b, H1K26. SirT1 deacetylates the residue and the PRC activity seems to methylate it. Most interestingly, the composition of the PRC complexes seems to alter with differentiation status. In embryonic stem cells the different isoforms of EED and SirT1 were maximal and their expression decreased as cells differentiate. Only one isoform of EED, EED3, was retained in fully differentiated tissue. Interestingly, all tissue-cultured cells showed high levels of EED and other subunits of PRC including SirT1 compared to differentiated mouse tissues. Similarly, prostrate cancer cells also showed the same trend with increased expression of PRC4. Experimentally, overexpression of Ezh2 can lead to the assembly of PRC4 in cultured cells. Taken together, these data show that SirT1-containing PRC are active in young actively dividing cells and decline with differentiation and development. This is in agreement with the theme that organismal aging is inversely proportional to SirT1 levels.

Polycomb System and DNA Repair

As discussed earlier in this chapter, one of the key findings in aged organisms is accumulation of damaged molecules, especially nuclear DNA. This damage may be induced by exposure to environmental factors like UV, chemicals, etc., or maybe triggered by internal factors like DNA replication, reactive oxygen species generated through metabolism, and spontaneous hydrolysis. Damaged DNA is effectively repaired by several DNA repair processes like homologous recombination-mediated repair or non-homologous end joining in young cells. However, older cells seem to be able to do so much less well than younger cells. Excellent recent reviews discuss the changes in the DNA repair mechanisms and machinery in aging (Lombard et al., 2005; Park and Gerson, 2005; Vijg, 2008).

When human cells are inflicted with DNA damage, they are repaired by either homologous recombination or non-homologous end joining pathway. Rad51, homologue of bacterial RecA, plays a major role in the homologous recombination process. It has several paralogs like XRCC2, XRCC3, Rad51B, Rad51C, and Rad51D in mammals (Takata et al., 2001) and knockout of these leads to impaired DNA repair. Although their exact molecular roles are unclear, they are critical for the assembly of the repair foci and preparing the chromosome for repair. It has recently been shown that EZH2, a PcG protein, directly regulates the expression RAD51 and its paralogs (Zeidler et al., 2005). EZH2 is the mammalian homologue of Drosophila *enhancer of zeste2*, which interacts directly with histone deacetylases and establishes gene repression. Overexpression of EZH2 is observed in aggressive breast tumors, leading to downregulation of the genes of Rad51 class that are essential for DNA repair. This suggests that downregulation of PcG leads to upregulation of homologous recombination-based DNA repair. Interestingly, it has been shown in *Drosophila* male germ cell line that as the organism ages, the homologous recombination pathway is used more frequently than non-homologous pathways (Preston et al., 2006). Is this an active mechanism to promote more fidel repair in aging organisms or is it a consequence of a failing PcG system?

Conclusions

PcG proteins are involved in regulation of several loci and only recently their link with cell-cycle control and senescence has started to emerge. Repression of these loci by PcG promotes cell division and renewal which are needed for normal life span. There are tantalizing clues to the possible involvement of PcG proteins in Sir2-mediated life span control, but possible 'life span phenotype' of PcG mutations has not been analyzed. Sir2 mediates longevity in the fly through a pathway related to calorie restriction where the Sir2 repressive functions are dependent on the PcG of genes. These observations indicate that PcG mutations may show premature aging although it needs to be tested directly. It is likely that Sir2-containing PcG complex may be targeting genetic loci for silencing that is needed for preventing aging, Fig. 1. These loci are likely to be environmental sensing and related to longevity that need to be maintained in repressed state in response to environment/caloric restriction when NAD levels are high, as weakening of this repressive system will lead to pleiotropic effects that will culminate into 'aging' phenotype, Fig. 2. Similarly, DNA repair and epigenetic mechanisms involved in chromatin-mediated regulation

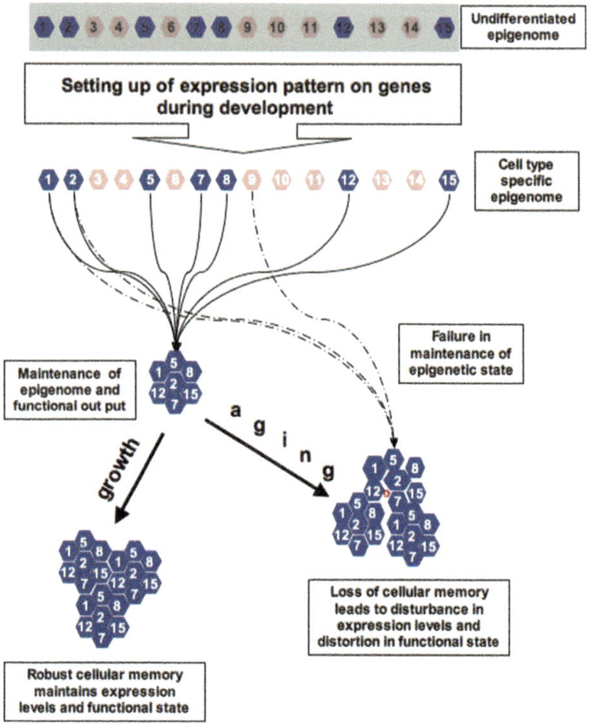

Fig. 2 A model linking PcG-mediated epigenetic memory system to aging process. During early development, expression pattern of genes is established. This expression pattern needs to be maintained to protect the identity of cell type. The complex regulatory events that lead to differentiation transfer the maintenance function to PcG-mediated memory system that works through chromatin modification. This process is depicted as conversion of undifferentiated epigenome to cell type-specific epigenomes. Several cell types, indicated as differently colored and numbered hexagons, function in a coordinated fashion in a given tissue/organ. During the life span of the organism, maintenance of the epigenetic state is essential for the functioning of the organ. Variety of PcG complexes may carry out specialized function of maintenance in various cell types. During the life span of an organism, accumulation of damage or spontaneous fault in PcG/trxG gene products can cause up- or downregulation of genes that can alter and deform the proteome of the cell type. This failure in the maintenance of the epigenetic state is shown by broken arrows. Such deformation will then lead to mis-regulation of a large number of genes at a mild level. While such mild mis-regulation of a large number of loci may not be lethal or disease prone, it may lead to general weakening of the cellular processes as seen in the aging process

are linked. As organisms age their genome is subjected to extensive damage which needs to be repaired and, importantly, each time epigenetic marks in the repaired portion of the genome also need to be reestablished. Since PcG system is well studied, its link with aging will help in understanding this complex process at genetic and molecular level. Furthermore, link to other distinct repressive chromatin features like heterochromatin and telomere has been studied in the context of aging. How many of these features are interlinked and how much is independent remains to be investigated. Progress in genomics and technical advances have brought these fields to an exciting stage where epigenetics of development, aging, longevity, environmental response, and connection of these processes to disease states is becoming evident.

So these epigenome-creating proteins have to be in perfect order or else after a damage, the new DNA may enter into a different expression state.

Acknowledgements KM acknowledges financial support from Council for Scientific and Industrial Research (India) and Department of Science & Technology (Govt. of India). RKM acknowledges financial support from Department of Biotechnology (Govt. of India), Human Frontier Science Program (HFSP), Indo–French Centre for the Promotion of Advanced Research (IFC-PAR/CEFIPRA), and Department of Science & Technology (Govt. of India).

References

Blastyak, A., Mishra, R.K., Karch, F., and Gyurkovics, H. (2006). Efficient and specific targeting of Polycomb group proteins requires cooperative interaction between Grainyhead and Pleiohomeotic. Mol. Cell Biol. **26:** 1434.

Brachmann, C.B., Sherman, J.M., Devine, S.E., Cameron, E.E., Pillus, L., and Boeke, J.D. (1995). The SIR2 gene family, conserved from bacteria to humans, functions in silencing, cell cycle progression, and chromosome stability. Genes Dev **9:** 2888–2902.

Brock, H.W., and Fisher, C.L. (2005). Maintenance of gene expression patterns. Dev.Dyn. **232:** 633.

Cao, R., and Zhang, Y. (2004). The functions of E(Z)/EZH2-mediated methylation of lysine 27 in histone H3. Curr Opin Genet Dev **14:** 155.

Chopra, V.S., and Mishra, R.K. (2005). To SIR with Polycomb: linking silencing mechanisms. Bioessays **27:** 119–121.

Collado, M., Gil, J., Efeyan, A., Guerra, C., Schuhmacher, A.J., Barradas, M., Benguria, A., Zaballos, A., Flores, J.M., Barbacid, M., Beach, D., and Serrano, M. (2005). Tumour biology: senescence in premalignant tumours. Nature **436:** 642.

Core, N., Joly, F., Boned, A., and Djabali, M. (2004). Disruption of E2F signaling suppresses the INK4a-induced proliferative defect in M33-deficient mice. Oncogene **23:** 7660–7668.

Core, N., Bel, S., Gaunt, S.J., Aurrand-Lions, M., Pearce, J., Fisher, A., and Djabali, M. (1997). Altered cellular proliferation and mesoderm patterning in Polycomb-M33-deficient mice. Development **124:** 721.

Dietrich, N., Bracken, A.P., Trinh, E., Schjerling, C.K., Koseki, H., Rappsilber, J., Helin, K., and Hansen, K.H. (2007). Bypass of senescence by the polycomb group protein CBX8 through direct binding to the INK4A-ARF locus. Embo J **26:** 1637–1648.

Diop, S.B., Bertaux, K., Vasanthi, D., Sarkeshik, A., Goirand, B., Aragnol, D., Tolwinski, N.S., Cole, M.D., Pradel, J., Yates, J.R., 3rd, Mishra, R.K., Graba, Y., and Saurin, A.J. (2008). Reptin and Pontin function antagonistically with PcG and TrxG complexes to mediate Hox gene control. EMBO Rep **9:** 260–266.

Francis, N.J., and Kingston, R.E. (2001). Mechanisms of transcriptional memory. Nat Rev Mol Cell Biol **2:** 409–421.

Furuyama, T., Banerjee, R., Breen, T.R., and Harte, P.J. (2004). SIR2 is required for polycomb silencing and is associated with an E(Z) histone methyltransferase complex. Curr Biol **14:** 1812–1821.

Gil, J., and Peters, G. (2006). Regulation of the INK4b-ARF-INK4a tumour suppressor locus: all for one or one for all. Nat Rev Mol Cell Biol **7:** 667–677.

Gil, J., Bernard, D., Martinez, D., and Beach, D. (2004). Polycomb CBX7 has a unifying role in cellular lifespan. Nat Cell Biol **6:** 67–72.

Gould, A. (1997). Functions of mammalian Polycomb group and trithorax group related genes. Curr Opin Genet Dev **7:** 488.

Guo, W.J., Datta, S., Band, V., and Dimri, G.P. (2007). Mel-18, a polycomb group protein, regulates cell proliferation and senescence via transcriptional repression of Bmi-1 and c-Myc oncoproteins. Mol Biol Cell **18:** 536–546.

Haupt, Y., Bath, M.L., Harris, A.W., and Adams, J.M. (1993). bmi-1 transgene induces lymphomas and collaborates with myc in tumorigenesis. Oncogene **8:** 3161–3164.

Imai, S., Armstrong, C.M., Kaeberlein, M., and Guarente, L. (2000). Transcriptional silencing and longevity protein Sir2 is an NAD-dependent histone deacetylase. Nature **403:** 795–800.

Isono, K., Fujimura, Y., Shinga, J., Yamaki, M., O-Wang, J., Takihara, Y., Murahashi, Y., Takada, Y., Mizutani-Koseki, Y., and Koseki, H. (2005). Mammalian polyhomeotic homologues Phc2 and Phc1 act in synergy to mediate polycomb repression of Hox genes. Mol Cell Biol **25:** 6694–6706.

Itahana, K., Campisi, J., and Dimri, G.P. (2004). Mechanisms of cellular senescence in human and mouse cells. Biogerontology **5:** 1–10.

Itahana, K., Zou, Y., Itahana, Y., Martinez, J.L., Beausejour, C., Jacobs, J.J., Van Lohuizen, M., Band, V., Campisi, J., and Dimri, G.P. (2003). Control of the replicative life span of human fibroblasts by p16 and the polycomb protein Bmi-1. Mol Cell Biol **23:** 389–401.

Jacobs, J.J., Kieboom, K., Marino, S., DePinho, R.A., and van Lohuizen, M. (1999). The oncogene and Polycomb-group gene bmi-1 regulates cell proliferation and senescence through the ink4a locus. Nature **397:** 164–168.

Kaeberlein, M. (2007). Molecular basis of ageing. EMBO Rep **8:** 907–911.

Kamijo, T., Weber, J.D., Zambetti, G., Zindy, F., Roussel, M.F., and Sherr, C.J. (1998). Functional and physical interactions of the ARF tumor suppressor with p53 and Mdm2. Proc Natl Acad Sci USA **95:** 8292–8297.

Kennison, J.A. (1995). The Polycomb and trithorax group proteins of Drosophila: trans-regulators of homeotic gene function. Annu Rev Genet **29:** 289.

Krimpenfort, P., Quon, K.C., Mooi, W.J., Loonstra, A., and Berns, A. (2001). Loss of p16Ink4a confers susceptibility to metastatic melanoma in mice. Nature **413:** 83–86.

Krishnamurthy, J., Torrice, C., Ramsey, M.R., Kovalev, G.I., Al-Regaiey, K., Su, L., and Sharpless, N.E. (2004). Ink4a/Arf expression is a biomarker of aging. J Clin Invest **114:** 1299–1307.

Kusama, S., Ueda, R., Suda, T., Nishihara, S., and Matsuura, E.T. (2006). Involvement of Drosophila Sir2-like genes in the regulation of life span. Genes Genet Syst **81:** 341–348.

Kuzmichev, A., Margueron, R., Vaquero, A., Preissner, T.S., Scher, M., Kirmizis, A., Ouyang, X., Brockdorff, N., Abate-Shen, C., Farnham, P., and Reinberg, D. (2005). Composition and histone substrates of polycomb repressive group complexes change during cellular differentiation. Proc Natl Acad Sci USA **102:** 1859–1864.

Lazzerini Denchi, E., Attwooll, C., Pasini, D., and Helin, K. (2005). Deregulated E2F activity induces hyperplasia and senescence-like features in the mouse pituitary gland. Mol Cell Biol **25:** 2660–2672.

Levine, S.S., King, I.F., and Kingston, R.E. (2004). Division of labor in polycomb group repression. Trends Biochem Sci **29:** 478–485.

Levine, S.S., Weiss, A., Erdjument-Bromage, H., Shao, Z., Tempst, P., and Kingston, R.E. (2002). The core of the polycomb repressive complex is compositionally and functionally conserved in flies and humans. Mol Cell Biol **22:** 6070–6078.

Liu, S., Dontu, G., Mantle, I.D., Patel, S., Ahn, N.S., Jackson, K.W., Suri, P., and Wicha, M.S. (2006). Hedgehog signaling and Bmi-1 regulate self-renewal of normal and malignant human mammary stem cells. Cancer Res **66:** 6063–6071.

Lombard, D.B., Chua, K.F., Mostoslavsky, R., Franco, S., Gostissa, M., and Alt, F.W. (2005). DNA repair, genome stability, and aging. Cell **120:** 497–512.

Loo, S., and Rine, J. (1995). Silencing and heritable domains of gene expression. Annu Rev Cell Dev Biol **11:** 519–548.

Masoro, E.J. (2003). Subfield history: caloric restriction, slowing aging, and extending life. Sci Aging Knowledge Environ **2003:** RE2.

Michaloglou, C., Vredeveld, L.C., Soengas, M.S., Denoyelle, C., Kuilman, T., van der Horst, C.M., Majoor, D.M., Shay, J.W., Mooi, W.J., and Peeper, D.S. (2005). BRAFE600-associated senescence-like cell cycle arrest of human naevi. Nature **436:** 720–724.

Miki, J., Fujimura, Y., Koseki, H., and Kamijo, T. (2007). Polycomb complexes regulate cellular senescence by repression of ARF in cooperation with E2F3. Genes Cells **12:** 1371–1382.

Min, J., Zhang, Y., and Xu, R.M. (2003). Structural basis for specific binding of Polycomb chromodomain to histone H3 methylated at Lys 27. Genes Dev. **17:** 1823.

Mishra, K., Chopra, V.S., Srinivasan, A., and Mishra, R.K. (2003). Trl-GAGA directly interacts with lola like and both are part of the repressive complex of Polycomb group of genes. Mech.Dev. **120:** 681.

Mulholland, N.M., King, I.F., and Kingston, R.E. (2003). Regulation of Polycomb group complexes by the sequence-specific DNA binding proteins Zeste and GAGA. Genes Dev **17:** 2741–2746.

Muller, J., Hart, C.M., Francis, N.J., Vargas, M.L., Sengupta, A., Wild, B., Miller, E.L., O'Connor, M.B., Kingston, R.E., and Simon, J.A. (2002). Histone methyltransferase activity of a Drosophila Polycomb group repressor complex. Cell **111:** 197–208.

Narita, M., Narita, M., Krizhanovsky, V., Nunez, S., Chicas, A., Hearn, S.A., Myers, M.P., and Lowe, S.W. (2006). A novel role for high-mobility group a proteins in cellular senescence and heterochromatin formation. Cell **126:** 503–514.

Narita, M., Nunez, S., Heard, E., Narita, M., Lin, A.W., Hearn, S.A., Spector, D.L., Hannon, G.J., and Lowe, S.W. (2003). Rb-mediated heterochromatin formation and silencing of E2F target genes during cellular senescence. Cell **113:** 703–716.

Otte, A.P., and Kwaks, T.H. (2003). Gene repression by Polycomb group protein complexes: a distinct complex for every occasion? Curr Opin Genet Dev **13:** 448–454.

Park, I.K., Qian, D., Kiel, M., Becker, M.W., Pihalja, M., Weissman, I.L., Morrison, S.J., and Clarke, M.F. (2003). Bmi-1 is required for maintenance of adult self-renewing haematopoietic stem cells. Nature **423:** 302–305.

Park, Y., and Gerson, S.L. (2005). DNA repair defects in stem cell function and aging. Annu Rev Med **56:** 495–508.

Pirrotta, V. (1997a). Chromatin-silencing mechanisms in Drosophila maintain patterns of gene expression. Trends Genet. **13:** 314.

Pirrotta, V. (1997b). PcG complexes and chromatin silencing. Curr Opin Genet Dev **7:** 249.

Pomerantz, J., Schreiber-Agus, N., Liegeois, N.J., Silverman, A., Alland, L., Chin, L., Potes, J., Chen, K., Orlow, I., Lee, H.W., Cordon-Cardo, C., and DePinho, R.A. (1998). The Ink4a tumor suppressor gene product, p19Arf, interacts with MDM2 and neutralizes MDM2's inhibition of p53. Cell **92:** 713–723.

Preston, C.R., Flores, C., and Engels, W.R. (2006). Age-dependent usage of double-strand-break repair pathways. Curr Biol **16:** 2009–2015.

Reinisch, C., Kandutsch, S., Uthman, A., and Pammer, J. (2006). BMI-1: a protein expressed in stem cells, specialized cells and tumors of the gastrointestinal tract. Histol Histopathol **21:** 1143–1149.

Ringrose, L., and Paro, R. (2004). Epigenetic regulation of cellular memory by the Polycomb and Trithorax group proteins. Annu Rev Genet **38:413–43.:** 413.

Rogina, B., and Helfand, S.L. (2004). Sir2 mediates longevity in the fly through a pathway related to calorie restriction. Proc Natl Acad Sci USA **101:** 15998–16003.

Rozovskaia, T., Tillib, S., Smith, S., Sedkov, Y., Rozenblatt-Rosen, O., Petruk, S., Yano, T., Nakamura, T., Ben-Simchon, L., Gildea, J., Croce, C.M., Shearn, A., Canaani, E., and Mazo, A. (1999). Trithorax and ASH1 interact directly and associate with the trithorax group-responsive bxd region of the Ultrabithorax promoter. Mol Cell Biol **19:** 6441–6447.

Russo, A.A., Tong, L., Lee, J.O., Jeffrey, P.D., and Pavletich, N.P. (1998). Structural basis for inhibition of the cyclin-dependent kinase Cdk6 by the tumour suppressor p16INK4a. Nature **395:** 237–243.

Sasaki, M., Ikeda, H., Sato, Y., and Nakanuma, Y. (2006). Decreased expression of Bmi1 is closely associated with cellular senescence in small bile ducts in primary biliary cirrhosis. Am J Pathol **169:** 831–845.

Schuettengruber, B., Chourrout, D., Vervoort, M., Leblanc, B., and Cavalli, G. (2007). Genome regulation by polycomb and trithorax proteins. Cell. **128:** 735.

Shao, Z., Raible, F., Mollaaghababa, R., Guyon, J.R., Wu, C.T., Bender, W., and Kingston, R.E. (1999). Stabilization of chromatin structure by PRC1, a Polycomb complex. Cell **98:** 37–46.

Shapiro, G.I., Edwards, C.D., Kobzik, L., Godleski, J., Richards, W., Sugarbaker, D.J., and Rollins, B.J. (1995). Reciprocal Rb inactivation and p16INK4 expression in primary lung cancers and cell lines. Cancer Res **55:** 505–509.

Takata, M., Sasaki, M.S., Tachiiri, S., Fukushima, T., Sonoda, E., Schild, D., Thompson, L.H., and Takeda, S. (2001). Chromosome instability and defective recombinational repair in knockout mutants of the five Rad51 paralogs. Mol Cell Biol **21:** 2858–2866.

Tanner, K.G., Landry, J., Sternglanz, R., and Denu, J.M. (2000). Silent information regulator 2 family of NAD- dependent histone/protein deacetylases generates a unique product, 1-O-acetyl-ADP-ribose. Proc Natl Acad Sci USA **97:** 14178–14182.

Tillib, S., Petruk, S., Sedkov, Y., Kuzin, A., Fujioka, M., Goto, T., and Mazo, A. (1999). Trithorax- and Polycomb-group response elements within an Ultrabithorax transcription maintenance unit consist of closely situated but separable sequences. Mol Cell Biol **19:** 5189–5202.

van der Lugt, N.M., Domen, J., Linders, K., van Roon, M., Robanus-Maandag, E., te Riele, H., van der Valk, M., Deschamps, J., Sofroniew, M., van Lohuizen, M., and et al. (1994). Posterior transformation, neurological abnormalities, and severe hematopoietic defects in mice with a targeted deletion of the bmi-1 proto-oncogene. Genes Dev **8:** 757–769.

Vasanthi, D., and Mishra, R.K. (2008). Epigenetic regulation of genes during development: A conserved theme from flies to mammals. J Genet Genomics **35:** 413–429.

Vijg, J. (2008). The role of DNA damage and repair in aging: New approaches to an old problem. Mech Ageing Dev **129:** 498–502.

Vonlanthen, S., Heighway, J., Altermatt, H.J., Gugger, M., Kappeler, A., Borner, M.M., van Lohuizen, M., and Betticher, D.C. (2001). The bmi-1 oncoprotein is differentially expressed in non-small cell lung cancer and correlates with INK4A-ARF locus expression. Br J Cancer **84:** 1372–1376.

Wang, L., Brown, J.L., Cao, R., Zhang, Y., Kassis, J.A., and Jones, R.S. (2004). Hierarchical recruitment of polycomb group silencing complexes. Mol.Cell. **14:** 637.

Zeidler, M., Varambally, S., Cao, Q., Chinnaiyan, A.M., Ferguson, D.O., Merajver, S.D., and Kleer, C.G. (2005). The Polycomb group protein EZH2 impairs DNA repair in breast epithelial cells. Neoplasia **7:** 1011–1019.

Zhang, Y., Xiong, Y., and Yarbrough, W.G. (1998). ARF promotes MDM2 degradation and stabilizes p53: ARF-INK4a locus deletion impairs both the Rb and p53 tumor suppression pathways. Cell **92:** 725–734.

Chromosomal Position Effect and Aging

Eric Gilson and Frédérique Magdinier

Abstract Aging is a complex biological process that manifests through cellular, physical, and metabolic changes in all types of tissues. Changes in chromosome structure and function throughout life play a pivotal role in this irreversible physiological process by affecting gene expression, replication, recombination, DNA repair, and epigenetic programming. Upon exogenous stress or simply as the result of cellular metabolism and cellular division, DNA rearrangements or redistribution of epigenetic factors might alter the chromatin environment and expression of genes regulating the aging process by the so-called chromosomal position effect. These changes can vary from cell to cell and can be associated with variegated pattern of gene expression. We will discuss here what is known on the mechanisms of various types of chromosomal position effects and their consequences on the aging process.

Keywords Chromosomal position effect · Telomeric position effect · Epigenetics · Aging · Telomeres · Chromatin

Introduction

The eukaryotic genome is organized into domains of individual genes or clusters with distinct patterns of expression. It is now evident that chromatin structure plays an important role in regulating gene transcription by providing a proper subnuclear environment to ensure correct spatial and temporal gene expression. The eukaryotic chromatin is highly dynamic and its state varies along the chromosomes and during cell cycle and development. Each chromatin state can be defined by its level of compaction, the positioning and the spacing of nucleosomes, its histone code, the covalent modification of the underlying DNA, its non-histone binding factors, the spatial localization within the nucleoplasm, and its dynamics during

F. Magdinier (✉)
Laboratoire de Biologie Moléculaire de la Cellule, CNRS UMR5239, Ecole Normale Supérieure de Lyon, UCBL1, IFR128. 46 allée d'Italie, 69364 Lyon Cedex 07, France
e-mail: frederique.magdinier@ens-lyon.fr

T.O. Tollefsbol (ed.), *Epigenetics of Aging*, DOI 10.1007/978-1-4419-0639-7_9,

cell cycle. Historically, open chromatin, where most of the transcription occurs, is referred to as "euchromatin" whereas condensed chromatin, where transcription is generally inhibited, is referred to as "heterochromatin," although various types of chromatin structure are evoked under these denominations. In fact, heterochromatin was originally described as a portion of the genome deeply stained from metaphase to interphase associated with the pericentric regions, telomeres, and some interstitial domains. The result of chromatin mosaicism along chromosomes is the neighboring between condensed and open regions. To an extent, the identity of these domains is maintained by different factors such as *cis*-regulators, fuzzy boundaries, or insulators that limit the influence of one region on the adjacent one (Fourel et al. 2004; Gaszner and Felsenfeld 2006; Valenzuela and Kamakaka 2006).

In higher eukaryotes, constitutive heterochromatin is enriched in methylated DNA, histone H3K9 di- and tri-methylation, H3K27 methylation, HP1 binding and can spread over flanking regions inducing thereby transcriptional silencing (Peng and Karpen 2008; Trojer and Reinberg 2007). When a gene is rearranged and is relocated near heterochromatin, its transcription becomes silent in a subset of cells, leading to a characteristic variegated pattern of expression in cells and tissues. This phenomenon was originally discovered in flies through an inversion of the X chromosome leading to the relocalization of the *white* gene close to pericentromeric heterochromatin and dubbed position effect variegation (PEV) (Eissenberg et al. 1990; Festenstein et al. 1999; James and Elgin 1986). It is now well established that the proximity of many types of chromosomal regions but also other types of DNA transaction, such as recombination, replication initiation, and DNA repair, can influence gene expression either positively (enhancer proximity) or negatively (silencer proximity) (Girton and Johansen 2008; Ottaviani et al. 2008; Rabbitts et al. 1983). Changes in chromosomal structure can also have long-range influence on DNA transaction by triggering the delocalization of specific chromatin factors. For instance, in budding yeast, the delocalization of heterochromatin factors due to telomere dysfunction stimulates the repressive function of *cis*-acting silencers located far away from telomeres (Marcand et al. 1996). These different types of position effects are globally referred herein as chromosomal position effect or CPE. Among them, telomere proximity resembles classical PEV by triggering the silencing of genes located at their proximity, defining the telomeric position effect (or TPE) (Gottschling et al. 1990). Most telomeres from *Saccharomyces cerevisiae* to *Homo sapiens* silence neighboring genes. TPE is modulated by the length and the structure of telomeres together with classical remodeling of the chromatin fiber (Ottaviani et al. 2008). The nature of telomeric and subtelomeric chromatin differs from global constitutive heterochromatin due to the specificity of its DNA sequences, the particular structure and dynamics of its chromatin, and the binding of specific factors (Blasco 2007b; Gilson and Geli 2007).

A common feature among eukaryotes is the progressive decline in vitality over the time. Aging encompasses a wide spectrum of degenerative processes such as for instance the accumulation of carbonylated proteins, damaged enzymes, protein misfolding, lipid peroxidation, or activation of inflammatory response pathways. Among them, the accumulation at the cytogenetic level of DNA lesions and chromatin changes will influence cellular function. A causal link between CPE and aging is not clearly established but rather derives from comparisons between young and old cells. For instance, the cytogenetically visible lesions such as translocations, insertions, dicentric, and acentric fragments inherent of the aging process might lead to altered expression of pro- or anti-aging genes by CPE. In addition, telomere shortening, which can lead to cellular senescence and appears to contribute to some aging processes in mammals, might alleviate the repression of pro-aging genes, which otherwise would have succumbed to TPE. It is worth noting that the strength or cell-type specificity of CPEs can be modulated by local or global changes in DNA methylation and chromatin structure. For instance, while DNA methylation pattern of certain genes in certain types of differentiated tissues changes during the aging process (Feil 2006; Fraga and Esteller 2007), modulation of the chromatin condensation may also influences the temporal and spatial expression pattern of individual or clusters of genes (see below for further details). Thus, long-range chromatin dysregulation as a consequence of chromosomal rearrangement could contribute to aging, drive the loss of overall cellular functions, and lay grounds for secondary genetic events (Fig. 1). In this chapter, we will discuss what is currently known on CPE and their associated changes in higher-order chromatin structure and the causes and consequences of such remodeling and epigenetic alterations during the aging process.

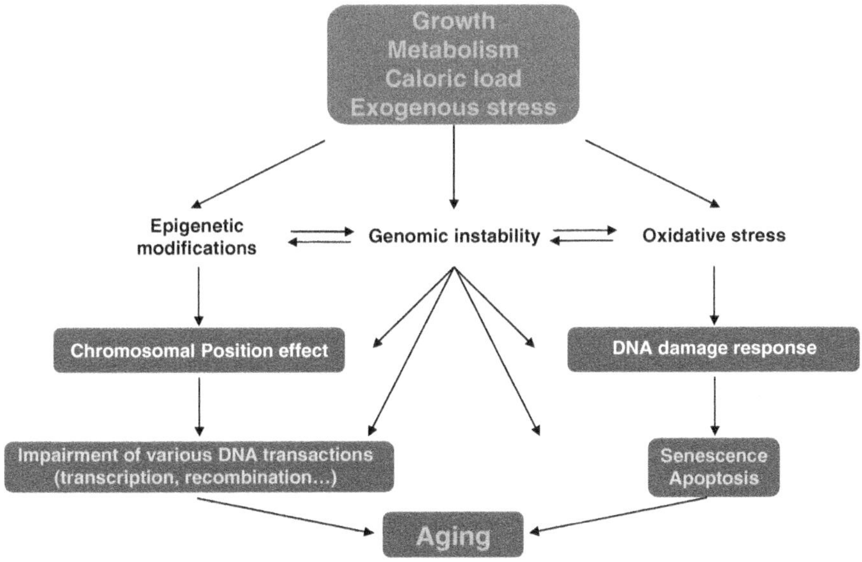

Fig. 1 At the organismal level, aging is likely the outcome of a complex interplay between genetic alterations and the pressure that the environment exerts on an organism. In this representation, epigenetic changes, genomic instability, and oxidative stress constitute a first step in the order of events leading to the variegation of gene expression through position effect mechanisms or telomeric position effect. As a consequence, these numerous changes affect cellular homeostasis and cellular response to different types of stress is impaired

Aging and Senescence in Budding Yeast

The mammalian aging process probably exceeds the yeast model in complexity. Nevertheless, this model organism provides important insights in the understanding of the chromatin changes involved in life span. In *S. cerevisiae,* aging, as an organismal process, refers to the fixed number of cell division that a daughter cell undergoes before stopping divisions. This limited life span is an intrinsic property of this organism and is genetically determined. By analogy to the mammalian situation, where the term senescence is used to refer to the limited proliferative capacities of somatic cells in culture, senescence in yeast corresponds to the delayed growth arrest of telomerase-negative cells. Since telomerase appears to be constitutively expressed in various yeast species, it seems unlikely that senescence occurs in naturally living cells, although one cannot rule out that telomerase is repressed under certain circumstances. Importantly, aging and senescence in yeast appear to obey to different mechanisms despite the existence of some connections at the chromatin level. As described below in details, aging is based on the accumulation of toxic products in mothers after asymmetrical cell division, a process that can be indirectly regulated by telomeric changes, while senescence is a direct consequence of telomere dysfunction. It is tempting to consider that yeast aging is a paradigm of aging in pluricellular organisms and that yeast senescence reflects the way in which several types of human somatic cells senesce in culture, since these cells do not express enough telomerase to replenish their telomeric DNA content at each cell division.

Yeast Aging Revealed the Role of the Sir2 NAD-Dependent Deacetylase

Heterochromatin in yeast stabilizes the repetitive components of the genome like the ribosomal DNA and maintains certain genes such as the mating-type loci in a silent state through cell division. A key regulator of the rDNA and mating-type loci is the Sir2 NAD+ histone deacetylase (HDAC) that prevents these regions from recombination and fusion throughout cell division. In agreement with a general role for heterochromatin in yeast aging, various aging genes are downregulated by the Sir complex (Daniel 2005). Sir proteins or Sirtuins are a conserved family found in numerous species. Sir2 (Silent information regulator 2) was the first sirtuin identified in yeast (Klar et al. 1979) as a regulator of ribosomal DNA recombination, gene silencing, DNA repair, genome stability, and longevity. Later, four additional genes with high homology to Sir2 were identified (Rine and Herskowitz 1987). None of the four are essential for cell survival but they are all involved in silencing at the mating-type loci and telomeres, cell cycle progression, and genomic integrity (Michan and Sinclair 2007). The first clear link between silencing and aging came from a genetic screen that isolated a gain-of-function mutation in SIR4, which abolishes silencing at telomere and at the mating-type loci (Kennedy et al.

1995; Kennedy et al. 1997). This mutant targets a greater amount of Sir2 and Sir3 to the nucleolus and increases longevity by 40%. Cells lacking Sir2 display accelerated aging whereas cells with extra copies have a reduced replicative life span (Kaeberleion et al., 1999; Sinclair and Guarente, 1997). Additional observations on other mutants such as SGS1, which encodes the RecQ DNA helicases, later converged toward a role of the relocalization of heterochromatin proteins to the nucleolus in aging. In the absence of RecQ helicases, the genome is highly unstable especially at repetitive loci and results in hyper recombination at the rDNA loci (Karow et al. 2000). This event leads to the relocalization of Sir3 to the nucleolus, which becomes enlarged and fragmented (Sinclair and Guarente 1997). As a consequence, life span is affected partly because of the dramatic effect on genomic stability in particular at the rDNA locus. Upon Sir2 relocalization, recombination between rDNA sequences results in the excision of a circular molecule of rDNA and the formation of an extrachromosomal circle (ERC) that are replicated and segregated exclusively to mother cells thanks to their anchorage to nuclear pores (Shcheprova et al. 2008). Introduction of a single ERC or any type of non-centromeric episome into young cells shortens lifetime and accelerates the onset of age-associated phenotypes (Falcon and Aris 2003). After several cell cycles, up to 1000 ERC can accumulate in a single cell, probably titrating out essential factors from the rest of the genome and leading to cell death (Sinclair and Guarente 1997) (Fig. 2). However, a complete abolition of ERC formation leads to only a moderate extension of life span (Kaeberlein and Kennedy 2005), suggesting that other processes contribute to yeast aging. For instance, during yeast cytokinesis, oxidatively damaged proteins are inherited asymmetrically after budding suggesting that the Sir2-dependent accumulation of oxidatively damaged proteins in mother cells might be a new mechanism for dealing with the accumulation of oxidative damage from aged mother to new born cells (Aguilaniu et al. 2003).

Other reports also stressed the importance of Sir2 relocalization upon DNA damage suggesting that beside the reorganization of the DNA molecule itself, the relocalization of chromatin-modifying factors might locally or globally influence the maintenance of chromosomal regions (Lee et al. 1999; Martin et al. 1999; McAinsh et al. 1999; Mills et al. 1999). The consequence of such redistribution is that previously silent genes may become transcriptionally active while active genes are repressed through remodeling process and subsequent chromatin compaction. The redistribution of Sir proteins during aging led to speculate that such a redistribution is an active defense process that the cell initiate to stabilize its genome (Oberdoerffer and Sinclair 2007).

Yeast Senescence: A Link with TPE?

TPE was first demonstrated by insertion of a construct containing a *URA3* marker next to an array of telomeric repeats. Integration of this construct close to the VII-L telomere deletes the terminal 15 kb of the chromosome and positions the *URA3*

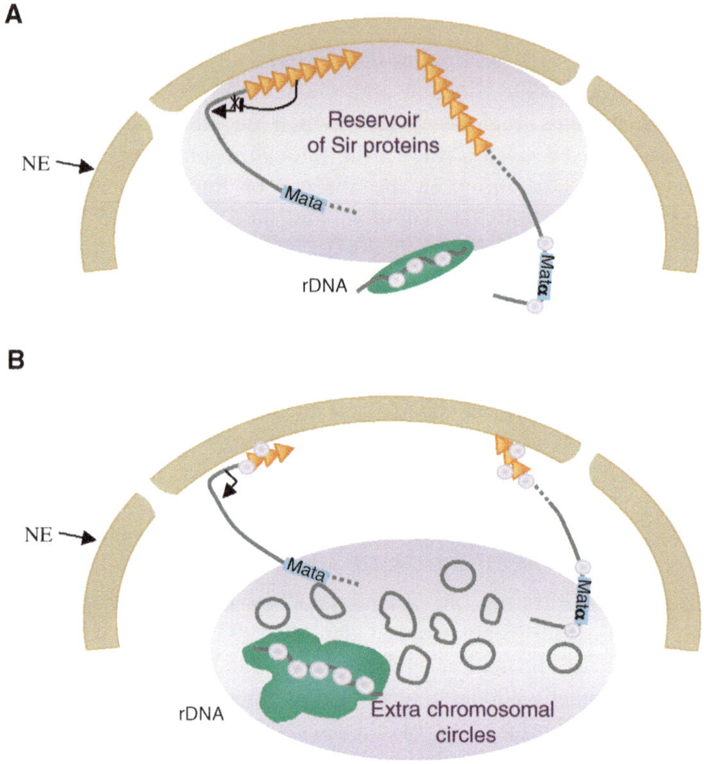

Fig. 2 *Saccharomyces cerevisiae* as a model organism for aging. **A**. Schematic representation of the yeast nucleus in young cells. Telomeres (*orange arrows*), a, and α mating-type loci and rDNA are enriched in Sir proteins. The high concentration of Sirs at the periphery of the nucleus acts as a reservoir that controls telomere structure and subsequently, the expression of subtelomeric genes. **B**. Upon aging, telomeres are shortened and recruit less Sir proteins. Fluctuation of the pool of Sirs present at chromosome ends directly controls the expression of subtelomeric genes. Sir proteins are relocalized to the nucleolus, which becomes enlarged and fragmented. After several cell cycles, the number of ERC increases and titrates all the Sir proteins. As a consequence, silenced genes become inactivated leading to age-related changes such as sterility and later, to cell death. This relocalization of Sir proteins increases the recombination rate between rDNA sequences and results in the formation of extrachromosomal circles (ERC) that are replicated and bind Sirs. In young cells, Sir proteins work against the accumulation of ERC and therefore delay the aging process

promoter 1.1 kb from the newly formed telomere. Cells expressing *URA3* grow on plates lacking uracil. However, on plates containing a drug toxic for *URA3* expressing cells (5-fluoro-orotic acid or 5-FOA), 20–60% of the cells were still able to grow, suggesting that *URA3* was silenced in the vicinity of the telomere (Gottschling et al. 1990). Following these observations, some of the features of TPE were concomitantly described, such as the stochastic reversibility or promoter independence and expression variegation.

Increasing the length of telomeres improves TPE while telomere shortening limits the silencing of subtelomeric genes (Eugster et al. 2006; Kyrion et al. 1993; Renauld et al. 1993). A plausible explanation is that longer telomeres bind more Rap1p that subsequently recruit more Sir complexes, thereby facilitating the formation of a heterochromatin complex able to polymerize and spread within the subtelomeric regions. Thus, the influence of telomere length on TPE is not merely the length itself but rather the changes in the recruitment of silencing factors. TPE induced by heterochromatin spreading hinges on Sir2p NAD+ dependent HDAC activity (Hoppe et al. 2002) and can be counteracted by Sas2p-dependant acetylation of H4-K16 (Kimura et al. 2002) placing the Sir proteins in a central position for the regulation of this mechanism (Fig. 2).

In budding yeast, the 32 telomeres are clustered into 4–6 *foci* which are primarily associated with the nuclear envelope (Gotta et al. 1996). This peripheral localization of telomeres is dependent on redundant pathways (Maillet et al. 2001). One acts through Ku and the second through Sir4-Esc1 (Hediger and Gasser 2002; Taddei and Gasser 2004). Relocation to this peripheral nuclear compartment probably does not cause repression per se (Tham et al. 2001) and silencing can be maintained without perinuclear anchoring (Gartenberg et al. 2004). All of the data on silencing and anchoring at the nuclear periphery converge toward a reservoir model where telomere clusters act as a subnuclear compartment concentrating key heterochromatin factors like the Sir proteins (Maillet et al. 1996b). Although the perinuclear location is not strictly required to maintain the silent state, silencers and telomeres would need to be somehow associated to this compartment to be in a local microenvironment containing enough silencing factors to shut down gene expression. These results are in favor of a model in which the Sir-mediated silent chromatin emanating from telomeres blocks transcription initiation in the subtelomeric regions suggesting that upon aging, the fluctuation of the pool of Sir proteins would directly control the expression of subtelomeric genes (Fig. 2).

The apparent bypass of senescence observed in Sir mutants is due to the simultaneous expression of a and alpha mating-type information, which indirectly causes an increase in the appearance of Rad52-dependent survivors (Lowell et al. 2003). Since rad52-negative Sir-negative cells senesce in the absence of telomerase, TPE is not required for the entry into senescence (Lowell et al. 2003), which, like in human cells, is associated with shortened telomeres being recognized as DNA damage (d'Adda di Fagagna et al. 2003) (Abdallah et al., in press). Nevertheless, even if TPE changes are not responsible of the growth arrest of senescent cells, it is possible that they contribute to the specific pattern of gene expression in senescent cells (Maillet et al. 1996a; Marcand et al. 1996). Indeed, the sequestering of silencing factors is expected to be telomere length dependent. In support of this, silencing at *HMR* is impaired in yeast cells carrying long telomeres (Buck and Shore 1995).

Consequences of Telomere Attrition in Senescence and Aging in Higher Eukaryotes

Yeast strains harboring mutations in genes required for telomerase function exhibit progressive shortening of telomeric DNA and growth arrest in the G2/M phase after 60–75 generations. This senescence process is MEC1 dependent (MEC1 is the yeast ortholog of ATR) (AS and Greider 2003; Enomoto et al. 2002). A minority of cells withstands loss of telomerase through RAD52-dependent amplification of telomeric and subtelomeric sequences; such survivors are now capable of long-term propagation with telomeres maintained by recombination rather than by telomerase (Lundblad and Blackburn 1993).

Human fibroblasts maintained in culture undergo many divisions before arresting growth. This limited life span of primary cells was first discovered by Hayflick and colleagues (Hayflick 1965). Later, it was discovered that the cells reach the so-called "Hayflick limit" because of telomere shortening and dysfunction in the absence of telomere maintenance catalyzed by the telomerase.

In vertebrates, telomeres are composed of stretches of TTAGGG repeats bound by a protein complex named shelterin or telosome involved in both telomere protection and length regulation (de Lange 2005). The cellular reverse transcriptase, telomerase, counteracts telomere shortening resulting from the incomplete telomere elongation after each round of DNA replication (Gilson and Geli 2007). The telomerase activity is easily detectable in germinal, embryonic, and stem cells where it contributes to maintain telomere length. This enzymatic complex extends the 3′ end of chromosomes by reverse transcription of the template region of its tightly associated RNA moiety. Telomerase expression is required for unlimited proliferation of yeast, protozoa, and immortal human tumor cells, as well as for the extended proliferation in some stem cells. In many adult stem cells or in activated lymphocytes, the presence of telomerase slows down but does not completely compensate replicative telomere erosion. Over the years, accumulating evidence has strengthened the view that an excessive telomere erosion is a tumor-suppressor mechanism and contributes to acquired and inherited aging processes, several premature aging syndromes. For instance, telomere length correlates with longevity and disease resistance in Human (Cawthon et al. 2003), whereas loss of telomerase causes accelerated aging in mice and several human syndromes (Blasco 2007a; Chang et al. 2004). Mutations reducing telomerase activity give rise to the premature-aging syndrome *dyskeratosis congenita* and aplastic anemia, as well as to idiopathic pulmonary fibrosis, all of which may be caused by telomere exhaustion and a reduced replicative potential of stem cells, a process which can be relevant for physiological aging. In somatic mammalian cells, the presence of a minimal set of very short telomeres is sufficient to trigger replicative senescence. Senescent cells are permanently arrested but still metabolically active. Based on the use of senescence markers, their number seems to increase in aged tissues although the specificity of these markers for the senescent state is still under debate. Senescence depends on the essential phosphoinositide (PI)-3-kinase-related protein kinase ATM and ATR involved in DNA damage checkpoint. If these checkpoints fail, cells resume division, develop genomic instability, and ultimately die during crisis. A few cells escape from crisis and their telomeres are maintained by recombination (alternative telomere lengthening, ALT).

Does Telomere Shortening Modulate Expression of Subtelomeric Genes?

Progressive telomere shortening is associated with changes in the chromatin structure at the tip of chromosomes but might also influence the architecture of subtelomeres. Subtelomeres are DNA sequences placed between chromosome-specific regions and chromosome ends with features that distinguish them from the rest of the genome (Mefford and Trask 2002; Riethman et al. 2001). In addition to the shelterin complex, both telomeres and subtelomeres contain nucleosomes (de Lange et al. 1990; Makarov et al. 1993; Pisano et al. 2008) that are enriched in chromatin marks found at constitutive heterochromatin regions and can spread toward the centromere (Benetti et al. 2007a, b; Garcia-Cao et al. 2004; Gonzalo et al. 2005; Gonzalo et al. 2006). Invalidation of the *Suv39h1* histone methyltransferase in the mouse correlates with extremely elongated telomeres (Garcia-Cao et al. 2004), while abrogation of the telomerase alters the chromatin status of telomeres and subtelomeric regions that are decondensed upon telomere size reduction in MEFs from late generation *Terc*$^{-/-}$ mice (Benetti et al. 2007a). Despite this heterochromatin-like organization, telomeres are not transcriptionally inert and polymerase II-dependent transcription originates in subtelomeres and processes through telomeric repeats, suggesting that the presence of repressive chromatin mark does not tightly pack telomeric and subtelomeric chromatin (Azzalin et al. 2007; Schoeftner and Blasco 2008).

Human subtelomeres vary in size from 10 to up to 300 kb in human cells. They contain repetitive sequences of different types and numerous genes but very little is known on their function in the regulation of cellular homeostasis (Mefford and Trask 2002; Riethman et al. 2001). However, these regions, prone to recombination and rearrangements, are associated with genome evolution, human disorders, but also aging possibly through TPE (Ottaviani et al. 2008). The capacity of mammalian telomeres to induce position effect has been controversial for many years. Compelling evidence for transcriptional silencing in the vicinity of human telomeres was provided more recently by using transgenes abutting telomeres, after telomere fragmentation similar to the approach used with yeast (Baur et al. 2001; Koering et al. 2002) and was later fueled by additional observations in cell culture and clinical samples (Ottaviani et al. 2008). By a telomere seeding procedure, natural telomeric regions have been replaced by artificial ones containing a reporter gene. Using this method, reporter genes in the vicinity of telomeric repeats are expressed on average ten-fold less than reporters at non-telomeric sites. Overexpression of the human telomerase reverse transcriptase (*hTERT*) resulted in telomere extension and decreased transgene expression (Baur et al. 2001), while overexpression of *TRF1*, involved in telomere length regulation, leads to the re-expression of the transgene (Koering et al. 2002), indicating the involvement of both telomere length and architecture in human TPE as observed in yeast. In addition, the treatment of cells with trichostatin A, an inhibitor of class I and II histone deacetylases, antagonizes TPE. In human cells, TPE is not sensitive to DNA methylation (Koering et al. 2002) while hypermethylation of the transgene appears as another mechanism of TPE in mouse ES cells (Pedram et al. 2006). In mammals, all three HP1 paralogs are found at telomeres, and loss of histone H3 methyltransferases leads to reduced levels of HP1 proteins at telomeres (Garcia-Cao et al. 2004; Koering et al. 2002; Sharma et al. 2003). Moreover, in human cells, TPE alleviation after TSA treatment correlates with HP1 delocalization (Koering et al. 2002). Taken together, these data suggest that in mammals, like in other simpler eukaryotic organisms, classical chromatin factors cooperate with telomere-associated proteins in the remodeling of the telomeric and subtelomeric regions and the propagation of the silencing at chromosome ends (Blasco 2007b).

The occurrence of telomeric position effect during senescence was recently investigated in human fibroblasts maintained in culture for an extended period of time (Ning et al. 2003). A total of 34 subtelomeric genes and the length of the corresponding telomeres were analyzed in young and senescent cells. Despite a differential expression for 17 out of these 34 genes, telomere length alone is not sufficient to determine the expression status of telomeric genes. Also, the analysis of eight telomeric genes on a single chromosome end showed that telomere shortening influences gene expression through the local alteration of chromatin structure. This observation fits the model proposed in yeast where TPE is influenced by the proteins bound to the telomeres rather than the telomere length per se (Ning et al. 2003).

In the human population the subtelomeric regions are highly polymorphic and length variation may be up to hundreds of kilobases among the different haplotypes. Therefore, transcriptional regulation of natural subtelomeric genes in human cells likely depends on telomere length, on the structure of the telomeric chromatin but also on the composition of the subtelomeric regions and the spatial organization of chromosome ends. Age-dependent telomere erosion might thus be a key player in the regulation of subtelomeric genes in elders as it was observed experimentally in artificial systems (Baur et al. 2001; Koering et al. 2002) and a number of factors that can influence directly or indirectly telomere structure may affect TPE by changing telomere conformation and maintenance and vice versa.

Recently, another player entered the field of telomere length regulation and aging through the characterization of the mammalian SIRT6 protein. In yeast, Sir2 functions as a histone deacetylase that regulates chromatin silencing, recombination genomic stability, and plays a central role in aging. Seven mammalian Sir2 homologs have been identified (SIRT1-SIRT7) (Dali-Youcef et al. 2007; Haigis and Guarente 2006). SIRT1 is closely related to Sir2 and represses expression of integrated reporter gene via histone deacetylation (Vaquero et al. 2004). Recently, it has been shown that SIRT6 deficiency in mice increases chromosomal aberrations such as fragmented chromosomes, detached centromeres and gaps, and leads to the development of a degenerative aging-like phenotype (Mostoslavsky et al. 2006). SIRT6 is a chromatin-associated protein that is required for normal base excision repair (BER) and appears to be critical for maintaining functional telomeres, allowing WRN interaction and preventing end-to-end fusion (Michishita et al. 2008). SIRT6 is required for proper interaction of WRN with telomeres during replication (Michishita et al. 2008). However, the consequence of *SIRT6* depletion on the chromatin structure of subtelomeres and the impact on gene expression remain to be determined. By analogy with the phenotype of mice lacking the *Suv39h1* histone methyltransferase, one can speculate that abnormalities of the telomeric chromatin might also impact on the organization of the subtelomeric regions in the absence of *SIRT6*.

In mammals, aging is associated with a multitude of changes in gene expression and increasing evidence supports the hypothesis of a link between modification of chromatin and senescence or aging. At this point, further studies are needed to elucidate the respective biological function of genes differentially expressed and influenced by the length of the distal TTAGGG repeats and the epigenetic status of subtelomeric chromatin. Interestingly, in baker's yeast, changes in telomere structures that perturb the distribution of heterochromatin factors within the nucleoplasm affect not only genes located in the subtelomeric regions but also some genes located far away from telomeres. It is thus expected, but not yet proven, that telomere shortening in mammalian cells triggers large-scale transcriptomic changes throughout the genome. In this case, decreasing the size of telomeric repeats might liberate factors that are normally telomere bound to modulate gene expression at internal chromatin sites. Thus, one can imagine that some aspect of the senescence phenotype might indeed reflect the misprogramming of transcription through disturbance of specific chromatin compartments (Fig. 1).

Chromosomal Instability and Position Effect

Detrimental effects of chromosome rearrangements followed by changes in chromatin structure have been observed for decades in human cells and model organisms. During aging, the progressive alteration of the control of DNA damage checkpoint increases the number of cells that can escape from cell cycle arrest and accumulates chromosomal abnormalities (Fig. 1). The outcome is an augmented genetic variability generated by the presence of unstable chromosomes and increasing frequency of polyploid cells. Also, there is a temporal relationship in the appearance of cells with chromosomal imbalance and the senescence state.

Heterochromatin Is Reorganized in Response to Senescence

Numerous evidences suggest that senescence plays a pivotal role in the balance between cancer and aging and it has been proposed that senescence in vivo contributes to aging through exhaustion of renewable stem cells and tissue dysfunction

(Campisi 2005; Wright and Shay 2002). Interestingly, in some cell types, senescence is associated with global changes in chromatin structure and accumulation of heterochromatin protein 1 (HP1) and histone H3 trimethylated on lysine 9 at specific foci (SAHF, senescence-associated heterochromatin foci) (Funayama et al. 2006; Narita et al. 2003; Zhang et al. 2005). In addition to these two well-known heterochromatin components, other proteins have been found to contribute to this massive heterochromatinization such as MacroH2A, HIRA, or ASF1 (Funayama and Ishikawa 2007; Zhang and Adams 2007; Zhang et al. 2007). Interestingly, orthologs of HIRA and ASF1 are associated with chromatin condensation and gene silencing in other species (Blackwell et al. 2004; Greenall et al. 2006; Kaufman et al. 1998; Sharp et al. 2001; Singer et al. 1998), in particular in yeast where it contributes to position effect. These compact chromatin foci are thought to silence expression by producing a repressive environment that prevents transcription of genes such as those controlling the E2F pathway (Narita et al. 2006).

Interestingly, regions of constitutive heterochromatin such as pericentromeres and telomeres are found at the periphery of these SAHF, suggesting that this massive reorganization of chromatin domains within the nuclear space will globally affect gene expression pattern (Funayama et al. 2006; Narita et al. 2003; Ye et al. 2007; Zhang et al. 2007).

In agreement with a role for SAHF during the aging process, increased heterochromatinization and activation of the HIRA/ASF1A pathway have also been observed in different tissues and species (Herbig et al. 2006; Jeyapalan et al. 2007; Sarg et al. 2002), and elderly people show increased heterochromatinization accompanied by transcriptional inactivation (Fig. 3). Hence, the chromatin remodeling observed in senescent cells might reflect the modifications of the nuclear architecture observed at the organismal level.

Loss of Silencing and Chromatin Changes

Besides the heterochromatinization of each individual chromosome through the formation of SAHF, chromosomal position effect in aging tissues might also be associated by the age-related loss of epigenetic silencing of repeated sequences as observed in the heart and in neurons. Specifically, the major satellite repeats that form the constitutive heterochromatin structure at centromeres were shown to become transcriptionally active (Gaubatz and Cutler 1990; Gaubatz and Flores 1990; Shen et al. 2008). One possible explanation could be the decrease in activity of chromatin-modifying enzyme and the ensuing alteration of the constitutive heterochromatin compartment after the relocalization of heterochromatin proteins as described above for the Sir proteins in yeast (Fig. 3).

Together with the chromatin factors mentioned earlier, SIRT1, a member of the SIR protein family in higher eukaryotes is known to be downregulated in senescent cells and during aging and a decreased histone H4-K16 acetylation has been observed at numerous loci (Sommer et al. 2006) while, intriguingly, trimethylation

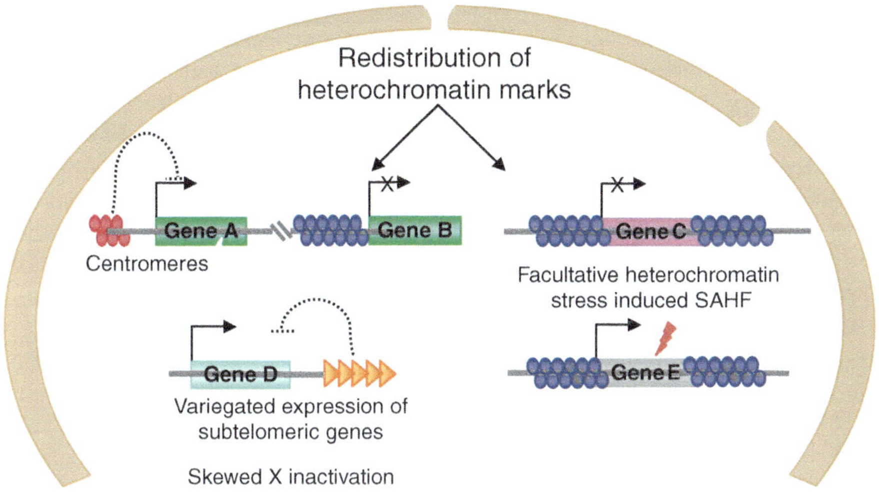

Fig. 3 Redistribution of the epigenetic marks in aging cells globally affects gene expression patterns. Aging is accompanied by alterations in the composition of the chromatin from site to site. On a global point of view, the main consequence appears to be the redistribution of heterochromatin marks with a decondensation of constitutive heterochromatin (pericentromeres and inactive X) and the condensation of euchromatic regions through the formation of heterochromatin foci. The integrity of the DNA is affected and a redistribution of chromatin marks at the site of DNA damage can be observed. At chromosome ends, the shortening of telomeres leads to a probable loss of heterochromatin marks at subtelomeric regions. This global reorganization of the chromatin architecture might be associated with a variegated expression of genes or clusters not only in the vicinity of these changes but also genomewide after the redistribution of key regulators

of the adjacent H4-K20 residue augments with age in rat liver and kidney, supporting the idea of a genome-wide switch from euchromatin to heterochromatin during tissue aging (Sarg et al. 2002).

At sites of damaged DNA, the recruitment of the DNA repair machinery involves histone modifiers that could spread over megabases around the DNA break, potentially affecting the epigenetic regulation of several genes at the site of breakage but also at the original sites of action of these factors as described for the Sir proteins in yeast (Botuyan et al. 2006; Huyen et al. 2004; Kim et al. 1999; Rogakou et al. 1999) pointing toward a global role for DNA damage in chromatin architecture at the point of breakage but also throughout the genome (Bahar et al. 2006). Some regions such as subtelomeres, fragile sites, or euchromatic loci are more prone to recombination, suggesting that changes are more often observed in certain regions and that consequences on gene expression might be directly influenced by local factors (Fig. 3).

Most of the genes differentially expressed during aging show poor conservation among tissues or species (Fraser et al. 2005; Park and Prolla 2005) underlining the apparent randomness of these changes. On the other hand, it has been shown in different tissues that genes physically clustered together change coordinately with age, suggesting that the borders of condensed chromatin are pushed toward euchromatin regions or that the decondensation of regions of constitutive heterochromatin such as the centromeres or the telomeres might influence the pattern of expression of numerous genes that would in turn contribute to the global modifications of the aging process. This hypothesis is further supported by work in flies showing that genes dysregulated during aging are often distributed on the same chromosomal region (Pletcher et al. 2002) putting position effect on the front seat for gene alteration in aging.

In conclusion, regional activation or inactivation of chromosomal domains suggest that some regions are more prone to global changes in their chromatin structure may be through the involvement of the long-distance regulation or the positioning to specific subnuclear domains.

Aging and X Inactivation

An active skewing of X inactivation has also been reported with increased age. Indeed, elderly women have a much higher frequency of skewed X inactivation in the myeloid cell lineage of peripheral blood cells compared to younger females (Busque et al. 1996; Gale et al. 1997; Sharp et al. 2000) with an increased tendency after 50–60 years, suggesting that hormonal changes could contribute to this age-related decondensation (Hatakeyama et al. 2004; Kristiansen et al. 2005). Age-related skewing is not solely a stochastic process as shown by analysis of mono- and dizygotic twins but may be due to complex mechanisms involving both random and genetic events (Kristiansen et al. 2005). Of note, several components of the SAHF are also implicated in the inactivation of the X chromosome suggesting that the delocalization of these factors to other chromosomes alters the proper regulation of the inactivated X chromosome and corresponding genes. The consequences of age-related skewing of X inactivation are not known. However, one possible consequence is the manifestation of X-linked disorders in elderly women such as X-linked hyper IgM syndrome, combined immunodeficiency, and chronic granulomatous disease (Au et al. 2004; Cazzola et al. 2000; Invernizzi et al. 2008).

Causes and Consequences of Environmental Adaptation on Aging Organisms

Aging is characterized as a progressive, generalized impairment of cellular functions resulting in an increasing susceptibility to environmental stress and a wide range of diseases. As we argue below, different mechanisms could be designed to accommodate the evolution of environmental conditions throughout life. One way might be the rapid regulation of subtelomeric genes by TPE. Indeed, a subtelomeric enrichment of genes related to stress response and metabolism in non-optimal growth conditions appears to be a conserved feature in many yeast species (Robyr et al. 2002) and clustering stress response genes at subtelomeres seems to be an evolutionarily conserved strategy that allows their reversible silencing and a fast response to changes in environmental conditions in various organisms (Barry et al. 2003; Borst and Ulbert 2001; Dreesen et al. 2007; Ottaviani et al. 2008). In budding yeast, most of these genes are silenced under optimal conditions but can be rapidly induced upon environmental changes. For instance, the *FLO* genes, involved in cellular adherence, are mostly contained within these domains and are silenced in a Sir-independent way which is, however, dependent upon Sir2p homologs, Hst1p and Hst2p (Halme et al. 2004). Also, stress-like nutrient starvation, heat shock, or chemical treatment can induce a hyperphosphorylation of Sir3p and decrease TPE at the truncated VII-L telomere (Stone and Pillus 1996). This also leads to an increase in the expression of natural subtelomeric genes such as the *PAU* genes which are involved in cell wall constitution and drug resistance (Ai et al. 2002). In *Schizosaccharomyces pombe*, many genes involved in response to nitrogen starvation are also clustered in subtelomeric regions and silenced by the Hda1p ortholog, Clr3 (Hansen et al. 2005).

One can imagine that a single mutation or epimutation altering TPE would allow the optimal expression of subtelomeric genes increasing the chances for the cell to express a gene that would be important for adaptation (Teixeira and Gilson 2005). On the contrary, alteration of chromatin and telomere associated with aging would limit the fast response to environmental stress and reduce longevity.

Interestingly, late generation m*Terc*$^{-/-}$ mice where telomerase is invalidated, exhibited a shorter life span associated with a greatly reduced capacity to overcome acute and chronic stress while age-matched control animals did not manifest such phenotypes, suggesting that the capacity to adapt to environmental changes is altered upon telomere dysfunction (Rudolph et al. 1999). Also several environmental factors that affect telomere length such as stress, smoking, obesity, and socioeconomic status might also accelerate aging (Canela et al. 2007; Cherkas et al. 2006; Epel et al. 2004; Valdes et al. 2005).

Other components of the chromatin conformation might also influence lifetime. For instance, overexpression of the SIRT1 histone deacetylase in the heart delays aging and protects against oxidative stress, suggesting that maintaining the global chromatin architecture upon stress-induced aging may represent a novel cardioprotection strategy in old individuals (Hsu et al. 2008). Increased expression of the mammalian SIRT1 promotes axon regrowth after injury and activation of this NAD-dependent deacetylase limits degeneration (Araki et al. 2004), suggesting that it may also protect neurons from age-related neurodegenerative diseases.

Interestingly, sensory perception may also affect life span in higher animals (Libert et al. 2007; Lindemann 2001). Many *Caenorhabditis elegans* mutants with reduced activity of putative chemosensory receptors have a significant increased life span (Alcedo and Kenyon 2004; Apfeld and Kenyon 1999) and some gustatory and olfactory neurons either promote or inhibit longevity (Alcedo and Kenyon 2004). Olfactory genes are preferentially positioned at subtelomeric positions; it would be interesting to correlate changes in olfactory stimuli in aged individuals and control of expression of subtelomeric clusters of genes by telomeres.

Another type of response to environmental conditions that affects aging is the response to nutrient restriction. In yeast, worms, flies, rodents, and possibly primates, diet extends life span (Cohen et al. 2004; Libert et al. 2007; Pletcher et al. 2002). One possible pathway would be the response to insulin/IGF1 and will not be discussed here (reviewed in Kenyon (2005)). Another pathway, directly associated with chromatin regulation involves the SIR proteins. The molecular consequences to dietary restriction in yeast grown in absence of glucose involve increasing activity of the Sir protein (Ai et al. 2002). The redistribution of this major regulator of gene silencing allows the reactivation of different genes such as subtelomeric genes as described above. Apparently, this ability for Sir2 to extend life span has been conserved during evolution since SIR2 orthologs increase survival in flies and worms. On the other hand, a connection between a shortened life span, olfaction, and food derived odors has been demonstrated (Libert et al. 2007). In diet-restricted long-lived flies, expression of the genes encoding the odorant receptors is strongly affected by both age and nutrient availability. Thus, as observed not only in yeast but also in worms, flies, rodents, or human cells, calorie restriction promotes survival by inducing the SIR histone deacetylase. The different pathways that might be involved in this extension of life span exceed the scope of this review but suggest that global changes linked to environmental conditions have a broad regulatory effect and likely involve global epigenetic regulation with a key role for the SIR proteins in this phenomenon.

Aging and Inflammation

Older subjects are more susceptible than younger ones to pathogenic stimuli (Krabbe et al. 2004). Moreover, one of the most dramatic transcriptional changes in aging flies is associated with response to microbial infection and involves increased expression of several anti-microbial genes suggesting that the response to infection is a primary concern in aging individuals (Pletcher et al. 2002). The immune response involves networks of regulation and a complex cascade of gene activation associated with epigenetic changes. This process occurs through a massive remodeling of the chromatin and subnuclear repositioning of genes and *cis*-regulating elements (Oberdoerffer and Sinclair 2007). Redistribution of the chromatin marks and alteration of the transcriptional accessibility through PEV or TPE might contribute to a modified sensibility or response to pathological stimuli in aged individuals. In agreement with this hypothesis, the rate of condensed chromatin increases in lymphocytes from older individuals, as observed for other tissues, suggesting that the expansion of silenced regions might affect the defense against pathogens (Lezhava 2001; Lezhava and Jokhadze 2007). During normal aging, depletion of lymphocytes has also been linked to a diminished response to circulating factors and in particular to reduced levels of IGF-I in the serum (Allman and Miller 2005a, b; Lombardi et al. 2005; Pifer et al. 2003; Stephan et al. 1997). Interestingly, the level of IGF-I is also diminished in mice defective for *SirT6*, suggesting a connection between the SIR and the IGF-I pathway as observed in other species such as *C. elegans* (Kenyon 2005; Michishita et al. 2008).

In aging hematopoietic stem cells, genes involved in inflammatory and stress response dominated the group of upregulated genes, whereas those participating in chromatin regulation and DNA repair were prominent among downregulated genes (Chambers et al. 2007; Rossi et al. 2005). The IgH and IgK gene clusters are upregulated in these cells and interestingly, the IgH genes cluster is localized at subtelomeres. Although this hypothesis has never been tested, it is tempting to speculate that telomeric position effect subsequent to telomere attrition might account for the specific changes at this cluster. Also, in human hepatic stellate cells undergoing senescence, increased expression of genes mediating inflammatory response has been observed (Schnabl et al. 2003). Consistent with these results, another analysis of senescent fibroblasts revealed that clusters of genes are indeed upregulated during senescence, independent of telomere length while no clustering of downregulated genes could be detected (Zhang et al. 2003). Interestingly, numerous genes encoding cytokines are located at subtelomeric loci and mice invalidated for the telomerase have an increased susceptibility to *C. albicans* and show immunosenescence (Blasco 2007a; Murciano et al. 2006).

Conclusions

Aging involves highly dynamic series of modifications at the cellular level that further lead to widespread changes at the level of the whole body. Importantly, the emergence of heterochromatin foci concomitant to telomere shortening reflects the large-scale shift of the constitutive heterochromatin compartment and raises the intriguing possibility of a global reprogramming of gene expression associated with position effect mechanisms. At the organismal level, variegated gene expression reflects this mosaic pattern of remodeled regions and the expression profiles determined in cells and tissues likely reflect the intrinsic complexity of this irremediable biological process.

Several other mechanisms governing the nuclear processes have not been evoked in this review however, it has to be kept in mind that other pathways might influence or be influenced by chromosomal position effect or telomeric position effect.

For instance, nuclear lamins are highly dynamic and play a role in the non-random positioning of chromosomes to nuclear subdomains and possibly the regulation of gene expression. The gene encoding the A-type lamins is mutated in the Hutchinson–Gilford premature aging syndrome (HGPS) (De Sandre-Giovannoli et al. 2003; Eriksson et al. 2003). Nuclei from patients with HGPS harbor a dysmorphic shape and a loss of heterochromatin-related proteins that are normally associated with the nuclear membrane and altered histone modification pattern (Scaffidi and Misteli 2006, 2008), suggesting that a dysfunctional nuclear envelope might participate in the aging process. The positioning of chromosome segments within the nuclear space might affect the regulation of individual or clusters of genes in aging cells. Research on the Werner premature aging syndrome supports the importance of the maintenance of genomic stability as a partial antidote to aging (Martin 2005). Also, the emergence of microRNAs or non-coding RNA in the regulation of chromatin, especially at telomeres (TERRA molecules), lays new ground for the deciphering of the epigenetic events involved in longevity.

Acknowledgments The work in Gilson lab is supported by the Ligue Nationale contre le Cancer (Equipe labellisée) and by the Association Française contre les Myopathies (AFM).

References

Abdallah P, Luciano P, Runge K, Lisby M, Geli V, Gilson E, Teixera T. A two step model for senescence triggered by a single critically short telomere, Nature Cell Biology. In press.

Aguilaniu H, Gustafsson L, Rigoulet M, Nystrom T (2003) Asymmetric inheritance of oxidatively damaged proteins during cytokinesis. Science 299: 1751–3

Ai W, Bertram PG, Tsang CK, Chan TF, Zheng XF (2002) Regulation of subtelomeric silencing during stress response. Mol Cell 10: 1295–305

Alcedo J, Kenyon C (2004) Regulation of C. elegans longevity by specific gustatory and olfactory neurons. Neuron 41: 45–55

Allman D, Miller JP (2005a) B cell development and receptor diversity during aging. Curr Opin Immunol 17: 463–7

Allman D, Miller JP (2005b) The aging of early B-cell precursors. Immunol Rev 205: 18–29

Apfeld J, Kenyon C (1999) Regulation of lifespan by sensory perception in Caenorhabditis elegans. Nature 402: 804–9

Araki T, Sasaki Y, Milbrandt J (2004) Increased nuclear NAD biosynthesis and SIRT1 activation prevent axonal degeneration. Science 305: 1010–3

AS IJ, Greider CW (2003) Short telomeres induce a DNA damage response in Saccharomyces cerevisiae. Mol Biol Cell 14: 987–1001

Au WY, Ma ES, Lam VM, Chan JL, Pang A, Kwong YL (2004) Glucose 6-phosphate dehydrogenase (G6PD) deficiency in elderly Chinese women heterozygous for G6PD variants. Am J Med Genet A 129A: 208–11

Azzalin CM, Reichenbach P, Khoriauli L, Giulotto E, Lingner J (2007) Telomeric repeat containing RNA and RNA surveillance factors at mammalian chromosome ends. Science 318: 798–801

Bahar R, Hartmann CH, Rodriguez KA, Denny AD, Busuttil RA, Dolle ME, Calder RB, Chisholm GB, Pollock BH, Klein CA, Vijg J (2006) Increased cell-to-cell variation in gene expression in ageing mouse heart. Nature 441: 1011–4

Barry JD, Ginger ML, Burton P, McCulloch R (2003) Why are parasite contingency genes often associated with telomeres? Int J Parasitol 33: 29–45

Baur JA, Zou Y, Shay JW, Wright WE (2001) Telomere position effect in human cells. Science 292: 2075–7

Benetti R, Garcia-Cao M, Blasco MA (2007a) Telomere length regulates the epigenetic status of mammalian telomeres and subtelomeres. Nat Genet 39: 243–50

Benetti R, Gonzalo S, Jaco I, Schotta G, Klatt P, Jenuwein T, Blasco MA (2007b) Suv4-20 h deficiency results in telomere elongation and derepression of telomere recombination. J Cell Biol 178: 925–36

Blackwell C, Martin KA, Greenall A, Pidoux A, Allshire RC, Whitehall SK (2004) The Schizosaccharomyces pombe HIRA-like protein Hip1 is required for the periodic expression of histone genes and contributes to the function of complex centromeres. Mol Cell Biol 24: 4309–20

Blasco MA (2007a) Telomere length, stem cells and aging. Nat Chem Biol 3: 640–9
Blasco MA (2007b) The epigenetic regulation of mammalian telomeres. Nat Rev Genet 8: 299–309
Borst P, Ulbert S (2001) Control of VSG gene expression sites. Mol Biochem Parasitol 114: 17–27
Botuyan MV, Lee J, Ward IM, Kim JE, Thompson JR, Chen J, Mer G (2006) Structural basis for the methylation state-specific recognition of histone H4-K20 by 53BP1 and Crb2 in DNA repair. Cell 127: 1361–73
Buck SW, Shore D (1995) Action of a RAP1 carboxy-terminal silencing domain reveals an underlying competition between HMR and telomeres in yeast. Genes Dev 9: 370–84
Busque L, Mio R, Mattioli J, Brais E, Blais N, Lalonde Y, Maragh M, Gilliland DG (1996) Nonrandom X-inactivation patterns in normal females: lyonization ratios vary with age. Blood 88: 59–65
Campisi J (2005) Senescent cells, tumor suppression, and organismal aging: good citizens, bad neighbors. Cell 120: 513–22
Canela A, Vera E, Klatt P, Blasco MA (2007) High-throughput telomere length quantification by FISH and its application to human population studies. Proc Natl Acad Sci USA 104: 5300–5
Cawthon RM, Smith KR, O′Brien E, Sivatchenko A, Kerber RA (2003) Association between telomere length in blood and mortality in people aged 60 years or older. Lancet 361: 393–5
Cazzola M, May A, Bergamaschi G, Cerani P, Rosti V, Bishop DF (2000) Familial-skewed X-chromosome inactivation as a predisposing factor for late-onset X-linked sideroblastic anemia in carrier females. Blood 96: 4363–5
Chambers SM, Shaw CA, Gatza C, Fisk CJ, Donehower LA, Goodell MA (2007) Aging hematopoietic stem cells decline in function and exhibit epigenetic dysregulation. PLoS Biol 5: e201
Chang S, Multani AS, Cabrera NG, Naylor ML, Laud P, Lombard D, Pathak S, Guarente L, DePinho RA (2004) Essential role of limiting telomeres in the pathogenesis of Werner syndrome. Nat Genet 36: 877–82
Cherkas LF, Aviv A, Valdes AM, Hunkin JL, Gardner JP, Surdulescu GL, Kimura M, Spector TD (2006) The effects of social status on biological aging as measured by white-blood-cell telomere length. Aging Cell 5: 361–5
Cohen HY, Miller C, Bitterman KJ, Wall NR, Hekking B, Kessler B, Howitz KT, Gorospe M, de Cabo R, Sinclair DA (2004) Calorie restriction promotes mammalian cell survival by inducing the SIRT1 deacetylase. Science 305: 390–2
d'Adda di Fagagna F, Reaper PM, Clay-Farrace L, Fiegler H, Carr P, Von Zglinicki T, Saretzki G, Carter NP, Jackson SP (2003) A DNA damage checkpoint response in telomere-initiated senescence. Nature 426: 194–8
Dali-Youcef N, Lagouge M, Froelich S, Koehl C, Schoonjans K, Auwerx J (2007) Sirtuins: the 'magnificent seven', function, metabolism and longevity. Ann Med 39: 335–45
Daniel J (2005) Sir-dependent downregulation of various aging processes. Mol Genet Genomics 274: 539–47
de Lange T (2005) Shelterin: the protein complex that shapes and safeguards human telomeres. Genes Dev 19: 2100–10
de Lange T, Shiue L, Myers RM, Cox DR, Naylor SL, Killery AM, Varmus HE (1990) Structure and variability of human chromosome ends. Mol Cell Biol 10: 518–27
De Sandre-Giovannoli A, Bernard R, Cau P, Navarro C, Amiel J, Boccaccio I, Lyonnet S, Stewart CL, Munnich A, Le Merrer M, Levy N (2003) Lamin a truncation in Hutchinson-Gilford progeria. Science 300: 2055
Dreesen O, Li B, Cross GA (2007) Telomere structure and function in trypanosomes: a proposal. Nat Rev Microbiol 5: 70–5
Eissenberg JC, James TC, Foster-Hartnett DM, Hartnett T, Ngan V, Elgin SC (1990) Mutation in a heterochromatin-specific chromosomal protein is associated with suppression of position-effect variegation in Drosophila melanogaster. Proc Natl Acad Sci USA 87: 9923–7

Enomoto S, Glowczewski L, Berman J (2002) MEC3, MEC1, and DDC2 are essential components of a telomere checkpoint pathway required for cell cycle arrest during senescence in Saccharomyces cerevisiae. Mol Biol Cell 13: 2626–38

Epel ES, Blackburn EH, Lin J, Dhabhar FS, Adler NE, Morrow JD, Cawthon RM (2004) Accelerated telomere shortening in response to life stress. Proc Natl Acad Sci USA 101: 17312–5

Eriksson M, Brown WT, Gordon LB, Glynn MW, Singer J, Scott L, Erdos MR, Robbins CM, Moses TY, Berglund P, Dutra A, Pak E, Durkin S, Csoka AB, Boehnke M, Glover TW, Collins FS (2003) Recurrent de novo point mutations in lamin A cause Hutchinson-Gilford progeria syndrome. Nature 423: 293–8

Eugster A, Lanzuolo C, Bonneton M, Luciano P, Pollice A, Pulitzer JF, Stegberg E, Berthiau AS, Forstemann K, Corda Y, Lingner J, Geli V, Gilson E (2006) The finger subdomain of yeast telomerase cooperates with Pif1p to limit telomere elongation. Nat Struct Mol Biol 13: 734–9

Falcon AA, Aris JP (2003) Plasmid accumulation reduces life span in Saccharomyces cerevisiae. J Biol Chem 278: 41607–17

Feil R (2006) Environmental and nutritional effects on the epigenetic regulation of genes. Mutat Res 600: 46–57

Festenstein R, Sharghi-Namini S, Fox M, Roderick K, Tolaini M, Norton T, Saveliev A, Kioussis D, Singh P (1999) Heterochromatin protein 1 modifies mammalian PEV in a dose- and chromosomal-context-dependent manner. Nat Genet 23: 457–61

Fourel G, Magdinier F, Gilson E (2004) Insulator dynamics and the setting of chromatin domains. Bioessays 26: 523–32

Fraga MF, Esteller M (2007) Epigenetics and aging: the targets and the marks. Trends Genet 23: 413–8

Fraser HB, Khaitovich P, Plotkin JB, Paabo S, Eisen MB (2005) Aging and gene expression in the primate brain. PLoS Biol 3: e274

Funayama R, Ishikawa F (2007) Cellular senescence and chromatin structure. Chromosoma 116: 431–40

Funayama R, Saito M, Tanobe H, Ishikawa F (2006) Loss of linker histone H1 in cellular senescence. J Cell Biol 175: 869–80

Gale RE, Fielding AK, Harrison CN, Linch DC (1997) Acquired skewing of X-chromosome inactivation patterns in myeloid cells of the elderly suggests stochastic clonal loss with age. Br J Haematol 98: 512–9

Garcia-Cao M, O'Sullivan R, Peters AH, Jenuwein T, Blasco MA (2004) Epigenetic regulation of telomere length in mammalian cells by the Suv39h1 and Suv39h2 histone methyltransferases. Nat Genet 36: 94–9

Gartenberg MR, Neumann FR, Laroche T, Blaszczyk M, Gasser SM (2004) Sir-mediated repression can occur independently of chromosomal and subnuclear contexts. Cell 119: 955–67

Gaszner M, Felsenfeld G (2006) Insulators: exploiting transcriptional and epigenetic mechanisms. Nat Rev Genet 7: 703–13

Gaubatz JW, Cutler RG (1990) Mouse satellite DNA is transcribed in senescent cardiac muscle. J Biol Chem 265: 17753–8

Gaubatz JW, Flores SC (1990) Tissue-specific and age-related variations in repetitive sequences of mouse extrachromosomal circular DNAs. Mutat Res 237: 29–36

Gilson E, Geli V (2007) How telomeres are replicated. Nat Rev Mol Cell Biol 8: 825–38

Girton JR, Johansen KM (2008) Chromatin structure and the regulation of gene expression: the lessons of PEV in Drosophila. Adv Genet 61: 1–43

Gonzalo S, Garcia-Cao M, Fraga MF, Schotta G, Peters AH, Cotter SE, Eguia R, Dean DC, Esteller M, Jenuwein T, Blasco MA (2005) Role of the RB1 family in stabilizing histone methylation at constitutive heterochromatin. Nat Cell Biol 7: 420–8

Gonzalo S, Jaco I, Fraga MF, Chen T, Li E, Esteller M, Blasco MA (2006) DNA methyltransferases control telomere length and telomere recombination in mammalian cells. Nat Cell Biol 8: 416–24

Gotta M, Laroche T, Formenton A, Maillet L, Scherthan H, Gasser SM (1996) The clustering of telomeres and colocalization with Rap1, Sir3, and Sir4 proteins in wild-type Saccharomyces cerevisiae. J Cell Biol 134: 1349–63

Gottschling DE, Aparicio OM, Billington BL, Zakian VA (1990) Position effect at S. cerevisiae telomeres: reversible repression of Pol II transcription. Cell 63: 751–62

Greenall A, Williams ES, Martin KA, Palmer JM, Gray J, Liu C, Whitehall SK (2006) Hip3 interacts with the HIRA proteins Hip1 and Slm9 and is required for transcriptional silencing and accurate chromosome segregation. J Biol Chem 281: 8732–9

Haigis MC, Guarente LP (2006) Mammalian sirtuins – emerging roles in physiology, aging, and calorie restriction. Genes Dev 20: 2913–21

Halme A, Bumgarner S, Styles C, Fink GR (2004) Genetic and epigenetic regulation of the FLO gene family generates cell-surface variation in yeast. Cell 116: 405–15

Hansen KR, Burns G, Mata J, Volpe TA, Martienssen RA, Bahler J, Thon G (2005) Global effects on gene expression in fission yeast by silencing and RNA interference machineries. Mol Cell Biol 25: 590–601

Hatakeyama C, Anderson CL, Beever CL, Penaherrera MS, Brown CJ, Robinson WP (2004) The dynamics of X-inactivation skewing as women age. Clin Genet 66: 327–32

Hayflick L (1965) The limited in vitro lifetime of human diploid cell strains. Exp Cell Res 37: 614–36

Hediger F, Gasser SM (2002) Nuclear organization and silencing: putting things in their place. Nat Cell Biol 4: E53–5

Herbig U, Ferreira M, Condel L, Carey D, Sedivy JM (2006) Cellular senescence in aging primates. Science 311: 1257

Hoppe GJ, Tanny JC, Rudner AD, Gerber SA, Danaie S, Gygi SP, Moazed D (2002) Steps in assembly of silent chromatin in yeast: Sir3-independent binding of a Sir2/Sir4 complex to silencers and role for Sir2-dependent deacetylation. Mol Cell Biol 22: 4167–80

Hsu CP, Odewale I, Alcendor RR, Sadoshima J (2008) Sirt1 protects the heart from aging and stress. Biol Chem 389: 221–31

Huyen Y, Zgheib O, Ditullio RA, Jr., Gorgoulis VG, Zacharatos P, Petty TJ, Sheston EA, Mellert HS,i Stavridi ES, Halazonetis TD (2004) Methylated lysine 79 of histone H3 targets 53BP1 to DNA double-strand breaks. Nature 432: 406–11

Invernizzi P, Pasini S, Selmi C, Miozzo M, Podda M (2008) Skewing of X chromosome inactivation in autoimmunity. Autoimmunity 41: 272–7

James TC, Elgin SC (1986) Identification of a nonhistone chromosomal protein associated with heterochromatin in Drosophila melanogaster and its gene. Mol Cell Biol 6: 3862–72

Jeyapalan JC, Ferreira M, Sedivy JM, Herbig U (2007) Accumulation of senescent cells in mitotic tissue of aging primates. Mech Ageing Dev 128: 36–44

Kaeberleion M, McVey M, Guareute L. The SIR2/3/4 complex and SIR2 alone promote longevity in Saccharomyces cerevisiae by two different mechanisms. Genes Dev, Vol 13 issue 9. Pages 2570–80. 1999

Kaeberlein M, Kennedy BK (2005) Large-scale identification in yeast of conserved ageing genes. Mech Ageing Dev 126: 17–21

Karow JK, Wu L, Hickson ID (2000) RecQ family helicases: roles in cancer and aging. Curr Opin Genet Dev 10: 32–8

Kaufman PD, Cohen JL, Osley MA (1998) Hir proteins are required for position-dependent gene silencing in Saccharomyces cerevisiae in the absence of chromatin assembly factor I. Mol Cell Biol 18: 4793–806

Kennedy BK, Austriaco NR, Jr., Zhang J, Guarente L (1995) Mutation in the silencing gene SIR4 can delay aging in S. cerevisiae. Cell 80: 485–96

Kennedy BK, Gotta M, Sinclair DA, Mills K, McNabb DS, Murthy M, Pak SM, Laroche T, Gasser SM, Guarente L (1997) Redistribution of silencing proteins from telomeres to the nucleolus is associated with extension of life span in S. cerevisiae. Cell 89: 381–91

Kenyon C (2005) The plasticity of aging: insights from long-lived mutants. Cell 120: 449–60
Kim S, Benguria A, Lai CY, Jazwinski SM (1999) Modulation of life-span by histone deacetylase genes in Saccharomyces cerevisiae. Mol Biol Cell 10: 3125–36
Kimura A, Umehara T, Horikoshi M (2002) Chromosomal gradient of histone acetylation established by Sas2p and Sir2p functions as a shield against gene silencing. Nat Genet 32: 370–7
Klar AJ, Fogel S, Macleod K (1979) MAR1-a regulator of the HMa and HMalpha Loci in Saccharomyces Cerevisiae. Genetics 93: 37–50
Koering CE, Pollice A, Zibella MP, Bauwens S, Puisieux A, Brunori M, Brun C, Martins L, Sabatier L, Pulitzer JF, Gilson E (2002) Human telomeric position effect is determined by chromosomal context and telomeric chromatin integrity. EMBO Rep 3: 1055–61
Krabbe KS, Pedersen M, Bruunsgaard H (2004) Inflammatory mediators in the elderly. Exp Gerontol 39: 687–99
Kristiansen M, Knudsen GP, Bathum L, Naumova AK, Sorensen TI, Brix TH, Svendsen AJ, Christensen K, Kyvik KO, Orstavik KH (2005) Twin study of genetic and aging effects on X chromosome inactivation. Eur J Hum Genet 13: 599–606
Kyrion G, Liu K, Liu C, Lustig AJ (1993) RAP1 and telomere structure regulate telomere position effects in Saccharomyces cerevisiae. Genes Dev 7: 1146–59
Lee SK, Johnson RE, Yu SL, Prakash L, Prakash S (1999) Requirement of yeast SGS1 and SRS2 genes for replication and transcription. Science 286: 2339–42
Lezhava T (2001) Chromosome and aging: genetic conception of aging. Biogerontology 2: 253–60
Lezhava T, Jokhadze T (2007) Activation of pericentromeric and telomeric heterochromatin in cultured lymphocytes from old individuals. Ann NY Acad Sci 1100: 387–99
Libert S, Zwiener J, Chu X, Vanvoorhies W, Roman G, Pletcher SD (2007) Regulation of Drosophila life span by olfaction and food-derived odors. Science 315: 1133–7
Lindemann B (2001) Receptors and transduction in taste. Nature 413: 219–25
Lombardi G, Di Somma C, Rota F, Colao A (2005) Associated hormonal decline in aging: is there a role for GH therapy in aging men? J Endocrinol Invest 28: 99–108
Lowell JE, Roughton AI, Lundblad V, Pillus L (2003) Telomerase-independent proliferation is influenced by cell type in Saccharomyces cerevisiae. Genetics 164: 909–21
Lundblad V, Blackburn EH (1993) An alternative pathway for yeast telomere maintenance rescues est1- senescence. Cell 73: 347–60
Maillet L, Boscheron C, Gotta M, Marcand S, Gilson E, Gasser SM (1996a) Evidence for silencing compartments within the yeast nucleus: a role for telomere proximity and Sir protein concentration in silencer-mediated repression. Genes Dev 10: 1796–811
Maillet L, Boscheron C, Gotta M, Marcand S, Gilson E, Gasser SM (1996b) Evidence for silencing compartments within the yeast nucleus: a role for telomere proximity and Sir protein concentration in silencer-mediated repression. Genes Develop 10: 1796–811
Maillet L, Gaden F, Brevet V, Fourel G, Martin SG, Dubrana K, Gasser SM, Gilson E (2001) Ku-deficient yeast strains exhibit alternative states of silencing competence. EMBO Rep 2: 203–10
Makarov VL, Lejnine S, Bedoyan J, Langmore JP (1993) Nucleosomal organization of telomere-specific chromatin in rat. Cell 73: 775–87
Marcand S, Buck SW, Moretti P, Gilson E, Shore D (1996) Silencing of genes at nontelomeric sites in yeast is controlled by sequestration of silencing factors at telomeres by Rap 1 protein. Genes Dev 10: 1297–309
Martin GM (2005) Genetic modulation of senescent phenotypes in Homo sapiens. Cell 120: 523–32
Martin SG, Laroche T, Suka N, Grunstein M, Gasser SM (1999) Relocalization of telomeric Ku and SIR proteins in response to DNA strand breaks in yeast. Cell 97: 621–33
McAinsh AD, Scott-Drew S, Murray JA, Jackson SP (1999) DNA damage triggers disruption of telomeric silencing and Mec1p-dependent relocation of Sir3p. Curr Biol 9: 963–6
Mefford HC, Trask BJ (2002) The complex structure and dynamic evolution of human subtelomeres. Nat Rev Genet 3: 91–102

Michan S, Sinclair D (2007) Sirtuins in mammals: insights into their biological function. Biochem J 404: 1–13
Michishita E, McCord RA, Berber E, Kioi M, Padilla-Nash H, Damian M, Cheung P, Kusumoto R, Kawahara TL, Barrett JC, Chang HY, Bohr VA, Ried T, Gozani O, Chua KF (2008) SIRT6 is a histone H3 lysine 9 deacetylase that modulates telomeric chromatin. Nature 452: 492–6
Mills KD, Sinclair DA, Guarente L (1999) MEC1-dependent redistribution of the Sir3 silencing protein from telomeres to DNA double-strand breaks. Cell 97: 609–20
Mostoslavsky R, Chua KF, Lombard DB, Pang WW, Fischer MR, Gellon L, Liu P, Mostoslavsky G, Franco S, Murphy MM, Mills KD, Patel P, Hsu JT, Hong AL, Ford E, Cheng HL, Kennedy C, Nunez N, Bronson R, Frendewey D, Auerbach W, Valenzuela D, Karow M, Hottiger MO, Hursting S, Barrett JC, Guarente L, Mulligan R, Demple B, Yancopoulos GD, Alt FW (2006) Genomic instability and aging-like phenotype in the absence of mammalian SIRT6. Cell 124: 315–29
Murciano C, Villamon E, Yanez A, O'Connor JE, Gozalbo D, Gil ML (2006) Impaired immune response to Candida albicans in aged mice. J Med Microbiol 55: 1649–56
Narita M, Krizhanovsky V, Nunez S, Chicas A, Hearn SA, Myers MP, Lowe SW (2006) A novel role for high-mobility group a proteins in cellular senescence and heterochromatin formation. Cell 126: 503–14
Narita M, Nunez S, Heard E, Lin AW, Hearn SA, Spector DL, Hannon GJ, Lowe SW (2003) Rb-mediated heterochromatin formation and silencing of E2F target genes during cellular senescence. Cell 113: 703–16
Ning Y, Xu JF, Li Y, Chavez L, Riethman HC, Lansdorp PM, Weng NP (2003) Telomere length and the expression of natural telomeric genes in human fibroblasts. Hum Mol Genet 12: 1329–36
Oberdoerffer P, Sinclair DA (2007) The role of nuclear architecture in genomic instability and ageing. Nat Rev Mol Cell Biol 8: 692–702
Ottaviani A, Gilson E, Magdinier F (2008) Telomeric position effect: From the yeast paradigm to human pathologies? Biochimie 90: 93–107
Park SK, Prolla TA (2005) Gene expression profiling studies of aging in cardiac and skeletal muscles. Cardiovasc Res 66: 205–12
Pedram M, Sprung CN, Gao Q, Lo AW, Reynolds GE, Murnane JP (2006) Telomere position effect and silencing of transgenes near telomeres in the mouse. Mol Cell Biol 26: 1865–78
Peng JC, Karpen GH (2008) Epigenetic regulation of heterochromatic DNA stability. Curr Opin Genet Dev 18: 204–11
Pifer J, Stephan RP, Lill-Elghanian DA, Le PT, Witte PL (2003) Role of stromal cells and their products in protecting young and aged B-lineage precursors from dexamethasone-induced apoptosis. Mech Ageing Dev 124: 207–18
Pisano S, Galati A, Cacchione S (2008) Telomeric nucleosomes: Forgotten players at chromosome ends. Cell Mol Life Sci
Pletcher SD, Macdonald SJ, Marguerie R, Certa U, Stearns SC, Goldstein DB, Partridge L (2002) Genome-wide transcript profiles in aging and calorically restricted Drosophila melanogaster. Curr Biol 12: 712–23
Rabbitts TH, Forster A, Baer R, Hamlyn PH (1983) Transcription enhancer identified near the human C mu immunoglobulin heavy chain gene is unavailable to the translocated c-myc gene in a Burkitt lymphoma. Nature 306: 806–9
Renauld H, Aparicio OM, Zierath PD, Billington BL, Chhablani SK, Gottschling DE (1993) Silent domains are assembled continuously from the telomere and are defined by promoter distance and strength, and by SIR3 dosage. Genes Dev 7: 1133–45
Riethman HC, Xiang Z, Paul S, Morse E, Hu XL, Flint J, Chi HC, Grady DL, Moyzis RK (2001) Integration of telomere sequences with the draft human genome sequence. Nature 409: 948–51
Rine J, Herskowitz I (1987) Four genes responsible for a position effect on expression from HML and HMR in Saccharomyces cerevisiae. Genetics 116: 9–22

Robyr D, Suka Y, Xenarios I, Kurdistani SK, Wang A, Suka N, Grunstein M (2002) Microarray deacetylation maps determine genome-wide functions for yeast histone deacetylases. Cell 109: 437–46
Rogakou EP, Boon C, Redon C, Bonner WM (1999) Megabase chromatin domains involved in DNA double-strand breaks in vivo. J Cell Biol 146: 905–16
Rossi DJ, Bryder D, Zahn JM, Ahlenius H, Sonu R, Wagers AJ, Weissman IL (2005) Cell intrinsic alterations underlie hematopoietic stem cell aging. Proc Natl Acad Sci USA 102: 9194–9
Rudolph KL, Chang S, Lee HW, Blasco M, Gottlieb GJ, Greider C, DePinho RA (1999) Longevity, stress response, and cancer in aging telomerase-deficient mice. Cell 96: 701–12
Sarg B, Koutzamani E, Helliger W, Rundquist I, Lindner HH (2002) Postsynthetic trimethylation of histone H4 at lysine 20 in mammalian tissues is associated with aging. J Biol Chem 277: 39195–201
Scaffidi P, Misteli T (2006) Lamin A-dependent nuclear defects in human aging. Science 312: 1059–63
Scaffidi P, Misteli T (2008) Lamin A-dependent misregulation of adult stem cells associated with accelerated ageing. Nat Cell Biol 10: 452–9
Schnabl B, Purbeck CA, Choi YH, Hagedorn CH, Brenner D (2003) Replicative senescence of activated human hepatic stellate cells is accompanied by a pronounced inflammatory but less fibrogenic phenotype. Hepatology 37: 653–64
Schoeftner S, Blasco MA (2008) Developmentally regulated transcription of mammalian telomeres by DNA-dependent RNA polymerase II. Nat Cell Biol 10: 228–36
Sharma GG, Hwang KK, Pandita RK, Gupta A, Dhar S, Parenteau J, Agarwal M, Worman HJ, Wellinger RJ, Pandita TK (2003) Human heterochromatin protein 1 isoforms HP1(Hsalpha) and HP1(Hsbeta) interfere with hTERT-telomere interactions and correlate with changes in cell growth and response to ionizing radiation. Mol Cell Biol 23: 8363–76
Sharp A, Robinson D, Jacobs P (2000) Age- and tissue-specific variation of X chromosome inactivation ratios in normal women. Hum Genet 107: 343–9
Sharp JA, Fouts ET, Krawitz DC, Kaufman PD (2001) Yeast histone deposition protein Asf1p requires Hir proteins and PCNA for heterochromatic silencing. Curr Biol 11: 463–73
Shcheprova Z, Baldi S, Frei SB, Gonnet G, Barral Y (2008) A mechanism for asymmetric segregation of age during yeast budding. Nature 454: 728–34
Shen S, Liu A, Li J, Wolubah C, Casaccia-Bonnefil P (2008) Epigenetic memory loss in aging oligodendrocytes in the corpus callosum. Neurobiol Aging 29: 452–63
Sinclair DA, Guarente L (1997) Extrachromosomal rDNA circles – a cause of aging in yeast. Cell 91: 1033–42
Singer MS, Kahana A, Wolf AJ, Meisinger LL, Peterson SE, Goggin C, Mahowald M, Gottschling DE (1998) Identification of high-copy disruptors of telomeric silencing in Saccharomyces cerevisiae. Genetics 150: 613–32
Sommer M, Poliak N, Upadhyay S, Ratovitski E, Nelkin BD, Donehower LA, Sidransky D (2006) DeltaNp63alpha overexpression induces downregulation of Sirt1 and an accelerated aging phenotype in the mouse. Cell Cycle 5: 2005–11
Stephan RP, Lill-Elghanian DA, Witte PL (1997) Development of B cells in aged mice: decline in the ability of pro-B cells to respond to IL-7 but not to other growth factors. J Immunol 158: 1598–609
Stone EM, Pillus L (1996) Activation of an MAP kinase cascade leads to Sir3p hyperphosphorylation and strengthens transcriptional silencing. J Cell Biol 135: 571–83
Taddei A, Gasser SM (2004) Multiple pathways for telomere tethering: functional implications of subnuclear position for heterochromatin formation. Biochim Biophys Acta 1677: 120–8
Teixeira MT, Gilson E (2005) Telomere maintenance, function and evolution: the yeast paradigm. Chromosome Res 13: 535–48
Tham WH, Wyithe JS, Ferrigno PK, Silver PA, Zakian VA (2001) Localization of yeast telomeres to the nuclear periphery is separable from transcriptional repression and telomere stability functions. Mol Cell 8: 189–99

Trojer P, Reinberg D (2007) Facultative heterochromatin: is there a distinctive molecular signature? Mol Cell 28: 1–13
Valdes AM, Andrew T, Gardner JP, Kimura M, Oelsner E, Cherkas LF, Aviv A, Spector TD (2005) Obesity, cigarette smoking, and telomere length in women. Lancet 366: 662–4
Valenzuela L, Kamakaka RT (2006) Chromatin insulators. Annu Rev Genet 40: 107–38
Vaquero A, Scher M, Lee D, Erdjument-Bromage H, Tempst P, Reinberg D (2004) Human SirT1 interacts with histone H1 and promotes formation of facultative heterochromatin. Mol Cell 16: 93–105
Wright WE, Shay JW (2002) Historical claims and current interpretations of replicative aging. Nat Biotechnol 20: 682–8
Ye X, Zerlanko B, Kennedy A, Banumathy G, Zhang R, Adams PD (2007) Downregulation of Wnt signaling is a trigger for formation of facultative heterochromatin and onset of cell senescence in primary human cells. Mol Cell 27: 183–96
Zhang H, Pan KH, Cohen SN (2003) Senescence-specific gene expression fingerprints reveal cell-type-dependent physical clustering of up-regulated chromosomal loci. Proc Natl Acad Sci USA 100: 3251–6
Zhang R, Adams PD (2007) Heterochromatin and its relationship to cell senescence and cancer therapy. Cell Cycle 6: 784–9
Zhang R, Chen W, Adams PD (2007) Molecular dissection of formation of senescence-associated heterochromatin foci. Mol Cell Biol 27: 2343–58
Zhang R, Poustovoitov MV, Ye X, Santos HA, Chen W, Daganzo SM, Erzberger JP, Serebriiskii IG, Canutescu AA, Dunbrack RL, Pehrson JR, Berger JM, Kaufman PD, Adams PD (2005) Formation of MacroH2A-containing senescence-associated heterochromatin foci and senescence driven by ASF1a and HIRA. Dev Cell 8: 19–30

Noncoding RNA for Presymptomatic Diagnosis of Age-Dependent Disease

Eugenia Wang

Abstract Age-dependent diseases generally involve long-term development of tissue system disorders; the late symptomatic manifestation is the culmination of numerous incremental programmatic shifts to dysfunctional states. Most current therapeutic treatments address illnesses characterizing the elderly symptomatically, largely attempting to reduce the severity of the disease rather than delaying or preventing its occurrence; this retrospective strategy is due to a serious lack of presymptomatic diagnosis. MicroRNAs have emerged recently as a major epigenetic factor controlling programmatic cell signaling controls and directing the optimal functionality of many tissues and their host cells. These small molecular species, while coding for no proteins, are key to coordinating systemic programming by controlling the expression levels of hundreds or thousands of genes via a negative regulatory mechanism, either by inhibiting their translation by binding at the 3′-untranslated region or degrading their mRNA messages by binding to their coding regions. These "genetic dimmer switches" are vital molecular program controls, exerting direct impact on target genes involved in cellular signaling pathways. Since microRNAs are necessary keys for the developmental construction of signaling pathways essential for neurons, cardiomyocytes, hepatocytes, etc., changes in microRNA expression, and the profiling of these changes qualitatively and quantitatively, may prove to be useful indices as presymptomatic diagnostic markers. The availability of these indices in the future should provide a jump start for preventive therapy for age-dependent diseases and frailty.

Keywords noncoding RNA · Post-transcriptional control · Transcription factor · Translation inhibition

E. Wang (✉)
Gheens Center on Aging and Department of Biochemistry and Molecular Biology, University of Louisville School of Medicine, Louisville, KY 40202, USA
e-mail: eugenia.wang@louisville.edu

T.O. Tollefsbol (ed.), *Epigenetics of Aging*, DOI 10.1007/978-1-4419-0639-7_10,

Introduction

Most age-dependent diseases are polygenic in nature and develop over a period of time. The ultimate symptomatic manifestation of such diseases is usually the "tip of the iceberg," with both genetic and environmental risk factors accumulating throughout the lifetime. Thus, early onset of a particular age-dependent debility may be the combination of "bad" genes and "bad" environmental factors. On the other hand, long-lived centenarians enjoy "good" genes and "good" environments, which may slow down and delay the onset of diseases till 100+ years old. A practical goal for healthy aging is to narrow the gap shown in Fig. 1 between normal aging and long-lived centenarians to the minimum and provide the maximal benefit to the majority of our population to live beyond 90 years disease-free.

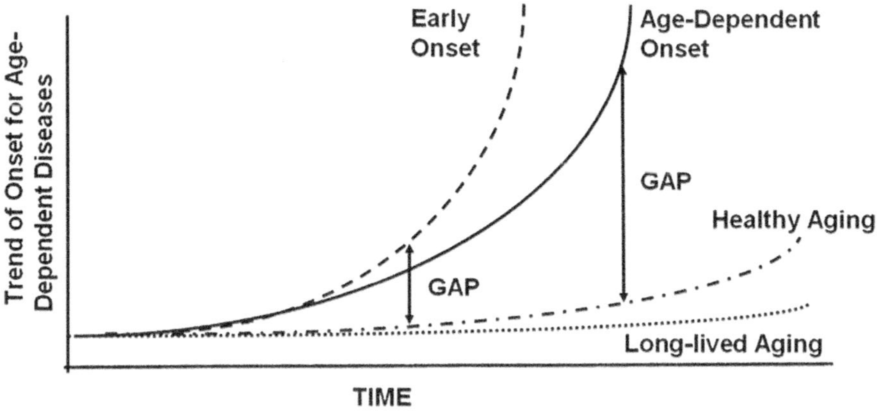

Fig. 1 "Gaps" between early and age-dependent disease onset, healthy aging, and individuals with extreme long-lived phenotype

At the system level, development and aging are at opposite ends of the spectrum; the former aims to establish an orderly hierarchy for constructing tissue and cellular functions, while the latter erodes this order by both programmatic and stochastic ("wear-and-tear" cumulative environmental insults) degradation. Obviously, the stronger the construction of the signaling pathways operating in individual tissues and cells, the more difficult the erosion. Likewise, when the impact from erosion is less, the original signaling pathways will be sustained and operational longer. Evaluating the strength of pathway construction and identifying the commencement of erosion are two daunting and challenging mysteries for us to solve.

Emerging results show that early childhood experience may determine late-life disease (Kuh 2006; Yi et al. 2007; Zhang et al. 2008). A special example of this is the discordance of colon cancer incidence among monozygotic twins reared apart: identical genetic endowment but different environments in the first 2 years of life may predestine one twin to be a cancer victim 60 years later, while the identical sibling remains healthy till 90 (Kaprio et al. 1987; Marcus et al. 1999; Fraga et al. 2005). This example may demonstrate a defective pathway construction in early childhood, setting the stage for cancer development later on. Once the signaling pathways are set in place during development, they are subjected to the impact of erosion during adult life. Obviously, qualitative and quantitative degrees of this erosion impact determine the threshold level tolerated by our system. Thus, an individual's life span may be illustrated by a series of overlapping equilibrium dynamics, with birth as the starting point, depicted by putatively perfect network equilibrium, and ending with the dismantling of equilibrium dynamics beyond repair at the end of the life span.

Midlife Deregulation of Programmatic Control

Although the fact may not be obvious or comforting, the root causes of age-dependent diseases such as Alzheimer's disease, osteoporosis, cancer, and cardiovascular disorders may be found decades earlier than the manifestation of symptoms. The initial early warning signs may be upstream malfunctions at physiological levels seemingly unrelated to the ultimate disease presentation. For example, kidney filtration rate, cardiac pumping efficiency, useful lung volume, and maximal breathing capacity all start to decline by the age of about 40 and continue to decline throughout later life (Schulz and Salthouse 1999). These physiological changes, which may contribute to the evolution of disease, are themselves caused by deleterious operations of many interacting biochemical reactions within the cells; it would be a daunting task to identify the initial molecular derailments leading to the eventual phenotypic disorders. For example, a recent finding shows that individuals manifesting high cholesterol levels in their forties are more likely to develop Alzheimer's disease (Solomon et al. 2008). Hypercholesterolemia can damage our arteries; the link to the etiology of Alzheimer's disease (AD) may reside in the production of amyloid plaques, although proving such a link may depend on future findings. Lowering cholesterol levels in the forties is probably beneficial not only for reducing the Alzheimer's disease risk but also for a whole host of elderly frailties, from AD to cardiovascular disorders, because of the complex underpinning contributing factors.

MicroRNA and Epigenetic Regulation

Noncoding RNAs include short RNA species that do not code for any specific proteins but function in post-transcriptional control of other genes' expression. A recent explosion of scientific attention is focused upon the microRNAs, in general only ~22 base pairs in length. Their unique sequences are complementary to either the 3′-untranslated (UTR) or the coding regions of their target genes; the former modality results in inhibition of translation, while the latter mode of binding results in mRNA degradation. In general, microRNAs are found in all living cells, from plants to humans, as negative regulators of gene expression at the post-transcriptional level (Finnegan and Matzke 2003; Lewis et al. 2003; Bartel 2004; Mansfield et al. 2004; Fazi and Nervi 2008). Most plant microRNAs function by binding to the target coding region, thus degrading the target gene's message, while animal microRNAs primarily bind to 3′-UTR regions with a signature "seed" sequence of eight nucleotide bases. It is predicted that some 1,200 human microRNAs may exist; more than 800 of them have actually been identified and sequence mapped already. Mammalian microRNAs are essential for the developmental regulation of cell lineages; several dominant microRNAs are keys to specific differentiation pathways. An example of microRNAs determining cell fate during development is microRNA *Let7*; a null mutant of *Let7* is embryonically lethal in *Caenorhabditis elegans* (Ruvkun et al. 2004). However, not all microRNAs function in this absolute fashion; thus we suggest that according to their functions, microRNAs may be classified into two groups: Case 1 microRNAs are those producing directly reciprocal levels of expression of their target genes; for example, microRNA 24 totally shuts down nonneuronal gene expression to allow neural cell differentiation. Case 2 microRNAs, on the other hand, are those operating at the level of equilibrium homeostasis, allowing balanced expression between themselves and their target, a "dimmer switch" effect.

During development, this type of microRNA function may be thought of as "canalization" (Waddington 1942; Siegal and Bergman 2002; Stearns 2002; Hornstein and Shomron 2006), allowing channels of cell lineage to be established and defined. Case 1 microRNAs allow the formation of the cell types needed for each tissue, while Case 2 microRNAs maintain system needs for each cell in its final differentiation destiny, controlling cell survival, oxidative defense, DNA repair, protein degradation, etc.

In principle, a single microRNA may have several dozens or hundreds of target genes, and vice versa, a single gene may be targeted by several microRNAs. This complexity constitutes microRNAs as the most versatile control of gene expression at the post-transcriptional level. Moreover, several microRNAs may be regulated as a group, thus exerting programmatic control for a given cellular state. For example, both *Let7* and *miR-34a* are downregulated in cancer cells, allowing the upregulation of their target gene, *H-Ras* and other oncogenes, etc (Yu et al. 2007; Büssing et al. 2008; Chan et al. 2008; Esquela-Kerscher et al. 2008; Kato and Slack 2008).

What controls microRNA up- or downregulation? Emerging findings show that microRNAs are activated by their own promoters and that often their target genes acting as the transcription activators are the exact factors binding these promoters and activating microRNA expression (Sylvestre et al., 2007). Thus, transcription factors targeted for silencing by specific microRNAs function as a feedback loop mechanism, suggesting "seesaw" regulation between these two molecular species. As depicted in Fig. 2 for a hypothetical apoptosis signaling pathway, three different transcriptional factors may control three different microRNAs, each responsible for a segment of this signaling pathway, and any two of them may form different compartments that eventually shape the apoptosis signaling pathway in its totality. Future experiments will reveal whether most microRNAs are activated via their own target's binding at the promoter region such that when too much transcriptional factor is present, it activates its own silencing microRNAs, with the end result of suppressing its own gene expression.

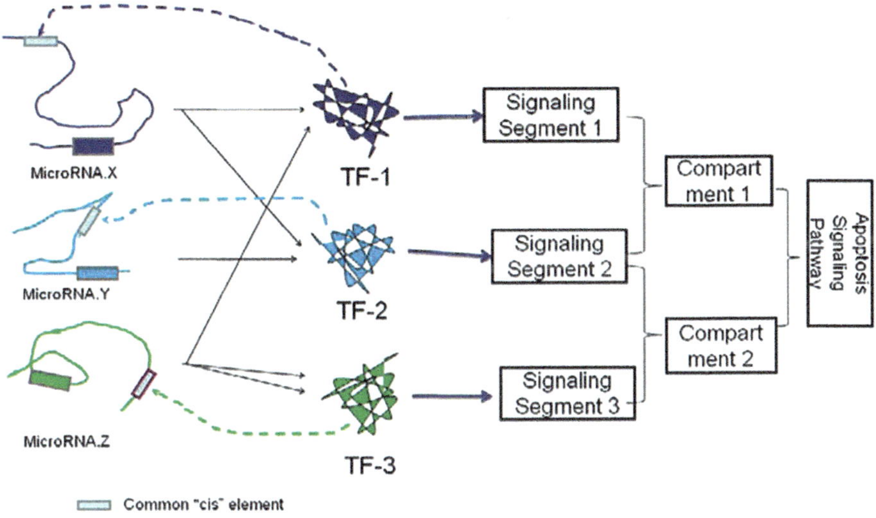

Fig. 2 Depiction of the feedback loop paradigm: *cis* elements for the promoter region of three microRNAs (X, Y, Z) and three transcriptional factors (TF 1, 2, and 3) together form an apoptosis pathway

Three forms of miRNAs exist in vivo: primary (pri-miRNA), precursor (pre-miRNA), and mature miRNAs. Pre-miRNAs are 60–110-nt-long molecules characterized by a folded-back hairpin structure; these precursors of mature miRNAs originate from longer Pri-RNAs, which are transcribed in the nucleus, cleaved by an RNase called Drosha, and then transported into the cytoplasm. Pre-miRNAs are further sheared by Dicer into 19–22-nt duplex molecules that function as mature miRNAs. The duplexes are subjected to further treatment in RNA-induced silencing complexes (RISC) and become two separated single strands, one of which is active and engaged in imperfect/perfect base pairing with specific sequences in target mRNAs (Lee et al. 1993; Reinhart et al. 2000).

MicroRNA genes are alleged with increasing evidence to regulate more than 25% of gene transcription (Lim et al. 2003). Since microRNA binding to specific target sites is sequence defined, it is especially vital in determining the exact action of microRNA functionality. Target binding sequence mutation may disable microRNAs from binding to their target gene's UTR region, thus reducing their repressing power. Many reports show that single nucleotide polymorphisms (SNPs) at either target binding sites or the microRNAs' own sequences may disable the post-transcriptional silencing of key oncogenes in cancer. Moreover, quantitative trait locus (QTL) mapping identifies mutations at some microRNA genomic locations, which result in changes in expression. For example, recent evidence shows that a miRNA mutation is responsible for the phenotype of muscularity in Texel sheep of Belgian origin (Clop et al. 2006); this observation provides a new way to search for mutations in numerous human genetic diseases that present no evidence of mutation in promoter regions, coding areas, or slicing sites of the genes themselves, but rather perhaps in the microRNAs controlling their regulation.

MicroRNAs as "Hubs" for Programmatic Control of Signaling Pathways

Much of the power of microRNAs stems from their multitarget function; like transcription factors, their molecular actions can regulate the expression of hundreds of genes. However, differing from transcription factors, microRNAs work at the post-transcriptional level by inhibiting translation or degrading messages. Moreover, target genes may be functionally related to a common signaling network. For example, as Fig. 3 depicts, the target genes of human miRNA *Let7*e can be organized into a signaling network for survival/apoptosis; the key nodes of this signaling pathway are represented by *Bcl2L1*, *Caspase 3*, *Stat3*, *IL6*, *IL10*, *PAX3*, *RB1*, etc. This figure represents a signaling pathway constructed by a partial list of genes identified by sequence homology as bearing *Let7e* target sites in their 3′-UTR regions; it is merely a theoretical model, by no means the actual working network for a particular cell's functional operation. To arrive at such a picture, each target must be verified not only by microRNA binding assays but also by determining the actual

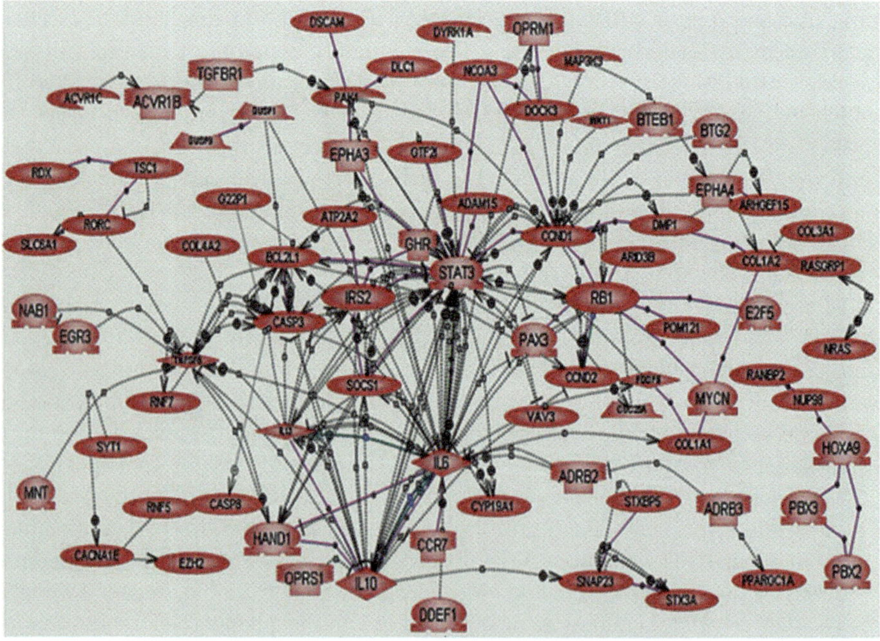

Fig. 3 Partial target genes of *Let-7e* and their predicted signaling network

silencing effects of *Let7*e on the potential targets. Moreover, as described above, individual microRNAs are likely controlled by feedback loops of their own targets, as with transcription factors. For example, *STAT3* is likely to be a candidate factor providing feedback to suppress *Let7e*'s own expression via its own promoter binding action. Finally, the signaling pathway presented in Fig. 3 is constructed with all of the target genes; needless to say, in any given cell at any time, not all of them are expressed, and therefore not all are present to participate in this network. Nevertheless, the complexity of microRNA regulation provides a picture that microRNA negatively regulates gene expression to accomplish control of a program for gene expression, rather than single gene action. This programmatic control may be illustrated by a circuit breaker controlling electrical supply to a section of a house, rather than an on/off switch controlling a single light bulb. This "circuit breaker" versus "single light bulb" analogy may provide the understanding that individual proteins, functioning in a signaling network catering to a specific cellular operation such as oxidative defense, may have underpinning regulation, enabling them to be expressed at either full or partial intensity, depending upon the level of defense required. Moreover, control via microRNAs provides a safeguard to prevent unwanted gene expression accidentally becoming activated by binding of a transcriptional factor. Thus the feedback loop between microRNA and its target transcriptional factors reflects nature's exquisite control of system-based regulation of individual gene expressions.

MicroRNAs as Biomarkers for Presymptomatic Diagnosis

As described earlier, individual microRNAs may be functionally categorized as underlying "hubs" controlling a group of genes operating in a single signaling pathway; gradually, microRNAs may be classified and associated with various disease states. Such a nomenclature is already in use; for example, *microRNA 34a* is largely considered as a "cancer microRNA," *microRNA 155* is considered as immune regulation based (Gottwein et al. 2007; Yin et al. 2008).

May there then be "longevity microRNAs?" The answer is certainly "yes!" But one must differentiate between life span and health span; it is not a generally desirable outcome to lengthen one without the other. In general, age-dependent frailty stems from micromolecular disorders involved in the regulation of DNA repair, oxidative defense, heat-shock response, protein degradation processing, control of intermediate metabolism, maintenance of genomic stability, cytoskeletal organization, cell-cycle regulation, etc. Signaling pathways controlling these cellular mechanisms may be interconnected by cross talk; numerous examples demonstrate that during aging, they become attenuated, misfire, or lose their ability to repair themselves. Until recently, research efforts have largely been directed to link these signaling disorders to one single "master" gene or one "dominant" pathway. Among these, the most noted is the IGF signaling pathway, remarkably identified as vital to the long-lived phenotype in diverse organisms from *C. elegans* to mouse and maybe even to nonhuman primates. Notwithstanding the success in this area, neither this pathway nor genes such as *IGF-1*, its receptors, and/or its many downstream factors operate in isolation. Rather, each may be part of a huge orchestration of complex signaling networks with cross-over, feedback loops, and overlapping operation. As complex as this scenario may be, it may be controlled by a few microRNAs serving as the underpinning "hubs" for most, if not all, of the signaling pathways for general cellular maintenance and well-being, as described above. Thus, specific microRNAs controlling DNA repair, oxidative defense, etc., may be viewed as longevity-determining microRNAs, as described above, maintaining the well-being of cells. Simplistically speaking, the microRNAs controlling these processes are universal to all cell types; other microRNAs may be added to this list depending on the specific functional needs of different cell types. In the case of neurons, cytoskeletal organization involving neurofilament and microtubule organization and synaptic junctions may involve the additional tissue-specific microRNAs required for their vital function. For colonic epithelial cells, cell-cycle control may be vital to regulate the precise proliferation program progressing from the colonic crypts to the apices of the villi; excessive proliferation would constitute hyperplasia of these cells, a prelude to cancer. *microRNA 34a* (Lodygin et al. 2008), controlling cell proliferation, is noted as a "tumor suppressor microRNA" (Medina and Slack 2008), and thus a "Case II" microRNA controlling the program of cell proliferation in colonic epithelial cells.

Since microRNAs are vital to the maintenance of cellular physiological status, their dysregulation becomes an obvious lead for identifying signaling disorders at the system level. Moreover, changes in microRNA expression may be the prelude to a programmatic shift of the entire system. Therefore, as described in Fig. 4, identifying microRNA changes in the presymptomatic stage may provide

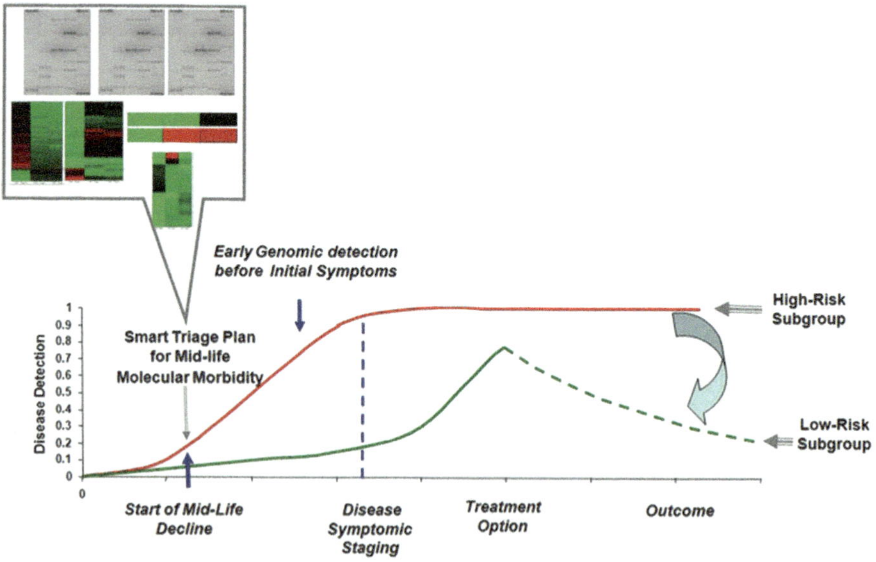

Fig. 4 MicroRNAs as a presymptomatic diagnostic for age-dependent diseases; microRNAs for high- and low-risk individuals' signatures

the advance diagnosis that we need to intervene in age-dependent diseases. Our recent results in aging mouse liver show that upregulation of several key microRNAs is reflected in corresponding downregulation of genes involved in DNA repair, intermediate metabolism, cell-cycle control, etc. (Maes, et al. 2008). Moreover, these changes are noticeable even at middle age and thus suggest that presymptomatic diagnosis is possible once we have identified lead microRNAs vital for signaling processes regulating DNA repair, oxidative stress, heat-shock response, etc. (Schipper et al. 2007; Maes et al. 2008; Medina and Slack 2008; Wang, Liang and Lu 2008).

Noncoding RNA: From Diagnosis to Therapeutic Treatment

Although qualitative and/or quantitative changes in microRNA expression may be used as presymptomatic diagnostics for systemic disorders of their target signaling pathway(s), and characterized as responsible for specific disease phenotypes, the challenge remains to investigate the functional links between specific miRNAs and their targets. Successful evidence of this sort of functional link will strengthen the notion that the lead microRNAs identified as diagnostic tools for a particular disease or disorder may themselves serve as precise leads to novel therapeutic approaches. Emerging success with animal models demonstrates this as the precise scenario involved; examples include *microRNA 375*, which regulates glucose metabolism and thus controls the diabetic course (El Ouaamari et al. 2008; Hennessy and O'Driscoll 2008), and *microRNA 21*, which regulates cardiac hypertrophy (Mann 2007; van Rooij and Olson 2007). Future work in transgenic mice, with key microRNAs conditionally activated (knocked in or out) during middle age, may provide essential animal models to investigate microRNA control of age-dependent diseases and allow us to study the presymptomatic application of microRNA control during midlife for late-life disease (Wang 2007).

Challenges of MicroRNAs as Presymptomatic Biomarkers: From Cultured Cell and Animal Model Studies to Human Application

The obvious path for translational study is to apply successes in laboratory discovery to clinical settings; this lengthy road is littered with frustrated, truncated developments. These incomplete journeys reaching clinical trials are due to several intrinsic problems built into our current technological approaches, including the following: (1) cultured cell models generally involve two-dimensional, monoculture experimental conditions testing systems composed of multiple interacting cell types in three-dimensional tissue operations; (2) most animal models are inbred strains for laboratory use, with an atypical genetic background; and (3) many human diseases do not occur in mice. Notwithstanding these problems, cell culture and animal models provide necessary tools, technology, and insights and will continue to reveal new avenues of diagnosis and therapeutic leads.

MicroRNA changes may prove to be the roots for prevalent, late-life, age-dependent diseases, the result of midlife dysregulation at the microRNA expression level. Slowing, reducing, and overriding this molecular degeneration in people in their forties and fifties may halt degeneration by the time they reach their sixties or seventies and beyond. Studying programmatic molecular changes at middle age may provide urgently needed presymptomatic diagnosis for late-life diseases, with the aim of gaining healthy elderly life for the baby boomer generation and their children in the decades to come.

Acknowledgments The author appreciates the comments and copy editing of Mr. Alan N. Bloch. Dr. Wang is supported by funding from the Gheens Foundation and a Kentucky "Bucks-for-Brains" endowed chair.

References

Bartel, D.P. 2004. MicroRNAs: genomics, biogenesis, mechanism, and function. Cell 116: 281–7.

Büssing, I., Slack, F.J., and Großhans, H. 2008. let-7 microRNAs in development, stem cells and cancer. Trends Mol Med 14(9): 400–9.E-pub Jul 30.

Chan, S.P., Ramaswamy, G., Choi, E.Y., and Slack, F.J. 2008. Identification of specific let-7 microRNA binding complexes in Caenorhabditis elegans. RNA 14 (10):2104–14. E-pub Aug 21.

Clop, A., Marcq, F., Takeda, H., Pirottin, D., Tordoir, X., Bibé, B., Bouix, J., Caiment, F., Elsen, J.M., Eychenne, F., Larzul, C., Laville, E., Meish, F., Milenkovic, D., Tobin, J., Charlier, C., and Georges, M. 2006. A mutation creating a potential illegitimate microRNA target site in the myostatin gene affects muscularity in sheep. Nat Genet 38 (7): 813–8.

El Ouaamari, A., Baroukh, N., Martens, G.A., Lebrun, P., Pipeleers, D., and Van Obberghen, E. 2008. miR-375 targets PDK1 and regulates glucose-induced biological responses in pancreatic β-cells. Diabetes 57 (10): 2708–17. E-pub Jun 30.

Esquela-Kerscher, A., Trang, P., Wiggins, J.F., Patrawala, L., Cheng, A., Ford, L., Weidhaas, J.B., Brown, D., Bader, A.G., and Slack, F.J. 2008. The let-7 microRNA reduces tumor growth in mouse models of lung cancer. Cell Cycle 7 (6):759–64.

Fazi, F. and Nervi, C. 2008. MicroRNA: basic mechanisms and transcriptional regulatory networks for cell fate determination. Cardiovasc Res 79 (4):553–61.

Finnegan, E.J. and Matzke, M.A. 2003. The small RNA world. J Cell Sci 116: 4689–93.

Fraga, M.F., Ballestar, E., Paz, M.F., Ropero, S., Setien, F., Ballestar, M.L., Heine-Suñer, D., Cigudosa, J.C., Urioste, M., Benitez, J., Boix-Chornet, M., Sanchez-Aguilera, A., Ling, C., Carlsson, E., Poulsen, P., Vaag, A., Stephan, Z., Spector, T.D., Wu, Y.Z., Plass, C., and Esteller, M. 2005. Epigenetic differences arise during the lifetime of monozygotic twins. Proc Natl Acad Sci USA 102 (30): 10604–9.

Gottwein, E., Mukherjee, N., Sachse, C., Frenzel, C., Majoros, W.H., Chi, J.T., Braich, R., Manoharan, M., Soutschek, J., Ohler, U., and Cullen, B.R. 2007. A viral microRNA functions as an orthologue of cellular miR-155. Nature 450 (7172): 1096–9.

Hennessy, E. and O'Driscoll, L. 2008. Molecular medicine of microRNAs: structure, function and implications for diabetes. Expert Rev Mol Med 10: e24.

Hornstein, E. and Shomron, N. 2006. Canalization of development by microRNAs. Nat Genet 38: S20–S24.

Kaprio, J., Alanko, A., Kivisaari, L., and Standertskjöld-Nordenstam, C.G. 1987. Mammographic patterns in twin pairs discordant for breast cancer. Br J Radiol 60 (713): 459–62.

Kato, M. and Slack, F.J. 2008. microRNAs: small molecules with big roles – C. elegans to human cancer. Biol Cell 100 (2): 71–81.

Kuh, D. 2006. A life course perspective on telomere length and social inequalities in aging. Aging Cell (6): 579–80.

Lee, R.C., Feinbaum, R.L., and Ambros, V. 1993. The *C. elegans* heterochronic gene lin-4 encodes small RNAs with antisense complementarity to lin-4. Cell 75: 843–54.

Lewis, B.P., Shih, I-H., Jones-Rhoades, M.W., Bartel, D.P., and Burge, C.B. 2003. Prediction of mammalian microRNA targets. Cell 115: 787–98.

Lim, L.P., Glasner, M.E., Yekta, S., Burge, C.B., and Bartel, D.P. 2003. Vertebrate microRNA genes. Science 299: 1540.

Lodygin, D., Tarasov, V., Epanchintsev, A., Berking, C., Knyazeva, T., Körner, H., Knyazev, P., Diebold, J., and Hermeking, H. 2008. Inactivation of miR-34a by aberrant CpG methylation in multiple types of cancer. Cell Cycle 7 (16): 2591–600. E-pub Aug. 1.

Maes, O.C., An. J., Sarojini, J., and Wang, E. 2008. Murine MicroRNAs implicated in liver functions and aging process. Mech Ageing Dev 129: 534–41.

Mann DL. 2007. MicroRNAs and the failing heart. N Engl J Med 356 (25): 2644–5.

Mansfield, J.H., Harfe, B.D., Nissen, R., Obenauer, J., Srineel, J., Chaudhuri, A., Farzan-Kashani, R., Zuker, M., Pasquinelli, A.E., Ruvkun, G., Sharp, P.A., Tabin, C.J., and McManus, M.T. 2004. MicroRNA-responsive 'sensor' transgenes uncover Hox-like and other developmentally regulated patterns of vertebrate microRNA expression. Nature Genetics 36 (10):1079–83 (Erratum, 36(11): 1238. Comment, 36(10): 1033–4.)

Marcus, D.M., Papastergiou, G.I., Patel, M., Pandya, A., and Brooks S.E. 1999. Discordant retinoblastoma in monozygotic twins. Am J Ophthalmol 128 (4): 524–6.

Medina, P.P. and Slack, F.J. 2008. microRNAs and cancer: An overview. Cell Cycle 7 (16): 2485–92.

Reinhart, B.J., Slack, F.J., Basson, M., Pasquinelli, A.E., Bettinger, J.C., Rougvie, A.E., Horvitz, H.R., and Ruvkun, G. 2000. The 21-nucleotide let-7 RNA regulates developmental timing in *Caenorhabditis elegans*. Nature 403: 901–6.

Ruvkun, G., Wightman, B., and Ha, I. 2004. The 20 years it took to recognize the importance of tiny RNAs. Cell S116: S93–S96.

Schipper, H.M., Maes, O.C., Chertkow, H.M., and Wang, E. 2007. MicroRNA expression in Alzheimer blood mononuclear cells. Gene Regul Sys Biol 1: 263–74.

Schulz, R. and Salthouse, T.A. 1999. Adult Development and Aging: Myths and Emerging Realities (3rd ed.) Upper Saddle River NJ: Prentice-Hall. (especially figure on pg 51, modified from Spirduso, W.W. 1995. Physical Dimensions of Aging. Champaign, IL: Human Kinetics.)

Siegal, M.L. and Bergman, A. 2002. Waddington's canalization revisited: developmental stability and evolution. Proc Natl Acad Sci USA 99 (16): 10528–32.

Solomon, A., Kivipelto, M., Zhou, J., and Whitmer, R.A. 2008. Midlife serum total cholesterol and risk of Alzheimer's disease and vascular dementia three decades later. Poster P04.067, American Academy of Neurology 60th Anniversary Annual Meeting, Chicago, April 12–19.

Stearns, S. C. 2002. Progress on canalization. Proc Natl Acad Sci USA 99(16): 10229–30.

Sylvestre, Y., deGuire, V., Querido, E., Mukhopadhyay, U.L., Bourdeau, V., Major, F., Ferbeyre, G., and Chartrand, P. 2007. An E2F/miR-20a autoregulatory feed-back loop. J Biol Chem 282(4): 2135–45.

van Rooij, E. and Olson, E.N. 2007. microRNAs put their signatures on the heart. Physiol Genomics 31(3): 365–6.

Waddington, C.H. 1942. Canalization of development and the inheritance of acquired characters. Nature 150(3811): 563–5.

Wang, E. 2007. MicroRNA, the putative molecular control for mid-life decline. Aging Res Rev 6(1):1–11.

Wang, Y., Liang, Y., and Lu, Q. 2008. MicroRNA epigenetic alterations: predicting biomarkers and therapeutic targets in human diseases. Clin Genet 74(4): 307–15. E-pub Aug 18.

Yi, Z., Gu, D., and Land, K.C. 2007. The association of childhood socioeconomic conditions with healthy longevity at the oldest-old ages in China. Demography 44(3):497–518. (Erratum in Nov;44(4) following table of contents.)

Yin, Q., Wang, X., McBride, J., Fewell, C., and Flemington E. 2008. B-cell receptor activation induces BIC/miR-155 expression through a conserved AP-1 element. J Biol Chem. 283(5): 2654–62.

Yu, F., Yao, H., Zhu, P., Zhang, X., Pan, Q., Gong, C., Huang, Y., Hu, X., Su, F., Lieberman, J., and Song, E. 2007. let-7 regulates self renewal and tumorigenicity of breast cancer cells. Cell 131(6):1109–23.

Zhang, Z., Gu, D., and Hayward, M.D. 2008. Early life influences on cognitive impairment among oldest old Chinese. J Gerontol B Psychol Sci Soc Sci 63(1):S25–33.

Part III
Impact of Epigenetics on Aging

Telomerase Control by Epigenetic Processes in Cellular Senescence

Huaping Chen and Trygve O. Tollefsbol

Abstract Cellular senescence is a controversial process that can prevent age-related diseases such as cancer while also promoting the aging process. A number of genes, such as those of oncogenes and tumor suppressors, have been shown to be of high importance in this process. Telomerase is a crucial enzyme that plays a pivotal role in cellular senescence by maintaining the stability of the genome. Five components of telomerase have been discovered so far, namely hTERT which is the catalytic subunit of telomerase, hTR which provides the RNA template for hTERT, a protein termed dyskerin which binds to hTR, and two additional proteins termed pontin and reptin which can assist the assembly of telomerase. Genetic control of these genes in cellular senescence has been investigated widely and many advances have been made in understanding the basis of the control of these genes. Epigenetic processes are important mechanisms that regulate the expression of genes and epigenetic control has also been shown to play an important role in cellular senescence. Our current insights on telomerase control by epigenetic processes in cellular senescence are reviewed in this chapter.

Keywords Telomerase · Epigenetic · Cellular senescence

Introduction

Like most other organisms, humans age and this is often accompanied by a series of age-related disease such as cancer. Many have long been fascinated with the causes for these processes. With decades of scientific research, a major biological process underlying this phenomenon has been revealed, which is a process termed cellular senescence. Every human is developed from a fertilized egg which undergoes

T.O. Tollefsbol (✉)
Center for Aging and Department of Biology, 175 Campbell Hall,
1300 University of Alabama at Birmingham, AL 35294–1170, USA
e-mail: trygve@uab.edu

T.O. Tollefsbol (ed.), *Epigenetics of Aging*, DOI 10.1007/978-1-4419-0639-7_11,

cell differentiation and division. It is inevitable that those differentiated cell types will suffer from factors involving the environment and potential mutation mistakes generated in DNA replication, which may potentially cause cancer transformation. However, cells have evolved a defense system to respond to those signals through DNA damage signals which can then arrest cell division or induce apoptosis, thereby preventing the occurrence of adverse biological processes such as cancer transformation. Cellular senescence is the mechanism that the defense system is based on; it is characterized by irreversible proliferative arrest. However, cellular senescence is a double-edged sword in that it has also been proposed as a direct cause of aging (Campisi and d'Adda di Fagagna, 2007). p16, a protein that can irreversibly induce cellular senescence, has been shown to be upregulated with increased age of stem and progenitor cells of the mouse brain, bone marrow, and pancreas (Campisi and d'Adda di Fagagna, 2007). Since senescent cells cannot be replaced in tissues, their number increases with age. In addition, they can secrete factors that may promote growth and angiogenic activity of nearby premalignant cells (Bavik et al., 2006). Moreover, these factors may inhibit the function of normal cells and limit the regenerative potential of stem cell pools, thereby contributing to the aging process of organisms (Campisi, 2005).

Pathways Involved in Cellular Senescence

Cellular senescence can be triggered by several different factors, which include dysfunctional telomeres, nontelomeric DNA damage signals, and stress and strong mitogenic signals (Fig. 1). Once those signals have been generated, corresponding cell signaling cascades will be activated, and the cells may enter into senescence or undergo apoptosis (Collado et al., 2007).

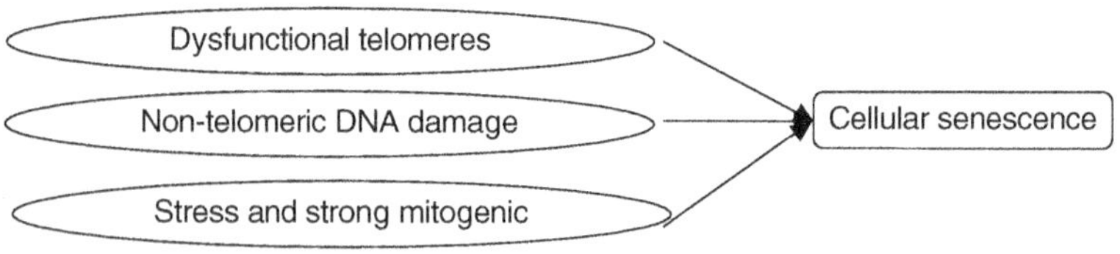

Fig. 1 Factors that can trigger cellular senescence. Dysfunctional telomeres, nontelomeric DNA damage, stress and strong mitogenic signals are three major factors that induce cellular senescence

Telomere attrition has been considered as the most important mechanism that leads to the senescence of normal somatic cells in terms of proliferation. Due to the end replication problem of linear chromosomes, cells will lose about 50–150 bp of sequence in the 5′-end of DNA strands (Harley et al., 1990; Hastie et al., 1990). After about 50 population doublings, telomeres will reach a critical length and the corresponding telomere structure will be changed, resulting in a classical DNA damage response (DDR) (d'Adda di Fagagna et al., 2003; Gire et al., 2004; Herbig et al.,

2004; Takai et al., 2003). A number of proteins have been indicated to be involved in this pathway. DNA damage foci found in those senescent cells contain telomeres, indicating the similar characteristic of dysfunctional telomeres and double-strand breaks (DSBs) (d'Adda di Fagagna et al., 2003). These signals may activate the *p53* gene, which can suppress tumor formation, and induce cells to enter into senescence. Inactivation of *p53* and *Rb* would allow further division of cells that may enter into crisis characterized by end-to-end chromosome fusions and apoptosis of cells. Very few cells ($\sim$1 in 10^7) may emerge from those cells by expressing telomerase to maintain telomeres.

For nontelomeric DNA damage signals, the same pathway may be activated to induce cells to enter into senescence. As to stress and strong mitogenic signals, overexpression of oncogenes such as *RAS* or suboptimal culture conditions could cause high expression of *p16*, further activating the tumor suppressor gene *RB*, which together with *E2F1* can repress a number of target genes that can cause chromatin reorganization and induce cells to enter into senescence.

Telomeres and Telomerase

Telomeres are special structures capped at the end of linear eukaryotic chromosomes, which can protect the natural ends of chromosomes from degradation or fusing with each other and can avoid recognition and processing of the end of chromosomes as DSBs. Mammalian telomeric DNA sequences contain tandem repeats of $(TTAGGG/CCCTAA)_n$ (Moyzis et al., 1988), which ends with a single-strand G-rich overhang at the 3′-end, and this overhang has been proposed to insert into the double-stranded region of telomeric DNA to form a displacement loop referred to as the "T-loop" (Griffith et al., 1999). The length of telomere repeats is highly variable among different species, and studies indicate that the number of this repeat sequence is rather different for chromosomes of individual cells in humans and mice (Baird et al., 2003; Griffith et al., 1999; Lansdorp et al., 1996; Zijlmans et al., 1997).

Mammalian TTAGGG telomere repeats are bound by a multiprotein complex termed telosome or shelterin (de Lange, 2005; Liu et al., 2004a). These proteins include the POT1/TPP1 heterodimer (which can bind to the G-strand overhang; Baumann and Cech, 2001; Liu et al., 2004b; Ye et al., 2004), TRF1 and TRF2 (which can bind to double-stranded repeats through their Myb-domain), RAP1 (repressor–activator protein 1), and TIN2 (TRF1-interacting nuclear factor 2, which can bind to TRF1 and TRF2; de Lange, 2005; Liu et al., 2004a). TANK1 and TANK2 poly(ADP)-ribosylates (also known as tankyrases) have also been indicated to interact with TRF1 (Smith et al., 1998). Among those proteins, TRF1 and TRF1-associated proteins can negatively regulate telomere length by controlling the access of telomerase to telomeres (de Lange, 2005), while TRF2 and POT1 can protect telomeres from end-to-end chromosome fusions by interacting with DNA damage signaling and repair factors (Celli and de Lange, 2005).

As has been mentioned above, the cell will lose up to 150 bp after each cell division due to the replication problem of linear chromosome at 5′ ends (Harley et al., 1990; Hastie et al., 1990); however, the repeated sequences in telomeres which do not encode genes can buffer this loss of genomic sequence at the end of chromosomes. Telomeres thus have become a mitotic clock for normal somatic cells. When telomeres have reached a critical length, DNA damage signals will be generated, and cellular senescence or apoptosis can be induced. For those cells such as stem cells, progenitor cells, and germline cells, which need to function by rapid unlimited division, telomerase adds hexamer repeats to the 3′-end of telomeres to compensate for the loss of telomeric DNA sequences (Autexier and Lue, 2006; Collins, 2006).

Five components of telomerase have been discovered so far, namely hTERT, which is the catalytic subunit of telomerase (Harrington et al., 1997; Meyerson et al., 1997; Nakamura et al., 1997) and is located on chromosome 5p15.33 (Bryce et al., 2000); hTERC, which provides an RNA template for hTERT (Feng et al., 1995); a protein termed dyskerin, which binds to hTERC; and two other proteins termed pontin and reptin, which can help the assembly of telomerase. Mutations of those components have been linked with congenital dyskeratosis characterized by mucocutaneous features and a number of other somatic abnormalities, including early graying, dental loss, bone marrow failure, osteoporosis, and an increased risk of malignancy (Vulliamy and Dokal, 2008).

Telomerase has been shown to be downregulated when cells undergo differentiation, and it is reactivated in almost all the cancer cells, which indicates telomerase has a close link to cellular senescence. Some cancer cells use another mechanism termed alternative lengthening of telomeres (ALT) to compensate for the loss of DNA in each cell cycle (Dunham et al., 2000). Telomerase has also been shown to have extra-telomeric effects that are important for the proliferation of stem cells and cancer cells (Lai et al., 2007; Stewart et al., 2002). Since most human somatic cells do not rely on the function of telomerase, telomerase-targeted therapy for cancer has received intense attention due to its specific effect on telomerase-positive cancer cells (Kim et al., 1994). Although stem cells in regenerative tissues also rely on telomerase to compensate for the attrition of telomeric DNA in each cell division, the telomerase inhibition effect on these normal cells should be minor due to their longer initial telomeres compared with cancer cells (Herbert et al., 1999; Zhang et al., 1999).

Besides the telomere maintenance effect of telomerase, telomerase may have a series of nontelomeric effects on cells, such as telomere capping, interacting with DNA damage signals, and modification of chromatin structure (Fig. 2).

Regulation mechanisms of telomerase have been investigated intensely in recent years. In fact, several levels of regulation have been revealed, which include regulation of transcription, alternative splicing, assembly, subcellular localization, post-translational modifications of various components, and accessibility to telomeres (Fig. 3). Regulation of transcription has been indicated as the major mechanism that regulates the activity of telomerase in cells (Nugent and Lundblad, 1998), and studies have revealed evidence that a complex regulation network is likely involved in this process.

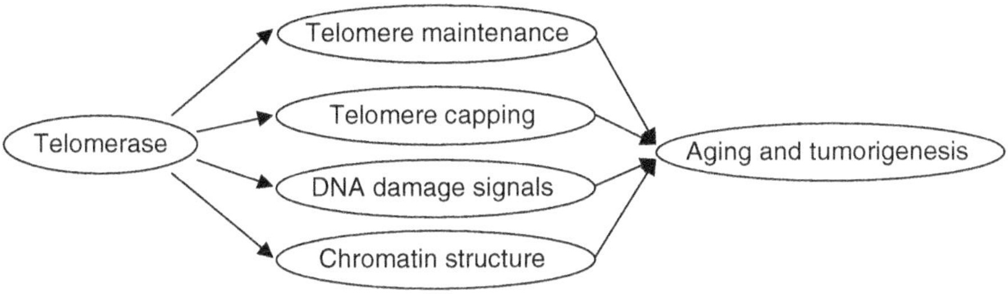

Fig. 2 Function of telomerase in aging and tumorigenesis. Telomerase can contribute to aging and tumorigenesis of humans by being involved in the following processes: telomere maintenance, telomere capping, interacting with DNA damage signals, and modification of chromatin structure to regulate gene expression

Fig. 3 Mechanisms that regulate function of telomerase. Function of telomerase can be regulated at a number of different levels. The two levels in shaded color refer to the involvement of epigenetic regulation mechanisms

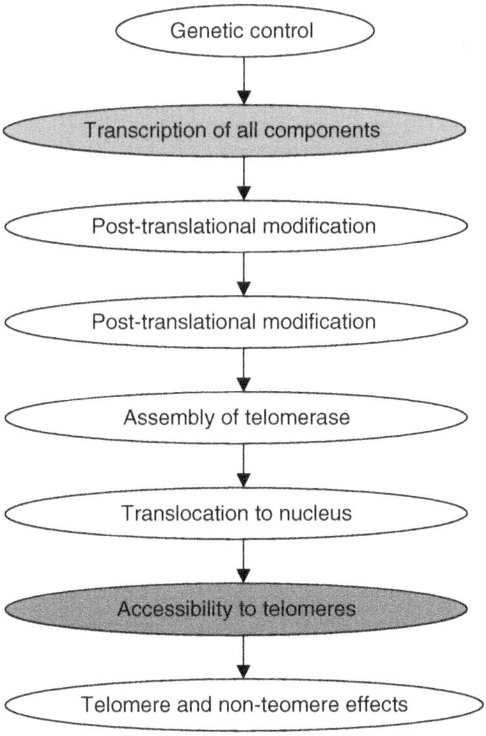

Epigenetic modifications of the regulation of gene expression have been considered as another important mechanism that contributes to the telomerase regulation network. Direct evidence has been found especially for processes such as aging and cancer. Discoveries from numerous laboratories have indicated that this mechanism is actively involved in the tight regulation of telomerase expression. In this chapter, epigenetic regulation of telomerase will be reviewed, especially with regard to the roles of hTERT, hTERC, and telomerase-associated proteins.

Telomerase Control by Genetic Processes

Genetic regulation of telomerase expression has been investigated widely, which can be reduced to the following three aspects.

Gains of Copy Number of TERT *and* TERC

Increases in the copy number of genes is a common mechanism used by human tumor or cancer cells to overexpress oncogenes (Rooney et al., 1999). *TERT* has been located to chromosome 5 at 5p15.33 (Bryce et al., 2000), *TERC* has been located to chromosome 3 at 3q26.3 (Soder et al., 1997), and dyskerin has been located to Xq28. All these regions have been shown to frequently undergo chromosomal gains (Knuutila et al., 1998; Rooney et al., 1999). Further investigation by using fluorescence in situ hybridization (FISH) and Southern blot analysis or PCR reveals that copy number of the *TERT* and *TERC* gene may increase in multiple tumors through chromosomal gains and gene amplifications in the process of cancer development due to clonal evolution (Cao et al., 2008).

Genetic Variation

Genetic variation has been considered as another mechanism that may contribute to the regulation of telomerase expression. Three correlated SNPs in *TERT* (–1381C>T, –244C>T, and Ex2-659G>A) have been found that may be associated with reduced risk of breast cancer among individuals with a family history of breast cancer (Savage et al., 2007).

Genetic Regulation on Transcription

Transcriptional regulation has been recognized as the most important mechanism that governs the expression of *hTERT*. The promoter region of the *hTERT* gene contains a series of binding sites for transcription factors and repressors, which include two E boxes that c-Myc may bind, several GC boxes that Sp1 may bind, and other potential cis-regulatory elements. Moreover, c-Myc competes with the Mad-Max network to control the transcription of *hTERT* (Xu et al., 2001). Estrogen has been shown to activate the expression of *hTERT* by binding to the estrogen-response element in ovarian epithelial cells and cancer cell lines (Bayne and Liu, 2005), while the transcription factor activator protein 1 represses *hTERT* expression by binding to its promoter (Takakura et al., 2005).

Telomerase Control by Epigenetic Processes in Cellular Senescence

Epigenetics is a field that has expanded rapidly in recent years; it has been indicated to play essential roles in cell differentiation, aging, and tumorigenesis and has been linked with gene silencing and reactivation as well as many other processes. Three basic directions of epigenetics have been investigated widely: DNA methylation, histone modification, and noncoding RNAs. These epigenetic processes can interplay with each other to regulate expression of genes. Understanding those processes from a perspective of epigenetics can not only clarify numerous puzzles related to the regulation of telomerase that have not been answered in former research but also provide valuable insights into developing novel medical approaches for aging and cancer-related diseases, such as discovery of potential targets in the pathology process of those disease by which intervention may be feasible.

Research in recent years revealed that epigenetic control is involved in the regulation of telomerase and this includes control of the gene expression of telomerase components and control of the accessibility of telomerase to telomeres, which is largely dependent on the three-dimensional structure of telomeres.

Methylation of CpG Islands

In the genomic sequence, there are areas that are rich in CpG dinucleotides and these areas are referred to as CpG islands. They generally exist in the promoter and first exon of genes. These CpG islands can undergo methylation, and this mechanism has been identified to participate in a number of basic biology processes such as development, aging, and age-related diseases by regulating the expression of genes involved in those processes. Methylation of CpG islands located in or near promoters of genes generally downregulates gene expression. DNA methylation is mediated by three DNA methyltransferases (DNMTs): DNMT1, DNMT3a, and DNMT3b. These enzymes can transfer methyl groups to cytosine located in the CpG dinucleotide (Jones and Baylin, 2002). DNMT1 is responsible for maintaining the methylation patterns following DNA replication, while DNMT3a and DNMT3b are responsible for de novo methylation (Okano et al., 1999). Another member of this gene family termed DNMT3L can interact with DNMT3a to promote de novo DNA methylation (Chedin et al., 2002). Mice with knockout of these genes are embryonic-lethal and have genomic hypomethylation (Okano et al., 1999). Moreover, changes in DNA methylation status have been associated with senescence and cancer. CpG islands are often unmethylated in normal cells while they can undergo methylation in cancer cells despite reduced global levels of DNA methylation (Jones, 1999). DNA methylation can induce gene silencing by either blocking the binding of transcription factors or binding proteins that can specifically bind to methylated DNA such as methyl CpG-binding protein 2 (MeCP2) and histone deacetylases (HDACs) which facilitate the formation of heterochromatin to prevent initiation of transcription (Nan et al., 1998).

In the promoter region of *hTERT*, clusters of CpG dinucleotides have been found (Horikawa et al., 1999). Several research groups have reported that the methylation status of CpG dinucleotides in the promoter of *hTERT* has a close link to the expression level of hTERT (Bechter et al., 2002; Dessain et al., 2000; Devereux et al., 1999; Guilleret et al., 2002a). However, conflicting data have been obtained concerning the methylation of CpG dinucleotides and the expression of *hTERT* in early studies. For example, hypomethylation of *hTERT* has been observed in undifferentiated and untransformed cells that do not express hTERT (Dessain et al., 2000), while methylation of the promoter has also been identified in differentiated and senescent cells (Lopatina et al., 2003; Shin et al., 2003). Through a comprehensive analysis of the *hTERT* promoter in a number of cancer cell lines with different tissue origins using methylation-specific PCR and bisulfite sequencing, Zinn et al. found that no methylation exists in the region near the transcription start site in telomerase-positive cell lines, while the region 600 bp upstream of the transcription start site is densely methylated (Zinn et al., 2007).

CpG islands have been identified in the promoter of *hTERC* in both humans and mice, which indicates that methylation may play a role in the regulation of its expression (Zhao et al., 1998). Apparently methylation of the promoter of *hTERC* in ALT cell lines has a close correlation with silencing of gene expression (Hoare et al., 2001). However, this relationship does not appear to exist in other cell lines, which indicates that the mechanisms that control the expression of *hTERC* are different between ALT cell lines and other cell lines (Hoare et al., 2001). A comprehensive study conducted on normal and tumor tissues and telomerase-positive and telomerase-negative cell lines indicates that *hTERC* expression is upregulated, while the cause of this does not appear to rely on DNA methylation of its promoter region (Guilleret et al., 2002b).

Histone Modification of Chromatin

Histone modification is another important epigenetic mechanism that regulates the expression of genes. Mammalian genomic DNA sequences can be packed into nucleosomes by wrapping around an octamer of core histone proteins, which include H2A, H2B, H3, and H4 (Luger et al., 1997). Another histone protein named H1 binds at the surface of nucleosome to fix the DNA sequence and histone proteins. Nucleosomes will then be packed into higher structures such as chromatin and finally into chromosomes. The N-terminal tails of core histones can be modified in a number of fashions, such as acetylation, biotinylation, methylation, phosphorylation, ubiquitination, SUMOylation, and ADP ribosylation. These modifications on histones can work with DNA methylation to control biological processes such as cell differentiation and aging by regulating the expression of key genes. Chromatin can be generally divided into two groups according to its biologic activity on gene expression, namely euchromatin, which can be transcribed, and heterochromatin, which cannot be transcribed. Each of these chromatins has specific modifications to maintain its status. Here we would like to focus on the expression of telomerase components under effects of histone methylation.

Histone Methylation and Acetylation

Histone methylation at lysines 4 or 79 of histone H3 (H3K4 or H3K79) associates with active transcription, and methylation at H3K9, H3K27, H4K20, and H1K26 correlates with gene silencing (Dillon et al., 2005). The number of methyl groups that can be added to lysine varies from one to three, which makes this methylation code more complex (Vakoc et al., 2006). The histone methylation is mediated by histone methyltransferases (HMTs).

Histone acetyltransferases (HATs) can transfer acetyl groups to lysine residues of histones. Since lysines in histone tails carry a positive charge, which can increase the interaction of histones with negatively charged DNA sequences to form a closed structure that does not favor gene transcription, acetylation of histones will generate a loose structure of chromatin which is usually associated with transcriptional activation of genes. This process can be reversed by histone deacetylases (HDACs).

As to components of telomerase, epigenetic mechanisms play pivotal roles in their expression process. For example, combination treatment of 5-azadeoxycytidine (a demethylation agent) with trichostatin A (a histone deacetylase inhibitor) can induce chromatin remodeling of both promoters and reactivation of *hTERC* and *hTERT* expression in ALT and normal cell lines (Atkinson et al., 2005).

The acetylation status of histones H3 and H4 and the methylation status of histone H3K9 have been shown to participate in the expression of *hTERC* and *hTERT*. For instance, histone H3 and H4 hypoacetylation and methylation of histone H3K9 have been linked with *hTERC* and *hTERT* silencing in ALT cell lines, while hyperacetylation of H3 and H4 and methylation of H3K4 are linked with hTERC and hTERT expression in telomerase-positive cell lines (Atkinson et al., 2005). Methylation of H4K20 is not associated with *hTERT* gene expression but may be specific to the promoters of *hTERC* and *hTERT* in ALT cell lines. Lysine-specific demethylase 1 (LSD1), which can catalyze demethylation of mono- and di-methylated histone H3K4 or K9, has been shown to repress *hTERT* transcription via demethylating H3K4 in normal and cancerous cells. HDACs also participate in the establishment of a stable repression state of the *hTERT* gene in normal or differentiated malignant cells (Zhu et al., 2008).

Trimethylation of H3K4 is another site-specific epigenetic mechanism that regulates the expression of *hTERT*. SET and MYND domain-containing protein 3 (SMYD3) is a histone methyltransferase. It has been indicated to induce *hTERT* expression by binding to the *hTERT* promoter and maintaining its histone H3K4 trimethylation. Knocking down SMYD3 in tumor cells will eliminate trimethylation of H3K4, decrease the binding of transcriptional factors like c-MYC and Sp1, and result in diminished histone H3 acetylation in the *hTERT* promoter region (Liu et al., 2007).

Epigenetic Control of Accessibility of Telomeres to Telomerase

Accessibility of telomerase to telomeres is another important mechanism by which cells control telomere length. Although former research indicates that telomerase can maintain the length of telomeres, a strict relationship between telomerase activity and telomere lengths is hard to establish (Bodnar et al., 1998). This is explained by the fact that a feedback loop is involved to regulate the accessibility of telomere ends to telomerase. When telomeres become short, it will generate a more accessible telomere structure which can increase the initiation frequency of telomerase (Smogorzewska and de Lange, 2004). It is reported that H3K9 of mammalian telomeres can undergo mono-, di-, and trimethylation, while proteins of heterochromatin protein 1 (HP1) family can be recruited to lysine residues undergoing di- and trimethylation. In this way, a closed chromatin state will form, and this may restrict accessibility of telomeres to telomerase (Blasco, 2005).

Conclusion

Cell senescence can benefit our health by preventing potential cancer transformation, while it can also promote the aging of humans on the other hand through the interaction between senescent cells and the surrounding normal cells. Telomerase is an important molecule that is involved in cell senescence. It can prevent the advent of replicative senescence in stem cells and germ cells, while it can also be reactivated when genomic instability is induced by environmental and genetic factors, thus promoting the development of cancer. Therefore, the regulation of telomerase has gained intense interest among scientists worldwide. Regulation of its function can be divided into several levels: accessibility to telomeres, regulation on expression through transcription factors, genetic regulation of its gene sequence, and epigenetic regulation. Actually, these mechanisms do not function independently but intertwine with each other so that some transcription factors can only bind to DNA elements in *hTERT* promoters that are unmethylated.

To clarify the interrelationship of those mechanisms and to resolve the sequence of those events, it is necessary to more fully understand the control of telomerase and its role in the process of aging. This would also help to design novel drugs targeted to telomerase for age-related diseases such as cancer.

Acknowledgment We apologize to those scientists whose work has not been cited in this chapter due to limited space. This work was supported by a grant from the National Cancer Institute (R01 CA129415) and Susan G. Komen for the Cure.

References

Atkinson, S., Hoare, S., Glasspool, R., and Keith, W. (2005). Lack of telomerase gene expression in alternative lengthening of telomere cells is associated with chromatin remodeling of the hTR and hTERT gene promoters. Cancer Res 65, 7585–7590.

Autexier, C., and Lue, N. F. (2006). The structure and function of telomerase reverse transcriptase. Annu Rev Biochem 75, 493–517.

Baird, D. M., Rowson, J., Wynford-Thomas, D., and Kipling, D. (2003). Extensive allelic variation and ultrashort telomeres in senescent human cells. Nat Genet 33, 203–207.

Baumann, P., and Cech, T. (2001). Pot1, the putative telomere end-binding protein in fission yeast and humans. Science 292, 1171–1175.

Bavik, C., Coleman, I., Dean, J., Knudsen, B., Plymate, S., and Nelson, P. (2006). The gene expression program of prostate fibroblast senescence modulates neoplastic epithelial cell proliferation through paracrine mechanisms. Cancer Res 66, 794–802.

Bayne, S., and Liu, J. (2005). Hormones and growth factors regulate telomerase activity in ageing and cancer. Mol Cell Endocrinol 240, 11–22.

Bechter, O., Eisterer, W., Dlaska, M., Kühr, T., and Thaler, J. (2002). CpG island methylation of the hTERT promoter is associated with lower telomerase activity in B-cell lymphocytic leukemia. Exp Hematol 30, 26–33.

Blasco, M. (2005). Mice with bad ends: mouse models for the study of telomeres and telomerase in cancer and aging. EMBO J 24, 1095–1103.

Bodnar, A., Ouellette, M., Frolkis, M., Holt, S., Chiu, C., Morin, G., Harley, C., Shay, J., Lichtsteiner, S., and Wright, W. (1998). Extension of life-span by introduction of telomerase into normal human cells. Science 279, 349–352.

Bryce, L., Morrison, N., Hoare, S., Muir, S., and Keith, W. (2000). Mapping of the gene for the human telomerase reverse transcriptase, hTERT, to chromosome 5p15.33 by fluorescence in situ hybridization. Neoplasia 2, 197–201.

Campisi, J. (2005). Suppressing cancer: the importance of being senescent. Science 309, 886–887.

Campisi, J., and d'Adda di Fagagna, F. (2007). Cellular senescence: when bad things happen to good cells. Nat Rev Mol Cell Biol 8, 729–740.

Cao, Y., Bryan, T., and Reddel, R. (2008). Increased copy number of the TERT and TERC telomerase subunit genes in cancer cells. Cancer Sci 99, 1092–1099.

Celli, G., and de Lange, T. (2005). DNA processing is not required for ATM-mediated telomere damage response after TRF2 deletion. Nat Cell Biol 7, 712–718.

Chedin, F., Lieber, M., and Hsieh, C. (2002). The DNA methyltransferase-like protein DNMT3L stimulates de novo methylation by Dnmt3a. Proc Natl Acad Sci USA 99, 16916–16921.

Collado, M., Blasco, M., and Serrano, M. (2007). Cellular senescence in cancer and aging. Cell 130, 223–233.

Collins, K. (2006). The biogenesis and regulation of telomerase holoenzymes. Nat Rev Mol Cell Biol 7, 484–494.

d'Adda di Fagagna, F., Reaper, P., Clay-Farrace, L., Fiegler, H., Carr, P., Von Zglinicki, T., Saretzki, G., Carter, N., and Jackson, S. (2003). A DNA damage checkpoint response in telomere-initiated senescence. Nature 426, 194–198.

de Lange, T. (2005). Shelterin: the protein complex that shapes and safeguards human telomeres. Genes Dev 19, 2100–2110.

Dessain, S., Yu, H., Reddel, R., Beijersbergen, R., and Weinberg, R. (2000). Methylation of the human telomerase gene CpG island. Cancer Res 60, 537–541.

Devereux, T., Horikawa, I., Anna, C., Annab, L., Afshari, C., and Barrett, J. (1999). DNA methylation analysis of the promoter region of the human telomerase reverse transcriptase (hTERT) gene. Cancer Res 59, 6087–6090.

Dillon, S., Zhang, X., Trievel, R., and Cheng, X. (2005). The SET-domain protein superfamily: protein lysine methyltransferases. Genome Biol 6, 227.

Dunham, M., Neumann, A., Fasching, C., and Reddel, R. (2000). Telomere maintenance by recombination in human cells. Nat Genet 26, 447–450.

Feng, J., Funk, W. D., Wang, S. S., Weinrich, S. L., Avilion, A. A., Chiu, C. P., Adams, R. R., Chang, E., Allsopp, R. C., Yu, J., et al. (1995). The RNA component of human telomerase. Science 269, 1236–1241.

Gire, V., Roux, P., Wynford-Thomas, D., Brondello, J., and Dulic, V. (2004). DNA damage checkpoint kinase Chk2 triggers replicative senescence. EMBO J 23, 2554–2563.

Griffith, J. D., Comeau, L., Rosenfield, S., Stansel, R. M., Bianchi, A., Moss, H., and de Lange, T. (1999). Mammalian telomeres end in a large duplex loop. Cell 97, 503–514.

Guilleret, I., Yan, P., Grange, F., Braunschweig, R., Bosman, F., and Benhattar, J. (2002a). Hypermethylation of the human telomerase catalytic subunit (hTERT) gene correlates with telomerase activity. Int J Cancer 101, 335–341.

Guilleret, I., Yan, P., Guillou, L., Braunschweig, R., Coindre, J., and Benhattar, J. (2002b). The human telomerase RNA gene (hTERC) is regulated during carcinogenesis but is not dependent on DNA methylation. Carcinogenesis 23, 2025–2030.

Harley, C., Futcher, A., and Greider, C. (1990). Telomeres shorten during ageing of human fibroblasts. Nature 345, 458–460.

Harrington, L., McPhail, T., Mar, V., Zhou, W., Oulton, R., Bass, M. B., Arruda, I., and Robinson, M. O. (1997). A mammalian telomerase-associated protein. Science 275, 973–977.

Hastie, N., Dempster, M., Dunlop, M., Thompson, A., Green, D., and Allshire, R. (1990). Telomere reduction in human colorectal carcinoma and with ageing. Nature 346, 866–868.

Herbert, B., Pitts, A. E., Baker, S. I., Hamilton, S. E., Wright, W. E., Shay, J. W., and Corey, D. R. (1999). Inhibition of human telomerase in immortal human cells leads to progressive telomere shortening and cell death. Proc Natl Acad Sci USA 96, 14276–14281.

Herbig, U., Jobling, W., Chen, B., Chen, D., and Sedivy, J. (2004). Telomere shortening triggers senescence of human cells through a pathway involving ATM, p53, and p21(CIP1), but not p16(INK4a). Mol Cell 14, 501–513.

Hoare, S., Bryce, L., Wisman, G., Burns, S., Going, J., van der Zee, A., and Keith, W. (2001). Lack of telomerase RNA gene hTERC expression in alternative lengthening of telomeres cells is associated with methylation of the hTERC promoter. Cancer Res 61, 27–32.

Horikawa, I., Cable, P., Afshari, C., and Barrett, J. (1999). Cloning and characterization of the promoter region of human telomerase reverse transcriptase gene. Cancer Res 59, 826–830.

Jones, P. (1999). The DNA methylation paradox. Trends Genet 15, 34–37.

Jones, P., and Baylin, S. (2002). The fundamental role of epigenetic events in cancer. Nat Rev Genet 3, 415–428.

Kim, N. W., Piatyszek, M. A., Prowse, K. R., Harley, C. B., West, M. D., Ho, P. L., Coviello, G. M., Wright, W. E., Weinrich, S. L., and Shay, J. W. (1994). Specific association of human telomerase activity with immortal cells and cancer. Science 266, 2011–2015.

Knuutila, S., Björkqvist, A., Autio, K., Tarkkanen, M., Wolf, M., Monni, O., Szymanska, J., Larramendy, M., Tapper, J., Pere, H., et al. (1998). DNA copy number amplifications in human neoplasms: review of comparative genomic hybridization studies. Am J Pathol 152, 1107–1123.

Lai, S., Cunningham, A., Huynh, V., Andrews, L., and Tollefsbol, T. (2007). Evidence of extratelomeric effects of hTERT and its regulation involving a feedback loop. Exp Cell Res 313, 322–330.

Lansdorp, P. M., Verwoerd, N. P., van de Rijke, F. M., Dragowska, V., Little, M. T., Dirks, R. W., Raap, A. K., and Tanke, H. J. (1996). Heterogeneity in telomere length of human chromosomes. Hum Mol Genet 5, 685–691.

Liu, C., Fang, X., Ge, Z., Jalink, M., Kyo, S., Björkholm, M., Gruber, A., Sjöberg, J., and Xu, D. (2007). The telomerase reverse transcriptase (hTERT) gene is a direct target of the histone methyltransferase SMYD3. Cancer Res 67, 2626–2631.

Liu, D., O'Connor, M., Qin, J., and Songyang, Z. (2004a). Telosome, a mammalian telomere-associated complex formed by multiple telomeric proteins. J Biol Chem 279, 51338–51342.

Liu, D., Safari, A., O'Connor, M., Chan, D., Laegeler, A., Qin, J., and Songyang, Z. (2004b). PTOP interacts with POT1 and regulates its localization to telomeres. Nat Cell Biol 6, 673–680.
Lopatina, N., Poole, J., Saldanha, S., Hansen, N., Key, J., Pita, M., Andrews, L., and Tollefsbol, T. (2003). Control mechanisms in the regulation of telomerase reverse transcriptase expression in differentiating human teratocarcinoma cells. Biochem Biophys Res Commun 306, 650–659.
Luger, K., Mäder, A., Richmond, R., Sargent, D., and Richmond, T. (1997). Crystal structure of the nucleosome core particle at 2.8 A resolution. Nature 389, 251–260.
Meyerson, M., Counter, C. M., Eaton, E. N., Ellisen, L. W., Steiner, P., Caddle, S. D., Ziaugra, L., Beijersbergen, R. L., Davidoff, M. J., Liu, Q., et al. (1997). hEST2, the putative human telomerase catalytic subunit gene, is up-regulated in tumor cells and during immortalization. Cell 90, 785–795.
Moyzis, R. K., Buckingham, J. M., Cram, L. S., Dani, M., Deaven, L. L., Jones, M. D., Meyne, J., Ratliff, R. L., and Wu, J. R. (1988). A highly conserved repetitive DNA sequence, (TTAGGG)n, present at the telomeres of human chromosomes. Proc Natl Acad Sci USA 85, 6622–6626.
Nakamura, T. M., Morin, G. B., Chapman, K. B., Weinrich, S. L., Andrews, W. H., Lingner, J., Harley, C. B., and Cech, T. R. (1997). Telomerase catalytic subunit homologs from fission yeast and human. Science 277, 955–959.
Nan, X., Ng, H., Johnson, C., Laherty, C., Turner, B., Eisenman, R., and Bird, A. (1998). Transcriptional repression by the methyl-CpG-binding protein MeCP2 involves a histone deacetylase complex. Nature 393, 386–389.
Nugent, C., and Lundblad, V. (1998). The telomerase reverse transcriptase: components and regulation. Genes Dev 12, 1073–1085.
Okano, M., Bell, D., Haber, D., and Li, E. (1999). DNA methyltransferases Dnmt3a and Dnmt3b are essential for de novo methylation and mammalian development. Cell 99, 247–257.
Rooney, P., Murray, G., Stevenson, D., Haites, N., Cassidy, J., and McLeod, H. (1999). Comparative genomic hybridization and chromosomal instability in solid tumours. Br J Cancer 80, 862–873.
Savage, S., Chanock, S., Lissowska, J., Brinton, L., Richesson, D., Peplonska, B., Bardin-Mikolajczak, A., Zatonski, W., Szeszenia-Dabrowska, N., and Garcia-Closas, M. (2007). Genetic variation in five genes important in telomere biology and risk for breast cancer. Br J Cancer 97, 832–836.
Shin, K., Kang, M., Dicterow, E., and Park, N. (2003). Hypermethylation of the hTERT promoter inhibits the expression of telomerase activity in normal oral fibroblasts and senescent normal oral keratinocytes. Br J Cancer 89, 1473–1478.
Smith, S., Giriat, I., Schmitt, A., and de Lange, T. (1998). Tankyrase, a poly(ADP-ribose) polymerase at human telomeres. Science 282, 1484–1487.
Smogorzewska, A., and de Lange, T. (2004). Regulation of telomerase by telomeric proteins. Annu Rev Biochem 73, 177–208.
Soder, A., Hoare, S., Muir, S., Going, J., Parkinson, E., and Keith, W. (1997). Amplification, increased dosage and in situ expression of the telomerase RNA gene in human cancer. Oncogene 14, 1013–1021.
Stewart, S., Hahn, W., O'Connor, B., Banner, E., Lundberg, A., Modha, P., Mizuno, H., Brooks, M., Fleming, M., Zimonjic, D., et al. (2002). Telomerase contributes to tumorigenesis by a telomere length-independent mechanism. Proc Natl Acad Sci USA 99, 12606–12611.
Takai, H., Smogorzewska, A., and de Lange, T. (2003). DNA damage foci at dysfunctional telomeres. Curr Biol 13, 1549–1556.
Takakura, M., Kyo, S., Inoue, M., Wright, W., and Shay, J. (2005). Function of AP-1 in transcription of the telomerase reverse transcriptase gene (TERT) in human and mouse cells. Mol Cell Biol 25, 8037–8043.
Vakoc, C., Sachdeva, M., Wang, H., and Blobel, G. (2006). Profile of histone lysine methylation across transcribed mammalian chromatin. Mol Cell Biol 26, 9185–9195.

Vulliamy, T., and Dokal, I. (2008). Dyskeratosis congenita: the diverse clinical presentation of mutations in the telomerase complex. Biochimie 90, 122–130.

Xu, D., Popov, N., Hou, M., Wang, Q., Björkholm, M., Gruber, A., Menkel, A., and Henriksson, M. (2001). Switch from Myc/Max to Mad1/Max binding and decrease in histone acetylation at the telomerase reverse transcriptase promoter during differentiation of HL60 cells. Proc Natl Acad Sci USA 98, 3826–3831.

Ye, J., Hockemeyer, D., Krutchinsky, A., Loayza, D., Hooper, S., Chait, B., and de Lange, T. (2004). POT1-interacting protein PIP1: a telomere length regulator that recruits POT1 to the TIN2/TRF1 complex. Genes Dev 18, 1649–1654.

Zhang, X., Mar, V., Zhou, W., Harrington, L., and Robinson, M. O. (1999). Telomere shortening and apoptosis in telomerase-inhibited human tumor cells. Genes Dev 13, 2388–2399.

Zhao, J., Hoare, S., McFarlane, R., Muir, S., Parkinson, E., Black, D., and Keith, W. (1998). Cloning and characterization of human and mouse telomerase RNA gene promoter sequences. Oncogene 16, 1345–1350.

Zhu, Q., Liu, C., Ge, Z., Fang, X., Zhang, X., Strååt, K., Björkholm, M., and Xu, D. (2008). Lysine-specific demethylase 1 (LSD1) Is required for the transcriptional repression of the telomerase reverse transcriptase (hTERT) gene. PLoS ONE 3, e1446.

Zijlmans, J. M., Martens, U. M., Poon, S. S., Raap, A. K., Tanke, H. J., Ward, R. K., and Lansdorp, P. M. (1997). Telomeres in the mouse have large inter-chromosomal variations in the number of T2AG3 repeats. Proc Natl Acad Sci USA 94, 7423–7428.

Zinn, R., Pruitt, K., Eguchi, S., Baylin, S., and Herman, J. (2007). hTERT is expressed in cancer cell lines despite promoter DNA methylation by preservation of unmethylated DNA and active chromatin around the transcription start site. Cancer Res 67, 194–201.

Telomeres, Epigenetics, and Aging

J. Arturo Londoño-Vallejo

Abstract Telomere shortening has been linked to aging and disease. On the other hand, recent experimental data point to epigenetics as a fundamental process in telomere function. How both aspects are connected and how they impact on organismal fitness will be the subject of intense research in the coming years. This chapter reviews different aspects of telomere biology and discusses their implications in the aging process.

Keywords Telomere length · Telomerase · Aging · Heterochromatin · Shelterin · Telomere maintenance

Introduction

Telomeres are essential structures for genome stability (Wong and Collins 2003). They protect the extremities of linear chromosomes from degradation and recombination. They also participate in the nuclear architecture and in the meiosis-specific genome reorganization and recombination. The primary structure of telomeres is composed of a short, 5′- to 3′ repeated G-rich sequence that ends in a single-stranded 3′ overhang, the G-tail (Rhodes et al. 2002). The double-stranded portion of telomeric sequences is bound by both histone and nonhistone protein complexes. In particular, a six-protein telomere-specific complex, called shelterin or telosome, plays an essential role in telomere protection and presumably is responsible for these chromosome ends to adopt singular structures, such as t-loops, in which the 3′ overhang folds back and invades the double-stranded portion of the repeats (de Lange 2005) (Fig. 1).

J.A. Londoño-Vallejo (✉)
UMR7147, Institut Curie-UPMC-CNRS, 26, rue d'Ulm, 75248 Paris, France
e-mail: arturo.Londono@curie.fr

T.O. Tollefsbol (ed.), *Epigenetics of Aging*, DOI 10.1007/978-1-4419-0639-7_12,

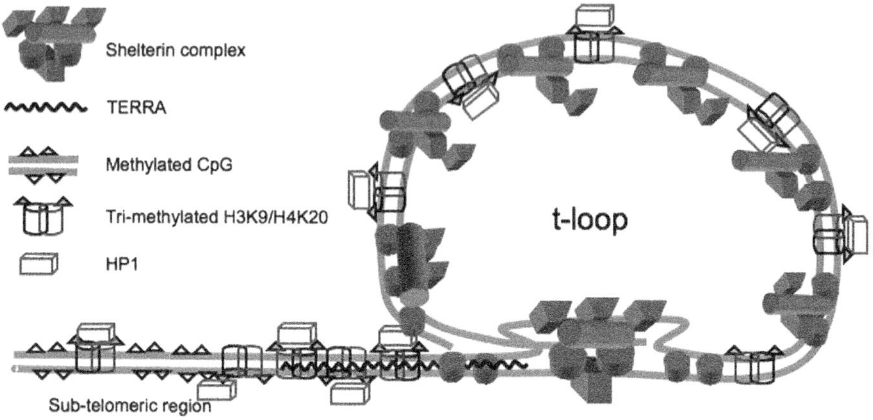

Fig. 1 The telomere is represented as a t-loop, where the G-tail invades the base of the double-stranded telomeric repeats. Telomere repeats are bound by both shelterin and histone complexes. Heterochromatic marks (trimethylation of H3K9 and H4K20 and recruitment of HP1) are present both at telomeres and at the subtelomeric regions, where CpG methylation also exists. The heterochromatic nature of telomeres, to which the telomere-associated RNA – TERRA – may contribute, influences length homeostasis and stability, likely by thwarting replication, limiting access to telomerase, and repressing recombination. Conversely, the length of telomeres influences the heterochromatic status of both telomeres and subtelomeric regions, at least in mouse cells. The intimate connection between histone and nonhistone complexes and its impact on the regulation of telomeric heterochromatin remains to be explored

The formation of the t-loop is thought to be a major mechanism by which the chromosome extremity is hidden away from the cell surveillance machinery (de Lange 2002). The disruption of the t-loop by telomere shortening or shelterin defaults provokes the recognition of the chromosome extremity as a double-strand break, triggering repair mechanisms that ultimately lead to chromosome end fusion or degradation (Blackburn 2001). On the other hand, a physiological disruption of the t-loop must take place during replication in order to allow the replisome to move forward toward the 3′ end (Gilson and Geli 2007; Verdun and Karlseder 2007). It has been proposed that the DNA damage signaling that results from this transient disruption triggers the recruitment of DNA repair proteins that, together with shelterin, will help to reestablish the t-loop structure (Verdun and Karlseder 2006). Importantly, the replication of the G-rich strand, which takes place through lagging-strand mechanisms, is always incomplete, due to the removal of the last RNA primer (Olovnikov 1971; Watson 1972). And since there is an absolute requirement to create an overhang on the sister chromatid that was completely replicated by leading mechanisms, there is also a 5′ to 3′ degradation of the parental C-rich strand (Makarov et al. 1997). As a consequence, telomere replication unavoidably leads to shortening by a few tens of repeats. Therefore, shortening of telomeres is the rule in dividing somatic cells that express no or very low levels of telomerase, the dedicated reverse transcriptase that specifically incorporates new telomeric repeats at the 3′ overhang (Shay and Wright 2005). When telomeres reach a critical length, cells stop dividing and enter a permanent state of senescence, which is considered a major antitumor mechanism (Shawi and Autexier 2008). When enough telomerase activity is present, telomere length is maintained and cells may then replicate indefinitely (Shay and Wright 2005).

Telomere length is therefore a highly dynamic variable of telomere function and is under tight control, directly reflecting the efficiency of telomere maintenance mechanisms (Hug and Lingner 2006). The best characterized telomere length homeostasis mechanism is the one exerted by the shelterin complex on telomerase activity. Long telomeres tend to bind more shelterin and to be less accessible to telomerase, whereas short telomeres tend to be more "open" and therefore are more likely to undergo elongation, thus preventing further shortening (Teixeira et al. 2004). This negative feedback that results in a net preference of telomerase for short telomeres determines an equilibrium point of telomere length within cells and eventually allows the rescue of potentially deleterious troubles in telomere replication or repair that provoke sudden excessive telomere shortening (Choi et al. 2001; Crabbe et al. 2004; Wyllie et al. 2000). Again, in the absence of telomerase, such accidents may rapidly lead to cell senescence, thus limiting cell renewal capacity.

Telomere Length: A Biomarker for Human Aging

In many organisms, telomere length decreases with increasing age, suggesting that telomerase activity is limiting, even in stem cell compartments (Canela et al. 2007; Harley et al. 1990; Hastie et al. 1990; Rufer et al. 1998; Vaziri et al. 1994). In humans, in particular, telomere shortening has been shown to occur not only during normal aging but also during many aging-related pathological conditions, such as cancer, arteriosclerosis, or cardiovascular diseases. Moreover, premature aging syndromes such as Werner syndrome, in which the gene coding for the WRN helicase is mutated (see page 14), are characterized by accelerated shortening of telomeres (Schulz et al. 1996; Tahara et al. 1997). Since telomere shortening triggers cell senescence in vitro, it has been proposed that telomere shortening is a major cause of the tissue dysfunction that characterizes the aging process in vivo, and thus telomere length is increasingly considered an aging biomarker (Bekaert et al. 2005a; von Zglinicki and Martin-Ruiz 2005). However, despite a clear association between physiological aging and telomere shortening, it has been difficult to distinguish between consequence and causality. Nevertheless, in pathological contexts such as dyskeratosis congenita (DC), another premature aging syndrome where there is a telomerase defect (Bessler et al. 2004), the phenomenon of anticipation (manifestation of the disease at earlier age in successive generations) strongly suggests that short telomeres are directly contributing to aging manifestations and may actually limit human life span (Armanios et al. 2005; Goldman et al. 2005; Vulliamy et al. 2004).

Genetics of Telomere Length

The mean telomere length found in an individual depends on length homeostatic mechanisms. There is marked variation in mean telomere length among individuals of the same age class and the available evidence strongly suggests that these differences are genetically determined. Monozygotic twins have very similar mean telomere lengths, whereas dizygotic twins show significant differences (Slagboom et al. 1994). Moreover, inherited paternal factors have a patent influence on the mean telomere lengths of their offspring (Njajou et al. 2007; Nordfjall et al. 2005). On the other hand, a positive correlation has been found between the age of fathers and telomere lengths in their children, suggesting that a vertical transmission of telomere lengths may contribute to length variations in the population (Unryn et al. 2005). X-linked factors have also been proposed (Nawrot et al. 2004) but have not been confirmed in independent studies. Studies in humans using marker linkage strategies have identified a few loci with significant influence on telomere lengths, although no obvious candidate genes have been found (Andrew et al. 2006; Vasa-Nicotera et al. 2005).

Since telomere length is the result of the equilibrium between lengthening and shortening mechanisms, genes responsible for setting telomere lengths may do so by promoting (or thwarting) efficiency of replication, telomerase activity, and/or telomere repair. For instance, the efficiency of telomere replication may be directly affected by the enzymatic activities of certain helicases such as WRN (Crabbe et al. 2004), RTEL1 (Ding et al. 2004), BLM (Opresko et al. 2002), or FANCJ (Wu et al. 2008) and the presence of particular allelic variants of any of these genes may lead to a more or less incomplete telomere replication. On the other hand, the levels of telomerase activity seem to be a major determinant of telomere length (Hug and Lingner 2006), these levels being directly dependent on the level of expression of both the catalytic subunit hTERT and the RNA moiety hTERC. Indeed, the level of telomerase in lymphocytes in response to a stimulus is genetically determined (Kosciolek and Rowley 1998) and single-nucleotide polymorphisms identified in the hTERT promoter in certain human populations have been associated with longer telomeres (Matsubara et al. 2006). hTERC, on the other hand, is expressed in most tissues but levels may be modulated according to the proliferation state (Weng et al. 1996). Experimentally, it is possible to increase telomere length by just overexpressing hTERC (Cristofari and Lingner 2006), indicating that its levels are limiting under normal conditions. This is again illustrated by defects in telomerase activity found in DC patients carrying one single normal *hTERC* gene (haploinsufficiency) (Vulliamy et al. 2001). Therefore, subtle differences in hTERC levels of expression among individuals may contribute to telomere length differences. Finally, oxidative damage constitutes a major extrinsic factor for telomere shortening (von Zglinicki 2000). Since resistance to oxidative stress is under genetic control (Powell et al. 2005), it is likely that genes in this pathway have a great impact on telomere length homeostasis.

Telomere length control is also exerted differently at different chromosome extremities. Single telomere lengths are heterogeneous within cells (Henderson et al.

1996; Lansdorp et al. 1996) and the chromosome distribution of this heterogeneity is characteristic of each individual and stably maintained during aging (Baird et al. 2003; Britt-Compton et al. 2006; Londoño-Vallejo et al. 2001). This single telomere length heterogeneity is most likely explained by allelic relative length polymorphisms (Graakjaer et al. 2004), which are defined in the zygote and depend directly on the relative contributions of single telomere lengths by the parents (Graakjaer et al. 2006). How a chromosome arm-specific telomere length is defined is not known, but it is reasonable to suppose that naturally occurring nucleotide variations in the subtelomeric regions (Baird et al. 2000; Coleman et al. 1999) are directly or indirectly responsible for such differences, perhaps through epigenetic mechanisms akin to imprinting (methylation-dependent, see below). Whatever be the case, the fact that just one or a few very short telomeres are enough to trigger mitotic senescence in vitro, and possibly in vivo, suggests that inheriting chromosome extremities with relatively shorter telomeres may impact, later in life, replication (or regeneration) capacity in aging individuals (Gilson and Londoño-Vallejo 2007). It also underlines the need of conducting population studies in which single, and not only overall, telomere lengths are measured.

Epigenetics of Telomeres

Telomeres Bear Heterochromatic Marks

The telomere DNA sequence found in humans (T2AG3, which is the same for most vertebrates) lacks CpGs and therefore cannot be the target of the same type of DNA methylation known to occur elsewhere in the genome. On the other hand, telomeric H3 and H4 histones undergo the same kind of epigenetic modifications associated with heterochromatin (for example, the pericentric chromatin), that is, trimethylation of lysine 9 on histone 3 (H3K9) and of lysine 20 on histone 4 (H4K20) and recruitment of HP1 (Blasco 2007) (Fig. 1). On the other hand, human and mouse telomeres are characterized by low levels of acetylated H3 and H4, features that also correspond to their heterochromatic status (Blasco 2007).

These modifications are likely important for telomere function since their modulation in mouse models results in alterations in telomere homeostasis. In particular, abolition of either the genes responsible for H3K9 trimethylation (Suv39h1 and Suv39h2) (Garcia-Cao et al. 2004), or the *Rb* genes that regulate H4K20 trimethylation (Gonzalo and Blasco 2005), or the enzymes responsible for the latter modification (the Suv420H1 and Suv420H2 histone methyltransferases or HMTases) (Benetti et al. 2007b) all result in telomere elongation and increased telomere sister chromatid recombination (T-SCE). These findings strongly suggest that disruption of heterochromatin perturbs telomere length homeostasis perhaps by imposing a more open configuration and/or by interfering with telomere length regulators. Conversely, telomere length may affect the epigenetic status of telomeres as demonstrated by the decrease in H3K9 and H4K20 methylation and the increase in H3 and H4 acetylation at telomeres of *terc*$^{-/-}$ mouse cells (Benetti et al. 2007a). These changes may also be present in naturally occurring short telomeres and perhaps contribute to telomere homeostasis by facilitating telomerase access to the telomere.

The heterochromatin status of telomeres also has great impact on telomere maintenance through the recruitment of factors essential for telomere replication. For instance, SIRT6, a member of the Sir2 family of histone deacetylases, is required to maintain correct (lower) levels of H3K9Ac, specifically during the S-phase (Michishita et al. 2008). In the absence of SIRT6, there is accumulation of H3K9Ac at telomeres, which results in inefficient recruitment of WRN helicase, leading to defects in telomere replication and to telomere loss (Michishita et al. 2008). It is then proposed that efficient recruitment of WRN to telomeres requires a transient altered chromatin state, but the molecular bases for this mechanism remain to be elucidated.

The epigenetic status of subtelomeric regions in mice and humans is connected to telomeres in that there seems to be a correlation in the histone methylation and acetylation patterns in both regions. However, subtelomeric DNA may also be subjected to DNA methylation and this modification seems to be independent of other heterochromatic marks since, in mice, lack of enzymes responsible for CpG methylation at subtelomeres does not affect other heterochromatic marks (no change in H3K9/H420 trimethylation at subtelomeres) while causing these marks to increase at telomeres (Gonzalo et al. 2006). Interestingly, loss of subtelomeric DNA methylation is accompanied by both increased T-SCE and telomere elongation (Gonzalo et al. 2006), the latter not necessarily being associated with a heterochromatic change at telomeres.

The consequences of subtelomeric chromatin modifications in humans appear to differ from what has been observed in mice. In a recent study, Yehezkel et al. examined the subtelomeric DNA in cells from patients with mutations in the gene *DNMT3B*, whose homologue is responsible for the de novo subtelomeric DNA methylation in mice (Yehezkel et al. 2008). As predicted, subtelomeric regions were hypomethylated in these patients but, in contrast to mice, telomeres were abnormally short and did not show signs of increased recombination.

In mice, for all cases of association of epigenetic modifications and telomere lengthening, the increase in telomere length seems to be dependent on telomerase activity. Therefore, it is likely that the contribution of recombination to telomere elongation is marginal. However, allowing recombination may be a first step in the activation of alternative mechanisms of telomere maintenance (ALT), i.e., independent of telomerase (Bryan et al. 1997; Cesare and Reddel 2008). These mechanisms are utilized by a small fraction of human cancer cells (Neumann and Reddel 2002) and it is likely that they require extensive and stable modifications of telomeric heterochromatin, although a clear link between these modifications and ALT in human cells is still missing. On the other hand, T-SCE does not exist (or else at very low levels) in normal human cells (Londoño-Vallejo et al. 2004) and there is no indication that ALT may occur under physiological conditions. In mice, however, both telomere recombination and lengthening are detected during the first cell divisions after fertilization, concurrently with a genome-wide decrease in DNA methylation and when very low (or no) telomerase activity is present (Liu et al. 2007). It is thus possible that a physiological ALT-like mechanism may contribute to telomere length setting at this early stage of mouse development, although it is not clear whether this is a regulated process or just an unintended manifestation of a transition between telomeric states (see below).

Telomere-Associated RNA

Telomeres have for a long time been considered as silent regions. Recently, however, it has been discovered that RNAs containing G-rich telomeric repeats are expressed in human and mouse cells (Azzalin et al. 2007; Schoeftner and Blasco 2008) (Fig. 1). At least a fraction of these molecules, called TERRA, contain subtelomeric sequences, indicating that transcription starts at subtelomeres of different chromosomes and proceeds toward the end (Azzalin et al. 2007).

The amount of TERRA varies in cell lines and seems to be positively correlated with telomere length, at least in human cells, and is present in both telomerase-positive and telomerase-negative cell lines (Azzalin and Lingner 2008). Importantly, these molecules were found to interact with telomeres even during transcriptionally inactive phases of the cell cycle, suggesting that TERRA is a constitutive component of telomeric heterochromatin and may indeed participate in maintaining the heterochromatic structure (Azzalin et al. 2007). Strikingly, TERRA abundance increases in cells lacking HMTases but diminishes in cells lacking DNA methyltransferases (DNMTases) (Schoeftner and Blasco 2008).

Although the functions of TERRA remain to be described, at least some of the pathways controlling its abundance and localization have started to be explored. These pathways implicate several suppressors with morphogenetic defects in genitalia (SMG) proteins, which are essential in nonsense-mediated mRNA decay (NMD) (Yamashita et al. 2005). Some of these proteins (SMG1, EST1A, and UPF1) are required to remove TERRA from telomeres and their absence leads to abrupt telomere loss, suggesting a default in telomere replication (Azzalin et al. 2007). On the other hand, EST1A interacts with telomerase (Redon et al. 2007), suggesting coordination between TERRA removal, telomere replication, and telomerase action.

Whether the abundance of TERRA in human cells is controlled during development, as seems to be the case in mice (Schoeftner and Blasco 2008), or varies with age remains to be determined. Also, the potential impact of the absence of TERRA on the heterochromatin architecture of telomeres and/or their metabolism remains to be determined. Nevertheless, similar to what occurs in other organisms (Sugiyama et al. 2005), it is possible that the presence of TERRA, or of small interfering RNAs derived from it, at telomeres is important to induce the heterochromatinization of these regions. Interestingly, TERRA is decreased in Dicer mutants, suggesting that disruption of the RNAi pathway affects TERRA stability (Schoeftner and Blasco 2008). At the same time, Dicer deficiency leads to hypomethylation of subtelomeric regions in mouse cells, caused by a decreased expression of DNMTases (Benetti et al. 2008), without affecting other heterochromatin marks at telomeres. As expected, Dicer-deficient cells bear longer telomeres and high levels of T-SCE (Benetti et al. 2008).

Once again, there seems to be another difference between mice and humans. In the study by Yehezkel et al. (2008) mentioned above, loss of methylation at subtelomeric regions of human cells mutated for DNMT3B was accompanied by an increase in TERRA association with telomeres. This accumulation of TERRA may contribute to telomere shortening through an increased heterochromatinization of telomeres. That RNA-mediated heterochromatinization may lead to telomere shortening is suggested by the observation that telomeres on the inactivated human X chromosome tend to shorten faster with age than its homologue (Surralles et al. 1999). Hyper-heterochromatinization of telomeres might impede replication fork progression, a correct access to telomerase, or both.

Shelterin: A Target for Epigenetic Modification

Finally, it must be stressed that telomeric sequences are bound by shelterin and that alterations in the structure or the composition of this complex have a profound impact on the function of telomeres, through modifications of either its length or its structure (Chan and Blackburn 2004; de Lange 2005). It is also clear that posttranslational modifications of shelterin components impact on telomere function. Such modifications, if persistent through mitosis or, a fortiori, through meiosis, may be considered as epigenetic (heritable) changes.

Probably the best known of posttranslational modifications of shelterin proteins is the poly-ADP-ribosylation of TRF1 (Rippmann et al. 2002), a protein that binds the double-stranded portion of telomeres and regulates telomere length (van Steensel and de Lange 1997). The accumulation of TRF1 on long telomeres prevents further elongation – through the action of POT1 (the shelterin component that binds the G-tail) – by telomerase (Loayza and De Lange 2003). However, the poly-ADP-ribosylation of TRF1 by tankyrase (in particular, TNKS1) diminishes the affinity of TRF1 for telomeric DNA, allowing the relief of the negative feedback (Smith et al. 1998). Furthermore, TIN2 (another shelterin protein) interacts with TRF1 and inhibits the poly-ADP-ribosylation activity of TNKS1 (Ye and De Lange 2004), thus impinging on telomere length homeostasis. While these changes are thought to be highly dynamic and transient in normal cells, in response to some unknown stimulus, they may become more permanent under physiological (during development, for instance) or pathological (such as cancer or DC) conditions.

Other modifications may include phosphorylation, for example, of TRF2 (Tanaka et al. 2005) or other components of shelterin, but such modifications also likely correspond to transient states in response to, for instance, telomere damage. However, it cannot be excluded that stable posttranslational modifications of shelterin components exist and that they may be recreated during telomere replication/remodeling, very much like the histone code, thus transmitting to the daughter cells a "telomere state" which will ultimately influence telomere function.

Epigenetics and the Telomere State

The observation that telomerase activity is enough to reach "immortalization" in certain types of human cells (Bodnar et al. 1998) has led to the supposition that length is the major determinant of telomere function. However, it is increasingly clear that telomere function is a matter not only of length but also of both structure and protein–protein interactions (Songyang and Liu 2006). The quality of this structure, be it the t-loop, the chromatin state, the relative concentrations of shelterin subcomplexes, and their relationship to other telomere ancillary proteins, ought to be the determinant in governing the telomere state and its impact on the physiology of subtelomeric regions as well as that of the cell.

As noted above, a transient telomere state characterized by low levels of H3K9Ac is required to allow efficient recruitment of WRN and normal telomere replication (Michishita et al. 2008). After replication, a "normal" telomere state should be reconstituted and transmitted to the daughter cell in order to ensure stability. In the presence of telomerase, this reconstitution involves a reposition of the number of repeats also. It also implicates the restitution of structural characteristics that do not depend on telomere elongation, such as the reformation of the t-loop and the replacement of all heterochromatin characteristics, including the association with TERRA. It is conceivable that changes in cell state (during differentiation, for instance) are accompanied by changes in telomere state. Levels of TERRA seem to vary during development, at least in mice (Schoeftner and Blasco 2008), and this may signify an important change in telomere function that must be maintained down in the line. Also, the levels of different shelterin components, or that of their isoforms (Lages et al. 2004; Yang et al. 2007), or else the levels of proteins important for telomere replication/repair – such as the WRN helicase (Motonaga et al. 2002) – may well vary between tissues and/or during aging, suggesting that different telomere states may accompany different pathways of differentiation.

The telomere state may also be transmitted through meiosis. For instance, it is not known whether telomere length heterogeneity in human cells is related to variable levels of heterochromatic marks either at telomeres or at subtelomeres. However, both the fact that variations in subtelomeric sequences are linked to specific telomeres lengths (Baird et al. 2003; Graakjaer et al. 2006) and the fact that the DNA methylation status of subtelomeric regions has a great impact on telomere length and recombination behavior in mice (Gonzalo et al. 2006) suggest that methylation of such regions may be, at least in part, responsible for the variations between single chromosome extremities within human cells. Moreover, since length variations exist among alleles of the same chromosome extremity and that these variations are detectable in germ cells and are transmitted to the offspring (Baird et al. 2003; Britt-Compton et al. 2006; Graakjaer et al. 2006; Londoño-Vallejo et al. 2001), it is conceivable that some kind of parental imprinting is the major determinant of telomere length and that this imprinting depends, like in other parts of the genome, either on the methylation status of subtelomeric DNA (Recillas-Targa 2002) or, perhaps, on other chromatin status impinging on, for instance, TERRA expression (Paldi 2003).

Finally, as mentioned above, the ALT-like mechanism that seems to operate during the first cell divisions after fertilization of mouse oocytes is concurrent to transient epigenetic modifications (Liu et al. 2007). Although this recombination pathway seems to be definitively repressed down the road of development, its reactivation is possible under particular circumstances, such as tumor transformation, in which telomere maintenance is essential but reactivation of telomerase is no longer possible. Experiments in yeast strongly suggest that the reactivation of such pathways does not depend on the acquisition of particular gene mutations but rather on the acquisition of particular telomere states through the modification of their epigenetic status (Makovets et al. 2008). In mammalian cells, it is possible that an epigenetic switch is required to turn ALT on, and the observations made in mice models bearing alterations in telomere chromatin strongly support this idea. Nevertheless, contrary to the yeast model, reintroduction of telomerase most of the time does not abolish ALT in human cells (Cerone et al. 2001), suggesting that elongating critically short telomeres or providing other capping functions dependent on telomerase is not enough to revert the telomere state that is permissive for recombination.

Telomere Position Effect (TPE)

TPE results from the influence of telomeres on the heterochromatinization of subtelomeric regions, which leads to repression (or modulation) of gene expression. This phenomenon has been well characterized in yeast but remains somewhat hard to pin down in mammals (Tham and Zakian 2002). Experimental evidence that TPE does exist in mammalian cells has been obtained using reporter transgenes integrated near telomeres (Baur et al. 2001; Koering et al. 2002; Pedram et al. 2006). In those experiments, it was shown that, as in yeast, a gene is expressed at lower levels when it is in the vicinity of telomeres and that the level of expression varies according to telomere length: the longer the telomere, the lower the gene expression level. On the other hand, overexpression of TRF1 favors gene expression, suggesting that both telomere length and structure influence TPE (Koering et al. 2002). Repression of the gene inserted near a telomere involves spreading of heterochromatin marks to the subtelomeric region and this may involve interactions between telomeric proteins, such as TRF1 or TIN2, and remodeling factors such as SALL1 (Netzer et al. 2001) or the heterochromatic protein HP1 (Kaminker et al. 2005).

In spite of the mentioned evidence, the proof that telomeres may influence expression of genes at their natural position is still to be provided. Observations in cells carrying deletions or translocations that put a telomere in the vicinity of genes indicate that expression of the latter is not affected, nor is there spreading of heterochromatic marks into the new subtelomeric region (Ofir et al. 1999). On the other hand, the patterns of expression of subtelomeric genes seem not to be particularly influenced by the length of the corresponding telomeres, even during the process of telomere shortening that accompanies proliferation (Ning et al. 2003). However, limited chromatin alterations at particular chromosome ends may explain observed changes in gene expression. If so, it would suggest that TPE, as in yeast, may be brought about through interactions with telomeric proteins rather than through telomere length (Ning et al. 2003; Ottaviani et al. 2008), but this remains speculative.

Also as in yeast, TPE may be highly dependent on the type (identity) of the chromosome extremity (Ottaviani et al. 2008). Most human chromosome extremities contain large blocks of segments that connect chromosome-specific sequences to telomeric sequences (Der-Sarkissian et al. 2002; Mefford and Trask 2002; Riethman 2008). These subtelomeric blocks are usually found duplicated at several chromosome extremities and may be present or absent at allelic positions, thus defining large segmental polymorphisms that are stably transmitted through meiosis (Der-Sarkissian et al. 2002). It is conceivable that TPE at particular chromosome extremities may be modulated by the presence/absence of such blocks and that polymorphisms in these regions could be responsible for potential differences in the way TPE influences gene expression among individuals; no data, however, have been provided so far in support of this hypothesis.

In all, it remains possible, but still to be demonstrated, that TPE contributes to the global gene expression changes that intervene during telomere-driven senescence in vitro and thereby to the aging process in vivo.

Impact of Telomeres on Aging and Disease

Aging, the process that renders an organism increasingly susceptible to death with increasing age (Kowald 2002), may be regarded as a consequence of the expression of deleterious mutations accumulated during life (Kirkwood 2002). Therefore, loss of function in genes implicated in DNA maintenance and repair may play a major role (Partridge 2001), which is also suggested by the observation that mutations having an impact on aging often affect genes whose products are implicated in these processes (for a recent review, see Lombard et al. 2005). Since telomeres are directly implicated in cell proliferation and tissue renewal capacity, it is expected that genes linked to telomere maintenance are tightly linked to aging and to diseases that lead to premature aging.

The strongest experimental arguments connecting telomere length and organismal aging have been obtained in mice that lack telomerase activity (Blasco et al. 1997; Rudolph et al. 1999). A detailed analysis of aging phenotypes in successive generations of *terc*$^{-/-}$ mice revealed that telomere length was inversely correlated with the incidence of aging manifestations such as skin lesions, alopecia, and hair graying, together with a shortened life span (Rudolph et al. 1999). When the *terc* knockout was introduced in mice with short telomeres, aging phenotypes appeared even earlier (Herrera et al. 1999). Alternatively, the reintroduction of telomerase in late-generation *terc*$^{-/-}$ mice rescued the chromosome instability and aging phenotypes (Samper et al. 2001), but only one copy of *terc* was not always sufficient to reestablish telomere homeostasis (haploinsufficiency) (Hathcock et al. 2002). This indicated that in mice, like in humans, levels of telomerase RNA are normally limiting.

In humans, premature aging or progeroid syndromes have provided important insights into the way telomeres impact on human aging. For instance, Werner's syndrome, which is an autosomal recessive genetic disorder caused by loss-of-function mutations in a gene coding for a helicase of the RecQ family (Yu et al. 1996), is associated with accelerated telomere shortening, although formal proof that telomere shortening causes Werner's syndrome manifestations is missing. This shortening may be explained by the fact that the WRN helicase has been implicated in telomere replication, its defect being responsible for the loss of the lagging-strand replicated telomere (Crabbe et al. 2004).

A tighter connection between telomere shortening and aging has been provided from patients with DC. Many of these patients carry a mutation either in *DKC1*, in *hTERC*, or more rarely in *hTERT* and bear shortened telomeres, and their lymphocytes show lower in vitro telomerase activity (Armanios et al. 2005; Marrone et al. 2004; Vulliamy et al. 2001). Clinical manifestations typically affect organs with high proliferation capacity, such as the hematopoietic system, thus pointing to telomere-dependent replicative senescence as the driving force of the disease (Shay and Wright 2004). More recently, mutations in *TINF2* (the gene coding for TIN2) have been found associated with severe forms of DC and this severity correlates with telomere shortness (Savage et al. 2008; Walne et al. 2008). As already mentioned, anticipation, a phenomenon by which the disease is manifested at earlier ages and in more deadly ways in later generations (Armanios et al. 2005; Goldman et al. 2005; Vulliamy et al. 2004), strongly supports the idea that short telomeres are the cause of the disease. Anticipation has been interpreted as the consequence of short telomeres being transmitted to the offspring together with a mutation in *hTERC*, and this hypothesis has also received strong support from mouse models of telomerase haploinsufficiency (Erdmann et al. 2004; Hathcock et al. 2002). Therefore, it is clear that inheriting short telomeres in the context of perturbed telomere homeostasis does accelerate aging. However, the consequences of inheriting short telomeres in the context of normal telomere maintenance mechanisms are not known.

Inheriting very short telomeres in a context of normal telomere homeostasis, as in siblings of DC patients of second and third generations in families with the autosomal form of DC, does not lead to premature aging (Goldman et al. 2005), although the effect on aging-related disease or life span remains to be determined. Interestingly, a recent study on telomere lengths of cerebral gray and white matter, obtained from autopsied patients of different ages, suggested that telomere lengths are well maintained under postmitotic conditions and that longer telomere lengths are associated with longer life span (Nakamura et al. 2007). If confirmed in larger and different ethnic groups, this would be the first indication that longer telomeres may retard aging manifestations and promote longevity. Still, it is possible that it is the shortening rate with age and not the initial telomere length at birth that is important in aging and longevity.

Alternatively, inheriting just one or a few telomeres – alleles – that are naturally very short may also have an impact on aging, since only one or a few critically short telomeres will be enough, after cell proliferation, to trigger mitotic senescence. This has been observed in vitro (Zou et al. 2004) and has also been suggested by the experiments in which the number of nuclei containing dysfunctional telomeres, but not the number of dysfunctional telomeres per nucleus, was found to increase with age in the skin of baboons (Jeyapalan et al. 2007). Since population studies of telomere lengths exclusively utilize techniques unable to evaluate single chromosome extremities, the impact of the presence of a few ones with very short telomeres on aging may have been neglected.

Independent of length, telomere structure may affect aging. As already mentioned, lasting modifications in shelterin components may modify the telomere structure in a permanent way and therefore influence telomere function and cell state. For example, the presence of short telomeres is not enough to recapitulate all the clinical features of DC in mice. However, when combined with mutations in POT1b, characteristic DC manifestations appear, suggesting that telomere dysfunction is caused not only by shortness but also by an abnormal structure (Hockemeyer et al. 2008). Similarly, as already noted, mutations in *TINF2* cause DC (Walne et al. 2008), likely through shortening of telomeres, but, since this protein is at the heart of the shelterin complex (de Lange 2005), it may also do so through alteration of the whole telomere structure.

Mice lacking SIRT6 also display a premature aging phenotype and have a short life span (Mostoslavsky et al. 2006). The observation that SIRT6 is required to modify the chromatin status of telomeres to allow the correct recruitment of WRN during telomere replication (Michishita et al. 2008) not only provides a remarkable molecular explanation for this model but also may become the paradigm for the relationship between chromatin regulation at telomeres, cellular senescence, and organismal aging.

Regarding the impact of telomeres on common aging-related diseases, it is worth noting that no other chromosome structure has been linked to major human health issues as tightly as telomeres. Telomere length has become an obligatory biomarker for anyone analyzing the impact of any factor (either environmental or genetic) on human fitness, even more so in aged populations. Short telomeres have then been described as associated with obesity, bone demineralization and other aging manifestations, and high risk of premature death and of development of cancer and vascular or colon diseases (Bekaert et al. 2005b; Benetos et al. 2004; Broberg et al. 2005; Cawthon et al. 2003; Epel et al. 2006; O'Sullivan et al. 2002, 2006; Shen et al. 2007; von Zglinicki et al. 2000; Wu et al. 2003), although some of these observations have not been reproduced in independent studies (Adams et al. 2007; Bischoff et al. 2006; Martin-Ruiz et al. 2005).

Nevertheless, deciding whether short telomeres are the cause and not just the consequence of an altered health status, which in turn is associated with aging or short life span, is not easy. For example, cancer incidence increases abruptly after the fourth decade of life (Balducci 2005) and more than two-thirds of all malignancies occur in people aged 65 and older. Intriguingly, patients with bladder, head and neck, lung, or renal cell carcinomas tend to bear shorter telomeres, in their peripheral blood lymphocytes, than do age-matched individuals without cancer but with similar risk factors (Wu et al. 2003). As in the case of chronic inflammatory diseases, such as arteriosclerosis (also connected to short telomeres), one could argue that immune responses elicited by cancer processes may impinge upon hematopoietic turnover, thus affecting the telomere length of nucleated blood cells. Whatever be the case, more studies are needed to pinpoint the real nature of the relationship between telomeres and age-related diseases.

Concluding Remarks

A lot remains to be explored about the relationship between telomeres and the aging process. Nevertheless, a great wealth of experimental and descriptive data indicate that telomere length has a great impact on health fitness. Polymorphisms in genes directly involved in telomere maintenance (replication, repair, elongation), and therefore likely to impact on telomere length homeostasis, potentially contribute to differences in telomere length among individuals and perhaps in the way they age. Telomere lengths also vary among chromosome extremities and this variation is definitely genetically determined, perhaps through imprinting mechanisms. Inheriting particularly short alleles may therefore influence the proliferation potential of cells in tissues with high turnover. On the other hand, programmed or stochastic modifications in the characteristics of telomeric heterochromatin or the shelterin complex may have a great impact on aging. It is clear, for example, that widespread changes in DNA methylation and histone modifications (Jones and Baylin 2007), as well as changes in the levels of shelterin components (Bellon et al. 2006; Lin et al. 2006; Ning et al. 2006; Oh et al. 2005; Yamada et al. 2002a, b), occur during cancer development, a major aging-related disease. Whether such changes also concur with normal aging or other aging-related diseases remains to be determined. On the other hand, telomere shortening with age may impact on the heterochromatic state of many chromosome extremities, thus influencing the gene expression landscape through TPE. Hopefully, future studies will come to precisely recognize and assess the influence of genetic and epigenetic factors, not only globally but also at specific chromosome extremities, on telomere function and their impact on aging.

Acknowledgments I thank all the people from my lab for fruitful discussions on this theme. Work in my laboratory is supported by grants from ARC, INCa, ANR, and La Ligue.

References

Adams, J., C. Martin-Ruiz, M. S. Pearce, M. White, L. Parker et al., 2007. No association between socio-economic status and white blood cell telomere length. Aging Cell **6:** 125–128.

Andrew, T., A. Aviv, M. Falchi, G. L. Surdulescu, J. P. Gardner et al., 2006. Mapping genetic loci that determine leukocyte telomere length in a large sample of unselected female sibling pairs. Am J Hum Genet **78:** 480–486.
Armanios, M., J. L. Chen, Y. P. Chang, R. A. Brodsky, A. Hawkins et al., 2005. Haploinsufficiency of telomerase reverse transcriptase leads to anticipation in autosomal dominant dyskeratosis congenita. Proc Natl Acad Sci USA **102:** 15960–15964.
Azzalin, C. M., and J. Lingner, 2008. Telomeres: the silence is broken. Cell Cycle **7:** 1161–1165.
Azzalin, C. M., P. Reichenbach, L. Khoriauli, E. Giulotto and J. Lingner, 2007. Telomeric repeat containing RNA and RNA surveillance factors at mammalian chromosome ends. Science **318:** 798–801.
Baird, D. M., J. Coleman, Z. H. Rosser and N. J. Royle, 2000. High levels of sequence polymorphism and linkage disequilibrium at the telomere of 12q: implications for telomere biology and human evolution. Am J Hum Genet **66:** 235–250.
Baird, D. M., J. Rowson, D. Wynford-Thomas and D. Kipling, 2003. Extensive allelic variation and ultrashort telomeres in senescent human cells. Nat Genet **33:** 203–207.
Balducci, L., 2005. Epidemiology of cancer and aging. J Oncol Manag **14:** 47–50.
Baur, J. A., Y. Zou, J. W. Shay and W. E. Wright, 2001. Telomere position effect in human cells. Science **292:** 2075–2077.
Bekaert, S., T. De Meyer and P. Van Oostveldt, 2005a. Telomere attrition as ageing biomarker. Anticancer Res **25:** 3011–3021.
Bekaert, S., I. Van Pottelbergh, T. De Meyer, H. Zmierczak, J. M. Kaufman et al., 2005b. Telomere length versus hormonal and bone mineral status in healthy elderly men. Mech Ageing Dev **126:** 1115–1122.
Bellon, M., A. Datta, M. Brown, J. F. Pouliquen, P. Couppie et al., 2006. Increased expression of telomere length regulating factors TRF1, TRF2 and TIN2 in patients with adult T-cell leukemia. Int J Cancer **119:** 2090–2097.
Benetos, A., J. P. Gardner, M. Zureik, C. Labat, L. Xiaobin et al., 2004. Short telomeres are associated with increased carotid atherosclerosis in hypertensive subjects. Hypertension **43:** 182–185.
Benetti, R., M. Garcia-Cao and M. A. Blasco, 2007a. Telomere length regulates the epigenetic status of mammalian telomeres and subtelomeres. Nat Genet **39:** 243–250.
Benetti, R., S. Gonzalo, I. Jaco, P. Munoz, S. Gonzalez et al., 2008. A mammalian microRNA cluster controls DNA methylation and telomere recombination via Rbl2-dependent regulation of DNA methyltransferases. Nat Struct Mol Biol **15:** 268–279.
Benetti, R., S. Gonzalo, I. Jaco, G. Schotta, P. Klatt et al., 2007b. Suv4-20 h deficiency results in telomere elongation and derepression of telomere recombination. J Cell Biol **178:** 925–936.
Bessler, M., D. B. Wilson and P. J. Mason, 2004. Dyskeratosis congenita and telomerase. Curr Opin Pediatr **16:** 23–28.
Bischoff, C., H. C. Petersen, J. Graakjaer, K. Andersen-Ranberg, J. W. Vaupel et al., 2006. No association between telomere length and survival among the elderly and oldest old. Epidemiology **17:** 190–194.
Blackburn, E. H., 2001. Switching and signaling at the telomere. Cell **106:** 661–673.
Blasco, M. A., 2007. The epigenetic regulation of mammalian telomeres. Nat Rev Genet **8:** 299–309.
Blasco, M. A., H. W. Lee, M. Rizen, D. Hanahan, R. DePinho et al., 1997. Mouse models for the study of telomerase. Ciba Found Symp **211:** 160–170; discuss.
Bodnar, A. G., M. Ouellette, M. Frolkis, S. E. Holt, C. P. Chiu et al., 1998. Extension of life-span by introduction of telomerase into normal human cells. Science **279:** 349–352.
Britt-Compton, B., J. Rowson, M. Locke, I. Mackenzie, D. Kipling et al., 2006. Structural stability and chromosome-specific telomere length is governed by cis-acting determinants in humans. Hum Mol Genet **15:** 725–733.
Broberg, K., J. Bjork, K. Paulsson, M. Hoglund and M. Albin, 2005. Constitutional short telomeres are strong genetic susceptibility markers for bladder cancer. Carcinogenesis **26:** 1263–1271.

Bryan, T. M., A. Englezou, L. Dalla-Pozza, M. A. Dunham and R. R. Reddel, 1997. Evidence for an alternative mechanism for maintaining telomere length in human tumors and tumor-derived cell lines. Nat Med **3:** 1271–1274.

Canela, A., E. Vera, P. Klatt and M. A. Blasco, 2007. High-throughput telomere length quantification by FISH and its application to human population studies. Proc Natl Acad Sci USA **104:** 5300–5305.

Cawthon, R. M., K. R. Smith, E. O'Brien, A. Sivatchenko and R. A. Kerber, 2003. Association between telomere length in blood and mortality in people aged 60 years or older. Lancet **361:** 393–395.

Cerone, M. A., J. A. Londoño-Vallejo and S. Bacchetti, 2001. Telomere maintenance by telomerase and by recombination can coexist in human cells. Hum Mol Genet **10:** 1945–1952.

Cesare, A. J., and R. R. Reddel, 2008. Telomere uncapping and alternative lengthening of telomeres. Mech Ageing Dev **129:** 99–108.

Chan, S. R., and E. H. Blackburn, 2004. Telomeres and telomerase. Philos Trans R Soc Lond B Biol Sci **359:** 109–121.

Choi, D., P. S. Whittier, J. Oshima and W. D. Funk, 2001. Telomerase expression prevents replicative senescence but does not fully reset mRNA expression patterns in Werner syndrome cell strains. FASEB J **15:** 1014–1020.

Coleman, J., D. M. Baird and N. J. Royle, 1999. The plasticity of human telomeres demonstrated by a hypervariable telomere repeat array that is located on some copies of 16p and 16q. Hum Mol Genet **8:** 1637–1646.

Crabbe, L., R. E. Verdun, C. I. Haggblom and J. Karlseder, 2004. Defective telomere lagging strand synthesis in cells lacking WRN helicase activity. Science **306:** 1951–1953.

Cristofari, G., and J. Lingner, 2006. Telomere length homeostasis requires that telomerase levels are limiting. Embo J **25:** 565–574.

de Lange, T., 2002. Protection of mammalian telomeres. Oncogene **21:** 532–540.

de Lange, T., 2005. Shelterin: the protein complex that shapes and safeguards human telomeres. Genes Dev **19:** 2100–2110.

Der-Sarkissian, H., G. Vergnaud, Y. M. Borde, G. Thomas and J. A. Londono-Vallejo, 2002. Segmental polymorphisms in the proterminal regions of a subset of human chromosomes. Genome Res **12:** 1673–1678.

Ding, H., M. Schertzer, X. Wu, M. Gertsenstein, S. Selig et al., 2004. Regulation of murine telomere length by Rtel: an essential gene encoding a helicase-like protein. Cell **117:** 873–886.

Epel, E. S., J. Lin, F. H. Wilhelm, O. M. Wolkowitz, R. Cawthon et al., 2006. Cell aging in relation to stress arousal and cardiovascular disease risk factors. Psychoneuroendocrinology **31:** 277–287.

Erdmann, N., Y. Liu and L. Harrington, 2004. Distinct dosage requirements for the maintenance of long and short telomeres in mTert heterozygous mice. Proc Natl Acad Sci USA **101:** 6080–6085. Epub 2004 Apr 6012.

Garcia-Cao, M., R. O'Sullivan, A. H. Peters, T. Jenuwein and M. A. Blasco, 2004. Epigenetic regulation of telomere length in mammalian cells by the Suv39h1 and Suv39h2 histone methyltransferases. Nat Genet **36:** 94–99. Epub 2003 Dec 2014.

Gilson, E., and V. Geli, 2007. How telomeres are replicated. Nat Rev Mol Cell Biol **8:** 825–838.

Gilson, E., and A. Londono-Vallejo, 2007. Telomere length profiles in humans: all ends are not equal. Cell Cycle **6:** 2486–2494.

Goldman, F., R. Bouarich, S. Kulkarni, S. Freeman, H. Y. Du et al., 2005. The effect of TERC haploinsufficiency on the inheritance of telomere length. Proc Natl Acad Sci USA **102:** 17119–17124.

Gonzalo, S., and M. A. Blasco, 2005. Role of Rb family in the epigenetic definition of chromatin. Cell Cycle **4:** 752–755.

Gonzalo, S., I. Jaco, M. F. Fraga, T. Chen, E. Li et al., 2006. DNA methyltransferases control telomere length and telomere recombination in mammalian cells. Nat Cell Biol **8:** 416–424.

Graakjaer, J., H. Der-Sarkissian, A. Schmitz, J. Bayer, G. Thomas et al., 2006. Allele-specific relative telomere lengths are inherited. Hum Genet **119:** 344–350.

Graakjaer, J., L. Pascoe, H. Der-Sarkissian, G. Thomas, S. Kolvraa et al., 2004. The relative lengths of individual telomeres are defined in the zygote and strictly maintained during life. Aging Cell **3:** 97–102.

Harley, C. B., A. B. Futcher and C. W. Greider, 1990. Telomeres shorten during ageing of human fibroblasts. Nature **345:** 458–460.

Hastie, N. D., M. Dempster, M. G. Dunlop, A. M. Thompson, D. K. Green et al., 1990. Telomere reduction in human colorectal carcinoma and with ageing. Nature **346:** 866–868.

Hathcock, K. S., M. T. Hemann, K. K. Opperman, M. A. Strong, C. W. Greider et al., 2002. Haploinsufficiency of mTR results in defects in telomere elongation. Proc Natl Acad Sci USA **99:** 3591–3596.

Henderson, S., R. Allsopp, D. Spector, S. S. Wang and C. Harley, 1996. In situ analysis of changes in telomere size during replicative aging and cell transformation. J Cell Biol **134:** 1–12.

Herrera, E., E. Samper and M. A. Blasco, 1999. Telomere shortening in $mTR^{-/-}$ embryos is associated with failure to close the neural tube. EMBO J. **18:** 1172–1181.

Hockemeyer, D., W. Palm, R. C. Wang, S. S. Couto and T. de Lange, 2008. Engineered telomere degradation models dyskeratosis congenita. Genes Dev **22:** 1773–1785.

Hug, N., and J. Lingner, 2006. Telomere length homeostasis. Chromosoma **115:** 413–425.

Jeyapalan, J. C., M. Ferreira, J. M. Sedivy and U. Herbig, 2007. Accumulation of senescent cells in mitotic tissue of aging primates. Mech Ageing Dev **128:** 36–44.

Jones, P. A., and S. B. Baylin, 2007. The epigenomics of cancer. Cell **128:** 683–692.

Kaminker, P., C. Plachot, S. H. Kim, P. Chung, D. Crippen et al., 2005. Higher-order nuclear organization in growth arrest of human mammary epithelial cells: a novel role for telomere-associated protein TIN2. J Cell Sci **118:** 1321–1330.

Kirkwood, T. B., 2002. Evolution of ageing. Mech Ageing Dev **123:** 737–745.

Koering, C. E., A. Pollice, M. P. Zibella, S. Bauwens, A. Puisieux et al., 2002. Human telomeric position effect is determined by chromosomal context and telomeric chromatin integrity. EMBO Rep **3:** 1055–1061.

Kosciolek, B. A., and P. T. Rowley, 1998. Human lymphocyte telomerase is genetically regulated. Genes Chromosomes Cancer **21:** 124–130.

Kowald, A., 2002. Lifespan does not measure ageing. Biogerontology **3:** 187–190.

Lages, C. S., O. Etienne, J. Comte, L. R. Gauthier, C. Granotier et al., 2004. Identification of alternative transcripts of the TRF1/Pin2 gene. J Cell Biochem **93:** 968–979.

Lansdorp, P. M., N. P. Verwoerd, F. M. van de Rijke, V. Dragowska, M. T. Little et al., 1996. Heterogeneity in telomere length of human chromosomes. Hum Mol Genet **5:** 685–691.

Lin, X., J. Gu, C. Lu, M. R. Spitz and X. Wu, 2006. Expression of telomere-associated genes as prognostic markers for overall survival in patients with non-small cell lung cancer. Clin Cancer Res **12:** 5720–5725.

Liu, L., S. M. Bailey, M. Okuka, P. Munoz, C. Li et al., 2007. Telomere lengthening early in development. Nat Cell Biol **9:** 1436–1441.

Loayza, D., and T. De Lange, 2003. POT1 as a terminal transducer of TRF1 telomere length control. Nature **25:** 25.

Lombard, D. B., K. F. Chua, R. Mostoslavsky, S. Franco, M. Gostissa et al., 2005. DNA repair, genome stability, and aging. Cell **120:** 497–512.

Londoño-Vallejo, J. A., H. Der-Sarkissian, L. Cazes, S. Bacchetti and R. Reddel, 2004. Alternative lengthening of telomeres is characterized by high rates of inter-telomeric exchange. Cancer Res **64:** 2324–2327.

Londono-Vallejo, J. A., H. DerSarkissian, L. Cazes and G. Thomas, 2001. Differences in telomere length between homologous chromosomes in humans. Nucleic Acids Res **29:** 3164–3171.

Makarov, V. L., Y. Hirose and J. P. Langmore, 1997. Long G tails at both ends of human chromosomes suggest a C strand degradation mechanism for telomere shortening. Cell **88:** 657–666.

Makovets, S., T. L. Williams and E. H. Blackburn, 2008. The telotype defines the telomere state in *Saccharomyces cerevisiae* and is inherited as a dominant non-Mendelian characteristic in cells lacking telomerase. Genetics **178:** 245–257.

Marrone, A., D. Stevens, T. Vulliamy, I. Dokal and P. J. Mason, 2004. Heterozygous telomerase RNA mutations found in dyskeratosis congenita and aplastic anemia reduce telomerase activity via haploinsufficiency. Blood **104:** 3936–3942.

Martin-Ruiz, C. M., J. Gussekloo, D. van Heemst, T. von Zglinicki and R. G. Westendorp, 2005. Telomere length in white blood cells is not associated with morbidity or mortality in the oldest old: a population-based study. Aging Cell **4:** 287–290.

Matsubara, Y., M. Murata, T. Yoshida, K. Watanabe, I. Saito et al., 2006. Telomere length of normal leukocytes is affected by a functional polymorphism of hTERT. Biochem Biophys Res Commun **341:** 128–131.

Mefford, H. C., and B. J. Trask, 2002. The complex structure and dynamic evolution of human subtelomeres. Nat Rev Genet **3:** 91–102.

Michishita, E., R. A. McCord, E. Berber, M. Kioi, H. Padilla-Nash et al., 2008. SIRT6 is a histone H3 lysine 9 deacetylase that modulates telomeric chromatin. Nature **452:** 492–496.

Mostoslavsky, R., K. F. Chua, D. B. Lombard, W. W. Pang, M. R. Fischer et al., 2006. Genomic instability and aging-like phenotype in the absence of mammalian SIRT6. Cell **124:** 315–329.

Motonaga, K., M. Itoh, Y. Hachiya, A. Endo, K. Kato et al., 2002. Age related expression of Werner's syndrome protein in selected tissues and coexpression of transcription factors. J Clin Pathol **55:** 195–199.

Nakamura, K., K. Takubo, N. Izumiyama-Shimomura, M. Sawabe, T. Arai et al., 2007. Telomeric DNA length in cerebral gray and white matter is associated with longevity in individuals aged 70 years or older. Exp Gerontol **42:** 944–950.

Nawrot, T. S., J. A. Staessen, J. P. Gardner and A. Aviv, 2004. Telomere length and possible link to X chromosome. Lancet **363:** 507–510.

Netzer, C., L. Rieger, A. Brero, C. D. Zhang, M. Hinzke et al., 2001. SALL1, the gene mutated in Townes-Brocks syndrome, encodes a transcriptional repressor which interacts with TRF1/PIN2 and localizes to pericentromeric heterochromatin. Hum Mol Genet **10:** 3017–3024.

Neumann, A. A., and R. R. Reddel, 2002. Telomere maintenance and cancer – look, no telomerase. Nat Rev Cancer **2:** 879–884.

Ning, H., T. Li, L. Zhao, J. Li, J. Liu et al., 2006. TRF2 promotes multidrug resistance in gastric cancer cells. Cancer Biol Ther **5:** 950–956.

Ning, Y., J. F. Xu, Y. Li, L. Chavez, H. C. Riethman et al., 2003. Telomere length and the expression of natural telomeric genes in human fibroblasts. Hum Mol Genet **12:** 1329–1336.

Njajou, O. T., R. M. Cawthon, C. M. Damcott, S. H. Wu, S. Ott et al., 2007. Telomere length is paternally inherited and is associated with parental lifespan. Proc Natl Acad Sci U S A.

Nordfjall, K., A. Larefalk, P. Lindgren, D. Holmberg and G. Roos, 2005. Telomere length and heredity: Indications of paternal inheritance. Proc Natl Acad Sci USA **102:** 16374–16378.

O'Sullivan, J., R. A. Risques, M. T. Mandelson, L. Chen, T. A. Brentnall et al., 2006. Telomere length in the colon declines with age: a relation to colorectal cancer? Cancer Epidemiol Biomarkers Prev **15:** 573–577.

O'Sullivan, J. N., M. P. Bronner, T. A. Brentnall, J. C. Finley, W. T. Shen et al., 2002. Chromosomal instability in ulcerative colitis is related to telomere shortening. Nat Genet **32:** 280–284.

Ofir, R., A. C. Wong, H. E. McDermid, K. L. Skorecki and S. Selig, 1999. Position effect of human telomeric repeats on replication timing. Proc Natl Acad Sci USA **96:** 11434–11439.

Oh, B. K., Y. J. Kim, C. Park and Y. N. Park, 2005. Up-regulation of telomere-binding proteins, TRF1, TRF2, and TIN2 is related to telomere shortening during human multistep hepatocarcinogenesis. Am J Pathol **166:** 73–80.

Olovnikov, A. M., 1971. Principle of marginotomy in template synthesis of polynucleotides. Dokl Akad Nauk SSSR **201:** 1496–1499.

Opresko, P. L., C. von Kobbe, J. P. Laine, J. Harrigan, I. D. Hickson et al., 2002. Telomere-binding protein TRF2 binds to and stimulates the Werner and Bloom syndrome helicases. J Biol Chem **277:** 41110–41119.

Ottaviani, A., E. Gilson and F. Magdinier, 2008. Telomeric position effect: from the yeast paradigm to human pathologies? Biochimie **90:** 93–107.

Paldi, A., 2003. Genomic imprinting: could the chromatin structure be the driving force? Curr Top Dev Biol **53:** 115–138.

Partridge, L., 2001. Evolutionary theories of ageing applied to long-lived organisms. Exp Gerontol **36:** 641–650.

Pedram, M., C. N. Sprung, Q. Gao, A. W. Lo, G. E. Reynolds et al., 2006. Telomere position effect and silencing of transgenes near telomeres in the mouse. Mol Cell Biol **26:** 1865–1878.

Powell, C. L., J. A. Swenberg and I. Rusyn, 2005. Expression of base excision DNA repair genes as a biomarker of oxidative DNA damage. Cancer Lett **229:** 1–11.

Recillas-Targa, F., 2002. DNA methylation, chromatin boundaries, and mechanisms of genomic imprinting. Arch Med Res **33:** 428–438.

Redon, S., P. Reichenbach and J. Lingner, 2007. Protein RNA and protein–protein interactions mediate association of human EST1A/SMG6 with telomerase. Nucleic Acids Res **35:** 7011–7022.

Rhodes, D., L. Fairall, T. Simonsson, R. Court and L. Chapman, 2002. Telomere architecture. EMBO Rep **3:** 1139–1145.

Riethman, H., 2008. Human telomere structure and biology. Annu Rev Genomics Hum Genet **9:** 1–19.

Rippmann, J. F., K. Damm and A. Schnapp, 2002. Functional characterization of the poly(ADP-ribose) polymerase activity of tankyrase 1, a potential regulator of telomere length. J Mol Biol **323:** 217–224.

Rudolph, K. L., S. Chang, H. W. Lee, M. Blasco, G. J. Gottlieb et al., 1999. Longevity, stress response, and cancer in aging telomerase-deficient mice. Cell **96:** 701–712.

Rufer, N., W. Dragowska, G. Thornbury, E. Roosnek and P. M. Lansdorp, 1998. Telomere length dynamics in human lymphocyte subpopulations measured by flow cytometry [see comments]. Nat Biotechnol **16:** 743–747.

Samper, E., J. M. Flores and M. A. Blasco, 2001. Restoration of telomerase activity rescues chromosomal instability and premature aging in Terc$^{-/-}$ mice with short telomeres. EMBO Rep **2:** 800–807.

Savage, S. A., N. Giri, G. M. Baerlocher, N. Orr, P. M. Lansdorp et al., 2008. TINF2, a component of the shelterin telomere protection complex, is mutated in dyskeratosis congenita. Am J Hum Genet **82:** 501–509.

Schoeftner, S., and M. A. Blasco, 2008. Developmentally regulated transcription of mammalian telomeres by DNA-dependent RNA polymerase II. Nat Cell Biol **10:** 228–236.

Schulz, V. P., V. A. Zakian, C. E. Ogburn, J. McKay, A. A. Jarzebowicz et al., 1996. Accelerated loss of telomeric repeats may not explain accelerated replicative decline of Werner syndrome cells. Hum Genet **97:** 750–754.

Shawi, M., and C. Autexier, 2008. Telomerase, senescence and ageing. Mech Ageing Dev **129:** 3–10.

Shay, J. W., and W. E. Wright, 2004. Telomeres in dyskeratosis congenita. Nat Genet **36:** 437–438.

Shay, J. W., and W. E. Wright, 2005. Senescence and immortalization: role of telomeres and telomerase. Carcinogenesis **26:** 867–874.

Shen, J., M. B. Terry, I. Gurvich, Y. Liao, R. T. Senie et al., 2007. Short telomere length and breast cancer risk: a study in sister sets. Cancer Res **67:** 5538–5544.

Slagboom, P. E., S. Droog and D. I. Boomsma, 1994. Genetic determination of telomere size in humans: a twin study of three age groups. Am J Hum Genet **55:** 876–882.

Smith, S., I. Giriat, A. Schmitt and T. de Lange, 1998. Tankyrase, a poly(ADP-ribose) polymerase at human telomeres. Science **282:** 1484–1487.

Songyang, Z., and D. Liu, 2006. Inside the mammalian telomere interactome: regulation and regulatory activities of telomeres. Crit Rev Eukaryot Gene Expr **16:** 103–118.

Sugiyama, T., H. Cam, A. Verdel, D. Moazed and S. I. Grewal, 2005. From the cover: RNA-dependent RNA polymerase is an essential component of a self-enforcing loop coupling heterochromatin assembly to siRNA production. Proc Natl Acad Sci USA **102:** 152–157.

Surralles, J., M. P. Hande, R. Marcos and P. M. Lansdorp, 1999. Accelerated telomere shortening in the human inactive X chromosome. Am J Hum Genet **65:** 1617–1622.

Tahara, H., Y. Tokutake, S. Maeda, H. Kataoka, T. Watanabe et al., 1997. Abnormal telomere dynamics of B-lymphoblastoid cell strains from Werner's syndrome patients transformed by Epstein-Barr virus. Oncogene **15:** 1911–1920.

Tanaka, H., M. S. Mendonca, P. S. Bradshaw, D. J. Hoelz, L. H. Malkas et al., 2005. DNA damage-induced phosphorylation of the human telomere-associated protein TRF2. Proc Natl Acad Sci USA **102:** 15539–15544.

Teixeira, M. T., M. Arneric, P. Sperisen and J. Lingner, 2004. Telomere length homeostasis is achieved via a switch between telomerase-extendible and -nonextendible states. Cell **117:** 323–335.

Tham, W. H., and V. A. Zakian, 2002. Transcriptional silencing at Saccharomyces telomeres: implications for other organisms. Oncogene **21:** 512–521.

Unryn, B. M., L. S. Cook and K. T. Riabowol, 2005. Paternal age is positively linked to telomere length of children. Aging Cell **4:** 97–101.

van Steensel, B., and T. de Lange, 1997. Control of telomere length by the human telomeric protein TRF1. Nature **385:** 740–743.

Vasa-Nicotera, M., S. Brouilette, M. Mangino, J. R. Thompson, P. Braund et al., 2005. Mapping of a major locus that determines telomere length in humans. Am J Hum Genet **76:** 147–151.

Vaziri, H., W. Dragowska, R. C. Allsopp, T. E. Thomas, C. B. Harley et al., 1994. Evidence for a mitotic clock in human hematopoietic stem cells: loss of telomeric DNA with age. Proc Natl Acad Sci USA **91:** 9857–9860.

Verdun, R. E., and J. Karlseder, 2006. The DNA damage machinery and homologous recombination pathway act consecutively to protect human telomeres. Cell **127:** 709–720.

Verdun, R. E., and J. Karlseder, 2007. Replication and protection of telomeres. Nature **447:** 924–931.

von Zglinicki, T., 2000. Role of oxidative stress in telomere length regulation and replicative senescence. Ann NY Acad Sci **908:** 99–110.

von Zglinicki, T., and C. M. Martin-Ruiz, 2005. Telomeres as biomarkers for ageing and age-related diseases. Curr Mol Med **5:** 197–203.

von Zglinicki, T., V. Serra, M. Lorenz, G. Saretzki, R. Lenzen-Grossimlighaus et al., 2000. Short telomeres in patients with vascular dementia: an indicator of low antioxidative capacity and a possible risk factor? [In Process Citation]. Lab Invest **80:** 1739–1747.

Vulliamy, T., A. Marrone, F. Goldman, A. Dearlove, M. Bessler et al., 2001. The RNA component of telomerase is mutated in autosomal dominant dyskeratosis congenita. Nature **413:** 432–435.

Vulliamy, T., A. Marrone, R. Szydlo, A. Walne, P. J. Mason et al., 2004. Disease anticipation is associated with progressive telomere shortening in families with dyskeratosis congenita due to mutations in TERC. Nat Genet **36:** 447–449. Epub 2004 Apr 2018.

Walne, A. J., T. J. Vulliamy, R. Beswick, M. Kirwan and I. Dokal, 2008. TINF2 mutations result in very short telomeres: Analysis of a large cohort of patients with dyskeratosis congenita and related bone marrow failure syndromes. Blood **112:** 3594–3600.

Watson, J. D., 1972. Origin of concatemeric T7 DNA. Nat New Biol **239:** 197–201.

Weng, N. P., B. L. Levine, C. H. June and R. J. Hodes, 1996. Regulated expression of telomerase activity in human T lymphocyte development and activation. J Exp Med **183:** 2471–2479.

Wong, J. M., and K. Collins, 2003. Telomere maintenance and disease. Lancet **362:** 983–988.

Wu, X., C. I. Amos, Y. Zhu, H. Zhao, B. H. Grossman et al., 2003. Telomere dysfunction: a potential cancer predisposition factor. J Natl Cancer Inst **95:** 1211–1218.

Wu, Y., K. Shin-ya and R. M. Brosh, Jr., 2008. FANCJ helicase defective in Fanconia anemia and breast cancer unwinds G-quadruplex DNA to defend genomic stability. Mol Cell Biol **28:** 4116–4128.

Wyllie, F. S., C. J. Jones, J. W. Skinner, M. F. Haughton, C. Wallis et al., 2000. Telomerase prevents the accelerated cell ageing of Werner syndrome fibroblasts. Nat Genet **24:** 16–17.

Yamada, K., A. Yagihashi, M. Yamada, K. Asanuma, R. Moriai et al., 2002a. Decreased gene expression for telomeric-repeat binding factors and TIN2 in malignant hematopoietic cells. Anticancer Res **22:** 1315–1320.

Yamada, M., N. Tsuji, M. Nakamura, R. Moriai, D. Kobayashi et al., 2002b. Down-regulation of TRF1, TRF2 and TIN2 genes is important to maintain telomeric DNA for gastric cancers. Anticancer Res **22:** 3303–3307.

Yamashita, A., I. Kashima and S. Ohno, 2005. The role of SMG-1 in nonsense-mediated mRNA decay. Biochim Biophys Acta **1754:** 305–315.

Yang, Q., R. Zhang, I. Horikawa, K. Fujita, Y. Afshar et al., 2007. Functional diversity of human protection of telomeres 1 isoforms in telomere protection and cellular senescence. Cancer Res **67:** 11677–11686.

Ye, J. Z., and T. De Lange, 2004. TIN2 is a tankyrase 1 PARP modulator in the TRF1 telomere length control complex. Nat Genet **9:** 9.

Yehezkel, S., Y. Segev, E. Viegas-Pequignot, K. Skorecki and S. Selig, 2008. Hypomethylation of subtelomeric regions in ICF syndrome is associated with abnormally short telomeres and enhanced transcription from telomeric regions. Hum Mol Genet **17:** 2776–2789.

Yu, C. E., J. Oshima, Y. H. Fu, E. M. Wijsman, F. Hisama et al., 1996. Positional cloning of the Werner's syndrome gene [see comments]. Science **272:** 258–262.

Zou, Y., A. Sfeir, S. M. Gryaznov, J. W. Shay and W. E. Wright, 2004. Does a sentinel or a subset of short telomeres determine replicative senescence? Mol Biol Cell **15:** 3709–3718.

Contributions of Tumor Suppressors to the Epigenetic Regulation of Aging Cells

Pinaki Bose, Amudha Ganapathy, and Karl Riabowol

Abstract Epigenetics refers to the initiation and maintenance of heritable patterns of gene expression and function without changes in the primary DNA sequence. Covalent modifications of histones and DNA methylation are the most common epigenetic modifications that influence the activity of various growth stimulatory and inhibitory genes. Cellular senescence might be viewed as a state when the "balance" of gene activity is skewed in the favour of growth inhibitory genes. By virtue of its influence on gene activity, the epigenetic state of chromatin profoundly regulates the establishment and perpetuation of senescence. Chromatin remodelling complexes such as histone acetyl transferases and histone deacetylases are active players in engineering the epigenetic landscape that leads to senescence induction. A variety of tumor suppressors are important components of various chromatin remodelling complexes and effect their growth inhibitory activity by bringing chromatin modifying proteins to DNA. In this chapter we discuss how major tumor suppressor proteins like p53, p16, PTEN, and ING engineer a change in the "epigenome" and enforce the senescent phenotype, mainly by regulating the state of facultative euchromatin.

Keywords Senescence · Epigenetics · Chromatin · Tumor suppressors · p53 · p16 · Rb · PTEN · ING

Introduction

The replicative life span of somatic human cells is limited to a finite number of population doublings in culture. This seminal discovery made by Leonard Hayflick in the early 1960s and now called the "Hayflick limit" heralded a field of study that

K. Riabowol (✉)
Departments of Biochemistry & Molecular Biology, Southern Alberta Cancer Research Institute, University of Calgary, 3330 Hospital Drive NW, Calgary, Alberta, Canada 42 N 4N1;
Departments of Oncology, Southern Alberta Cancer Research Institute, University of Calgary, 3330 Hospital Drive NW, Calgary, Alberta, Canada 42N 4N1
e-mail: karl@ucalgary.ca

T.O. Tollefsbol (ed.), *Epigenetics of Aging*, DOI 10.1007/978-1-4419-0639-7_13,

has addressed a question relevant to all: how do organisms age, particularly at the cellular level?

The "Hayflick limit" or the number of divisions that replicative cell types will undergo in culture is characteristic of the species, cell type, and donor age, and is enforced in human cells by the erosion of telomeres that occurs with each cell division. At the end of their replicative life span, senescent cells remain metabolically active but are unable to respond to mitogenic stimuli and cannot reenter the cell cycle. In contrast to senescence, replication-competent cells frequently exit the cell cycle and become quiescent, but they can be induced to divide again through mitogen stimulation. Thus, a major hallmark of senescence is irreversible cell cycle arrest and what appear to be permanent changes in chromatin organization and cytoskeletal structure (see Fig. 1). Changes in the state of chromatin organization are emerging as a possible causal agent of irreversible cell cycle arrest in senescence, perhaps due to heterochromatinization or condensation of certain chromatin domains that is believed to repress the expression of genes required for cell proliferation. These chromatin modifications are brought about by the alteration of key residues on histones and other proteins involved in chromatin assembly.

Tumor suppressors are a class of proteins that negatively regulate cell proliferation when activated in response to diverse stimuli such as DNA damage, telomere shortening, and other stresses and were initially called tumor suppressors since their inactivation promoted the emergence and growth of tumors. Tumor suppressors prevent malignancy by inducing cell cycle arrest and/or apoptosis depending on the type and degree of damage inflicted on the cell machinery. Since irreversible cell cycle arrest is a characteristic feature of senescence, tumor suppressor proteins are believed to have profound effects on the establishment of the senescent state and its perpetuation. In this chapter we will discuss the role of a subset of tumor suppressors in modulating the onset and maintenance of senescence. We will also

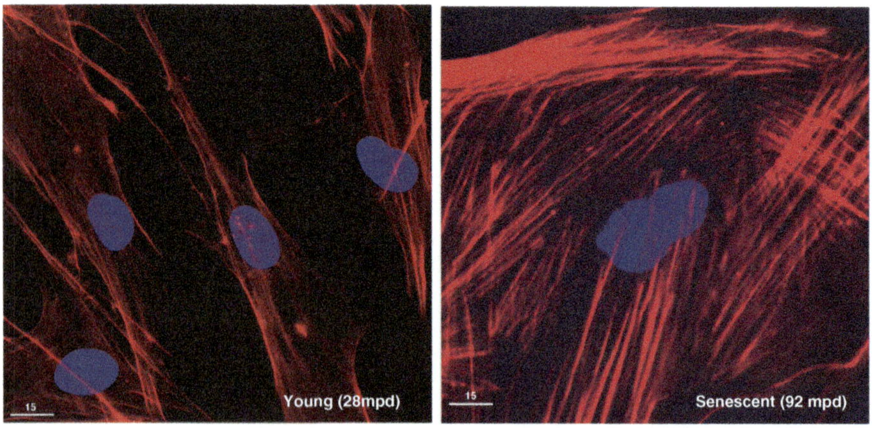

Fig. 1 Primary human fibroblasts increase in size and develop stress fibers during replicative senescence. The bars indicate 15 μM

discuss how major tumor suppressor proteins are involved in determining chromatin structure through epigenetic modifications that contribute to cellular senescence.

Linkage Between Replicative Senescence and Organismal Aging

Evidence recently accumulated from different experimental systems has now very strongly supported the idea that replicative senescence plays a causative role in organismal aging. It has also been shown that cellular senescence and apoptosis are important determinants in the development of certain age-related diseases (Zhang and Herman, 2002). Senescent cells have been detected and are found to accumulate with age in primate skin (Herbig et al., 2006; Jeyapalan et al., 2007), human vascular tissue (Minamino and Komuro, 2007; Matthews et al., 2006), and rodent and human kidneys (Melk et al., 2003; Melk et al., 2004; Krishnamurthy et al., 2004) among others.

Telomere erosion is considered to be the initiating signal responsible for replicative senescence in human cells. Telomere shortening occurs with increasing age in a variety of tissues including skin, liver, kidney, and lymphocytes (Hastie et al., 1990; Allsopp et al., 1992) and in animal models (Martin et al., 2002; Baerlocher et al., 2003). It has been shown that telomere dysfunction induces senescence in the fibroblasts of primate dermis (Herbig et al., 2006). Telomere loss, therefore, limits cellular replication and leads to the accumulation of senescent cells in old animals (Herbig et al., 2006; Jeyapalan et al., 2007).

Increased expression of heterochromatin-associated factors, including activation of the histone regulating A/antisilencing factor 1 (HIRA/ASF1) pathway, has been reported in the skin of aging primates (Herbig et al., 2006). It has also been observed that in baboon dermal fibroblasts, a correlation exists between the level of HIRA expression and the age of the donor (Jeyapalan et al., 2007). These observations link HIRA/ASF1 accumulation to the induction of senescence and suggest that senescence in vivo may correlate with age-related decline in physiological functions. The abundance of histone H4 methylated on lysine 20 has also been reported to increase with age in rat liver and kidney (Sarg et al., 2002), and this modification that is linked to transcriptional repression (Kousarides, 2007) supports the idea of heterochromatin accumulating as various tissues age in vivo.

Recent advances in the understanding of the biological role of the p53 family protein, p63, have unraveled one of the most tangible links between replicative senescence and organismal aging to date. For example, p63 heterozygous mutant mice have shortened life spans and an accelerated aging phenotype (Keyes et al., 2005). These mice were also less tumor-prone when compared to wild-type animals. Knockdown of p63 in primary keratinocytes (a cell type that expresses high levels of p63) caused growth arrest, morphological changes consistent with senescence, and elevation of endogenous β-gal activity. Also, conditional p63 deficiency in a keratin-5-expressing cell-specific background increased expression of senescence markers such as β-gal activity and p16INK4a in these cells (Keyes et al., 2006). A similar phenotype was observed when p63 deficiency was induced during embryogenesis. Finally, the interesting observation that selective p63 ablation in proliferative epithelia of adult mice recapitulated features of accelerated aging and concomitant induction of senescence lends credence to the theory that senescence is a causative factor in the aging process.

Caloric restriction has also been invoked to explain the link between cellular senescence and organismal aging. Li et al. (1997) studied the effects of age and long-term caloric restriction on the proliferation capacity of murine lens epithelial cells in vitro and in vivo. They showed that, as expected, old mice had more senescent lens epithelial cells than did young mice and also that diet-restricted animals had lower levels of senescent cells than did age-matched mice fed ad libitum. This confirmed a previous study reporting that caloric restriction not only extends life span and reduces age-related changes (Lane et al., 1992) but also reduces the rate of accumulation of senescent cells, thus further suggesting a relationship between replicative senescence and organismal aging. However, the mechanisms by which caloric restriction extends replicative and organismal life spans remain unclear. Several groups have reported a decline in metabolic activity with caloric restriction which in turn might reduce the production of reactive oxygen species, thus mitigating cellular DNA damage and prolonging life span (Masoro, 1996; Sohal and Weindruch, 1996). Since changes in chromatin structure are being linked increasingly to cell senescence, it will be interesting to test whether caloric restriction affects chromatin structure to any significant degree.

A better appreciation of the role of cellular senescence in various premature aging syndromes is also gradually emerging, and cells from individuals with Hutchinson–Gilford progeria syndrome (HGPS) or Werner's syndrome (WS) show a reduced replicative life span in culture and faster acquisition of senescence-like phenotypes, including changes in chromatin structure. In fact, changes in chromatin structure in HGPS have been linked to at least two tumor suppressors, Rb and ING1, as noted below.

Chromatin Structure Modification: Epigenetics at Work

Nucleosomes form the basic repeat unit of chromatin; 146 bp of DNA is wrapped around an octamer of core histones. Chromatin can be broadly classified into euchromatin and heterochromatin. Euchromatin is the transcription-competent type that is decondensed during interphase. Heterochromatin, on the other hand, can be further subdivided into constitutive and facultative types. Constitutive heterochromatin, such as that found in the pericentromeric regions of chromosomes, remains transcriptionally silent throughout the cell cycle and is unresponsive to external and internal proliferative stimuli. The conformation of constitutive heterochromatin is not cell type specific and is essentially fixed throughout the lifetime of an organism. In contrast, facultative heterochromatin is dynamic and oscillates between euchromatic and heterochromatic states in a tightly regulated manner. Whether facultative heterochromatin is euchromatic or heterochromatic is determined by epigenetic modifications of DNA and histones that initiate and maintain heritable patterns of gene expression and gene function without affecting changes in DNA sequence. Epigenetic signaling has crucial roles in tumorigenesis (Jones and Baylin, 2002) and cellular senescence. The best known examples of epigenetic mechanisms are covalent modifications of histones by acetylation, ubiquitination, methylation, ADP-ribosylation, and phosphorylation (Kouzarides, 2007). These modifications, performed by histone acetyltransferases (HATs), deacetylases (HDACs), methyltransferases (HMTs), kinases (HKs), and ubiquitin ligases, offer a mechanism through which upstream signaling pathways converge on common targets to regulate gene expression. HATs and HDACs play important roles in other cellular processes including transcription, DNA replication, and cell cycle progression. HATs acetylate key amino acid residues on histone tails, relaxing chromatin and generally promoting gene expression, while HDACs deacetylate the residues and work antagonistically to HATs. Also, many tumor suppressors including p53, are themselves posttranslationally modified by acetylation, which affects their activity. The links between these epigenetic changes and how they correspond to the process of cell aging are discussed in detail below.

Aging: A Change in the Epigenome

Consistent with the idea that there is a general repression of gene expression during cellular senescence, chromatin compaction is one of the main features that accompanies the onset of senescence (Barbie et al., 2004). As will be evident from further discussion and other chapters in this book, senescence-associated heterochromatic foci (SAHF) formation is an important phenomena that has been linked to cell aging and the repression of proliferation-inducing genes.

Certain regions of DNA, particularly some gene promoters, have clusters of cytosine and guanine residues termed CpG islands. CpG island methylation has long been associated with stable gene silencing and heterochromatin formation (Eden et al., 1998; Jones and Laird, 1999). Counterintuitively, it was discovered that CpG island methylation decreased as a function of population doublings in normal cells (Wilson and Jones, 1983) and during organismal aging (Hornsby et al., 1992). Enzymes responsible for this methylation, DNA methyltransferases (DNMTs), also showed a decrease in expression with age (Vertino et al., 1994). Although this argues against DNA methylation having a role in cell aging, the observation might instead be consistent with DNA methylation repressing a set of growth-inhibitory genes and this repression being attenuated with the onset of senescence, thus pushing cells toward proliferative arrest (Baylin and Herman, 2000; Jones and Baylin, 2002). The latter possibility is supported by the fact that with age inhibition of DNMT results in the transcriptional activation of the *p21* gene in human fibroblasts (Young and Smith, 2001). Recently, it has been shown that, although the activity of maintenance DNMTs is decreased in senescent fibroblasts, de novo methylation activity is actually increased during senescence (Lopatina et al., 2002). This is consistent with the fact that global genomic methylation decreases in senescence, with a concomitant rise in gene-localized hypermethylation due to increased de novo DNMT activity.

While it is a potentially attractive mechanism to explain altered gene expression during cell aging and the permanent exit from the cell cycle seen in senescent cells, DNA methylation also appears to drift with time (Shmookler Reis and Goldstein, 1982) and so may often represent an effect rather than a cause of senescence.

SAHF Formation

The establishment of senescence is increasingly believed to be associated with gross reorganization of chromatin structure, leading to transcriptional silencing of growth-promoting genes. Compelling evidence now links punctate, highly condensed domains of facultative heterochromatin, called senescence-associated heterochromatin foci (SAHF), to the onset of senescence (Wright and Shay, 2002; Narita et al., 2003; Campisi, 2005; Herbig and Sedivy, 2006). The importance of chromatin domains like SAHF is further illustrated by the fact that many genes that are up- or downregulated during senescence appear to be physically clustered (Zhang et al., 2003). The two chromatin regulators, HIRA and ASF1, discussed earlier in this chapter, have been implicated in driving the formation of SAHF in human cells (Zhang et al., 2005; Zhang et al., 2007). These proteins promote the accumulation of heterochromatin proteins such as lysine 9 methylated histone 3, the histone H2A variant macro-H2A (Zhang et al., 2005), and HP1α, β, and γ (Narita et al., 2003) in SAHF. The accumulation of these heterochromatin proteins in SAHF may contribute to the stable growth arrest seen in senescent cells. The high-mobility group A proteins (HMGA 1 or HMGA 2) are also considered integral structural components of SAHF in normal human fibroblasts (Narita et al., 2006). Knockdown of HMGA by RNAi prevents RasG12V-induced SAHF formation and proliferation arrest, indicating that HMGA proteins act as essential components of SAHF formation which in turn might limit proliferative capacity. Both increased association of HMGA2 with chromatin and loss of histone H1 have been reported to be essential for SAHF formation (Funayama et al., 2006). The loss of histone H1 not only enhances the incorporation of HMGA2 into chromatin and the alteration of chromatin structure but also induces some senescence mediators including p53 and p21 with the associated growth arrest and cellular changes predicted (Funayama and Ishikawa, 2007).

The expression of the p16INK4a tumor suppressor has been shown to be markedly elevated in senescent cells (Alcorta et al., 1996; Wong and Riabowol, 1996; Stein et al., 1999). Enforced expression of p16INK4a was reported to be sufficient to induce SAHF (Narita et al., 2003). Despite its beneficial role in SAHF formation, depletion of p16 in senescent cells has little impact on SAHF or E2F-target gene expression, which might emphasize its importance in the establishment but perhaps not in the maintenance of SAHF (Wang et al., 1999; Narita et al., 2006). Additional proteins affecting the Rb pathway include prohibitin, a potent tumor suppressor/growth regulatory protein, which appears to be linked to the regulation of senescence (Liu et al., 1994). Similar to Rb, prohibitin interacts with and represses E2F family members (Rastogi et al., 2006). Decreased cellular levels of prohibitin lead to a reduction in the number of SAHF and an impaired ability of cells to undergo senescence, again suggesting an integral role for epigenetics and chromatin structure modification in cellular senescence (Rastogi et al., 2006).

Downregulation of the Wnt pathway in senescent cells also serves as a trigger for SAHF formation (Ye et al., 2007). Perhaps in a manner opposite to that seen by another signal transducer, ras, when it is overexpressed (Serrano et al., 1997), repression of the Wnt family member, Wnt2, occurs independently of p53 and Rb in early senescence and is found to be involved in driving the relocalization of HIRA to PML bodies (Ye et al., 2007). Cycling of HIRA through PML bodies has been shown to be important for the formation of SAHF (Zhang et al., 2005); thus, Wnt function is intimately linked to SAHF formation and the assembly of facultative heterochromatin might help to establish cell senescence. In contrast to in vitro cell reports, studies done in whole animal models have indicated that Wnt signaling is augmented in aging. Two recent papers have reported that increased Wnt signaling also results in stem cell senescence in various tissues of aging mice (Brack et al., 2007; Liu et al., 2007). An additional tumor suppressor linked to the formation of SAHF is a splicing isoform of ING1. In this study, discussed below, expression of ING1a is sufficient, by itself, to efficiently induce the formation of SAHF (Soliman et al., 2008).

Major Tumor Suppressor Pathways Contributing to the Senescence Phenotype

The importance of senescence as a tumor suppressive mechanism has been emphasized by recent studies showing that reactivation of p53 in murine tumors causes cell senescence and associated tumor regression (Ventura et al., 2007; Xue et al., 2007). Tumor suppressors limit cellular proliferation in response to a variety of stimuli including DNA damage, telomere shortening, and other stresses. Consistent with a role in suppressing cancer, cellular senescence is positively regulated by several tumor suppressors (reviewed in Campisi, 2005) including p53 and Rb (Campisi, 2001), p21WAF1 (Stein et al., 1999), p16INK4a, and $p19^{ARF}$ (Alcorta et al., 1996). The activities of the two major tumor suppressors, Rb and p53, are tightly regulated by posttranslational modifications such as phosphorylation, acetylation, and ubiquitination (Classon and Harlow, 2002; Sharpless and DePinho, 2002). p53 serves as a transcription factor directly controlling the expression of genes responsible for inducing cell cycle arrest or apoptosis in response to genomic damage. In contrast, Rb regulates transcription indirectly by interacting with other transcription factors and recruiting chromatin-remodeling proteins to genes that control cell cycle progression and differentiation. Signaling pathways controlled by p53 and Rb are essential for the establishment and maintenance of senescence-induced/associated growth arrest. While these pathways have been well described in the past regarding their roles in senescence (Campisi and d'Adda di Fagagna, 2007), we will focus primarily upon how they may contribute to senescence via their effects on chromatin.

p16/Rb Pathway

Rb was one of the first tumor suppressor genes to be cloned (Friend et al., 1986; Lee et al., 1987) and plays a key role in the regulation of cellular senescence in certain model systems. Several models have shown that Rb in its active hypophosphorylated form acts as a repressor of S-phase gene expression and thereby hinders cell cycle progression through the G1/S boundary (Sellers et al., 1995; Weintraub et al., 1995); it is also involved in the maintenance of cell cycle arrest during senescence of murine cells (Sage et al., 2003). Unphosphorylated or hypophosphorylated Rb associates with several transcription factors including members of the E2F family, abrogating their transactivation functions. When Rb is phosphorylated by CDKs, it releases E2F allowing transcription of growth-stimulatory genes, promoting the initiation of DNA replication (reviewed in Trimarchi and Lees, 2002; Stevaux and Dyson, 2002). A large body of evidence supports the idea that Rb family proteins (pRb1/105, p107, pRb2/p130) associate with a wide variety of transcription factors and chromatin-remodeling enzymes to control gene expression (Macaluso et al., 2006; Parakati and DiMario, 2005), and given the constitutive activation of Rb in senescence, this pathway likely plays an essential role in regulating gene expression during senescence.

Cyclin-dependent kinase inhibitors (CDKIs), including the p16 protein, the first to be identified to interact with CDK4 in cells transformed with SV40 virus, are a family of low-molecular-weight proteins that function to inhibit CDKs (Li et al., 1994). p16 is inactivated in a large proportion of human tumors (reviewed in Ruas and Peters, 1998) and is upregulated severalfold in cells entering senescence (Wong and Riabowol, 1996; Hara et al., 1996). Consistent with a role of the p16/Rb pathway in senescence, it has been observed that overexpression of telomerase alone is not sufficient to immortalize keratinocytes and human mammary epithelial cells; however, immortalization can be achieved by inactivation of the Rb/p16 pathway or downregulation of p16 expression, in combination with telomerase activity (Kiyono et al., 1998). Inactivation of p16 alone causes extension of life span or immortalization, which agrees with the observation that there is a high frequency of p16 gene inactivation in immortal cell lines (Noble et al., 1996; Loughran et al., 1996). Telomere shortening is likely responsible for p16 gene expression in normal cell senescence (Vaziri and Benchimol, 1996). However, other stresses like overexpression of oncogenic ras (Serrano et al., 1997; Malumbres et al., 2000) or ING1 (Soliman et al., 2008) also induce the expression of p16 without telomere shortening in both humans and rodents, with the different products of the p16 locus appearing to be coordinately regulated (Collado et al., 2007). Consistent with a causal role in cell aging, knockdown of p16 using short interfering RNAs (siRNAs) has been reported to inhibit ras-induced senescence in human epithelial cells (Bond et al., 2004) and expression of p16 is known to increase with age in mammals (reviewed in Kim and Sharpless, 2006). In the same vein, loss of telomeric DNA and structure that occurs with senescence (Harley et al., 1990) induces a stress signal that augments p16 expression. How this signal is transduced may be related to the observation that the p38MAP kinase can contribute to senescence-associated upregulation of p16 (Deng et al., 2004; Ito et al., 2006). Although the molecular mechanism linking p38MAP kinase to p16 is not yet clearly understood, p38MAP kinase targets MAPKAP3 (Gaestel, 2006), and several studies have reported that phosphorylation of polycomb proteins by MAPKAP3 represses the expression of p16 (Bernard et al., 2005; Voncken et al., 2005). Consistent with regulation via polycomb proteins, senescence-associated inactivation of polycomb proteins leads to the upregulation of p16 (Bracken et al., 2007). Very recently, it was also shown that enhanced p38MAP kinase activity functions as a signaling effector in oxidative stress-induced inhibition of steroidogenesis during aging (Abidi et al., 2008). These findings suggest that the upregulation of p16 observed during aging may occur via a ROS/oxidative stress-mediated p38MAPK pathway that operates in part through polycomb proteins. Another key transducer in this pathway may also be the ING PHD proteins (discussed in detail below), since ING proteins have been recently shown to interact with p38MAP kinase in several different species (Gordon et al., 2008) and ING1 can also independently induce p16 expression (Soliman et al., 2008). Additional studies also suggest that polycomb group silencing and SWI/SNF activation epigenetically regulate the INK4b-ARF-INK4a in MRT cells (Kia et al., 2008).

Consistent with a role for stress pathways in transducing senescence-inducing signals, treatment with genotoxic stresses induces the expression of p16 in hematopoietic stem cells (HSC) (Meng et al., 2003; Wang et al., 2006). Oxidative stress has also been shown to contribute to decline in HSC function in a p16/ARF-dependent manner and a causal relationship between p16 expression and the age-induced decline in HSC function or other self-renewing tissues in vivo, such as neural stem cells and pancreatic islets, has also been suggested (Janzen et al., 2006; Krishnamurthy et al., 2006; Molofsky et al., 2006). These observations link a variety of stresses to the induction of p16 expression which, in turn, may regulate stem cell function and when activated induce a senescent phenotype.

p53/p21 Pathway

p53 has been called the "guardian of the genome" (Lane, 1992) and is found to be inactivated in approximately half of all human cancers. Normally, p53 is expressed as a latent transcription factor with a very short half-life. There are several targets of p53 involved in a variety of processes, such as cell cycle arrest and apoptosis.

It is well known that p53 levels are mainly regulated at the posttranslational level, by the rate of ubiquitination and subsequent proteolytic degradation. The levels of p53 protein are controlled in part by the MDM2 protein, an oncogene frequently amplified or overexpressed in sarcomas and other human cancers (reviewed in Momand et al., 1998). Once p53 gets activated by ATM/ATR-Chk2-mediated phosphorylation in response to DNA damage (Canman et al., 1998; Tibbetts et al., 1999), MDM2 gets sequestered by ARF, which allows for further stabilization of p53 (Sherr, 2001). In addition to activation in response to different acute stresses such as oxidative or radiation-induced damage, several studies have suggested that the transcriptional activity of p53 is increased in a chronic manner with the accumulation of cell doublings as cells become senescent (Atadja et al., 1995; Bond et al., 1996; Vaziri et al., 1997).

It has been shown that activation of p53 also plays a major role in the generation of early age-associated phenotypes in mice and also in induced premature senescence (Tyner et al., 2002). One of the obvious ways in which activation of p53 has been linked to the induction of senescence is through the various genes regulated by p53 (Buckbinder et al., 1994). One that has received considerable study is p21 which was first identified in a screen for senescence-inducing genes (Noda et al., 1994). The *p21* gene product, a member of the "Cip/Kip" family of cyclin-dependent kinase inhibitors, was also found to increase when fibroblasts and keratinocytes approached senescence. Furthermore, it was reported that the elimination of p21 through homologous recombination extended the life span of human diploid fibroblasts in culture (Brown et al., 1998), suggesting that p21 indeed has a major effect upon proliferative life span. Clear links also exist between the Rb and p53 pathways. For example, the *INK4a* gene locus which encodes the p16 regulator of Rb also contains an overlapping gene which encodes a protein called ARF (alternative reading frame) that is 19 kDa in mice ($p19^{Arf}$) and 14 kDa ($p14^{Arf}$) in human cells that impinge upon p53 function.

The relationship between the p53 and Rb pathways is shown in Fig. 2. Overexpression of $p19^{Arf}$ inhibits MDM2, resulting in the stabilization of p53 and activation of its downstream targets (reviewed in Bringold and Serrano, 2000). Further, increased transcription of the $p19^{Arf}$ gene was seen in senescent murine fibroblasts (Kamijo et al., 1997; Zindy et al., 1998; Jacobs et al., 1999) but surprisingly not in human keratinocytes (Munro et al., 1999). Although initially described as primarily a regulator of p53, triple knockout mice lacking functional p53, MDM2, and $\mathrm{p19^{Arf}}$ develop tumors at a greater frequency than do mice lacking p53 and MDM2 or p53 alone, suggesting that $p19^{Arf}$ has tumor suppressor activity that is independent of MDM2 and p53 and so may contribute to senescence via an additional mechanism (Weber et al., 2000).

Overexpression of oncogenic ras induces acute stress that triggers premature senescence (Serrano et al., 1997), which is accompanied by enhanced expression of p16 and $\mathrm{p19^{Arf}}$ and therefore leads to both p53 and Rb activation (Serrano et al., 1997; Zhu et al., 1998; Lin et al., 1998). It has been reported that in some cell systems, acute expression of both ras and raf can induce p21 independently of p53 (Sewing et al., 1997; Woods et al., 1997), suggesting that ras-induced premature senescence may be enforced in a fundamentally distinct manner from replicative senescence. Furthermore, senescence in murine versus human cells shows clear differences since the activity of p53 was reported to actually decline in tissues of aging mice following response to ionizing radiation (Feng et al., 2007). Since both age and other stresses induce p53 in human cells, this further emphasizes the difference between rodent and human cell cycle regulation in response to stress and age.

The Phosphatase and Tensin Homologue Deleted on Chromosome 10 (PTEN)

PTEN is a dual specificity phosphatase and tumor suppressor that negatively regulates the activity of the highly oncogenic phosphatidylinositol 3-kinase (PI3K)/AKT

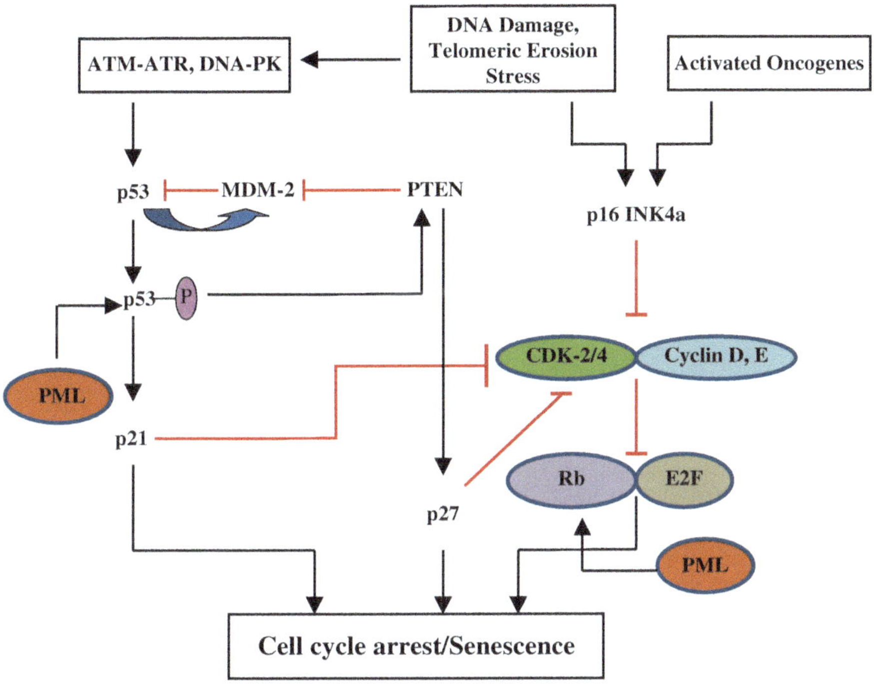

Fig. 2 Signaling pathways controlled by p53 and pRb are the major regulators of senescence. When hypophosphorylated, Rb binds to and inactivates the transcription of E2F-responsive genes. Rb releases E2F when phosphorylated by cyclins which, in turn, induces the transcription of proteins involved in cell proliferation. On the other hand, p21 prevents Rb phosphorylation via inhibition of the CDK4/2 kinases. p53 activation is achieved by phosphorylation by the ATM/ATR and CHK1/CHK2 proteins, thereby leading to the upregulation of p21, which again induces cell cycle arrest. PTEN induces p27, another cell cycle inhibitor that promotes senescence. PTEN can also be induced by p53 through a positive feedback mechanism in which p53 is stabilized by the inactivation of MDM2 by PTEN. PML cooperates with p53 and Rb in the induction of cell cycle arrest

prosurvival pathway (Stambolic et al., 1998; Cantley and Neel, 1999). While homozygous knockouts of PTEN are embryonic lethal, heterozygous knockouts ($PTEN^{+/-}$) acquire autoimmunity and develop various cancer types (Suzuki et al., 2008). It has been shown that downregulation of PTEN occurs in a variety of human cancers and that upregulation blocks cell growth, perhaps due to the induction of the *p27* gene (Li and Sun, 1998). PTEN, by virtue of its inhibitory regulation by PI3Ks, is also linked to cell senescence, since treatment of human fibroblasts with PI3K inhibitors reduces the life span of cells and accelerates the onset of senescence. In this study (Tresini et al., 1998), inhibition of PI3K was associated with elevated levels of *p27* and the induction of senescence markers. A link between PTEN and chromatin structure via the Rb pathway was suggested by the observation that PTEN does not induce G1 arrest in either Saos-2 or C33A cells that lack Rb (Paramio et al., 1999). Links between PTEN and other tumor suppressor pathways have also been found, including the observation that PTEN is required for p53-induced apoptosis in murine embryonic fibroblasts. In these studies, PTEN was able to physically bind to p53 and prevent its MDM2-mediated degradation (Freeman et al., 2003; Zhou et al., 2003), which also increased p53 DNA-binding ability (Freeman et al., 2003). Also, combined inactivation of both PTEN and p53 in mouse prostate was required for the development of invasive prostate cancer (Chen et al., 2005), since inactivation of PTEN alone induces growth arrest through the p53-dependent cellular senescence pathway both in vitro and in vivo (Chen et al., 2005). These observations suggest that PTEN may cooperate with both the Rb and p53 tumor suppressor pathways (Fig. 2). Recently, it has also been shown that downregulation of PTEN in hepatocellular carcinoma is not only due to mutations but also in part due to PTEN promoter methylation and regulation through other epigenetic modifications (Wang et al., 2007), thus linking PTEN function to chromatin remodeling.

Promyelocytic Leukemia (PML)

Replicative senescence and ras-induced premature senescence have been shown to augment levels of PML, a tumor suppressor and regulator of both p53- and Rb-mediated senescence-inducing pathways. PML is a RING finger protein that localizes to large nuclear structures called promyelocytic leukemia oncogenic domains (PODs) or PML nuclear bodies (reviewed in Salomoni et al., 2008). PML nuclear bodies are implicated in multiple cellular processes including cell proliferation, apoptosis, and senescence (Salomoni and Pandolfi, 2002). Increased PML expression was initially observed in a microarray study of ras-induced senescence (Ferbeyre et al., 2000). Since then, it has been shown that increasing PML expression can trigger many features of senescence in both human and mouse fibroblasts, and a dramatic increase in the size and number of PML nuclear bodies is seen with cell aging (Ferbeyre et al., 2000; Pearson et al., 2000). Several components of PML bodies have been identified, including SUMO-1, Sp100, Sp140, CREB-binding protein (CBP), p53, and spectrin. Among these components, p53 has been shown to physically interact with PML (Pearson and Pelicci, 2001), and p53 acetylation at lysine 382 has been reported to occur in replicative senescence and may be essential for optimal activation of p53 by PML (Pearson et al., 2000). p53 localization to PML nuclear bodies in senescent cells has also been implicated in regulating the activity of p53 by inhibiting its ubiquitin-mediated proteosomal degradation (Bernardi et al., 2004).

SIRT1, a sirtuin, is a NAD^+-dependent type III HDAC that interacts with histone H1 and p53, deacetylates H1 on K26 (Vaquero et al., 2004) and p53 on lys 382, which is also a site for ING2-mediated acetylation (Kataoka et al., 2003). SIRT1 has been implicated in regulating premature cellular senescence induced by the tumor suppressors PML and p53 (Langley et al., 2002). PML also induces cellular senescence by remodeling chromatin structure. This is part of a sequential process which starts with the incorporation of HIRA in PML nuclear bodies where it transiently colocalizes with heterochromatin protein 1 (HP1) before HP1 is incorporated into

SAHF. This might represent one of the initial events taking place during senescence induction. Also, a physical association of HIRA and another chromatin regulator ASF1a may be required for SAHF formation and efficient senescence-associated cell cycle exit (Zhang et al., 2005). Interestingly, it has been reported that giant PML nuclear bodies observed in lymphocytes of patients with immunodeficiency, centromeric instability, and facial dysmorphy (ICF) in the G2 phase of the cell cycle function in epigenetic regulation of chromatin during this phase of the cell cycle (Luciani et al., 2006). Several groups have reported that translocation of Rb to PML nuclear bodies in senescent cells permits posttranslational modifications of Rb that are necessary for the repression of E2F transcription factors and subsequent chromatin alterations (Khan et al., 2001; Mallette et al., 2004). These findings all indicate that PML nuclear bodies act as important target sites for the modification of other tumor suppressors and various proteins involved in senescence and also modulate events that reshape the chromatin environment of a presenescent cell.

ING Proteins Effect Epigenetic Changes Occurring During Senescence

The inhibitor of growth (ING) family of plant homeodomain (PHD) containing tumor suppressors (ING1–5) is an evolutionarily conserved group of proteins that affect a variety of cellular processes including apoptosis, DNA damage signaling, cell cycle progression, and replicative senescence. They are downregulated in a wide variety of cancer types including breast, blood, lung, brain, and liver cancer, supporting their classification as type II tumor suppressors which are frequently downregulated but seldom mutated as in the case of type I tumor suppressors (Sager, 1997). Functional cooperation between ING and p53 has been reported to induce senescence and apoptosis. Inhibition of ING1 or ING2 expression has been shown to cause a moderate extension of the replicative life span of diploid human fibroblasts (Garkavtsev et al., 1996; Pedeux et al., 2005), which suggests that these members of the ING family may play a role in the induction or the maintenance of replicative senescence.

A number of studies have reported the involvement of ING proteins in chromatin remodeling through their interaction with different HAT and HDAC complexes (Loewith et al., 2000; Vieyra et al., 2002; Pedeux et al., 2005; Doyon et al., 2006). Studies done in yeast showed for the first time that Yng2, the yeast homologue of ING1, was able to interact with Tra1, a component of the SAGA and NuA4 HAT complexes as shown in Fig. 3 (Loewith et al., 2000). Endogenous human ING1 protein has also been shown to coimmunoprecipitate with HAT activity and complexes containing proteins such as CBP, PCAF, p300, and TRRAP (Vieyra et al., 2002). ING1 has also been shown to be a component of the Sin3–HDAC1 complex through direct interaction of the 125 N-terminal amino acids of ING1 with the SAP30 subunit (Kuzmichev et al., 2002), and it has been implicated in promoting acetylation of particular residues of p53 that are also targets of deacetylation by SIRT1 (Kataoka et al., 2003). Further, a recent study reported that ING1 and

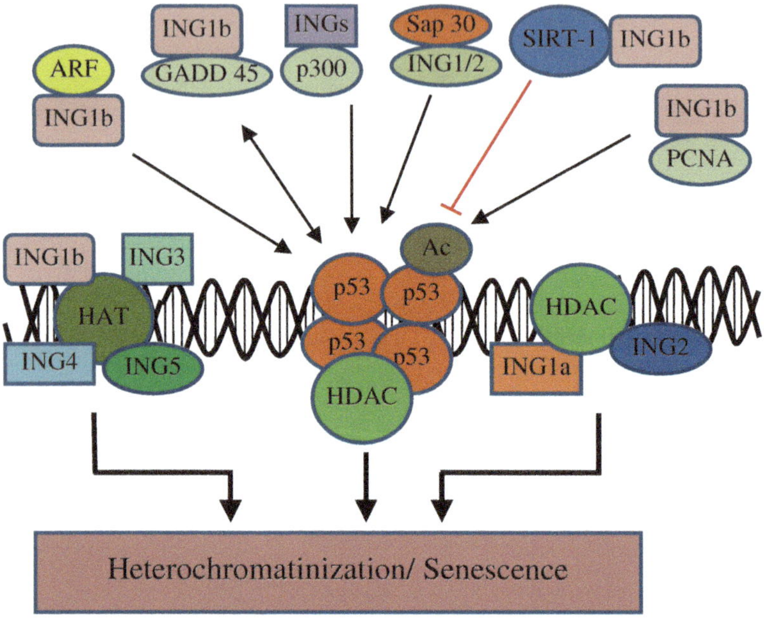

Fig. 3 ING-interacting proteins and their roles in heterochromatinization. ING family members interact with tumor suppressors, transcription factors, and HAT and HDAC complexes, facilitating heterochromatin formation. In association with p300, PCNA, and ARF, ING proteins promote acetylation of p53, leading to the upregulation of cell cycle inhibitors resulting in cell cycle arrest. Thus, ING proteins can be postulated to act in concert with other tumor suppressors and transcription factors to bring about and/or maintain permanent cell cycle arrest and cellular senescence through heterochromatin formation and/or maintenance

ING2 proteins interact directly with SIRT1 (Binda et al., 2008). Thus, ING1 and ING2 proteins may serve as links between the Sin3–HDAC complex and SIRT1. ING1 and ING2 recruit SIRT1 to the transcriptional repression domain of the protein RBP1, another component of the Sin3–HDAC1 complex, inhibiting its transcriptional repression activity (Binda et al., 2008).

ING1 has also been proposed to play a critical role in oncogene-induced senescence. Although recent reports indicate that p53 activity is not dependent upon ING1 (Coles et al., 2007), impaired binding of the heterochromatin-associated protein HP1 was observed in ING1-deficient murine fibroblasts after expression of RasV12 which activates p53. This suggests a role for ING1, possibly in collaboration with p53, in heterochromatin formation (Abad et al., 2007).

ING proteins may affect senescence through p53, by regulating p53 acetylation during replicative senescence. While the activity of p53 has long been known to increase during senescence (Atadja et al., 1995; Bond et al., 1996; Vaziri et al., 1997), ING2 expression has more recently been reported to increase in late passage human primary cells where it colocalizes with Ser15-phosphorylated p53 (Pedeux et al., 2005). The HAT, p300, was also detected in these immunocomplexes, suggesting that ING2 may regulate p53 acetylation via p300. Indeed the association between p300 and p53 was enhanced in cells overexpressing ING2. Furthermore, both ING1 and ING2 strongly enhance the transcriptional transactivation activity of p53 and thus negatively regulate cell proliferation (Garkavtsev et al., 1998; Nagashima et al., 2001; Kataoka et al., 2003). The effects of ING proteins on Rb and p53 may be mediated through linking these tumor suppressors to particular sites on chromatin via direct or regulated binding to histones. Trimethylated histone 3 Lys-4 (H3K4me3) has been reported to be associated with active gene expression and as an epigenetic mark that serves as a docking pad for transcriptional activators. Recently, the PHD domains of ING proteins have been found to specifically and robustly bind to histones, with ING1 and ING2 being sensitive to the methylation status of H3K4 (Martin et al., 2006; Palacios et al., 2008; Pena et al., 2006; Shi et al., 2006). Since ING1 and ING2 are components of the Sin3–HDAC1 complex (Doyon et al., 2006), this interaction should serve to recruit HDAC chromatin-remodeling activity to regions containing H3K4me3 and thus repress gene expression via localized and reversible heterochromatinization of DNA regions.

The two major isoforms of ING1, p47ING1a (ING1a) and p33ING1b (ING1b), have been reported to have antagonistic functions (Vieyra et al., 2002). ING1b is the major isoform expressed in proliferating human cells where ING1a is expressed at relatively low levels (Feng et al., 2002). ING1b exhibits UV-induced binding to proliferating cell nuclear antigen (PCNA) but ING1a does not (Scott et al., 2001). The ING1b isoform shows preferential association with HATs, whereas ING1a binds more avidly to HDACs. Also, overexpression of ING1b, but not ING1a, induces apoptosis (Vieyra et al., 2002). A very recent report (Soliman et al., 2008) further links ING1 isoforms to the processes of senescence and heterochromatinization. In this study, both mRNA and protein levels of ING1a were seen to significantly increase in senescent cells with a concomitant decrease in ING1b expression. A 15- to 20-fold difference in the ratio between ING1b and ING1a was noted in senescent versus replication-competent primary fibroblasts, and the interaction between ING1a and HDAC1 increased severalfold in senescent cells, suggesting a functional link between the expression of this isoform of ING1 and gene repression observed during senescence. A causal role of ING1a in inducing senescence was also supported by studies in which ING1a was overexpressed in replication-competent, young fibroblasts. ING1a-transfected cells showed the formation of SAHF, distinct HP1γ foci, and several other indices of senescence, such as elevated Rb and p16 expression, senescence-associated β-gal expression, and enlarged flattened morphology accompanied by cell cycle arrest, further highlighting the role of ING1 in chromatin remodeling during senescence (Soliman et al., 2008).

The above-mentioned findings allude to the importance of ING proteins in modulating chromatin dynamics through direct interaction with chromatin-remodeling proteins, epigenetic marks on histones, and other tumor suppressors like p53, Rb, and p16. These events reshape chromatin architecture and might reflect an important role of ING proteins in tumor suppression and senescence induction via other major tumor suppressor pathways.

Table 1 Tumor suppressors associated with various chromatin-remodeling complexes

Tumor Suppressors	Biological Effects	Associated Chromatin-Remodeling Complexes	References
p53	Chromatin regulator	p400 E1A (SWI2/SNF2 remodelling proteins),	Chan *et al.*, 2005
	Gene expression regulator	HSP 27	O'Callaghan-Sunol *et al.*, 2007;
	Mitochondrial mediator	EZH2 (HMTase)	Tang *et al.*, 2004;
	of apoptosis	PML	Langley *et al.*, 2002; Ye *et al.*, 2007
		BRAC1	reviewed in Deng and Brodie, 2000
Rb	G1-S transition	p16,HDAC	reviewed in Macaluso *et al.*, 2006
	Gene expression regulator	MRG	Garcia *et al.*, 2008
	SAHF formation	BrG1	Kang *et al.*, 2004
		LAP2α	Naetar *et al.*, 2007
		Suv39H1 HMT	Vandel *et al.*, 2001
		PML	Ye *et al.*, 2007
		E2F	Narita *et al.*, 2003
		BRAC1	reviewed in Deng and Brodie, 2000
ING1	Growth arrest	mSin 3A HDAC1/2	Kuzmichev *et al.*, 2002
	Senescence	PCNA	Scott *et al.*, 2001
	Apoptosis	ARF	González *et al.*, 2006
		P300/CBP, PCAF, GCN5	Vieyra *et al.*, 2002
		GADD45	Cheung *et al.*, 2001
		HP-1, Sap30, Brg1, TRAPP	reviewed in Campos *et al.*, 2004
ING2		HDAC1/2	Doyon *et al.*, 2006
		CBP/P300	Pedeux *et al.*, 2005
		SnoN	Sarker *et al.*, 2008
		PIP	Gozani *et al.*, 2003
		H3K4me2/3	Pena *et al.*, 2006
ING3		mSin 3A HDAC,Tip60/Nu A4 HAT complexes	Doyon *et al.*, 2006
		H3K4me3	Pena *et al.*, 2006

Table 1 (Continued)

Tumor Suppressors	Biological Effects	Associated Chromatin-Remodeling Complexes	References
ING4		HBO1 HAT, Tip60/Nu A4 HAT, TRAFF,	Doyon *et al.*, 2006
		BAF53a, DAMP1, Actin	Reviewed in Campos *et al.*, 2004
		Liprin 1 alpha	Shen *et al.*, 2007
		HIF prolylhydroxylase	Ozer *et al.*, 2005
		p53/p300	Shiseki *et al.*, 2003
		P65/NF-kappa B	Garkavtsev *et al.*, 2004
		H3K4me1/2/3	Palacios *et al.* 2006
ING5		HBO1 HAT	Doyon *et al.*, 2006
		p300, p53/p300	Shiseki *et al.*,
		MOZ/MORF HAT	Doyon *et al.*, 2006
		H3K4me3	Pena *et al.*, 2006
PML	SAHF formation	HIRA/ASF-1	Ye *et al.*, 2007
	Chromatin reorganization	p53	Pearson *et al.*, 2001
	Regulation of transcription genes	TTRAP	Xu *et al.*, 2008
		SIRT-1	Langley *et al.*, 2002
		HDAC	Wu *et al.*, 2001
Prohibitin	Cell growth	RNF-2 (polycomb protein)	Choi *et al.*, 2008
	Cell signaling	E2F	Rastogi *et al.*, 2006
		P53	Fusaro *et al.*, 2003

HDACs and Senescence

As discussed above, chromatin architecture is an important determinant of the senescent phenotype. Various chromatin-remodeling complexes effect changes in chromatin organization that, in turn, are believed to form the basis for heterochromatinization and thereby cellular senescence. The differential activity of HAT and HDAC complexes determines the steady state level of histone acetylation (reviewed in Berger, 2007), one of the main determinants of heterochromatinization observed during senescence. Histone deacetylation leads to the condensation of chromatin which plays a pivotal role in gene silencing. The enzymes responsible for deacetylation of histones, HDACs, are critical players in various biological processes, including cell cycle progression, cellular differentiation, apoptosis, and senescence. A variety of proteins, including a large number of tumor suppressors, are found in complexes with HDAC proteins. These associated factors generally serve to directly or indirectly tether HDAC complexes to DNA, thus bringing chromatin-modifying activity close to histones. A large body of evidence has now linked senescence induction to changes in epigenetic "marks" on histones and has led to the belief that these changes, which profoundly affect the chromatin environment, might influence gene expression patterns involved in the onset of senescence or a bypass of the senescence-inducing signal, leading to apoptosis or tumorigenesis.

Three classes of HDACs have been characterized thus far (Thiagalingam et al., 2003). Class I and II HDACs share such common properties as binding to zinc, sensitivity to trichostatin A, and a similar mechanism of action (Finnin et al., 1999). Rpd3 proteins and HDA1 represent the yeast counterparts of Class I and Class II HDACs, respectively. Class III HDACs share homology within their catalytic core domain with the yeast NAD^+-dependent histone deacetylase silent information regulator 2 (SIR2) and are generally referred to as sirtuins (reviewed in Blander and Guarente, 2004). Here we will limit our discussion to Class I HDACs.

Conflicting reports exist regarding the expression levels of HDAC1 in senescence in vitro and in vivo. Reduced expression of HDAC1 has been reported in senescent WI-38 cells and human foreskin fibroblasts (Wagner et al., 2001; Place et al., 2005). Both mRNA and protein levels of HDAC1 were shown to be reduced as cells approached senescence. A senescence-specific isoform of HDAC2, which is not detected in young fibroblasts, has also been reported (Wagner et al., 2001). On the other hand, HDAC1 has been reported to increase in senescent melanocytes and senescent intradermal melanocytic nevi. Also, ectopic expression of HDAC1 in melanocytes induces senescence (Bandyopadhyay et al., 2007). HDAC1 overexpression leads to the formation of heterochromatin protein 1 beta (HP1β) foci, stable recruitment of Rb to chromatin, methylation of H3 K9, and significant increase in heterochromatinization. A recent study has also reported an increase in HDAC1 levels in quiescent livers of old mice (Wang et al., 2008a). In livers of old mice, HDAC1 interacts with the transcription factor C/EBPalpha and is recruited by this protein to E2F-dependent promoters. The recruitment of HDAC1 to c-Myc and FoxM1B promoters leads to deacetylation of histone H3 at K9 on these E2F-dependent promoters, leading to inhibition of cell proliferation. Thus HDAC1 augments the growth-inhibitory activity of C/EBPalpha. In a different study, the same group also reported that HDAC1 levels are elevated in human liver tumors and in young mouse livers, proliferating after partial hepatectomy (PH) (Wang et al., 2008). This might be due to the fact that HDAC1 has an inhibitory effect on the C/EBPalpha promoter which in turn exerts an inhibitory effect on liver proliferation. A similar mechanism is also thought to be responsible for the hyperproliferation of liver tumors.

Histone deacetylase inhibitors have begun to emerge quite recently as a promising class of antineoplastic agents since they show selective activity against diverse malignancies and have notable effects on tumor cell proliferation and apoptosis in vitro and in vivo (reviewed in Minucci and Pelicci, 2006). However, despite the obvious effect of these compounds on generally increasing euchromatin and disregulating gene expression, the molecular mechanisms behind the biological effects of HDAC inhibitors are not yet fully understood. For example, it has been shown that the substrates affected by HDAC inhibitors are not restricted to histones but also include various transcriptional regulators, such as p53 and E2F-1, which may in turn further extend to general transcription factors, such as TFIIB or TFIIF (Marks et al., 2001, Choi et al., 2004) and cell cycle regulatory proteins (Gui et al., 2004).

HDAC inhibitors likely exert their effects on cells by a general disregulation of gene expression via alteration of chromatin structure. Why cancer cells appear to be more sensitive to these agents than normal cells is unclear, but it may relate to the possibility that in aneuploid and rapidly growing cancer cells, chromatin regions may be less well defined and a large complement may exist in the form of facultative heterochromatin. These regions of facultative heterochromatin may contain proapoptotic and/or antiproliferative genes (imparting apoptotic resistance to these cells) which, when activated by HDAC inhibitors, can induce apoptosis in these cells (Peart et al., 2005). Also, HDAC activity is required for the activation of certain transcription factors like STAT5 (Rascle et al., 2003). Thus, inhibition of HDAC activity can prevent expression of genes for which STAT5 is required as a transcription factor.

Conclusions

Several proteins characterized as tumor suppressors, generally due to their growth-inhibitory roles and inactivation of tumorigenesis, have been found to impact the structure of chromatin. Some act as transcription factors (p53), others directly regulate transcription factors (Rb/p16/PTEN), while most appear to be able to regulate gene expression through alteration of chromatin structure (p53, Rb, ING) via diverse mechanisms that impinge upon HAT, HDAC, and other activities that modify histones. These effects on heterochromatin then lead to a generalized increase in heterochromatin and to the formation of various forms of heterochromatic foci that are thought to impose the senescent phenotype. One paradox related to this idea is that the incidence of cancer increases dramatically in the elderly who have accumulated a significant number of senescent cells. However, it has been proposed that senescence may play a role in blocking tumorigenesis since senescent cells are very inefficient in acquiring an immortal phenotype (Campisi, 2001). Therefore, how senescence might contribute to cancer is less obvious, but one possibility is that senescent cells might affect the probability of transformation in nearby, nonsenescent cell types (Krotolika and Campisi, 2002). Another possibility is that presenescent cells which have shorter telomeres may become genetically unstable and preneoplastic. Also, the chronic accumulation of DNA damage that leads to senescence induction might be a trigger for carcinogenesis. Consistent with this idea, it has long been proposed that tumor suppressor proteins function at the "crossroads" of cancer and aging. Indeed active and opposite roles for the p53, Rb, and ING, and other tumor suppressors have been reported in both cancer and senescence. A better understanding of how each of the major tumor suppressor families contributes to altering gene expression via altering chromatin structure in both acute and chronic time frames will significantly clarify how senescence may serve to block cancer cell proliferation in some cases while enhancing the rate of tumorigenesis in others.

References

Abad, M., Menéndez, C., Füchtbauer, A., Serrano, M., Füchtbauer, E.M., Palmero, I. 2007. Ing1 mediates p53 accumulation and chromatin modification in response to oncogenic stress. J Biol Chem. 282:31060–7.

Abidi, P., Leers-Sucheta, S., Cortez, Y., Han, J., Azhar, S. 2008. Evidence that age-related changes in p38 MAP kinase contribute to the decreased steroid production by the adrenocortical cells from old rats. Aging Cell. 7:168–78.

Alcorta, D.A., Xiong, Y., Phelps, D., Hannon, G., Beach, D., Barrett, J.C. 1996. Involvement of the cyclin-dependent kinase inhibitor p16 (INK4a) in replicative senescence of normal human fibroblasts. Proc Natl Acad Sci. 93:13742–7.

Allsopp, R.C., Vaziri, H., Patterson, C., Goldstein, S., Younglai, E.V., Futcher, A.B., Greider C.W., Harley, C.B. 1992. Telomere length predicts replicative capacity of human fibroblasts. Proc Natl Acad Sci. 89:10114–8.

Atadja, P., Wong, H., Garkavtsev, I., Veillette, C., Riabowol, K. 1995. Increased activity of p53 in senescing fibroblasts. Proc Natl Acad Sci. 2:8348–52.

Baerlocher, G.M., Mak, J., Röth, A., Rice, K.S., Lansdorp, P.M. 2003. Telomere shortening in leukocyte subpopulations from baboons. J Leukoc Biol. 73:289–96.

Bandyopadhyay, D., Curry, J.L., Lin, Q., Richards, H.W., Chen, D., Hornsby, P.J., Timchenko, N.A., Medrano, E.E. 2007. Dynamic assembly of chromatin complexes during cellular senescence: implications for the growth arrest of human melanocytic nevi. Aging Cell. 6(4): 577–91.

Barbie, D.A., Kudlow, B.A., Frock, R., Zhao, J., Johnson, B.R., Dyson, N., Harlow, E., Kennedy, B.K. 2004. Nuclear reorganization of mammalian DNA synthesis prior to cell cycle exit. Mol Cell Biol. 24:595–607.

Baylin, S.B., Herman, J.G. 2000. DNA hypermethylation in tumorigenesis: epigenetics joins genetics. Trends Genet. 16:168–74.

Berger, S.L. 2007. The complex language of chromatin regulation during transcription. Nature. 447:407–12.

Bernard, D., Martinez-Leal, J.F., Rizzo, S., Martinez, D., Hudson, D., Visakorpi, T., Peters, G., Carnero, A., Beach, D., Gil, J. 2005. CBX7 controls the growth of normal and tumor-derived prostate cells by repressing the Ink4a/Arf locus. Oncogene. 24:5543–51.

Bernardi, R., Scaglioni, P.P., Bergmann, S., Horn, H.F., Vousden, K.H., Pandolfi, P.P. 2004. PML regulates p53 stability by sequestering Mdm2 to the nucleolus. Nat Cell Biol. 6:665–72.
Binda, O., Nassif, C., Branton, P.E. 2008. SIRT1 negatively regulates HDAC1-dependent transcriptional repression by the RBP1 family of proteins. Oncogene. 27:3384–92.
Blander, G., Guarente, L. 2004. The Sir2 family of protein deacetylases. Annu Rev Biochem. 73:417–35.
Bond, J., Haughton, M., Blaydes, J., Gire, V., Wynford-Thomas, D., Wyllie, F. 1996. Evidence that transcriptional activation by p53 plays a direct role in the induction of cellular senescence. Oncogene. 13:2097–104.
Bond, J., Jones, C., Haughton, M., DeMicco, C., Kipling, D., Wynford-Thomas, D. 2004. Direct evidence from siRNA-directed "knock down" that p16(INK4a) is required for human fibroblast senescence and for limiting ras-induced epithelial cell proliferation. Exp Cell Res. 292: 151–6.
Brack, A.S., Conboy, M.J., Roy, S., Lee, M., Kuo, C.J., Keller, C., Rando, T.A. 2007. Increased Wnt signaling during aging alters muscle stem cell fate and increases fibrosis. Science. 317: 807–10.
Bracken, A.P., Kleine-Kohlbrecher, D., Dietrich, N., Pasini, D., Gargiulo, G., Beekman, C., Theilgaard-Mönch, K., Minucci, S., Porse, B.T., Marine, J.C., Hansen, K.H., Helin, K. 2007. The Polycomb group proteins bind throughout the INK4A-ARF locus and are disassociated in senescent cells. Genes Dev. 21:525–30.
Bringold, F., Serrano, M. 2000. Tumor suppressors and oncogenes in cellular senescence. Exp Gerontol. 35:317–29.
Brown, D.R., Thomas, C.A., Deb, S.P. 1998. The human oncoprotein MDM2 arrests the cell cycle: elimination of its cell-cycle-inhibitory function induces tumorigenesis. EMBO J. 17: 2513–25.
Buckbinder, L., Talbott, R., Seizinger, B.R., Kley, N. 1994. Gene regulation by temperature-sensitive p53 mutants: identification of p53 response genes. Proc Natl Acad Sci. 91:10640–4.
Campisi J. 2001. Cellular senescence as a tumor-suppressor mechanism. Trends Cell Biol. 11(11):S27–31.
Campisi, J. 2005. Senescent cells, tumor suppression, and organismal aging: good citizens, bad neighbors. Cell. 120:513–22.
Campisi, J., d'Adda di Fagagna, F. 2007. Cellular senescence: when bad things happen to good cells. Nat Rev Mol Cell Biol. 8:729–40.
Campos, E.I., Chin, M.Y., Kuo, W.H., Li, G. 2004. Biological functions of the ING family tumor suppressors. Cell Mol Life Sci. 61:2597–613.
Canman, C.E., Lim, D.S., Cimprich, K.A., Taya, Y., Tamai, K., Sakaguchi, K., Appella, E., Kastan, M.B., Siliciano, J.D. 1998. Activation of the ATM kinase by ionizing radiation and phosphorylation of p53. Science. 281(5383):1677–9.
Cantley, L.C., Neel, B.G. 1999. New insights into tumor suppression: PTEN suppresses tumor formation by restraining the phosphoinositide 3-kinase/AKT pathway. Proc Natl Acad Sci. 96: 4240–5
Chan, H.M., Narita, M., Lowe, S.W., Livingston, D.M. 2005. The p400 E1A-associated protein is a novel component of the p53 → p21 senescence pathway. Genes Dev. 19:196–201.
Chen, Z., Trotman, L.C., Shaffer, D., Lin, H.K., Dotan, Z.A., Niki, M., Koutcher,. J.A., Scher, H.I., Ludwig, T., Gerald, W., Cordon-Cardo, C., Pandolfi, P.P. 2005. Crucial role of p53-dependent cellular senescence in suppression of Pten-deficient tumorigenesis. Nature. 436:725–30.
Cheung, K.J Jr, Mitchell, D., Lin, P., Li, G. 2001. The tumor suppressor candidate p33(ING1) mediates repair of UV-damaged DNA. Cancer Res. 61:4974–7.
Choi, C.H., Burton, Z.F., Usheva, A. 2004. Auto-acetylation of transcription factors as a control mechanism in gene expression. Cell Cycle. 3:114–5.
Choi, D., Lee, S.J., Hong, S., Kim, I.H., Kang, S. 2008. Prohibitin interacts with RNF2 and regulates E2F1 function via dual pathways. Oncogene. 27:1716–25.
Classon, M., Harlow, E. 2002. The retinoblastoma tumour suppressor in development and cancer. Nat Rev Cancer. 2:910–7.

Coles, A.H., Liang, H., Zhu, Z., Marfella, C.G., Kang, J., Imbalzano, A.N., Jones, S.N. 2007. Deletion of p37Ing1 in mice reveals a p53-independent role for Ing1 in the suppression of cell proliferation, apoptosis, and tumorigenesis. Cancer Res. 67:2054–61.
Collado, M., Blasco, M.A., Serrano, M. 2007. Cellular senescence in cancer and aging. Cell. 130: 223–33.
Deng, C.X., Brodie, S.G. 2000. Roles of BRCA1 and its interacting proteins. Bioessays. 22: 728–37.
Deng, Q., Liao, R., Wu, B.L., Sun, P. 2004. High intensity ras signaling induces premature senescence by activating p38 pathway in primary human fibroblasts. J Biol Chem. 279: 1050–9.
Doyon, Y., Cayrou, C., Ullah, M., Landry, A.J, Côté, V., Selleck, W., Lane, W.S, Tan, S., Yang, X.J., Côté, J. 2006. ING tumor suppressor proteins are critical regulators of chromatin acetylation required for genome expression and perpetuation. Mol Cell. 21:51–64.
Eden, S., Hashimshony, T., Keshet, I., Cedar, H., Thorne, A.W. 1998. DNA methylation models histone acetylation. Nature. 394:842.
Feng, X., Hara, Y., Riabowol, K. 2002. Different HATS of the ING1 gene family. Trends Cell Biol. 12:532–8.
Feng, Z., Hu, W., Teresky, A.K., Hernando, E., Cordon-Cardo, C., Levine, A.J. 2007. Declining p53 function in the aging process: a possible mechanism for the increased tumor incidence in older populations. Proc Natl Acad Sci. 104:16633–8.
Ferbeyre, G., de Stanchina, E., Querido, E., Baptiste, N., Prives, C., Lowe, S.W. 2000. PML is induced by oncogenic ras and promotes premature senescence. Genes Dev. 14:2015–27.
Finnin, M.S., Donigian, J.R., Cohen, A., Richon, V.M., Rifkind, R.A., Marks, P.A., Breslow, R., Pavletich, N.P. 1999. Structures of a histone deacetylase homologue bound to the TSA and SAHA inhibitors. Nature. 401(6749):188–93.
Freeman, D.J., Li, A.G., Wei, G., Li, H.H., Kertesz, N., Lesche, R., Whale, A.D., Martinez-Diaz, H., Rozengurt, N., Cardiff, R.D., Liu, X., Wu, H. 2003. PTEN tumor suppressor regulates p53 protein levels and activity through phosphatase-dependent and -independent mechanisms. Cancer Cell. 3:117–30.
Friend, S.H., Bernards, R., Rogelj, S., Weinberg, R.A., Rapaport, J.M., Albert, D.M., Dryja, T.P. 1986. A human DNA segment with properties of the gene that predisposes to retinoblastoma and osteosarcoma. Nature. 323: 643–6.
Funayama, R., Ishikawa, F. 2007. Cellular senescence and chromatin structure. Chromosoma. 116: 431–40.
Funayama, R., Saito, M., Tanobe, H., Ishikawa, F. 2006. Loss of linker histone H1 in cellular senescence. J Cell Biol. 175:869–80.
Fusaro, G., Dasgupta, P., Rastogi, S., Joshi, B., Chellappan, S. 2003. Prohibitin induces the transcriptional activity of p53 and is exported from the nucleus upon apoptotic signaling. J Biol Chem. 278:47853–61.
Gaestel, M. 2006. MAPKAP kinases – MKs – two's company, three's a crowd. Nat Rev Mol Cell Biol. 7:120–30.
Garcia, S.N., Pereira-Smith, O. 2008. MRGing chromatin dynamics and cellular senescence. Cell Biochem Biophys. 50:133–41.
Garkavtsev, I., Kazarov, A., Gudkov, A., Riabowol, K. 1996. Suppression of the novel growth inhibitor p33ING1 promotes neoplastic transformation. Nat Genet. 14: 415–20.
Garkavtsev, I., Grigorian, I.A., Ossovskaya, V.S., Chernov, M.V., Chumakov, P.M., Gudkov, A.V. 1998. The candidate tumour suppressor p33ING1 cooperates with p53 in cell growth control. Nature. 391:295–8.
Garkavtsev, I., Kozin, S.V., Chernova, O., Xu, L., Winkler, F., Brown, E., Barnett, G.H., Jain, R.K. 2004. The candidate tumour suppressor protein ING4 regulates brain tumour growth and angiogenesis. Nature. 428:328–32.
González, L., Freije, J.M., Cal, S., López-Otín, C., Serrano, M., Palmero, I. 2006. A functional link between the tumour suppressors ARF and p33ING1. Oncogene. 25:5173–9.

Gordon, P.M.K., Soliman, M.A., Bose, P., Trinh, Q., Sensen, C.W., Riabowol, K. 2008. Interspecies data-mining to predict novel protein–protein interactions in humans. BMC Genomics. 9: 426.
Gozani, O., Karuman, P., Jones, D.R., Ivanov, D., Cha, J., Lugovskoy, A.A., Baird, C.L., Zhu, H., Field, S.J., Lessnick, S.L., Villasenor, J., Mehrotra, B., Chen, J., Rao, V.R., Brugge, J.S., Ferguson, C.G., Payrastre, B., Myszka, D.G., Cantley, L.C., Wagner. G., Divecha, N., Prestwich, G.D., Yuan, J. The PHD finger of the chromatin-associated protein ING2 functions as a nuclear phosphoinositide receptor. Cell. 114:99–111.
Gui, C.Y., Ngo, L., Xu W.S., Richon, V.M., Marks, P.A. 2004. Histone deacetylase (HDAC) inhibitor activation of p21WAF1 involves changes in promoter-associated proteins, including HDAC1. Proc Natl Acad Sci USA.101:1241–6.
Hara, E., Smith, R., Parry, D., Tahara, H., Stone, S., Peters, G. 1996. Regulation of p16CDKN2 expression and its implications for cell immortalization and senescence. Mol Cell Biol. 16: 859–67.
Harley, C.B., Futcher, A.B., Greider, C.W. 1990. Telomeres shorten during ageing of human fibroblasts. Nature. 345:458–60.
Hastie, N.D., Dempster, M., Dunlop, M.G., Thompson, A.M., Green, D.K., Allshire, R.C. 1990. Telomere reduction in human colorectal carcinoma and with ageing. Nature. 346: 866–8.
Herbig, U., Ferreira, M., Condel, L., Carey, D., Sedivy, J.M. 2006. Cellular senescence in aging primates. Science. 311:1257.
Herbig, U., Sedivy, J.M. 2006. Regulation of growth arrest in senescence: telomere damage is not the end of the story. Mech Ageing Dev. 127:16–24.
Hornsby, P.J., Yang, L., Gunter, L.E. 1992. Demethylation of satellite I DNA during senescence of bovine adrenocortical cells in culture. Mutat Res. 275:13–9.
Ito, K., Hirao, A., Arai, F., Takubo, K., Matsuoka, S., Miyamoto, K., Ohmura, M., Naka, K., Hosokawa, K., Ikeda, Y., Suda, T. 2006. Reactive oxygen species act through p38 MAPK to limit the lifespan of hematopoietic stem cells. Nat Med. 12:446–51.
Jacobs, J.J., Kieboom, K., Marino, S., DePinho, R.A., van Lohuizen, M. 1999. The oncogene and Polycomb-group gene bmi-1 regulates cell proliferation and senescence through the ink4a locus. Nature. 397:164–8.
Janzen, V., Forkert, R., Fleming, H.E., Saito, Y., Waring, M.T., Dombkowski, D.M., Cheng, T., DePinho, R.A., Sharpless, N.E, Scadden DT. 2006. Stem-cell ageing modified by the cyclin-dependent kinase inhibitor p16INK4a. Nature. 443(7110):421–6.
Jeyapalan, J.C., Ferreira, M., Sedivy, J.M., Herbig, U. 2007. Accumulation of senescent cells in mitotic tissue of aging primates. Mech Ageing Dev. 128:36–44.
Jones, P.A., Laird, P.W. 1999. Cancer epigenetics comes of age. Nat Genet. 21:163–7.
Jones, P.A., Baylin, S.B. 2002. The fundamental role of epigenetic events in cancer. Nat Rev Genet. 3:415–28.
Kamijo, T., Zindy, F., Roussel, M.F., Quelle, D.E., Downing, J.R., Ashmun, R.A., Grosveld, G., Sherr C.J. 1997. Tumor suppression at the mouse INK4a locus mediated by the alternative reading frame product p19ARF. Cell. 91:649–59.
Kang, H., Cui, K., Zhao, K. 2004. BRG1 controls the activity of the retinoblastoma protein via regulation of p21CIP1/WAF1/SDI. Mol Cell Biol. 24:1188–99.
Kataoka, H., Bonnefin, P., Vieyra, D., Feng, X., Hara, Y., Miura, Y., Joh, T., Nakabayashi, H., Vaziri, H., Harris, C.C., Riabowol, K. 2003. ING1 represses transcription by direct DNA binding and through effects on p53. Cancer Res. 63:5785–92.
Keyes, W.M., Mills, A.A. 2006. p63: a new link between senescence and aging. Cell Cycle. 5: 260–5.
Keyes, W.M., Wu, Y., Vogel, H., Guo, X., Lowe, S.W., Mills, A.A. 2005. p63 deficiency activates a program of cellular senescence and leads to accelerated aging. Genes Dev. 19:1986–99.
Khan, M.M., Nomura, T., Kim, H., Kaul, S.C., Wadhwa, R., Shinagawa, T., Ichikawa-Iwata, E., Zhong, S., Pandolfi, P.P., Ishii, S.2001. Role of PML and PML-RARalpha in Mad-mediated transcriptional repression. Mol Cell. 7:1233–43.

Kia, S.K., Gorski, M.M., Giannakopoulos, S., Verrijzer, C.P. 2008. SWI/SNF mediates polycomb eviction and epigenetic reprogramming of the INK4b-ARF-INK4a locus. Mol Cell Biol. 28:3457–64.

Kim, W.Y., Sharpless, N.E. 2006. The regulation of INK4/ARF in cancer and aging. Cell. 127: 265–75.

Kiyono, T., Foster, S.A., Koop, J.I., McDougall, J.K., Galloway, D.A., Klingelhutz, A.J. 1998. Both Rb/p16INK4a inactivation and telomerase activity are required to immortalize human epithelial cells. Nature. 396:84–88.

Kouzarides, T. 2007. Chromatin modifications and their function. Cell. 128:693–705.

Krishnamurthy, J., Ramsey, M.R., Ligon, K.L., Torrice, C., Koh, A., Bonner-Weir, S., Sharpless, N.E. 2006. p16INK4a induces an age-dependent decline in islet regenerative potential. Nature. 443: 453–7.

Krishnamurthy, J., Torrice, C., Ramsey, M.R., Kovalev, G.I., Al-Regaiey, K., Su, L., Sharpless, N.E. 2004. Ink4a/Arf expression is a biomarker of aging. J Clin Invest. 114:1299–307.

Kuzmichev, A., Zhang, Y., Erdjument-Bromage, H., Tempst, P., Reinberg, D. 2002. Role of the Sin3-histone deacetylase complex in growth regulation by the candidate tumor suppressor p33(ING1). Mol Cell Biol. 22:835–48.

Krotolika, A. and Campisi, J. 2002. Cancer and aging: a model for the cancer promoting effects of the aging stroma. Int J Biochem Cell Biol. 34:1401–14.

Lane, D.P. 1992. Cancer. p53, guardian of the genome. Nature. 358:15–6.

Lane, M.A., Ingram, D.K., Cutler, R.G., Knapka, J.J., Barnard, D.E., Roth. G.S. 1992. Dietary restriction in nonhuman primates: progress report on the NIA study. Ann NY Acad Sci. 673:36–45.

Langley, E., Pearson, M., Faretta, M., Bauer, U.M., Frye, R.A., Minucci, S., Pelicci, P.G., Kouzarides, T. 2002. Human SIR2 deacetylates p53 and antagonizes PML/p53-induced cellular senescence. EMBO J. 21:2383–96.

Lee, W.H., Bookstein, R., Hong, F., Young, L.J., Shew, J.Y., Lee, E.Y. 1987. Human retinoblastoma susceptibility gene: cloning, identification, and sequence. Science. 235:1394–9.

Li, D.M., Sun, H. 1998. PTEN/MMAC1/TEP1 suppresses the tumorigenicity and induces G1 cell cycle arrest in human glioblastoma cells. Proc Natl Acad Sci. 95:15406–11.

Li, T., Santockyte, R., Shen, R.F., Tekle, E., Wang, G., Yang, D.C., Chock, P.B. 2006. Expression of SUMO-2/3 induced senescence through p53- and pRB-mediated pathways. J Biol Chem. 281:36221–7.

Li, Y., Yan, Q., Wolf, N.S. 1997. Long term calorie restriction delays age-related decline in proliferation capacity of murine lens epithelial cells in vitro and in vivo. Invest Ophthalmol Vis Sci. 38: 100–107.

Lin, A.W., Barradas, M., Stone, J.C., van Aelst, L., Serrano, M., Lowe, S.W. 1998. Premature senescence involving p53 and p16 is activated in response to constitutive MEK/MAPK mitogenic signaling. Genes Dev. 12:3008–19.

Liu, H., Fergusson, M.M., Castilho, R.M., Liu, J., Cao, L., Chen, J., Malide, D., Rovira, I.I., Schimel, D., Kuo, C.J., Gutkind, J.S., Hwang, P.M., Finkel, T. 2007. Augmented Wnt signaling in a mammalian model of accelerated aging. Science. 317:803–6.

Liu, X.T., Stewart, C.A., King, R.L., Danner, D.A., Dell'Orco, R.T., McClung, J.K. 1994. Prohibitin expression during cellular senescence of human diploid fibroblasts. Biochem Biophys Res Commun. 201(1):409–14.

Loewith, R., Meijer, M., Lees-Miller, S.P., Riabowol, K., Young, D. 2000. Three yeast proteins related to the human candidate tumor suppressor p33(ING1) are associated with histone acetyltransferase activities. Mol Cell Biol. 20:3807–16.

Lopatina, N., Haskell, J.F., Andrews, L.G., Poole, J.C., Saldanha, S., Tollefsbol, T. 2002. Differential maintenance and de novo methylating activity by three DNA methyltransferases in aging and immortalized fibroblasts. J Cell Biochem. 84:324–34.

Loughran, O., Malliri, A., Owens, D., Gallimore, P.H., Stanley, M.A., Ozanne, B., Frame, M.C., Parkinson, E.K. 1996. Association of CDKN2A/p16INK4A with human head and neck

keratinocyte replicative senescence: relationship of dysfunction to immortality and neoplasia. Oncogene. 13:561–8.
Luciani, J.J., Depetris, D., Usson, Y., Metzler-Guillemain, C., Mignon-Ravix, C., Mitchell, M.J., Megarbane, A., Sarda, P., Sirma, H., Moncla, A., Feunteun, J., Mattei, M.G. 2006. PML nuclear bodies are highly organised DNA–protein structures with a function in heterochromatin remodelling at the G2 phase. J Cell Sci. 119:2518–31.
Macaluso, M., Montanar, M., Giordano, A. 2006. Rb family proteins as modulators of gene expression and new aspects regarding the interaction with chromatin remodeling enzymes. Oncogene. 25:5263–7.
Mallette, F.A., Goumard, S., Gaumont-Leclerc, M.F., Moiseeva, O., Ferbeyre, G. 2004. Human fibroblasts require the Rb family of tumor suppressors, but not p53, for PML-induced senescence. Oncogene. 23:91–9.
Malumbres, M., Pérez De Castro, I., Hernández, M.I., Jiménez, M., Corral, T., Pellicer, A. 2000. Cellular response to oncogenic ras involves induction of the Cdk4 and Cdk6 inhibitor p15(INK4b). Mol Cell Biol. 20:2915–25.
Marks, P., Rifkind, R.A., Richon, V.M., Breslow, R., Miller, T., Kelly, W.K. 2001. Histone deacetylases and cancer: causes and therapies. Nat Rev Cancer. 1:194–202.
Martin, D.G., Baetz, K., Shi, X., Walter, K.L., MacDonald, V.E., Wlodarski, M.J., Gozani, O., Hieter, P., Howe, L. 2006. The Yng1p plant homeodomain finger is a methyl-histone binding module that recognizes lysine 4-methylated histone H3. Mol Cell Biol. 26(21): 7871–9.
Martin, L.J., Mahaney, M.C., Bronikowski, A.M., Dee Carey, K., Dyke, B., Comuzzie, A.G. 2002. Lifespan in captive baboons is heritable. Mech Ageing Dev. 123:1461–7.
Masoro, E.J. 1996. Possible mechanisms underlying the antiaging actions of caloric restriction. Toxicol Pathol. 24:738–41.
Matthews, C., Gorenne, I., Scott, S., Figg, N., Kirkpatrick, P., Ritchie, A., Goddard, M., Bennett, M. 2006. Vascular smooth muscle cells undergo telomere-based senescence in human atherosclerosis: effects of telomerase and oxidative stress. Circ Res. 99:156–64.
Melk, A., Kittikowit, W., Sandhu, I., Halloran, K.M., Grimm, P., Schmidt, B.M., Halloran, P.F. 2003. Cell senescence in rat kidneys *in vivo* increases with growth and age despite lack of telomere shortening. Kidney Int. 63:2134–43.
Melk, A., Schmidt, B.M., Takeuchi, O., Sawitzki, B., Rayner, D.C., Halloran, P.F. 2004. Expression of p16INK4a and other cell cycle regulator and senescence associated genes in aging human kidney. Kidney Int. 65: 510–20.
Meng, A., Wang, Y., Van Zant, G., Zhou, D. 2003. Ionizing radiation and busulfan induce premature senescence in murine bone marrow hematopoietic cells. Cancer Res. 63:5414–9.
Minamino, T., Komuro, I. 2007. Vascular cell senescence: contribution to atherosclerosis. Circ Res. 100:15–26.
Minucci, S., Pelicci, P.G. 2006. *Histone deacetylase inhibitors and the promise of epigenetic (and more) treatments for cancer.* Nat Rev Cancer. 6:38–51.
Molofsky, A.V., Slutsky, S.G., Joseph, N.M., He, S., Pardal, R., Krishnamurthy, J., Sharpless, N.E., Morrison, S.J. 2006. Increasing p16INK4a expression decreases forebrain progenitors and neurogenesis during ageing. Nature. 443:448–52.
Momand, J., Jung D., Wilczynski, S., Niland, J. 1998. The MDM2 gene amplification database. Nucleic Acids Res. 26: 3453–9.
Munro, J., Stott, F.J., Vousden, K.H., Peters, G., Parkinson, E.K. 1999. Role of the alternative INK4A proteins in human keratinocyte senescence: evidence for the specific inactivation of p16INK4A upon immortalization. Cancer Res. 59: 2516–21.
Naetar, N., Hutter, S., Dorner, D., Dechat, T., Korbei, B., Gotzmann, J., Beug, H., Foisner, R. 2007. LAP2alpha-binding protein LINT-25 is a novel chromatin-associated protein involved in cell cycle exit. J Cell Sci. 120:737–47.
Nagashima, M., Shiseki, M., Miura, K., Hagiwara, K., Linke, S.P., Pedeux, R., Wang, X.W., Yokota, J., Riabowol, K., Harris, C.C. 2001. DNA damage-inducible gene p33ING2

negatively regulates cell proliferation through acetylation of p53. Proc Natl Acad Sci. 98: 9671–6.

Narita, M., Narita, M., Krizhanovsky, V., Nuñez, S., Chicas, A., Hearn, S.A., Myers, M.P., Lowe, S.W. 2006 A novel role for high-mobility group a proteins in cellular senescence and heterochromatin formation. Cell.126: 503–14.

Narita, M., Nunez, S., Heard, E., Narita, M., Lin, A.W., Hearn, S.A., Spector, D.L., Hannon, G.J., Lowe, S.W. 2003. Rb-mediated heterochromatin formation and silencing of E2F target genes during cellular senescence. Cell.113:703–16.

Noble, J.R., Rogan, E.M., Neumann, A.A., Maclean, K., Bryan, T.M., Reddel, R.R. 1996. Association of extended *in vitro* proliferative potential with loss of p16INK4 expression. Oncogene. 13:1259–68.

Noda, A., Ning, Y., Venable, S.F., Pereira-Smith, O.M., Smith, J.R. 1994. Cloning of senescent cell-derived inhibitors of DNA synthesis using an expression screen. Exp Cell Res. 211:90–8.

O'Callaghan-Sunol, C., Gaba,i V.L., Sherman, M.Y. 2007. Hsp27 modulates p53 signaling and suppresses cellular senescence. Cancer Res. 67:11779–88.

Ozer, A., Wu, L.C., 2005. Bruick RK. The candidate tumor suppressor ING4 represses activation of the hypoxia inducible factor (HIF). Proc Natl Acad Sci. 102:7481–6.

Palacios, A., Muñoz, I.G., Pantoja-Uceda, D., Marcaida, M.J., Torres, D., Martín-García, J.M., Luque, I., Montoya, G., Blanco, F.J. 2008. Molecular basis of histone H3K4me3 recognition by ING4. J Biol Chem. 283:15956–64.

Parakati, R., DiMario, J.X. 2005. Dynamic transcriptional regulatory complexes, including E2F4, p107, p130, and Sp1, control fibroblast growth factor receptor 1 gene expression during myogenesis. J Biol Chem. 280:21284–94.

Paramio, J.M., Navarro, M., Segrelles, C., Gómez-Casero, E., Jorcano, J.L. 1999. PTEN tumour suppressor is linked to the cell cycle control through the retinoblastoma protein. Oncogene.18:7462–8.

Pearson, M., Carbone, R., Sebastiani, C., Cioce, M., Fagioli, M., Saito, S., Higashimoto, Y., Appella, E., Minucci, S., Pandolfi, P.P., Pelicci, P.G. 2000. PML regulates p53 acetylation and premature senescence induced by oncogenic Ras. Nature. 406:207–10.

Pearson, M., Pelicci, P.G. 2001. PML interaction with p53 and its role in apoptosis and replicative senescence. Oncogene. 20:7250–6.

Peart, M.J., Smyth, G.K., van Laar, R.K., Bowtell, D.D., Richon, V.M., Marks, P.A., Holloway, A.J., Johnstone, R.W. 2005. Identification and functional significance of genes regulated by structurally different histone deacetylase inhibitors. Proc Natl Acad Sci. 102: 3697–3702.

Pedeux, R., Sengupta, S., Shen, J.C., Demidov, O.N., Saito, S., Onogi, H., Kumamoto, K., Wincovitch, S., Garfield, S.H., McMenamin, M., Nagashima, M., Grossman, S.R, Appella, E., Harris, C.C. 2005. ING2 regulates the onset of replicative senescence by induction of p300-dependent p53 acetylation. Mol Cell Biol. 25:6639–48.

Pena, P.V., Davrazou, F., Shi, X., Walter, K.L., Verkhusha, V.V., Gozani, O., Zhao, R., Kutateladze, T.G. 2006. Molecular mechanism of histone H3K4me3 recognition by plant homeodomain of ING2. Nature. 442:100–3.

Place, R.F., Noonan, E.J., Giardina, C. 2005. HDACs and the senescent phenotype of WI-38 cells. BMC Cell Biol. 6:37.

Rascle, A., Johnston, J.A., Amati, B. 2003. Deacetylase activity is required for recruitment of the basal transcription machinery and transactivation by STAT5. Mol Cell Biol. 23: 4162–73.

Rastogi, S., Joshi, B., Dasgupta, P., Morris, M., Wright, K., Chellappan, S. 2006. Prohibitin facilitates cellular senescence by recruiting specific corepressors to inhibit E2F target genes. Mol Cell Biol. 26:4161–71.

Ruas, M., Peters, G. 1998. The p16INK4a/CDKN2A tumor suppressor and its relatives. Biochim Biophys Acta. 1378:F115–177.

Sage, J., Miller, A.L., Perez-Mancera, P.A., Wysocki, J.M., Jacks, T. 2003. Acute mutation of retinoblastoma gene function is sufficient for cell cycle re-entry. Nature. 424: 223–228.

Sager, R. 1997. Expression genetics in cancer: shifting the focus from DNA to RNA. Proc Natl Acad Sci USA. 94:952–955.
Salomoni, P., Ferguson, B.J., Wyllie, A.H., Rich, T. 2008. New insights into the role of PML in tumour suppression. Cell Res. 18: 622–40.
Salomoni, P., Pandolfi, P.P. 2002. The role of PML in tumor suppression. Cell. 108: 165–70.
Sarg, B., Koutzamani, E., Helliger, W., Rundquist, I., Lindner, H.H. 2002. Postsynthetic trimethylation of histone H4 at lysine 20 in mammalian tissues is associated with aging. J Biol Chem. 277: 39195–201.
Sarker, K.P., Kataoka, H., Chan, A., Netherton, S.J., Pot, I., Huynh, M.A., Feng, X., Bonni, A., Riabowol, K., Bonni, S. 2008. ING2 as a novel mediator of transforming growth factor-beta-dependent responses in epithelial cells. J Biol Chem. 283:13269–79.
Scott, M., Bonnefin, P., Vieyra, D., Boisvert, F.M., Young, D., Bazett-Jones, D.P., Riabowol, K. 2001. UV-induced binding of ING1 to PCNA regulates the induction of apoptosis. J Cell Sci. 114:3455–62.
Sellers, W.R., Rodgers, J.W., Kaelin, W.G. Jr. 1995. A potent transrepression domain in the retinoblastoma protein induces a cell cycle arrest when bound to E2F sites. Proc Natl Acad Sci. 92:11544–8.
Serrano, M., Lin, A.W., McCurrach, M.E., Beach, D., Lowe, S.W. 1997. Oncogenic ras provokes premature cell senescence associated with accumulation of p53 and p16INK4a. Cell. 88: 593–602.
Sewing, A., Wiseman, B., Lloyd, A.C., Land, H. 1997. High-intensity Raf signal causes cell cycle arrest mediated by p21Cip1. Mol Cell Biol. 17:5588–97.
Sharpless, N.E., DePinho, R.A. 2002. p53: good cop/bad cop. Cell.110: 9–12.
Shen, J.C., Unoki, M., Ythier, D., Duperray, A., Varticovski, L., Kumamoto, K., Pedeux, R., Harris, C.C. 2007. Inhibitor of growth 4 suppresses cell spreading and cell migration by interacting with a novel binding partner, liprin alpha1. Cancer Res. 67:2552–8.
Sherr, C.J., 2001. The INK4a/ARF network in tumour suppression. Nat Rev Mol Cell Biol. 2: 731–737.
Shi, X., Hong, T., Walter, K.L., Ewalt, M., Michishita, E., Hung, T., Carney, D., Peña, P., Lan, F., Kaadige, M.R., Lacoste, N., Cayrou, C., Davrazou, F., Saha, A., Cairns, B.R., Ayer, D.E., Kutateladze, T.G., Shi, Y., Côté, J., Chua, K.F., Gozani, O. 2006. ING2 PHD domain links histone H3 lysine 4 methylation to active gene repression. Nature. 442:96–9.
Shiseki, M., Nagashima, M., Pedeux, R.M., Kitahama-Shiseki, M., Miura, K., Okamura, S., Onogi, H., Higashimoto, Y., Appella, E., Yokota, J., Harris, C.C. 2003. p29ING4 and p28ING5 bind to p53 and p300, and enhance p53 activity. Cancer Res. 63:2373–8.
Shmookler Reis, R.J., Goldstein, S. 1982. Interclonal variation in methylation patterns for expressed and non-expressed genes. Nucleic Acids Res. 10:4293–304.
Sohal, R.S., Weindruch, R. 1996. Oxidative stress, caloric restriction, and aging. Science. 273: 59–63.
Soliman, M.A., Berardi, P., Pastyryeva, S., Bonnefin, P., Feng, X., Colina, A., Young, D., Riabowol, K. 2008. ING1a expression increases during replicative senescence and induces a senescent phenotype. Aging Cell. 2008 Aug 7.
Stambolic, V., Suzuki, A., de la Pompa, J.L., Brothers, G.M., Mirtsos, C., Sasaki, T., Ruland, J., Penninger, J.M., Siderovski, D.P., Mak, T.W. 1998. Negative regulation of PKB/Akt-dependent cell survival by the tumor suppressor PTEN. Cell. 95:29–39.
Stein, G.H., Drullinger, L.F., Soulard, A., Dulić, V. 1999. Differential roles for cyclin-dependent kinase inhibitors p21 and p16 in the mechanisms of senescence and differentiation in human fibroblasts. Mol Cell Biol. 19:2109–17.
Stevaux, O., Dyson, N.J. 2002. A revised picture of the E2F transcriptional network and RB function. Curr Opin Cell Biol. 14:684–91.
Suzuki, A., Nakano, T., Mak, T.W., Sasaki, T. 2008. Portrait of PTEN: messages from mutant mice. Cancer Sci. 99:209–13.

Tang, X., Milyavsky, M., Shats, I., Erez, N., Goldfinger, N., Rotter, V. 2004. Activated p53 suppresses the histone methyltransferase EZH2 gene. Oncogene. 23:5759–69.

Thiagalingam, S., Cheng, K.H., Lee, H.J., Mineva, N., Thiagalingam, A., Ponte, J.F. 2003. Histone deacetylases: unique players in shaping the epigenetic histone code. Ann NY Acad Sci. 983: 84–100.

Tibbetts, R.S., Brumbaugh, K.M., Williams, J.M., Sarkaria, J.N., Cliby, W.A., Shieh, S.Y., Taya, Y., Prives, C., Abraham, R.T. 1999. A role for ATR in the DNA damage-induced phosphorylation of p53. Genes Dev. 13:152–7.

Tresini, M., Mawal-Dewan, M., Cristofalo, V.J., Sell, C. 1998. A phosphatidylinositol 3-kinase inhibitor induces a senescent-like growth arrest in human diploid fibroblasts. Cancer Res. 58: 1–4.

Trimarchi, J.M., Lees, J.A. 2002. Sibling rivalry in the E2F family. Nat Rev Mol Cell Biol. 3: 11–20.

Tyner, S.D., Venkatachalam, S., Choi, J., Jones, S., Ghebranious, N., Igelmann, H., Lu, X., Soron, G., Cooper, B., Brayton, C., Hee Park, S., Thompson, T., Karsenty, G., Bradley, A., Donehower, L.A. 2002. p53 mutant mice that display early ageing-associated phenotypes. Nature. 415: 45–53.

Vandel, L., Nicolas, E., Vaute, O., Ferreira, R., Ait-Si-Ali, S., Trouche, D. 2001. Transcriptional repression by the retinoblastoma protein through the recruitment of a histone methyltransferase. Mol Cell Biol. 21:6484–94.

Vaquero, A., Scher, M., Lee, D., Erdjument-Bromage, H., Tempst, P., Reinberg, D. 2004. Human SirT1 interacts with histone H1 and promotes formation of facultative heterochromatin. Mol Cell. 16: 93–105.

Vaziri, H., Benchimol, S. 1996. From telomere loss to p53 induction and activation of a DNA-damage pathway at senescence: the telomere loss/DNA damage model of cell aging. Exp Gerontol. 31:295–301.

Vaziri, H., West, M.D., Allsopp, R.C., Davison, T.S., Wu, Y.S., Arrowsmith, C.H., Poirier, G.G., Benchimol, S. 1997. ATM-dependent telomere loss in aging human diploid fibroblasts and DNA damage lead to the post-translational activation of p53 protein involving poly(ADP-ribose) polymerase. EMBO J. 16:6018–33.

Ventura, A., Kirsch, D.G., McLaughlin, M.E., Tuveson, D.A., Grimm, J., Lintault, L., Newman, J., Reczek, E.E., Weissleder, R., Jacks, T. 2007. Restoration of p53 function leads to tumour regression in vivo. Nature. 445:661–5.

Vertino, P.M., Issa, J.P., Pereira-Smith, O.M., Baylin, S.B. 1994. Stabilization of DNA methyltransferase levels and CpG island hypermethylation precede SV40-induced immortalization of human fibroblasts. Cell Growth Differ. 5:1395–402.

Vieyra, D., Loewith, R., Scott, M., Bonnefin, P., Boisvert, F.M., Cheema, P., Pastyryeva, S., Meijer, M., Johnston, R.N., Bazett-Jones, D.P., McMahon, S., Cole, M.D., Young, D., Riabowol, K. 2002. Human ING1 proteins differentially regulate histone acetylation. J Biol Chem. 277: 29832–9.

Voncken, J.W., Niessen, H., Neufeld, B., Rennefahrt, U., Dahlmans, V., Kubben, N., Holzer, B., Ludwig, S., Rapp, U.R. 2005. MAPKAP kinase 3pK phosphorylates and regulates chromatin association of the polycomb group protein Bmi1. J Biol Chem. 280:5178–87.

Wagner M, Brosch G, Zwerschke W, Seto E, Loidl P, Jansen-Dürr P. 2001. Histone deacetylases in replicative senescence: evidence for a senescence-specific form of HDAC-2.FEBS Lett. 499: 101–6.

Wang, G.L., Salisbury, E., Shi, X., Timchenko, L., Medrano, E.E., Timchenko, N.A. 2008a. HDAC1 cooperates with C/EBPalpha in the inhibition of liver proliferation in old mice. J Biol Chem. 283:26169–78.

Wang, G.L., Salisbury, E., Shi, X., Timchenko, L., Medrano, E.E., Timchenko, N.A. 2008b. HDAC1 promotes liver proliferation in young mice via interactions with C/EBPbeta. J Biol Chem. 283:26179–87.

Wang, L., Wang, W.L., Zhang, Y., Guo, S.P., Zhang, J., Li, Q.L. 2007. Epigenetic and genetic alterations of PTEN in hepatocellular carcinoma. Hepatol Res. 37:389–396.
Wang, Y., Schulte, B.A., Zhou, D. 2006. Hematopoietic stem cell senescence and long-term bone marrow injury. Cell Cycle. 5:35–8.
Weber, J.D., Jeffers, J.R., Rehg, J.E., Randle, D.H., Lozano, G., Roussel, M.F., Sherr, C.J., Zambetti, G.P. 2000. p53-independent functions of the p19(ARF) tumor suppressor. Genes Dev. 14:2358–65.
Weintraub, S.J., Chow, K.N., Luo, R.X., Zhang, S.H., He, S., Dean, D.C. 1995. Mechanism of active transcriptional repression by the retinoblastoma protein. Nature. 375:812–5.
Wilson, V.L., Jones, P.A. 1983. DNA methylation decreases in aging but not in immortal cells. Science. 220:1055–7.
Wong, H., Riabowol, K. 1996. Differential CDK-inhibitor gene expression in aging human diploid fibroblasts. Exp Gerontol. 31:311–25.
Woods, D., Parry, D., Cherwinski, H., Bosch, E., Lees, E., McMahon, M. 1997. Raf-induced proliferation or cell cycle arrest is determined by the level of Raf activity with arrest mediated by p21Cip1. Mol Cell Biol. 17:5598–611.
Wright WE, Shay JW. Historical claims and current interpretations of replicative aging. Nat Biotechnol. 2002;20(7):682–8.
Wu, W.S., Vallian, S., Seto, E., Yang, W.M., Edmondson, D., Roth. S., Chang, K.S. 2001. The growth suppressor PML represses transcription by functionally and physically interacting with histone deacetylases. Mol Cell Biol. 21:2259–68.
Xu, G.L., Pan, Y.K., Wang, B.Y., Huang, L., Tian, L., Xue, J.L., Chen, J.Z., Jia, W. 2008. TTRAP is a novel PML nuclear bodies-associated protein. Biochem Biophys Res Commun. 375:395–8.
Xue, W., Zender, L., Miething, C., Dickins, R.A., Hernando, E., Krizhanovsky, V., Cordon-Cardo, C., Lowe, S.W. 2007. Senescence and tumour clearance is triggered by p53 restoration in murine liver carcinomas. Nature. 445:656–60.
Ye, X., Zerlanko, B., Kennedy, A., Banumathy, G., Zhang, R., Adams, P.D. 2007. Downregulation of Wnt signaling is a trigger for formation of facultative heterochromatin and onset of cell senescence in primary human cells. Mol Cell. 27:183–96.
Young, J.I., Smith, J.R. 2001. DNA methyltransferase inhibition in normal human fibroblasts induces a p21-dependent cell cycle withdrawal. J Biol Chem. 276:19610–6.
Zhang, H., Pan, K.H., Cohen, S.N. 2003. Senescence-specific gene expression fingerprints reveal cell-type-dependent physical clustering of up-regulated chromosomal loci. Proc Natl Acad Sci. 100:3251–6.
Zhang, R., Chen, W., Adams, P.D. 2007. Molecular dissection of formation of senescence-associated heterochromatin foci. Mol Cell Biol. 27:2343–58.
Zhang, R., Poustovoitov, M.V., Ye, X., Santos, H.A., Chen, W., Daganzo, S.M., Erzberger, J.P., Serebriiskii, I.G., Canutescu, A.A., Dunbrack, R.L., Pehrson, J.R., Berger, J.M., Kaufman, P.D., Adams, P.D. 2005. Formation of MacroH2A-containing senescence-associated heterochromatin foci and senescence driven by ASF1a and HIRA. Dev Cell. 8:19–30.
Zhang, Y., Herman, B. 2002 Ageing and apoptosis. Mech Ageing Dev. 123: 245–60.
Zhou, M., Gu, L., Findley, H.W., Jiang, R., Woods, W.G. 2003. PTEN reverses MDM2-mediated chemotherapy resistance by interacting with p53 in acute lymphoblastic leukemia cells. Cancer Res. 63: 6357–62.
Zhu, J., Woods, D., McMahon, M., Bishop, J.M. 1998. Senescence of human fibroblasts induced by oncogenic Raf. Genes Dev. 12: 2997–3007.
Zindy, F., Eischen, C.M., Randle, D.H., Kamijo, T., Cleveland, J.L., Sherr. C.J., Roussel, M.F. 1998. Myc signaling via the ARF tumor suppressor regulates p53-dependent apoptosis and immortalization. Genes Dev. 12:2424–33.

Epigenetic Drift and Aging

Ester Lara*, **Vincenzo Calvanese***, **and Mario F. Fraga**

Abstract Epigenetics of aging is an emerging field that promises exciting revelations in the near future. Epigenetic pathways, including DNA methylation and histone modification, are determinants of normal development and can change during aging. Some of the epigenetic alterations described during aging, as hypermethylation at specific promoters and decrease of global DNA methylation, are also associated with tumour development. The epigenetic changes occurring during development and aging can be stochastic or depend on environmental factors. Future challenges in the field involve the determination of the precise molecular mechanisms that create age-dependent epigenetic variation and how these epigenetic changes affect the aging phenotype.

Keywords methylation · histone modifications · aging · development

Introduction

Epigenetics was originally defined in 1942 by Conrad Waddington as the study of how genotypes give rise to phenotypes through programmed changes during development. So, at first, the field of epigenetics was strictly related to the regulation of gene expression during embryonic development. Arthur Riggs subsequently defined it more precisely as the study of heritable mitotic and meiotic changes in gene function that cannot be explained by changes in the DNA sequence. According to this definition, an epigenetic event is something that affects genes without changing the nucleotide sequence, in a way that can be inherited through cell division and, eventually, gamete formation. Substantial effort has been made in recent decades to

---

*Both authors contributed equally

M.F. Fraga (✉)
Department of Immunology and Oncology, National Center for Biotechnology, CNB-CSIC, Cantoblanco, Madrid E-28049, Spain; Laboratorio de Epigenética del Cáncer, Instituto Universitario de Oncología del Principado de Asturias (IUOPA), Universidad de Oviedo, Oviedo, Spain
e-mail: mffraga@cnio.es

T.O. Tollefsbol (ed.), *Epigenetics of Aging*, DOI 10.1007/978-1-4419-0639-7_14,

discover the molecular mechanisms involved in this additional way of storing information and regulating gene expression in the cell. With what we know today, we can say that these epigenetic modifications seem to be quite stable, but, at the same time, can be modulated by many factors, including physiological and pathological circumstances and the environment. While we know some of the mechanisms by which these modifications are conserved over many cell generations, we have little evidence of the mechanisms of inheritance in organism reproduction (for reviews, see Holliday 2006; Rakyan and Beck 2006; Whitelaw and Whitelaw 2006).

Epigenetic mechanisms involve covalent chemical modification of DNA (methylation) or chromatin (histone modification). Moreover, an increasing number of additional mechanisms, mostly related to the former two, regulate gene expression and chromatin structure (non-coding RNAs, among others). These mechanisms are not described in this chapter, although it should be appreciated that in the near future they will probably turn out to be more important than the first two.

DNA Methylation

DNA methylation in mammals consists of the addition of a methyl group to the aromatic ring of a single DNA base. This is a widespread phenomenon in the genome of many organisms, and in mammals, it is mostly restricted to the 5 carbon of the cytosine ring of a CpG dinucleotide. In normal human tissues, 5-methylcytosine accounts for 3–6% of the total cytosine. The CpG dinucleotide is found at a very low frequency in the genome but is concentrated in particular in gene promoters (or their surrounding areas), where it mostly regulates gene expression, blocking transcription when the methyl group is present. Most of these regions, called CpG islands, are subjected to dynamic methylation modifications during development that are linked to the rapid differentiation and formation of various tissues. Once differentiation is complete, a tissue-specific methylation is established in each cell type and is basically maintained for the rest of a cell's life.

Promoter methylation accounts for only a small part of global genome methylation: the bulk of CpG methylated in the genome is localized in repetitive sequences, most of which are derived from transposable elements. Methylation keeps these sequences silenced, making the event of amplification and new insertion in the genome extremely rare.

At least two more genetic mechanisms rely on DNA methylation in normal cells: genomic imprinting and X-inactivation. Genomic imprinting occurs in some genes whose expression is always restricted either to the maternal or to the paternal allele. It requires DNA methylation at one of the two parental alleles of a gene to ensure monoallelic expression. A similar gene-dosage reduction is involved in X-chromosome inactivation in females (Heard and Disteche 2006; Lyon 1961; Okamoto et al. 2004).

In relation to the mechanism by which mCpG drives transcriptional repression and produces a close chromatin structure, it was formerly proposed that the methyl group itself, exposed in the major groove of the double helix, interfered with transcription factor binding. However, the most likely mechanism has turned out to be the interaction of these methylated residues with proteins containing a methyl-binding domain (MBD). These proteins link the methyl group of CpGs to complex chromatin remodelling machinery that is able to turn off transcription and lock the chromatin in a condensed state (Hendrich and Bird 1998).

To understand how this post-synthetic modification is inherited, we merely have to look at the enzymatic system responsible for its establishment: an enzyme family called DNA methyltransferases (DNMTs). DNMT1 is responsible for the maintenance of this modification in DNA replication. In fact, when a new strand of DNA is synthesized, the mCpG site is copied to an antisense CpG on the other strand, creating a hemi-methylated site. DNMT1 specifically recognizes these hemi-methylated CpGs and transfers a methyl group to the unmethylated cytosine ring. According to Bostick et al. (2007), DNMT1 acts in conjunction with the UHRF1 protein, which contains a methyl DNA-binding protein that preferentially binds to hemi-methylated CpG sites. UHRF1 is required for the stable association of DNMT1 with chromatin to facilitate the maintenance of DNA methylation. In this way, the methylation can be maintained in the two daughter cells in mitosis. These features explain the stability of the modification, which enables it to be inherited in cell division. Thus, it is possible to define an epimutation as a stable change of methylation at a particular site that is perpetuated over generations. In fact, DNA methylation is more flexible than a genetic alteration. As it is generated and perpetuated by cellular enzymes, it can be modulated as easily as many cellular processes. In any case, the original idea of DNA methylation stability in somatic cells has changed in the light of recent studies. Metivier et al. (2008) and Kangaspeska et al. (2008) demonstrated a dual role of DNMTs in cyclical methylation/demethylation inherent to transcriptional cycling of the *pS2* and *ERα* genes. Nevertheless, even though there are a considerable number of exceptions, most genome methylation is not maintained during meiosis and gamete formation. The whole genome undergoes global demethylation in gametes and methylation is subsequently re-established in early embryonic stages by the de novo DNA methyltransferases DNMT3a and 3b, which are specifically expressed during the first phase of embryo development (Bestor 2000).

There are clear examples of the inheritance of epigenetic status from generation to generation in plants and a few in mammals, which we will discuss later. However, the inheritance of epigenetic modifications has different features from those of classical Mendelian inheritance of genetic entities.

There are many pathological circumstances under which the machinery maintaining the correct methylation pattern can be disrupted. The most paradigmatic and best studied example is cancer. The alterations of DNA methylation in cancerous tissues were first seen many decades ago, and since then epigenetics has become a crucial component in the study of cancer biology (Jones and Baylin 2002). The proof of its importance is illustrated by the frequent treatment of certain cancer types with antiblastic drugs (e.g., decitabine), which work by inhibiting DNA methylation. In cancer cells, the transcriptional silencing of tumour suppressor genes by CpG island-promoter hypermethylation is the key to the tumourigenic process and contributes to all of the typical hallmarks of a cancer cell that result from tumour suppressor inactivation. By contrast, repetitive genomic sequences, which are normally heavily methylated, lose a substantial proportion of their methylation, resulting in a global loss of methylcytosine content in the cell.

Histone Modification

The second and more complex aspect of epigenetics is chromatin modification. The protagonists here are histones, the small basic proteins forming the nucleosome. The core histones, H2A, H2B, H3 and H4, together with 147 base pairs of genomic DNA wrapped around them, comprise the nucleosomes, which are the basic units of chromatin. The close interactions between DNA and histone proteins lead to a high degree of structure condensation, which, by default, impedes gene transcription. Histone proteins have some positively charged tails that protrude from the core structure of the nucleosome and can be modified in many amino acid residues. Given that there are at least 30 sites of possible modification for each nucleosome and six types of modification (methylation, acetylation, phosphorylation, ubiquitylation, sumoylation and proline isomerization) and that some of the latter can occur in different configurations (for example, lysine can be mono-, di- or tri-methylated), a bewilderingly large number of combinations are possible (reviewed in Fraga and Esteller 2005).

Fortunately, as with all complex biological systems, we have begun to understand the basic rules governing the combinations of modifications that regulate many biological processes, such as gene expression, DNA repair, chromatin compaction and genome stability, as well as important genetic processes such as X-inactivation.

In eukaryotic cells, gene activation is closely associated with covalent modifications of histone N-terminal tails, which often differ between active and silenced chromatins. Acetylation is the most extensively studied modification that can effectively influence gene expression. Acetylation of lysine 14 or 9 in histone H3, and lysine 16 of histone H4, performed by histone acetyltransferase enzymes (HATs), is generally associated with active gene transcription. Other modifications, such as methylation of histone H3 at the arginine 17 position or phosphorylation of H3 at the serine 10 position, also confer an open chromatin conformation that facilitates transcription.

Gene activation by histone acetylation has a biophysical explanation. The lysine has a positive charge at its end and can bind tightly to the negatively charged DNA to form a closed chromatin structure that impedes the access of transcription factors. Acetylation of lysine residues removes their positive charge and attenuates the charge interaction between histone tails and DNA. An alternative mechanism has been proposed in which various histone modifications act as docking sites for other chromatin-modifying enzymes and basal transcription machinery. For example, a protein domain called bromodomain can specifically bind to acetylated lysines. The bromodomain is often found in enzymes that help activate transcription, including SWI/SNF, an ATP-dependent chromatin-remodelling complex. Lysine acetylation can recruit the SWI/SNF complex to facilitate transcription activation. Histone H3-K9 methylation is associated with transcriptional silencing. Proteins with a chromodomain, such as HP1, can specifically bind to methylated lysine. HP1 is a transcription-silencing protein that interacts with histone deacetylase (HDAC). Its binding to methylated H3-K9 results in histone deacetylation that eventually leads to gene silencing. On the other hand, H3-K4 methylation is recognized by the chromodomain protein CHD1, which can further recruit HATs to activate target gene transcription.

Increasing evidence indicates that DNA methylation and histone modifications are interrelated in gene regulation in many circumstances. DNMT1, for example, associates with HDAC2 and other proteins to form a silencing complex, which is recruited to heterochromatin. It is possible that HDAC2 is the major player involved in keeping the heterochromatin hypoacetylated. In addition, DNMT3a and DNMT1 bind directly to HDAC1, and histone methylation is also coupled with methylated repetitive sequences. Although many studies have suggested that DNA methylation may be a principal means of silencing target genes preceding other epigenetic pathways, there are also situations where DNA methylation seems to act as a secondary event to stabilize the gene-silencing status initiated by existing chromatin modification. Loss of histone acetylation and H3-K4 methylation is the first step towards reversible transcriptional repression, which is followed by H3-K9 methylation and promoter DNA methylation to stabilize the silenced chromatin state. These observations indicate that there is a close cooperation between DNA methylation and histone modifications in regulating gene activities during development.

The above are only a few of the epigenetic marks and pathways involved in chromatin modification, but the number of additional mechanisms, enzymes, combinations of marks and proteins recognizing them and linking them to the final function is increasing all the time. Furthermore, for many of these chemical modifications, we have identified the enzymatic systems that remove the mark. So, we return to the question of how a system of reversible marks that are modified and established during development can be inherited in meiosis and sexual reproduction. Even though we may currently have some idea, the question is far from being completely answered. Perhaps, a more profound knowledge of other epigenetic mechanisms, like those from the realm of non-coding RNA, whose specific moiety is determined by the nucleotide sequence, could be the key to solving this mystery.

Epigenetics and Aging

The aging process consists of a complex of anatomical, physiological, biochemical and genetic changes that all organisms undergo during their lifetime. Epigenetics is probably only one of several components of aging, but its features make it a very strong candidate for explaining these changes. Epigenetic players are enzymes and, as is the case with all enzymatic systems, they can be easily regulated and have a certain failure rate. DNA replication and repair are also enzymatic processes but evolution has optimized these systems so that their error rate is minimal. Selective pressure has not been as strong as on epigenetic enzymes and the fact that the epigenetic status of each cell has to change during development, differentiation and physiological responses makes the system intrinsically flexible. Thus, despite it being a heritable system, it can probably be modulated by external factors, as we will describe later.

For these reasons, we will try to dissect the various aspects of epigenetic involvement in aging, from the first evidence of epigenetic changes in aging to the recent genome-wide analysis of epigenetic changes in aging. We will focus our attention on the following points:

- epigenetic changes during aging;
- the role of environment in aging-dependent epigenetic changes;
- the impact of epigenetic alteration on the aging phenotype;
- intergenerational transmission through generations of epigenetic changes accumulated during development and aging.

Epigenetic Changes During Development and Aging

As soon as DNA methylation was discovered, scientists began investigating this phenomenon during the aging process. The first report was published more than 40 years ago by Berdyshev et al. (1967), who discovered that 5-methyldeoxycytidine levels decrease with age in spawning humpbacked salmon. Later, Vanyushin et al. (1973) performed a more detailed study in rats, in which they detected a global loss of cytosine methylation during aging in brain and heart, but not in liver and lung, and proposed that the former two organs are primarily affected by the aging process. More recently, Wilson et al. (1987) confirmed the gradual loss of DNA methylation in different mouse tissues with age and in human bronchial epithelial cells. They provided convincing evidence that the reduction in methylation is unrelated to the proliferation rate of the cells and so cannot be ascribed to the dilution effect of cell division.

When techniques for the study of promoter-specific DNA methylation first became available, the change of methylation status with aging of some specific promoters was readily studied. Several specific regions of the genomic DNA become hypermethylated during aging. For instance, an increase of methylation in ribosomal DNA clusters in livers and germ cells of senescent rats was noted by Oakes et al. (2003) using restriction landmark genomic scanning, a method that enables specific methylation patterns of CpG island sequences to be determined. This observation is important because it could be associated with the decrease of RNA levels during aging. Moreover, failure to maintain normal DNA methylation patterns in male germ cells could be one of the mechanisms underlying age-related abnormalities in fertility and progeny outcome.

Methylation of promoter CpG islands in non-tumourigenic tissues has been reported for several genes, including *estrogen receptor* (*ER*), *myogenic differentiation antigen 1* (*MYOD1*), *insulin-like growth factor II* (*IGF2*) and *tumour suppressor candidate 33* (*N33*). In some cases, such as *MLH1* and *p14ARF*, colon promoter hypermethylation was more common in aged tissues (reviewed in Issa 2003). A recent study found promoter hypermethylation of the tumour suppressor genes *lysyl oxidase* (*LOX*), *p16INK4a*, *runt-related transcription factor 3* (*RUNX3*) and *TPA-inducible gene 1* (*TIG1*) in non-neoplastic gastric mucosa to be positively and significantly associated with aging (So et al. 2006). Other examples of genes with increased promoter methylation during aging include those encoding *E-cadherin*, *c-fos* and *collagen* $\alpha 1$ (reviewed in Fraga and Esteller 2007). Although it is possible to associate the accumulation of methylation at the promoters of these tumour suppressor genes during aging with the predisposition to developing cancer, there is no experimental or mechanistic evidence of a direct relationship between these genes and aging.

Another group of studies centred their attention on a classic in vitro model for aging: the replicative senescence of primary cultured cells. The process of cellular senescence was first described by Hayflick and Moorhead (1961), who observed that normal human fibroblasts were able to enter a state of irreversible growth arrest after serial cultivation in vitro while cancer cells were able to proliferate indefinitely. They proposed that there were some factors whose gradual loss through cell proliferation limited the number of divisions and that this process could contribute to the organismal aging. It is still not completely clear how this cellular senescence contributes to aging, but two main processes have been suggested: the accumulation of senescent cells in tissues and the limitation of regenerative potential of adult stem cell pools.

As in every field of biology, an in vitro model, despite its obvious limitations due to the non-physiological environment, has been useful for deriving a molecular description of some of the processes associated with senescence and aging. Wilson and Jones (1983) first explained how global DNA methylation also decreases with the number of cell passages in cultures of normal diploid fibroblasts of mice, hamsters and humans, while immortal cell lines have stable levels of methylation. The greatest rate of loss of 5-methylcytosine residues was observed in mouse cells, which survived the fewest divisions, implying that the rate of methylation loss is correlated with functional senescence.

One of the principal factors involved in cell senescence is telomere shortening. Studies of telomere length regulation identified the first molecular mechanism capable of counting cell divisions and implementing cell cycle arrest (Harley et al. 1990). Telomeres consist of repetitive DNA elements at the end of linear chromosomes that protect the DNA ends from degradation and recombination. Due to the intrinsic inability of the replication machinery to copy the ends of linear molecules, telomeres become progressively shorter with every round of cell division. Eventually, telomeres reach a critically short length and behave like double-stranded DNA breaks that activate the p53 tumour suppressor protein, which results in telomere-initiated senescence or apoptosis. Telomerase is a ribonucleoprotein with DNA polymerase activity that elongates telomeres, but its level of activity in most adult tissues is not sufficient to compensate for the progressive telomere attrition that occurs with aging (reviewed in Collado et al. 2007). A strong relationship between DNA methylation and telomere length has recently been described. Mouse embryonic stem cells genetically deficient for DNMT1 or both DNMT3a and DNMT3b have dramatically elongated telomeres compared with wild-type controls. Even if telomere repeats (TTAGGG) lack the canonical CpG methylation site, mouse subtelomeric regions are heavily methylated. However, this modification is not as strong as in DNMT-deficient cells. Other heterochromatic marks, such as histone H3-*K*9 and histone H4-*K*20 trimethylation, remain at both subtelomeric and telomeric regions in these cells (Gonzalo et al. 2006).

In summary, these and many other reports show that two specific alterations of DNA methylation occur during aging: a decrease in global 5-methylcytosine and the hypermethylation of a specific promoter. Intriguingly, global DNA hypomethylation, aberrant promoter hypermethylation and modest DNMT overexpression are known epigenetic alterations in cancer. Thus, the accumulation of epigenetic alterations during aging is comparable to a milder form of the aberrant epigenetic alteration observed in cancer. It is therefore reasonable to think that it might contribute to tumourigenic transformation.

Thus, what are the possible causes of this DNA methylation drift? The loss of global DNA methylation during aging is probably the result of the passive demethylation of heterochromatic DNA arising from the progressive loss of DNMT1 efficacy, the erroneous targeting of the enzyme by other cofactors or both. It is also possible that the natural response of the cell to loss of DNA methylation in repeated DNA sequences is to overexpress the de novo DNA methylase DNMT3b, as previously found in cultured fibroblasts. A logical outcome of DNMT3b overexpression is that regions such as promoter CpG islands, which are commonly unmethylated in normal cells, become aberrantly hypermethylated.

Other epigenetic layers, such as histone modifications, also have a defined profile during aging and cell transformation. For example, the trimethylation of H4-K20, which is enriched in differentiated cells, increases with age and is commonly reduced in cancer cells. These variations can be due to deregulation of the principal H4-K20 methyltransferase Suv4–20 h (a homolog of *Drosophila* suppressor of variegation 4-20) of other histone-modifying enzymes. The increase of trimethylated H4-K20 in aged cells has been associated with defects in the nuclear lamina but little is known about the molecular mechanism linking the nuclear lamins and the histone-modifying machinery.

Furthermore, the interactions between polycomb group proteins in regulating histone modifications may be involved in gene control during the senescence process. For instance, the downregulation of EZH2 (a histone methyltransferase) due to BMI1 protein leads to the loss of methylation at H3-K27 and the activation of *INK4A* transcription, resulting in senescence (Bracken et al. 2007). Another histone methyltransferase called Suv39h1, associated with the methylation of histone H3 lysine 9 (H3-K9me), is involved in cellular senescence by mediating the silencing of growth-promoting genes by heterochromatin formation as H3-K9me creates binding sites for HP1 proteins (Braig et al. 2005). Likewise, Narita et al. (2003) observed that the amount of H3-K9me associated with *cyclin A* and *PCNA* promoters increased in senescent cells. They also described senescence-associated DNA foci defined as senescence-associated heterochromatic foci (SAHF) in which H3-K9me and HP1 proteins are concentrated.

In addition to H4-K20 and H3-K9, other epigenetic mechanisms potentially involved in both aging and cancer have recently been described. Of these, the sirtuins, which comprise the class III family of HDACs, deserve special attention. Histone acetylation is crucial for the control of chromatin structure and thus for the regulation of gene expression. Sirtuins are nicotinamide adenine dinucleotide (NAD) dependent and participate in a range of cellular events, including chromatin remodelling, transcriptional silencing, mitosis and control of life span. The founding member of the family, silent information regulator 2 (Sir2), was initially described in yeast. Sir2-like enzymes catalyse a reaction involving the cleavage of NAD. In yeast, deletion of Sir2 shortened the life span, whereas an extra copy of this gene increased it, implying that the Sir2 family has an important role in aging. The crucial importance of NAD^+ in many metabolic pathways and the fact that sirtuins can control the activity of many other proteins involved in cell growth suggest that the Sir2 family of proteins is involved in the elongation of life span mediated by caloric restriction. It has been suggested that carbon flow in glycolysis and the tricarboxylic acid cycle is much reduced under caloric restriction, and so less NAD is available for Sir2. In this way, the Sir2 proteins might link metabolic rate and aging through NAD-dependent gene regulation and chromatin remodelling. The extension of life span by caloric restriction requires Sir2 and is accompanied by increased respiration, which, in turn, increases Sir2 activity. This evidence has sparked considerable interest in the sirtuins, the mammalian orthologs of Sir2. Seven homologues of the yeast Sir2 protein, sirtuin 1–7 (SIRT1–7), are known in mammals, where they have roles in the regulation of gene expression, stress responses, DNA repair, apoptosis, cell cycle, genomic stability and insulin regulation. Thus, the sirtuins are fundamental to the control of crucial metabolic pathways and the regulation of cell growth and cancer. Among the family members, SIRT1 and SIRT2 are of particular interest because they are altered in cancer cells, their expression might be age dependent and they can act on histone tails (reviewed in Fraga and Esteller 2007).

The Role of Environment in Aging-Dependent Epigenetic Changes

We have described in detail a great deal of evidence of epigenetic changes during the aging process and we already have some idea about how they may be related to aging in terms of loss of tissue regeneration and greater predisposition to tumourigenesis. We need to know whether this epigenetic drift is a normal default process due to the imperfection of the machinery involved in maintaining the epigenetic status or whether it is dependent on environment and environmental stress or both, in which case what the relative proportions are. In other words, to what extent is it a stochastic and unchangeable process? By how much could manipulation of environment, diet and lifestyle arrest or slow the aging process? Here, we will present some evidence about how environment can affect aging-related epigenetic drift. Many studies have shown how environmental factors can affect epigenetic status of adult tissues. For example, Belinsky et al. (2002) demonstrated gene-specific methylation changes in bronchial epithelium that can be measured in the sputum of patients. In particular, aberrant promoter methylation, which is very common in lung cancer, was found in *p16* and, to a lesser extent, in *DAP kinase*, both of which are well-known tumour suppressor genes. It was detected in the bronchial epithelium of current smokers as frequently as in that of former smoker controls but not in the same tissue of never-smokers, indicating that the epigenetic inactivation of this gene is probably permissive for the acquisition of additional genetic and epigenetic changes leading to lung cancer.

Many other epigenetic effects have been observed in humans and animals, as a result of diet (for example, there are nutrients that are methyl donors or HDAC inhibitors) or environmental pollutants (most of which are reviewed in Tang and Ho 2007).

To discover how important these environmental factors may be to aging-related epigenetic drift, we need to refer to a classical tool in human genetics studies: monozygotic twin discordance.

Twin studies are extremely useful in human genetics and medicine for estimating the relative importance of genetic and non-genetic components in any phenotype or disease. The rationale underlying classical twin studies is the assumption that MZ twins are genetically identical, whereas DZ twins share 50% of their segregating genes on average and are as genetically different or similar as are ordinary siblings. By calculating concordance rates and correlation coefficients, it is possible to estimate degrees of similarity. Greater similarity in MZ than in DZ twins simply means that the investigated phenotype or disease probably has a genetic component. By contrast, if there is a significantly higher concordance between MZ twins than between DZ twins, the genetic component is likely to be predominant in this trait (reviewed in Poulsen et al. 2007).

The interaction between environmental factors and phenotypic discordance within MZ twins was first noticed many years ago. However, until recently, little was known about the molecular mechanisms by which environmental factors can influence gene function.

A recent study performed in our laboratory and those of other workers addressed the epigenetic contribution to twin discordance and tried to elucidate the effect of environmental characteristics on gene function. To answer our question about the extent to which the epigenetic changes during aging can be affected by environmental factors, we can adapt the monozygotic twin model. The hypothesis is that if the observed epigenetic drift is determined by genetic and heritable factors, we would expect no difference in the epigenetic variables between twins. If drift is totally stochastic and the environment has no effect on it, we would expect to see differences whose nature does not depend on whether or not the twins shared the same environment and lifestyle. Global changes would probably not be detected because changes in both directions would compensate each other. If there is an environmental component, we would expect to see a difference in the epigenetic drift that is larger in older twin pairs, and particularly in those who did not share the same environment and lifestyle.

An analysis of global- and locus-specific DNA methylation and histone-modification differences in blood samples from 80 MZ twins showed that young MZ twin pairs were essentially indistinguishable in their epigenetic marks, whereas a fraction of the elderly MZ twin pairs had substantial variations in several tissues distributed throughout their genomes, which affected repeat DNA sequences and single-copy genes. The degree of discordance in epigenetic pattern was related to environmental differences between twins because these differences were accentuated in the couples who had different lifestyles and who had spent less of their lives together. Moreover, differences in gene expression in the elder twin pairs were four times greater than those observed in the younger twin pairs (Fraga et al. 2005).

Smoking habit, physical activity and diet, among others, are external factors that have been proposed as having a long-term influence on epigenetic modifications. However, it is possible that small defects in transmitting epigenetic information through successive cell divisions, or maintaining it in differentiated cells, accumulate in a process that could be considered as aging-associated epigenetic drift. Epigenetic defects would be expected to accumulate faster than genetic mutations because their consequences for survival are probably less severe and because the cells have not developed a comparable number of mechanisms to correct them.

There is increasing evidence of epigenetic modulation in response to environmental factors from other sources. Examples include an abnormal intrauterine environment associated with epigenetic downregulation of genes involved in pancreatic β-cell function and a maternal diet associated with the DNA methylation profile of offspring. There are, however, several epigenetic changes that occur during ontogenic development that cannot be explained by environmental effects alone, either in twins or in inbred animal studies. A large group of experiments on laboratory animals showed that the reduction of genetic variability by using inbred strains and the reduction of environmental variability by highly standardized husbandry in laboratory animals does not remarkably reduce the range of random variability in quantitative biological traits. For this reason, the existence of a third component was hypothesized that significantly contributes for creating biological random variability (Gartner 1990). More recent studies have given a deeper insight into the problem using a robust and highly fecund parthenogenetic marbled crayfish as the model experimental animal. This animal can tolerate a very broad range of experimental conditions, so it is recommended for research on environmental epigenomics. Isogenic crayfish exhibited broad variation in colouration, growth, life span, reproduction, behaviour and number of sense organs, even when reared under identical conditions. Global DNA methylation varied among batchmates and among tissues, but this variation did not appear to be associated with variation in the life-history parameters (Vogt et al. 2008). Therefore, epigenetic modifications can result from stochastic events and external environmental factors, making them an obvious candidate of molecular mechanism for generating phenotypic variation. Gene function and chromatin structure can be modulated by means of plastic chemical modifications in the DNA and the accompanying histones, and these modifications can be affected by environmental factors. The idea that environmental effects can provoke epigenetic-mediated phenotypic responses is attractive. However, the precise mechanism by which the environment generates phenotypically adaptive responses is still unknown and represents a fascinating area for future research.

The Impact of Epigenetic Alteration on the Aging Phenotype

It is important to determine the genomic location of epigenetic changes, since those occurring in non-coding sequences will have fewer biological consequences than those occurring in coding sequences. Epigenetics provides the link between the environment and the development of diseases. Environmental factors affecting DNA methylation include diet, proteins, drugs and hormones.

The association between epigenetic changes and disease phenotypes has been well studied. One of the best examples relating epigenetics and disease phenotype is found in *agouti* mice. The mouse *agouti* alleles regulate the production of pigment in individual hair follicles. The *A* allele is responsible for the black, wild-type coat colour of mice, whereas the A^{vy}, A^{iapy} and A^{hvy} alleles are responsible for a range of phenotypes in these mice. This spectrum includes coat colour, which varies from entirely yellow to fully *agouti*. Yellow and mottled mice are obese and prone to diabetes and cancer, in contrast to fully *agouti* mice, known as pseudo*agouti*s, which are lean and non-diabetic. Expression of the A^{vy}, A^{iapy} and A^{hvy} alleles is controlled by an intracisternal A particle (IAP) retrotransposon, through spontaneous insertions of single IAP sequences in different regions of the *agouti* gene. Transcription originating in an IAP retrotransposon inserted upstream of the *agouti* gene (*A*) causes ectopic expression of *agouti* protein, resulting in yellow fur, obesity, diabetes and increased susceptibility to tumours. Isogenic A^{vy} mice are epigenetic mosaics and have coats that vary in a continuous spectrum from full yellow, through variegated yellow/*agouti*, to full *agouti* (pseudo*agouti*) due to IAP expression. The activity of the A^{vy} allele is associated with the epigenetic state. It is more frequently hypermethylated in pseudo*agouti* than in yellow mice (Morgan et al. 1999). Furthermore, the LTR in the IAP of A^{iapy} and A^{hvy} alleles is more highly methylated in pseudo*agouti* mice.

Similar to mouse studies, there are several reports of epigenetics and disease in humans. Studies of the role of epigenetic factors in psychiatric diseases have focused on DNA methylation. Those concerned with mental illnesses such as schizophrenia and bipolar disorders have focused on the epigenetic differences in discordant MZ twins. Genes involved in dopaminergic pathways are being studied as candidate genes for schizophrenia. Several studies have supported the involvement of the *dopamine receptor D2* (*DRD2*) in the pathogenesis of this disease. Different methylation patterns in the promoter region of *DRD2* have been detected in the PBL of two schizophrenic MZ twin pairs (Petronis et al. 2003). By contrast, Zhang et al. (2007) selected 48 discordant sib pairs with schizophrenia and found cytosine methylation both in patients and in normal subjects. Given the large sample, these results suggest that aberrant methylation is not a universal phenomenon in schizophrenia. Mill et al. (2006) demonstrated that MZ twins (discordant for birth weight) have methylation discordance in the promoter region of another widely studied psychiatric candidate gene *catechol-O-methyltransferase* (*COMT*). They also found no association between birth weight and the degree of *COMT* methylation or any evidence to suggest that MZ-twin methylation discordance can be explained by birth-weight differences.

Rosa et al. (2008) suggested that pairs of twins discordant for bipolar disease may be more discordant for X-inactivation patterns compared to the other pairs of twins they studied, so the X chromosome may be involved in this disease. Kuratomi et al. (2008) related a reduced DNA methylation status in the *peptidylprolyl isomerase E-like* (*PPIEL*) promoter in patients with bipolar disorder.

Aberrant hypermethylation of CpG islands located in the promoter regions of tumour suppressor genes is one of the most important mechanisms of improper gene inactivation in cancer. Nishida et al. (2008) reported that aberrant methylation of a limited number of loci is commonly seen in the normal aging liver and that these epigenetic alterations gradually progress and expand to a larger panel of methylation markers in hepatocellular carcinoma (HCC).

The genetics of Alzheimer's disease (AD) plays a role in a small percentage of patients, making the environment and other non-genetic factors responsible for the more common sporadic forms. AD appears to be caused by amyloid-β overproduction and accumulation, possibly arising from the loss of epigenetic control in the expression of the genes involved in amyloid-β protein precursor (AβPP) processing. The process is strictly related to *S*-adenosylmethionine (SAM) metabolism, which is required by methyltransferase as a methyl group donor. The methylation of *amyloid precursor protein* (*APP*) gene promoter can alter *APP* expression and affect amyloidogenesis. Researchers have found lower SAM levels in AD patients, suggesting that age-related demethylation of cytosines has some significance in amyloid-β protein deposition in the aged brain (Tohgi et al. 1999).

Type 2 diabetes has increased incidence with age and the severity of its metabolic defects increases with time. It is characterized by insulin resistance in the liver and peripheral target tissues and by hyperglycaemia. Reduced oxidative capacity of the mitochondria in skeletal muscle is associated with insulin resistance in this illness. Defects in the skeletal muscle respiratory chain may contribute to the pathogenesis of type 2 diabetes. *COX7A1* is a nuclear-encoded oxidative phosphorylation gene encoding a subunit of cytochrome *c* oxidase and is expressed in skeletal and heart muscle. *COX7A1* is significantly downregulated in the muscle of type 2 diabetic patients. The level of DNA methylation in the *COX7A1* promoter region was found to be higher in muscle from elderly twin pairs than from young twins, suggesting that differences in DNA methylation become more pronounced with increased age (Ronn et al. 2008).

Intergenerational Transmission of Epigenetic Changes Accumulated During Development and Aging

As epigenetic factors can be affected by the environment, there has recently been a debate about whether, when these alterations occur in germ cells, they can be transmitted from generation to generation. This possibility is simultaneously attractive, because it supports the idea that the environment can shape the evolution of the species, and provocative, because it revives the discounted evolutionary theories of Lamarck. In fact, much of the genomic DNA methylation is erased between fertilization and preimplantation. This is followed by a wave of methylation after implantation. Mammalian development is dependent on DNMT, which requires SAM and uses zinc as a cofactor. Synthesis of SAM is dependent on dietary folates, vitamin B_{12}, methionine, betaine and choline. In the maternal human diet, folic acid is important for the prevention of neural tube birth defects. Something similar occurs in mice. Wolff et al. (1998), Cropley et al. (2006) and Waterland (2006) reported that specific methyl supplements in the diets of pregnant mouse dams can affect the expression of *agouti* gene in the offspring. IAP expression is regulated by DNA methylation and chromatin packaging, and has adverse health effects, whereby it produces ectopic expression of *agouti* protein, resulting in yellow, obese and diabetic mice. For this reason, a methyl-supplemented diet may suppress IAP expression and have positive effects on the health and longevity of offspring. Morgan et al. (1999) studied epigenetic inheritance at the *agouti* locus in mice. They reported that the phenotype of a dam with the A^{vy} allele is related to the phenotypes of the offspring, so yellow dams produce yellow and mottled offspring, but not pseudo*agouti* offspring, whereas pseudo*agouti* dams produce 20% pseudo*agouti* offspring. They also noted a grandmaternal effect; the passage of the allele through two generations of pseudo*agouti* females produced significantly more pseudo*agouti* offspring than through only one generation of pseudo*agouti* dams. Epigenetic modifications of DNA produce alterations in transcription state. Defects in the epigenetic silencing machinery can lead to epimutation, the abnormal silencing of a gene. Abnormal *MLH1* (a DNA mismatch-repair gene) promoter methylation is related to HNPCC (Lynch syndrome). Morak et al. (2008) reported evidence of inheritance of epimutation, since *MLH1* promoter methylation was found in a patient and his mother, which indicated familial predisposition. Chan et al. (2006) documented a stably inherited, allele-specific and mosaic methylation in the promoter of another DNA mismatch-repair gene called *MSH2*, in a family affected with HNPCC.

Some studies have addressed how environmental factors can affect DNA methylation patterns transmitted from one generation to the next. For instance, Anway et al. (2005) analysed the effects of two endocrine disruptors that frequently occur in the environment on diseases that commonly involve epigenetic alterations: vinclozolin, a fungicide used in viticulture, and methoxychlor, a pesticide. The transient exposure of gestating female rats to these compounds promoted an adult testis phenotype featuring increased spermatogenic cell apoptosis in offspring from the F1 to the F4 generation. This transgenerational phenotype appears to be associated with altered DNA methylation of the male germ line.

Overall, epigenetic inheritance is a sporadic process in mammals, but it may be difficult to distinguish it from genetic inheritance, variable expressivity and incomplete penetrance. Epigenetic changes tend to be stochastic and reversible, so the occurrence and inheritance of epimutations are probably governed by completely different rules from those of Mendelian genetics.

Concluding Remarks

Epigenetics clearly has an important role in determining phenotypic differences during aging. Given that we know that changes occur in DNA methylation and histone modifications during aging, it is reasonable to expect that these will probably have repercussions for some age-associated diseases, such as some mental illnesses, diabetes, and the development of some types of cancer. Once we understand the nature of the close relationship between epigenetic modifications and the aging process, we will be able to explore new strategies for minimizing the impact of epigenetics on the development of aging-related diseases. However, this will not be straightforward because some epigenetic changes that occur during ontogenic development and aging cannot be explained by environmental effects alone: there is also a stochastic component to phenotypic variability. The molecular mechanisms by which particular environmental exposures induce specific epigenetic changes are still unknown and are a matter of interest for future research. We also need to establish what proportion of heritable phenotypic variability associated with development and aging can be ascribed to epigenetic factors. Our understanding of epigenetic variation and inheritance is still in its infancy, but ongoing studies will undoubtedly provide important information in the near future.

References

Anway MD, Cupp AS, Uzumcu M, Skinner MK. 2005. Epigenetic transgenerational actions of endocrine disruptors and male fertility. Science 308:1466–9

Belinsky SA, Palmisano WA, Gilliland FD, Crooks LA, Divine KK, Winters SA, Grimes MJ, Harms HJ, Tellez CS, Smith TM, Moots PP, Lechner JF, Stidley CA, Crowell RE. 2002. Aberrant promoter methylation in bronchial epithelium and sputum from current and former smokers. Cancer Res 62:2370–7

Berdyshev GD, Korotaev GK, Boyarskikh GV, Vanyushin BF. 1967. Nucleotide composition of DNA and RNA from somatic tissues of humpback salmon and its changes during spawning. Biokhimia 32

Bestor TH. 2000. The DNA methyltransferases of mammals. Hum Mol Genet 9:2395–402

Bostick M, Kim JK, Esteve PO, Clark A, Pradhan S, Jacobsen SE. 2007. UHRF1 plays a role in maintaining DNA methylation in mammalian cells. Science 317:1760–4

Bracken AP, Kleine-Kohlbrecher D, Dietrich N, Pasini D, Gargiulo G, Beekman C, Theilgaard-Monch K, Minucci S, Porse BT, Marine JC, Hansen KH, Helin K. 2007. The Polycomb group proteins bind throughout the INK4A-ARF locus and are disassociated in senescent cells. Genes Dev 21:525–30

Braig M, Lee S, Loddenkemper C, Rudolph C, Peters AH, Schlegelberger B, Stein H, Dorken B, Jenuwein T, Schmitt CA. 2005. Oncogene-induced senescence as an initial barrier in lymphoma development. Nature 436:660–5

Chan TL, Yuen ST, Kong CK, Chan YW, Chan AS, Ng WF, Tsui WY, Lo MW, Tam WY, Li VS, Leung SY. 2006. Heritable germline epimutation of MSH2 in a family with hereditary nonpolyposis colorectal cancer. Nat Genet 38:1178–83

Collado M, Blasco MA, Serrano M. 2007. Cellular senescence in cancer and aging. Cell 130: 223–33

Cropley JE, Suter CM, Beckman KB, Martin DI. 2006. Germ-line epigenetic modification of the murine A^{vy} allele by nutritional supplementation. Proc Natl Acad Sci USA 103:17308–12

Fraga MF, Ballestar E, Paz MF, Ropero S, Setien F, Ballestar ML, Heine-Suner D, Cigudosa JC, Urioste M, Benitez J, Boix-Chornet M, Sanchez-Aguilera A, Ling C, Carlsson E, Poulsen P, Vaag A, Stephan Z, Spector TD, Wu YZ, Plass C, Esteller M. 2005. Epigenetic differences arise during the lifetime of monozygotic twins. Proc Natl Acad Sci USA 102:10604–9

Fraga MF, Esteller M. 2005. Towards the human cancer epigenome: a first draft of histone modifications. Cell Cycle 4:1377–81

Fraga MF, Esteller M. 2007. Epigenetics and aging: the targets and the marks. Trends Genet 23:413–8

Gartner K. 1990. A third component causing random variability beside environment and genotype. A reason for the limited success of a 30 year long effort to standardize laboratory animals? Lab Anim 24:71–7

Gonzalo S, Jaco I, Fraga MF, Chen T, Li E, Esteller M, Blasco MA. 2006. DNA methyltransferases control telomere length and telomere recombination in mammalian cells. Nat Cell Biol 8:416–24

Harley CB, Futcher AB, Greider CW. 1990. Telomeres shorten during ageing of human fibroblasts. Nature 345:458–60

Hayflick L, Moorhead PS. 1961. The serial cultivation of human diploid cell strains. Exp Cell Res 25:585–621

Heard E, Disteche CM. 2006. Dosage compensation in mammals: fine-tuning the expression of the X chromosome. Genes Dev 20:1848–67

Hendrich B, Bird A. 1998. Identification and characterization of a family of mammalian methyl-CpG binding proteins. Mol Cell Biol 18:6538–47

Holliday R. 2006. Epigenetics: a historical overview. Epigenetics 1:76–80

Issa JP. 2003. Age-related epigenetic changes and the immune system. Clin Immunol 109:103–8

Jones PA, Baylin SB. 2002. The fundamental role of epigenetic events in cancer. Nat Rev Genet 3:415–28

Kangaspeska S, Stride B, Metivier R, Polycarpou-Schwarz M, Ibberson D, Carmouche RP, Benes V, Gannon F, Reid G. 2008. Transient cyclical methylation of promoter DNA. Nature 452:112–5

Kuratomi G, Iwamoto K, Bundo M, Kusumi I, Kato N, Iwata N, Ozaki N, Kato T. 2008. Aberrant DNA methylation associated with bipolar disorder identified from discordant monozygotic twins. Mol Psychiatry 13:429–41

Lyon MF. 1961. Gene action in the X-chromosome of the mouse (*Mus musculus* L.). Nature 190:372–3

Metivier R, Gallais R, Tiffoche C, Le Peron C, Jurkowska RZ, Carmouche RP, Ibberson D, Barath P, Demay F, Reid G, Benes V, Jeltsch A, Gannon F, Salbert G. 2008. Cyclical DNA methylation of a transcriptionally active promoter. Nature 452:45–50

Mill J, Dempster E, Caspi A, Williams B, Moffitt T, Craig I. 2006. Evidence for monozygotic twin (MZ) discordance in methylation level at two CpG sites in the promoter region of the catechol-O-methyltransferase (COMT) gene. Am J Med Genet B Neuropsychiatr Genet 141B: 421–5

Morak M, Schackert HK, Rahner N, Betz B, Ebert M, Walldorf C, Royer-Pokora B, Schulmann K, von Knebel-Doeberitz M, Dietmaier W, Keller G, Kerker B, Leitner G, Holinski-Feder E. 2008.

Further evidence for heritability of an epimutation in one of 12 cases with MLH1 promoter methylation in blood cells clinically displaying HNPCC. Eur J Hum Genet 16:804–11
Morgan HD, Sutherland HG, Martin DI, Whitelaw E. 1999. Epigenetic inheritance at the agouti locus in the mouse. Nat Genet 23:314–8
Narita M, Nunez S, Heard E, Narita M, Lin AW, Hearn SA, Spector DL, Hannon GJ, Lowe SW. 2003. Rb-mediated heterochromatin formation and silencing of E2F target genes during cellular senescence. Cell 113:703–16
Nishida N, Nagasaka T, Nishimura T, Ikai I, Boland CR, Goel A. 2008. Aberrant methylation of multiple tumor suppressor genes in aging liver, chronic hepatitis, and hepatocellular carcinoma. Hepatology 47:908–18
Oakes CC, Smiraglia DJ, Plass C, Trasler JM, Robaire B. 2003. Aging results in hypermethylation of ribosomal DNA in sperm and liver of male rats. Proc Natl Acad Sci USA 100:1775–80
Okamoto I, Otte AP, Allis CD, Reinberg D, Heard E. 2004. Epigenetic dynamics of imprinted X inactivation during early mouse development. Science 303:644–9
Petronis A, Gottesman, II, Kan P, Kennedy JL, Basile VS, Paterson AD, Popendikyte V. 2003. Monozygotic twins exhibit numerous epigenetic differences: clues to twin discordance? Schizophr Bull 29:169–78
Poulsen P, Esteller M, Vaag A, Fraga MF. 2007. The epigenetic basis of twin discordance in age-related diseases. Pediatr Res 61:38R–42R
Rakyan VK, Beck S. 2006. Epigenetic variation and inheritance in mammals. Curr Opin Genet Dev 16:573–7
Ronn T, Poulsen P, Hansson O, Holmkvist J, Almgren P, Nilsson P, Tuomi T, Isomaa B, Groop L, Vaag A, Ling C. 2008. Age influences DNA methylation and gene expression of COX7A1 in human skeletal muscle. Diabetologia 51:1159–68
Rosa A, Picchioni MM, Kalidindi S, Loat CS, Knight J, Toulopoulou T, Vonk R, van der Schot AC, Nolen W, Kahn RS, McGuffin P, Murray RM, Craig IW. 2008. Differential methylation of the X-chromosome is a possible source of discordance for bipolar disorder female monozygotic twins. Am J Med Genet B Neuropsychiatr Genet 147B:459–62
So K, Tamura G, Honda T, Homma N, Endoh M, Togawa N, Nishizuka S, Motoyama T. 2006. Quantitative assessment of RUNX3 methylation in neoplastic and non-neoplastic gastric epithelia using a DNA microarray. Pathol Int 56:571–5
Tang WY, Ho SM. 2007. Epigenetic reprogramming and imprinting in origins of disease. Rev Endocr Metab Disord 8:173–82
Tohgi H, Utsugisawa K, Nagane Y, Yoshimura M, Genda Y, Ukitsu M. 1999. Reduction with age in methylcytosine in the promoter region -224 approximately -101 of the amyloid precursor protein gene in autopsy human cortex. Brain Res Mol Brain Res 70:288–92
Vanyushin BF, Nemirovski LE, Klimenko VV, Vasiliev VK, Belozersky AN. 1973. The 5-Methylcytosine in DNA of rats. Gerontologia 19:138–152
Vogt G, Huber M, Thiemann M, van den Boogaart G, Schmitz OJ, Schubart CD. 2008. Production of different phenotypes from the same genotype in the same environment by developmental variation. J Exp Biol 211:510–23
Waterland RA. 2006. Assessing the effects of high methionine intake on DNA methylation. J Nutr 136:1706S–1710S
Whitelaw NC, Whitelaw E. 2006. How lifetimes shape epigenotype within and across generations. Hum Mol Genet 15 Spec No 2:R131–7
Wilson VL, Jones PA. 1983. DNA methylation decreases in aging but not in immortal cells. Science 220:1055–7
Wilson VL, Smith RA, Ma S, Cutler RG. 1987. Genomic 5-methyldeoxycytidine decreases with age. J Biol Chem 262:9948–51
Wolff GL, Kodell RL, Moore SR, Cooney CA. 1998. Maternal epigenetics and methyl supplements affect agouti gene expression in Avy/a mice. Faseb J 12:949–57
Zhang AP, Yu J, Liu JX, Zhang HY, Du YY, Zhu JD, He G, Li XW, Gu NF, Feng GY, He L. 2007. The DNA methylation profile within the 5′-regulatory region of DRD2 in discordant sib pairs with schizophrenia. Schizophr Res 90:97–103

Role of Epigenetics in Age-Related Long-Term Memory Loss

J. Tyson DeAngelis and Trygve O. Tollefsbol

Abstract The epigenetic mechanisms of DNA methylation and histone modifications have been shown to play a vital role in long-term memory formation by regulating the expression of genes involved in memory processes. In fear conditioned mice, CpG islands in the promoter regions of the memory-linked genes, *reelin* and *Protein Phosphatase 1 (PP1)*, undergo either hypomethylation *(reelin)* or hypermethylation *(PP1)*. By altering the expression of these genes, epigenetic mechanisms can affect the ability to form long-term memories. The pivotal role of histone modifications in long-term potentiation is made evident by such genetic disorders as Rubenstein-Taybi syndrome (RTS), a disease characterized in part by cognitive dysfunction. The histone acetyl transferase domain of CREB-binding protein is catalytically inert in RTS patients and is believed to be the cause of a limited ability for memory formation. The role of epigenetics in age-related memory loss is best seen by studying the classic age-related disease, Alzheimer's disease (AD), which is characterized by severe memory loss. Biological aging correlates with a loss of expression and activity of certain epigenetic modulating enzymes, such as DNA methyltransferase1. The loss of maintenance of the epigenome is believed to be key to understanding the role of epigenetics in age-related cognitive diseases, such as AD.

Keywords DNA methylation · Histone modifications · Long-term potentiation · Rubenstein–Taybi syndrome · *Reelin* · *PP1* · Alzheimer's disease

---

T.O. Tollefsbol (✉)
Center for Aging and Department of Biology, 175 Campbell Hall, 1300 University of Alabama, Birmingham, AL 35294-1170, USA
e-mail: trygve@uab.edu

T.O. Tollefsbol (ed.), *Epigenetics of Aging*, DOI 10.1007/978-1-4419-0639-7_15,

Introduction

The epigenetic mechanisms of DNA methylation and histone modifications are essential to almost every aspect of human development and have been shown to regulate a number of biological processes. In recent years, epigenetic processes have been shown to play a role in long-term memory formation as well as many other cognitive processes. In concert with histone acetylation, DNA methylation provides a means for environmental stimuli to assert heritable changes in gene expression through the covalent modification of the DNA itself and its surrounding chromatin. In essence, long-term memory is nothing more than the storage of environmental stimuli. Overall, DNA methylation patterns may serve to define the epigenetic state of a cell, which in turn can dictate the phenotypic state of the cell. This notion is made evident by the essential role DNA methylation has been shown to play in embryonic development and cellular differentiation, processes that both involve a highly coordinated transformation from one state to another.

The epigenetic process of DNA methylation is carried out by the DNA methyltransferase (DNMT) family of enzymes and involves the covalent addition of a methyl group, donated by *S*-adenosyl methionine (SAM), to cytosine bases in CpG dinucleotides (Bestor 2000). DNA hypermethylation of specific regions of CG-rich DNA, known as CpG islands, in a promoter region usually leads to a loss of gene expression. DNA hypermethylation can repress transcription by sterically preventing a methylation-sensitive transcription factor from binding and/or can repress transcription through the recruitment of methylcytosine-binding domain (MBD) proteins to areas of dense hypermethylation. MBD proteins have been shown to complex with histone-modifying enzymes, linking changes in DNA methylation to changes in, for example, histone acetylation (Ng et al. 1999).

Chromatin modifications involve the addition or the removal of functional groups from specific residues along the N-terminus tails of the histone proteins. The most studied histone alteration is histone acetylation. The acetylation of histone tails primarily occurs at lysine residues and decreases the affinity of the histone tails for DNA. Acetyl groups are added by histone acetyltransferases (HATs) and usually result in an upregulation in gene expression. The removal of acetyl groups is catalyzed by histone deacetylases (HDACs) which work in opposition of HATs. There are other major types of histone modifications, such as histone methylation and histone phosphorylation. Although these modifications are essential to transcriptional regulation, their role in memory formation is for the most part unknown. Histone phosphorylation has recently been shown to correlate with the formation of long-term memory and will be discussed briefly later in the chapter.

Synaptic plasticity is loosely defined as the strengthening of the connection between two neuronal cells. Long-term potentiation (LTP) is the increase of synaptic plasticity over time and it is the concept of LTP that forms the basis for memory formation. There are two different phases of LTP: the first being the early phase LTP (E-LTP), in which cellular proteins are immediately modified in response to an action potential, and the second phase of memory formation being the late-LTP (L-LTP). L-LTP, in contrast to E-LTP, involves the synthesis of new proteins (Nguyen et al. 1994) and is thought to be more closely linked to long-term memory formation. The role of epigenetic processes in memory formation has only recently been revealed and it is the ability of epigenetic mechanisms to elicit long-term changes in gene expression in response to external stimuli that may be the key to understanding how learning and memory function at the molecular level (Table 1).

Table 1 Examples of memory-linked genes known to participate in memory formation or long-term potentiation (LTP)

Gene	Link to long-term potentiation (LTP)	Reference
MeCP1	Mice lacking MeCP1, a methyl cytosine-binding protein, display decreases in LTP	Zhao et al. (2003)
Protein phosphatase 1 (PP1)	*PP1* is a memory suppressor gene that has been shown to undergo hypermethylation in its promoter region during LTP	Miller and Sweatt (2007)
Reelin	The promoter region of the *reelin* gene, a memory promoter gene linked to increases in LTP, undergoes hypomethylation during LTP	Levenson et al. (2008)
Presenilin 1 (PS1)	A gene linked to AD that is regulated by DNA methylation and may provide a possible target for epigenetic-based AD therapeutics involving the administration of *S*-adenosyl methionine	Scarpa et al. (2006)
CREB-binding protein (CBP)	Inactivating mutations in the gene of CBP, a HAT known to regulate memory formation, are associated with RTS patients which display a decrease in the ability to form memories	Kalkhoven et al. (2003)

DNA Methylation in Memory Formation

DNA methylation provides the means for a separate tier of information to be stored on the DNA itself and was originally thought to be more of a stable, persistent change; however, DNA methylation patterns have now been shown to be dynamically regulated, especially when it comes to memory formation. Although the precise role of DNA methylation in memory formation is not well understood, it has become apparent that DNA methylation plays a vital role in the formation of long-term memory. Some of the first studies that highlighted the role of DNA methylation in memory formation were the studies that inhibited DNA methylation or knocked out DNA methylation modulators. Levenson and colleagues illustrated that inhibition of DNA methylation by DNMT inhibitors, such as 5-azacytidine, resulted in a drastic decrease in LTP and radical changes in gene expression (Levenson et al.

2006). The importance of DNA methylation in LTP is further indicated by studies showing that knockout mice deficient in *MeCP1*, an MBD known to complex with HDACs, had a severe decrease in LTP in the hippocampus (Zhao et al. 2003).

The site-specific effects of DNA methylation on LTP have been investigated in a number of memory-linked genes (Table 1). Two genes in particular were chosen as examples to explain the role of DNA methylation in memory formation. Reelin is an extracellular matrix protein that is essential to the development of the higher order structure of the brain, assists in the formation of novel neuronal connections, and regulates LTP and synaptic plasticity (Quattrocchi et al. 2002; Beffert et al. 2005). *Reelin* expression in the hippocampus is highly dynamic, especially during LTP, and regulation of *reelin* expression has been shown to be influenced by both DNA methylation and histone acetylation (Levenson et al. 2008). *Protein phosphatase 1 (PP1)* is another memory-linked gene that is known as a memory suppressor. Increased *PP1* expression elicits a decrease in LTP and the ability to form long-term memories, whereas inhibition of *PP1* promotes learning and long-term memory formation (Gräff and Mansuy 2008).

In 2007, Miller and Sweatt published the findings that investigated the role of DNA methylation in LTP in fear-conditioned animals. Fear conditioning is a well-known psychiatric experiment that promotes LTP and provides a means to study the mechanisms of memory formation. They examined the regulation of both the *reelin* gene and the *PP1* gene in LTP with regard to DNA methylation. They showed that the promoter of the *PP1* gene underwent hypermethylation upon fear conditioning, resulting in a repression of *PP1* expression (Miller and Sweatt 2007). Interestingly, the *reelin* gene underwent hypomethylation, which resulted in an upregulation of reelin expression (Miller and Sweatt 2007). This suggests that there is an active mechanism for DNA demethylation, a process that has been thought to be a passive one, and also hints at the existence of a DNA-demethylating enzyme that may be involved in the epigenetic regulation of gene expression.

The transcriptional regulatory role of DNA methylation in memory formation appears to be very similar to its role in other biological processes; however, the mechanisms that are causing these drastic alterations to DNA methylation patterns in a site-specific fashion are still unknown. It is well known that mechanisms of DNA methylation and histone modifications are intricately entwined and it is this relationship with chromatin alterations that may be the driving force behind site-specific changes in DNA methylation. Given this relationship, the higher stability of DNA methylation, as compared to the much more dynamic process of modifying histone tails, may allow for a semipermanent storage of information regarding the alterations that histones have undergone over time in that region of DNA.

Histone Acetylation in Memory Formation

L-LTP describes the phase of LTP that deals with the formation of long-term memories and involves the upregulation in the expression of genes known to promote memory formation. The transition of a gene from a transcriptionally inactive to a transcriptionally active state is controlled in part by histone modifications. Histone modifications like acetylation have been shown to play an integral part in LTP, yet the exact mechanisms that drive the histone-modifying processes are still unknown. Increases in the levels of histone acetylation have been shown to coincide with LTP (Levenson et al. 2004). Interestingly, injections of HDAC inhibitors such as trichostatin A (TSA) directly into the hippocampus have been shown to increase LTP and the ability to form long-term memories (Miller et al. 2008).

Investigations of certain genetic disorders, such as Rubenstein–Taybi syndrome (RTS), have illustrated the importance of histone acetylation in long-term memory. RTS patients are characterized by physical abnormalities, cognitive dysfunction, and varying degrees of mental retardation. RTS is a disorder in which a mutation in the gene coding for the CREB-binding protein (CBP) causes the HAT domain of CBP to be catalytically inert (Kalkhoven et al. 2003). Most importantly, RTS results in a limited ability to form memories, a symptom that in mice can be reversed through the introduction of TSA (Alarcón et al. 2004).

Recent findings by Sweatt and colleagues illustrated that increases in histone acetylation during LTP were also accompanied by an increase in both histone phosphorylation and phosphorylation of the extracellular signal-regulated kinase/mitogen-activated protein kinase (ERK/MAPK) pathway (Levenson et al. 2004). These findings suggest that histone phosphorylation may be a link between cellular signaling pathways and histone modifications, furthering the notion that epigenetics provides a means for environmental stimuli to assert covalent modifications to DNA and its surrounding chromatin.

Age-Associated Memory Loss

The essential role of DNA methylation in memory formation may for the first time explain the loss of cognitive function, including memory loss, which is known to occur with aging. To date, there has yet to be a published study indicating the role of epigenetic mechanisms in age-related memory loss. Despite the lack of solid evidence implicating epigenetic mechanisms in age-related memory decline, there are multiple indicators that suggest that dysregulation of epigenetic mechanisms is a contributing factor.

It is well known that with biological aging comes a genome-wide DNA hypomethylation (Wilson et al. 1987). Cellular aging is associated with a decline in the expression and activity of DNMT1 (Casillas et al. 2003), which may be the cause for age-associated DNA hypomethylation. A loss of DNMT1 expression would have dire consequences on the ability to form long-term memories, especially given that DNMT1 has been shown to be highly expressed in brain tissue (Liu et al. 2007). For example, hypomethylation in the promoter regions of memory suppressor genes such as *PP1* would result in an upregulation in *PP1* expression and a subsequent decrease in LTP.

One of the best examples regarding age-related memory loss is Alzheimer's disease (AD). AD is a cognitive disorder characterized by severe memory dysfunction. *Presenilin 1* (*PS1*) is a gene known to affect the onset of Alzheimer's disease and has been shown to be regulated by DNA methylation (Scarpa et al. 2006). Interestingly, AD-associated increases in *PS1* expression lead to an increase in the acetylating activity of CBP, once again linking DNA methylation to histone modifications (Marambaud et al. 2003). Although increases in histone acetylation are known to occur with LTP, aberrant and unregulated increases in histone acetylation would more than likely prevent the mechanisms driving LTP from functioning properly. Given the presence of epigenetic mechanisms in the etiology of AD, Scarpa and colleagues proposed that the promoter of *PS1* could be induced to undergo hypermethylation by administering SAM to AD patients (Scarpa et al. 2006).

Conclusions

In recent years, the field of epigenetics has grown exponentially, with an increasing number of essential biological processes being shown to be regulated by epigenetic mechanisms. Besides epigenetics, no other biological mechanism displays the ability to assert heritable changes in gene expression in response to environmental stimuli. Although the molecular mechanisms that regulate LTP are far from understood, it has become clear that both DNA methylation and histone acetylation play a vital part in long-term memory formation.

Acknowledgments This work was supported in part by grants from the National Cancer Institute and Susan G. Komen for the Cure.

References

Alarcón, J., Malleret, G., Touzani, K., Vronskaya, S., Ishii, S., Kandel, E., et al. 2004. Chromatin acetylation, memory, and LTP are impaired in $CBP^{+/-}$ mice: a model for the cognitive deficit in Rubinstein–Taybi syndrome and its amelioration. Neuron 42: 947–959.

Beffert, U., Weeber, E., Durudas, A., Qiu, S., Masiulis, I., Sweatt, J., et al. 2005. Modulation of synaptic plasticity and memory by Reelin involves differential splicing of the lipoprotein receptor Apoer2. Neuron 47: 567–579.

Bestor, T. H. 2000. The DNA methyltransferases of mammals. Hum Mol Genet 9: 2395–2402.

Casillas, M. A., Jr., Lopatina, N., Andrews, L. G., and Tollefsbol, T. O. 2003. Transcriptional control of the DNA methyltransferases is altered in aging and neoplastically-transformed human fibroblasts. Mol Cell Biochem 252: 33–43.

Gräff, J., and Mansuy, I. 2008. Epigenetic codes in cognition and behaviour. Behav Brain Res 192: 70–87.

Kalkhoven, E., Roelfsema, J., Teunissen, H., den Boer, A., Ariyurek, Y., Zantema, A., et al. 2003. Loss of CBP acetyltransferase activity by PHD finger mutations in Rubinstein–Taybi syndrome. Hum Mol Genet 12: 441–450.

Levenson, J. M., O'Riordan, K. J., Brown, K. D., Trinh, M. A., Molfese, D. L., and Sweatt, J. D. 2004. Regulation of histone acetylation during memory formation in the hippocampus. J Biol Chem 279: 40545–40559.

Levenson, J., Roth, T., Lubin, F., Miller, C., Huang, I., Desai, P., et al. 2006. Evidence that DNA (cytosine-5) methyltransferase regulates synaptic plasticity in the hippocampus. J Biol Chem 281: 15763–15773.

Levenson, J., Qiu, S., and Weeber, E. 2008. The role of reelin in adult synaptic function and the genetic and epigenetic regulation of the reelin gene. Biochim Biophys Acta 1779: 422–431.

Liu, L., van Groen, T., Kadish, I., and Tollefsbol, T. 2007. DNA methylation impacts on learning and memory in aging. Neurobiol Aging. E-pub ahead of print. PMID:17850924.

Marambaud, P., Wen, P., Dutt, A., Shioi, J., Takashima, A., Siman, R., et al. 2003. A CBP binding transcriptional repressor produced by the PS1/epsilon-cleavage of N-cadherin is inhibited by PS1 FAD mutations. Cell 114: 635–645.

Miller, C. A., and Sweatt, J. D. 2007. Covalent modification of DNA regulates memory formation. Neuron 53: 857–869.

Miller, C. A., Campbell, S. L., and Sweatt, J. D. 2008. DNA methylation and histone acetylation work in concert to regulate memory formation and synaptic plasticity. Neurobiol Learn Mem 89: 599–603.

Ng, H. H.,Zhang, Y., Hendrich, B., Johnson, C. A., Turner, B. M., Erdjument-Bromage, H., et al. 1999. MBD2 is a transcriptional repressor belonging to the MeCP1 histone deacetylase complex. Nat Genet 23: 58–61.

Nguyen, P., Abel, T., and Kandel, E. 1994. Requirement of a critical period of transcription for induction of a late phase of LTP. Science 265: 1104–1107.

Quattrocchi, C., Wannenes, F., Persico, A., Ciafré, S., D'Arcangelo, G., Farace, M., et al. 2002. Reelin is a serine protease of the extracellular matrix. J Biol Chem 277: 303–309.

Scarpa, S., Cavallaro, R., D'Anselmi, F., and Fuso, A. 2006. Gene silencing through methylation: an epigenetic intervention on Alzheimer disease. J Alzheimer's Dis 9: 407–414.

Wilson, V., Smith, R., Ma, S., and Cutler, R. 1987. Genomic 5-methyldeoxycytidine decreases with age. J Biol Chem 262: 9948–9951.

Zhao, X., Ueba, T., Christie, B., Barkho, B., McConnell, M., Nakashima, K., et al. 2003. Mice lacking methyl-CpG binding protein 1 have deficits in adult neurogenesis and hippocampal function. Proc Natl Acad Sci USA 100: 6777–6782.

Part IV
Epigenetics of Age-Related Diseases

The Epigenetics of Age-Related Cancers

Kristen H. Taylor, Lynda B. Bennett, Gerald L. Arthur, Huidong Shi, and Charles W. Caldwell

Abstract Individuals of all ages can be detrimentally affected by cancer, albeit the disease is more prevalent in aging individuals. Epigenetic modulation is essential for normal development and becomes altered in cancer and aging. The most well-studied epigenetic elements include DNA methylation, histone modifications, miRNAs, and the binding of co-regulatory proteins such as Polycomb group and Trithorax proteins. Each of these "contributors" to the epigenetic landscape is subjected to aberrant events that are associated with aging and with cancer. We discuss the types of epigenetic modifications that are associated with cancers that affect the young and those that affect adults. It is known that erroneous DNA damage repair, inflammation, and proinflammatory signaling occur frequently in the elderly and that these processes can also result in aberrant epigenetic modifications. Acute lymphoblastic leukemia is one type of cancer that spans the entire lifetime from birth to the aged and demonstrates a number of age-related differences in clinical behavior. Therefore, ALL provides an excellent model to begin to decipher age-related epigenetic impacts on the disease. Finally, mechanistic scenarios are given which may explain the relationship between aging and the development of cancer, and future directions which will augment the elucidation of the complex relationship between cancer and aging are discussed.

Keywords epigenetics · aging · cancer

Introduction

One of the hallmarks of epigenetic alterations in cancer is the concomitant finding of genomic hypomethylation and regulatory region hypermethylation compared to normal cells. Reconciliation of these observations, along with other aberrant associations of additional epigenetic and genetic changes in cancers, is a major effort now

C.W. Caldwell (✉)
Department of Pathology and Anatomical Sciences, Ellis Fischel Cancer Center, University of Missouri, 115 Business Loop I-70 West, Columbia, MO 65203, USA
e-mail: caldwellc@health.missouri.edu

T.O. Tollefsbol (ed.), *Epigenetics of Aging*, DOI 10.1007/978-1-4419-0639-7_16,

under way, but we are still in the early stages of actually understanding and confirming many of these. There is little doubt that both the types and the clinical behaviors of cancers differ in an age-related manner. Thus, it is likely that age-related mechanisms contribute to these variable biological behaviors. However, since cancer is actually a very large number of diseases rather than a single disease, these broad assertions are difficult to reconcile. Furthermore, some forms of cancer, usually the most common, have been studied in more detail than others, thus the role(s) of genetics and epigenetics is better understood in these tumor types. To further complicate the elucidation of age-related mechanisms, cancer is common in aging individuals but is much less common in children, with many of the pediatric tumors considered to be very uncommon to rare in occurrence.

Associated with aging are an increase in DNA damage and unrepaired double-strand breaks, impaired cell cycle regulation, hormonal changes, and an increase in exposures to mutagens. There is considerable experimental support that aberrant repair of DNA strand breaks and dysregulation of the cell cycle are good mechanistic candidates for many types of cancers. At a very basic level, it is intuitive that under conditions where a given tissue or cell type is chronically stimulated and undergoing rapid mitosis, there will be increased opportunity for errors in DNA repair that may develop into cancers. This biological process can occur in a number of anatomic sites that are characterized by robust temporal proliferative responses to stimulatory agents. Examples here include the bone marrow, peripheral lymphoid organs, and parts of the gastrointestinal tract during immune responses of various types. Other anatomic sites, such as the breast and prostate, are more prone to hormonal changes which are induced by puberty and menopause. Additionally, exposures to mutagens such as those found in cigarette smoke and other chemicals may produce cellular responses which over time result in the development of cancer. While these processes and mechanisms may play a prominent role(s) in carcinogenesis of adult cancers, they likely have different roles in pediatric cancers. In a broad generalization, pediatric cancers tend to occur in less-differentiated cells and tissues, whereas those of adults typically occur in more differentiated somatic tissues.

This chapter describes epigenetic changes that are associated with aging, including embryogenesis, and provides examples of known epigenetic changes that have been identified in the most common cancers of the young and the old. Finally, we attempt to synthesize the development of cancer and aging and discuss future directions which will aid in further elucidating the complex relationship between cancer and aging.

Epigenetic Changes in Aging and Cancer

Completion of the Human Genome Project has allowed us to dissect genome organization, determine gene numbers, and search for regulatory sequences embedded in DNA. More recent research has also highlighted the critical and rapidly evolving role of epigenetics in regulation of gene expression patterns (Bernstein et al. 2007;

Estecio et al. 2007; Esteller 2007). As we now know, there are multiple known, and likely other undiscovered, epigenetic marks that work together to affect the chromatin state and gene expression in normal and neoplastic conditions. The major epigenetic marks currently identified include DNA methylation, histone modifications, and binding of co-regulatory proteins such as Polycomb group (PcG) and Trithorax proteins.

DNA Methylation and Histone Modifications

A number of possible mechanisms exist that can affect DNA methylation and gene expression including overexpression of DNMTs, hypermethylation of gene promoter regions, hypomethylation of genomic sequences, and selective demethylation of normally methylated gene promoters (Fig 1). During aging, certain gene-specific loci are known to acquire increased levels of DNA methylation in normal somatic tissues (Issa et al. 1994; Issa 2003; Shen et al. 2007). Conversely, one gene-specific example of age-related demethylation, killer immunoglobulin-like receptor (*KIR*) expression, is usually restricted to natural killer cells and is not expressed on T cells of neonates (Li et al. 2008). However, with increasing age, KIR proteins do become expressed on T cells and are thought to contribute to the development of age-related autoimmune diseases. This pattern of gene expression involves two epigenetic modifications, histone H3 demethylation and DNA demethylation. It is also evident that age-related changes involving epigenetic modifications occur in hematopoietic stem cells (HSCs) and may participate in regulation of the long-term fate of these HSCs (Wagner et al. 2008). Whether these processes contribute to the observation that bone marrow failure syndromes such as myelodysplasia occur most commonly in older adults is yet to be addressed. Not only do lymphoid and hematopoietic cells demonstrate increasing (or decreasing) DNA methylation with aging, but this phenomenon is also observed in normal epithelial tissue from the colon (Issa et al. 1994).

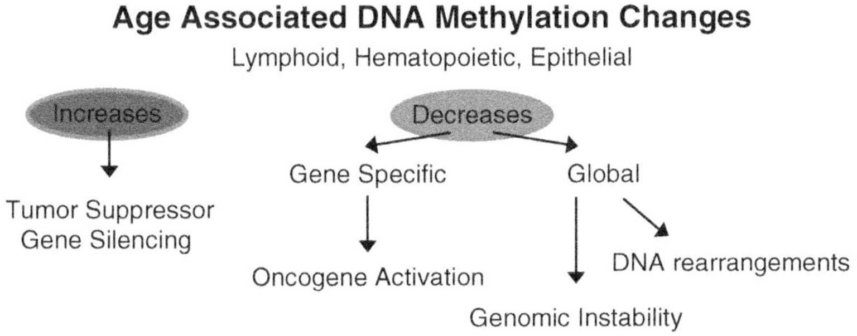

Fig. 1 Age-associated DNA methylation changes. Age-associated increases and decreases in DNA methylation are observed in lymphoid, hematopoietic, and epithelial tissues. These methylation changes are associated with tumor suppressor gene silencing, oncogene activation, and genomic instability

Polycomb Group Proteins

PcG proteins were first recognized for their role as epigenetic modulators of *Drosophila* homeotic (Hox) gene clusters, although additional roles in mammalian systems are now known to include targets relating mainly to transcription factors, developmental regulators, and maintenance of stem cell pluripotency (Ringrose 2007). It has been hypothesized that new proliferative demands occur as normal cells transform into cancers and that this may require a reversion of differentiated characteristics to allow more of an embryonic or stem cell-like phenotype (Monk and Holding 2001) where genes expressed in embryonic cells, but not in adult cells, become reactivated in tumors and subsequently affect the regulation of various processes in both embryos and adults, including the cell cycle. A growing body of work has linked human PcG genes to various hematological and epithelial cancers, identifying novel mechanisms of malignant transformation (Dukers et al. 2004; Pasini et al. 2004; Raaphorst 2005; Schlesinger et al. 2007). In humans, the Polycomb repressive complex 1 (PRC1) contains BMI1, RING, and CBX (CBX2, CBX4, CBX6, CBX7, CBX8) proteins (and others). The other main complex, PRC2, is composed of EED, EZH, SUZ12, and YY1. PcG complexes associate with, or contain, various enzymes that modify histone tails, including histone deacetylases (HDACs) and histone methyl-transferases (HMTs). Histone tail acetylation and methylation generally result in gene activation and silencing, respectively. In normal adult cells, expression of SUZ12, EZH2, and EED is very low, but these proteins have been found to be highly expressed in a variety of human cancers (reviewed in Raaphorst (2005)).

PcG genes can function as a novel class of oncogenes and anti-oncogenes (Pasini, Bracken and Helin 2004; van Galen et al. 2004; Raaphorst 2005). Not only are PcG genes such as *BMI1* and *EZH2* capable of cellular transformation, but they are also vital for cell survival. Abnormal PcG expression is widespread in human non-Hodgkin's lymphomas and differs by subtype (Bracken et al. 2003; Gyory and Minarovits 2005; Raaphorst 2005). Mantle cell lymphoma (MCL), an aggressive lymphoma mainly present in adults, abnormally expresses *BMI1* as well as several of its PRC1-binding partners (Visser et al. 2001). In the normal counterpart cells, the mature B cells of germinal centers (GC), expression of the majority of PRC1 and PRC2 genes is dichotomized in resting and activated cells (Gyory and Minarovits 2005). For instance, BMI1 and its binding partners in PRC1 are primarily present in resting B cells within the GC mantle zone and in non-dividing centrocytes found in the GC follicle (reviewed in van Galen et al. (2004)). However, these are absent in proliferating follicular centroblasts which express the PRC2 proteins EZH2 and EED instead. Neoplastic centrocytes and centroblasts in follicular lymphoma have lost this mutually exclusive expression and actually coexpress all PcG proteins of the PRC1 and PRC2 complexes, suggesting an inability to correctly regulate the PRC1 complex during cell divisions. Lymphomas are not the only tumors that aberrantly express PcG proteins. As reviewed by Raaphorst, many types of solid epithelial tumors display increased expression of PcG proteins and some of the patterns are associated with a loss of differentiation, metastatic behavior, and poor prognosis (Raaphorst 2005).

Chromatin States and Cellular Function

Epigenetic changes lie at the heart of how organisms generate different types of tissue under different circumstances in embryonic development, in regulating cell renewal in adults, and in the cellular responses of the organism to environmental stress and diseases such as cancer.

The role of nuclear chromatin structure in general and its functions in cancer is another area of research that is important to our understanding of this group of diseases. Han et al. (2008) recently reported on intriguing functions of a chromatin-organizing protein associated with progression of breast cancer. The AT-rich-binding protein 1 (SATB1) is aberrantly expressed in human metastatic breast cancer and coordinately regulates the expression of sets of genes that facilitate tumor growth and metastasis. Previous investigations demonstrated that SATB1 could facilitate an active chromatin structure through interactions with AT-rich DNA sequences that apparently can become unpaired through torsional stress-induced changes. SATB1 is generally localized to cage-like nuclear networks that anchor loops of chromatin and recruits chromatin-remodeling proteins to these anchorage sites (matrix attachment regions). Thus, SATB1 can regulate histone modifications and nucleosome positioning over long stretches of DNA (Gal-Yam et al. 2008). Functionally, SATB1 was first shown to coordinately regulate gene transcription during T-cell development (Cai et al. 2006). The Han study found that SATB1 protein levels are elevated in highly metastatic breast cancer cell lines and in 100% of the poorly differentiated infiltrating ductal carcinomas and in 7/12 moderately differentiated human primary tumors, but not in the surrounding normal tissue (Han et al. 2008). They also established a strong reciprocal relationship between increased SATB1 protein levels and decreased overall survival time. In vivo experiments with SATB1 depletion from aggressive MDA-MB-231 cells blocked formation of initial tumors and of lung metastases. Conversely, ectopic expression of SATB1 in a non-metastatic breast cancer cell line induced growth of large tumors and metastases. Gene expression profiles of MDA-MB-231 cells treated with a small hairpin RNA targeting *SATB1* demonstrated altered expression of >1,000 genes, mostly those encoding cell adhesion proteins. When the SATB1-dependent gene set was compared with genes whose dysregulation in breast tumors correlated with poor prognosis from a previously published study (van, V et al. 2002), 63 common genes were identified. Thus, SATB1 may coordinately regulate many genes that can inhibit the spread of breast, and perhaps other, cancers. This newly identified epigenetic mechanism in breast cancer highlights the very fertile field of research into the cancer epigenome.

Known Contributors to Epigenetic Alterations

DNA Damage Repair and Inflammation

DNA damage repair involves an evolutionarily conserved pathway that affects the life span of normal cells, as well as the development of various cancers.

Human cells are presented with more than 1×10^5 DNA lesions/day that must be repaired to avoid potential problems such as chromosomal instability, mutations, and cell death. Mammalian cells have therefore developed at least five major pathways, each responsible for repairing specific types of DNA lesions. Double-strand breaks (DSBs) are mainly repaired through homologous recombination (HR) and non-homologous end joining (NHEJ). Single-strand breaks (SSBs) are mainly repaired through excision repair pathways including base excision repair (BER), nucleotide excision repair (NER), and mismatch repair (MMR).

The DSBs can arise from exposures to ionizing radiation, mutagenic chemicals, and free radicals and set off a cascade of events known as the DNA damage response (DDR). The preferred repair mechanism when a second copy of the damaged DNA is present to serve as a template is HR, which occurs mainly during late phases of the cell cycle and results in DNA repair that is mostly error-free. The process of NHEJ occurs during early phases of the cell cycle when a sister DNA strand is unavailable to serve as a template for HR. In this case, the DSBs are repaired through end-ligation, a reaction that typically involves gain or loss of a few nucleotides (Calin and Croce 2006). As one example, NHEJ represents the main repair mechanism for normal, physiologic V(D)J somatic hypermutations and class switch recombination events during immune responses, generally located in peripheral lymphoid tissues. In cases where NHEJ is not able to participate in the DSB repair reaction, a microhomology-mediated end-joining reaction exists as a salvage pathway, although the factors involved in this particular reaction remain largely unknown (Ching et al. 2005; Cohen et al. 2008).

When SSBs are repaired through excision repair, BER is responsible for repairing small chemical alterations that frequently miscode and therefore are of special relevance in mutagenesis. BER is likely the main sentinel against damage from reactive oxygen species (ROS), methylation, and alkylation, and thus it is particularly important with regard to epigenetic alterations, aging, and cancer. In its core reaction, a set of DNA glycosylases that recognize a specific lesion cleaves the damaged base, producing an abasic site that is subsequently processed by the APE1 endonuclease to generate a nick. DNA polymerase β (polβ) then performs a single-nucleotide gap-filling reaction that removes the 5′ baseless sugar residue. In an alternative pathway, larger 2–10 nucleotide regions may be removed and synthesized de novo, in a reaction involving additional proteins, such as Polδ/ϵ, the co-factor PCNA, the FEN1 endonuclease, and Ligase1. Certain of these polymerases are error-prone and may induce additional lesions. The NER pathway is used to repair a large group of helix-distorting lesions that could interfere with base pairing. Most of these lesions arise from exogenous sources such as UV light. This pathway is thought to be the most versatile in terms of lesion recognition and can survey the entire genome for lesions. Finally, MMR is responsible for removing mispaired nucleotides caused by replication errors and aberrant HR. It is also involved in the repair of oxidative damage and has been suggested to act as a backup repair mechanism to BER. Cells deficient in MMR display microsatellite instability and have causative roles in several malignancies, but little is known about the relationship of MMR with aging.

As can be surmised, the various types of DDR are very important during virtually any type of reactive or inflammatory process that stimulates cells to divide rapidly, because each cell division requires one or more forms of DNA repair for successful cell division. Prolongation of such stimulatory conditions, as in chronic inflammation, then increases the probability of defects that may or may not be faithfully repaired and can result in genetic lesions, or possibly, epigenetic lesions. As one example, of all the non-Hodgkin's lymphoma subtypes, the most common forms arise in a lymphoid germinal center stage of maturation. This is an anatomical and functional site that is characterized by very rapid mitosis, somatic hypermutations, and class switch recombination. This then sets the stage for an overlay or interaction between aberrant DNA strand breaks and potential epigenetic alterations that may lead to altered DNA methylation.

One very intriguing clue into possible development of aberrant DNA methylation comes from investigations into relationships between inflammation, DNA damage, and cytosine methylation (Valinluck et al. 2004, 2005; Valinluck and Sowers 2007a,b). Many types of DNA damage can inhibit DNA–protein interactions involving methyl-binding proteins (MBPs) and DNMT1-mediated methylation of the target cytosine. Interrupting the cascade of events important in gene silencing via interference with binding of these proteins could lead to inappropriate activation (or suppression) of genes and proteins important to normal cells. Functionally, most forms of cytosine damage could interfere with both activities through interference with the critical physicochemical contact points that facilitate these DNA–protein interactions. Conversely, another group of cytosine damage products may actually facilitate both binding of MBPs and enzymatic methylation. As one example, halogenated cytosine residues may prove particularly important in this context and represent a link to aging, chronic inflammation, and development of cancers. Recent studies suggest that halogenation of nucleic acids is a significant form of DNA damage particularly in areas of tissue inflammation (Valinluck and Sowers 2007b). The hydroxychloride (HOCl) produced by activated neutrophils and the hydroxybromide (HOBr) from activated eosinophils were reported to react with DNA and form 5-chlorocytosine and 5-bromocytosine. When this happens, the MBPs cannot distinguish between methylated and halogenated DNA and actually bind to both with high affinity. Additionally, DNMT1 could not distinguish between 5-methylcytosine and certain 5-halocytosines (Valinluck and Sowers 2007a, b). Thus, the formation and persistence of 5-halocytosine in CpG dinucleotides at sites of chronic inflammation, particularly in anatomic sites such as peripheral lymphoid tissues, could lead to inappropriate de novo methylation that could subsequently be replicated in daughter cells through the normal hemimethylation functions of DNMT1. This inflammatory process then may well modify the epigenome during aging and contribute to development of cancers including lymphomas and other diseases.

Proinflammatory Signaling

The insulin/IGF1-like signaling (IIS) pathway represents a conserved pathway regulating metabolism in many organisms. Similar to DNA repair pathways, good evidence links the IIS pathway to both aging and cancer (reviewed in Fraga and Esteller (2007); Fraga et al. (2007); Mostoslavsky (2008)). The binding of insulin (or related peptides) to transmembrane receptors may lead to receptor autophosphorylation and phosphorylation of insulin receptor substrate (IRS) proteins, thus facilitating recruitment of several additional proteins and initiating multiple signaling pathways. One major function involves activation of phosphatidylinositol-3-OH kinase (PI3K), which then leads to the activation of PDK1 kinase and then to the activation of AKT/PKB kinase. AKT/PKB kinase is a critical mediator of multiple downstream processes involved in metabolic adaptation, stress responses, cell cycle regulation, and growth survival. The FOXO transcription factor family represents one of the main effectors of the PI3K–AKT signaling pathway. Activation of AKT leads to FOXO phosphorylation and cytoplasmic localization of the proteins. However, reducing IIS allows nuclear translocation of the FOXO proteins where they can activate or repress transcription and promote stress resistance, cell death, cell cycle arrest, and metabolic changes in a cell-context-dependent manner. The character of the specific response likely depends on the type and intensity of the original stimulus. For instance, the DNA damage response induces a strong IIS response to allow cellular adaptation.

Mammals also have distinct hormones that can provoke IIS to induce different outcomes. Insulin directs anabolic metabolism, whereas IGF1 (mostly by the liver in response to pituitary growth hormone (GH)) mainly directs somatic growth and differentiation. Levels of IGF1 may be most relevant for influencing life span in mammals since reduced IGF1 signaling extends longevity in mice. In contrast, reduction in insulin sensitivity or insulin secretion produces diabetes and shortens life span in both mice and humans. In humans, reduced IIS has not been clearly linked with longevity; indeed low levels of IGF1 have been linked to cardiovascular disease and diabetes, whereas high IGF1 levels may confer susceptibility to cancer.

Defects in several of the IIS pathway components also correlate with increased incidence of certain human cancers. For instance, PTEN encodes a major lipid phosphatase that functions in PI3K signaling and inhibits the PI3K/AKT pathway. PTEN is mutated in a wide variety of cancer types and germ line mutations in the PTEN gene are found in cancer-prone syndromes such as Bannayan–Zonana and Cowden syndromes. Consistent with this, knockout mice deficient in PTEN develop a range of tumors, including those of skin, prostate, colon, and breast carcinomas, as well as thymic lymphomas. Loss of PTEN leads to the activation of the insulin-signaling pathway and stimulates cellular proliferation.

Proinflammatory signals are also known to induce epigenetic changes. Activation of methionine synthase by *IGF-1* and dopamine has been shown to stimulate *PI3K*- and *MAPK*-dependent mechanisms that result in increased DNA methylation, and conversely, inhibition of these pathways led to decreased methylation (Waly et al. 2004). It was also reported that interleukin-6 (*IL-6*) can upregulate *DNMT1* gene expression in human erythroleukemia cells. Chronic activation of the proinflammatory pathway may also result in transcription factor-directed aberrant methylation in target promoters. It was shown recently that in malignant T cells, *DNMT1* and constantly activated *STAT3* cooperatively bind in vivo to *STAT3* SIE/GAS binding sites identified in the hypermethylated *SHP-1* promoter (Zhang et al. 2005). Removing *DNMT1* and/or *STAT3* by small-interfering RNA induces DNA demethylation and expression of the *SHP-1* gene. These data indicate that a transcription factor like *STAT3* may, in part, transform cells by inducing epigenetic silencing of its target genes in cooperation with *DNMT1*. Since *STAT3* is frequently activated by various cytokines and growth factors during chronic inflammation it might change the epigenetic profiles of *STAT3* target genes and generate predisposition to neoplasia.

Epigenetics in Embryogenesis

Embryonic and fetal development is a delicately balanced process involving multiple signaling pathways during very exact spatiotemporal windows. Small deviations can have significant effects on normal development. The fetal environment is clearly a site of epigenetic control of fetal gene expression (reviewed in Nafee et al. (2008)). During early stages, from conception to fetal maturity, changes in gene expression accompany development of over 200 cell types that comprise the various organs and tissues in humans. These changes are largely controlled dynamically through epigenetic mechanisms. Between fertilization and blastocyst formation, a first phase of active demethylation occurs that involves the paternal genome with the exception of paternally imprinted genes, pericentromeric heterochromatin, and some repetitive elements. A form of passive demethylation then ensues until the morula stage, leading to overall decreased DNA methylation in the nucleus. De novo methylation begins after the fifth cell cycle, coincident with the first differentiation events, establishment of the inner cell layer that will create all adult tissues, and the trophectoderm that forms the placenta. Beyond this, a series of dynamic changes in the epigenome will occur in a programmed manner and in fact can be altered at various times during intrauterine development through environmental signals and other as yet unidentified mechanisms that can lead to unintended re-setting of the epigenetic imprints that may lead to pathological processes (Anway and Skinner 2006; Jirtle and Skinner 2007; Anway et al. 2008; Anway and Skinner 2008a, b). This period of abundant epigenetic modification is an ideal time for processes to go awry and potentially leads to lethal or sublethal changes in the ultimate phenotype of the infant. It is not clear how, or if, these normal dynamic epigenetic changes may set the stage for acquisition of genetic aberrations such as *MLL* fusion genes or other acquired genetic abnormalities that might later contribute to disease. This is an exciting area for future research that could be exceedingly important in potentially preventing many diseases.

The placenta plays a crucial intermediary role in maintaining the proper microenvironment for fetal development, and events that alter the balance of metabolic parameters involving the placenta can have enormous consequences. In an interesting review on maternal transmission of risk for atherosclerosis, DeRuiter et al. (2008) describe the ways in which the health status of the mother at specific points in gestation can affect the epigenome (mainly through DNA methylation) in a variety of ways. One mechanism discussed is that of fetal undernutrition that can result in a deficiency in one-carbon metabolism, decreases in DNA methyltransferase activity, and gene-specific hypomethylation, events that can then alter chromatin structure. Another way in which the fetal microenvironment might play a role in disease development is through events that lead to intrauterine growth restriction with the characteristic postnatal overgrowth phase in which the rapid growth leads to an increase in DNA strand breaks that may also affect DNA methylation. There are likely a number of parallels between maternal transmission of atherosclerosis and that of certain cancers. Certainly, cardiovascular disease, diabetes, and cancer are all common diseases of an aged population, and there is evidence to suggest that these might have origins through epigenetic modifications that occur in utero through endocrine disruption (Anway et al. 2006; Anway and Skinner 2006; Skinner and Anway 2007; Jirtle and Skinner 2007; Anway and Skinner 2008a, b).

Differences in Cancers of Young and Adult Patients

For incompletely understood reasons, the cancers that occur in children and adults differ in a variety of ways. There actually seems to be a general bimodality to cancer development, with a group of diseases common to young patients and a different group more common in adults (Fig. 2).

Fig. 2 Common cancers in children versus adults. The most common cancers in children are leukemia, lymphoma, and tumors of the central nervous system. Conversely, adults and the elderly have a high incidence of breast, prostate, and lung cancers

Common Cancers in Children versus Adults

Pediatric (embryonal tissue)	**Adult** (epithelial tissue)	**Elderly**
Leukemias	Breast	Prostate
Lymphomas	Melanoma	Colorectal
Central Nervous System	Testes	Lung
	NHL	Breast
	Thyroid	

Studies of cancer epigenetics, particularly DNA methylation, have mainly been published using adult tumor tissue and less so for pediatric malignancies. However, there are a few reports that indicate this is an area in need of more detailed research. In infants less than 1 year of age, the most common tumors are embryonal tumors such as nephroblastoma (Wilms' tumor), neuroblastoma, retinoblastoma, rhabdomyosarcoma, medulloblastoma, and Ewing's sarcoma. Additionally, a very specific form of acute lymphoblastic leukemia (ALL) and acute megakaryoblastic leukemia occur in infants at birth or within weeks, and the latter frequently occurs in those with Down's syndrome and can spontaneously resolve without treatment (Hitzler and Zipursky 2005; Hitzler 2007). The most common tumors occurring in children between ages 2 and 10 include ALL, which will be discussed later, certain types of lymphomas, brain tumors, soft tissue tumors, and bone sarcomas.

Nine candidate genes including *p16(INK4A), MGMT, GSTP1, RASSF1A, APC, DAPK, RARβ, CDH1*, and *CDH13* were examined for promoter methylation in 175 primary pediatric tumors and 23 tumor cell lines using methylation-specific PCR in a study of the major pediatric tumors (Wilms' tumor, neuroblastoma, hepatoblastoma, medulloblastoma, rhabdomyosarcoma, osteosarcoma, Ewing's sarcoma, retinoblastoma, and acute leukemia) (Harada et al. 2002). The most frequently methylated gene in both primary tumors (40%) and cell lines (86%) was *RASSF1A*. This observation was also confirmed in another study (Wong et al. 2004) of pediatric tumors, including neuroblastoma, thyroid carcinoma, hepatocellular carcinoma, pancreatoblastoma, adrenocortical carcinoma, Wilms' tumor, Burkitt's lymphoma, and T-cell lymphoma. In this case, the majority (75%) of patients with *RASSF1A* methylation were male. Methylated *RASSF1A* alleles were also detected in 4/13 adjacent non-tumor tissues, raising the question if this may be an early event in pediatric cancer.

Neuroblastoma is the most common extracranial solid tumor in infancy and childhood. DNA methylation status of 45 candidate genes was reported using 10 neuroblastoma cell lines as well as 10 genes in tissue from 118 cases of primary neuroblastoma (Alaminos et al. 2004). Clustering of methylation data from the cell lines distinguished those with *MYCN* amplification (a negative prognostic factor) from the others. Hypermethylation of *HOXA9* was associated with mortality in non-infant (childhood) patients and in tumors lacking *MYCN* amplification. Hypermethylation of the proapoptotic genes *TMS1* and *CCND2* was associated with stage 4-progressing tumors, but the genes were never methylated in stage 4S tumors (commonly these undergo spontaneous regression). Further, methylation of *RARβ2* was associated with patient survival. Thus, in primary neuroblastoma, DNA methylation patterns among 10 genes revealed several clinically relevant groups. Certainly this type of preliminary, but very intriguing, data requires follow-up with more detailed epigenetic studies, as this is still a difficult tumor to manage clinically in some cases. Perhaps the epigenetic stratification of treatment could improve this situation.

DNA methylation of candidate genes has also been reported in human retinoblastoma (Cohen et al. 2008) showing 89% of tumors were methylated in *RASSF1A*, 52% in *NEUROG1*, 5% in *DAPK*, *RUNX3*, and *CACNA1G*, and no methylation of *RARβ2*, *SOCS1*, or *IGF-2*. The high methylation status of *NEUROG1* is interesting and may suggest an alternative neural pathway in the development and progression of retinoblastoma, but further studies are needed. The results of *RASSF1A* methylation are consistent with the other larger studies reported in pediatric cancers noted above.

Accumulating evidence indicates a significant involvement of epigenetic events in medulloblastoma development (Lindsey et al. 2004, 2005, 2007). In studies of candidate tumor suppressor genes (*RASSF1A, CASP8, S100*, and *HIC1*), each was inactivated in >30% of medulloblastomas by promoter hypermethylation, leading to the silencing of their gene expression. While preliminary, this study suggests again that epigenetic manipulations might be a useful therapeutic option, but much more research will be needed to confirm the overall epigenetic status of this tumor type before targeted intervention can become common.

Sarcomas are relatively uncommon (<10% of all human cancers) and are believed to originate from mesenchymal progenitor cells, although this is still under investigation (Riggi and Stamenkovic 2007). Nevertheless, they are very aggressive pediatric (and adult) cancers with a high metastatic potential and are typically refractory to most chemo- and radiation therapy. Ewing's sarcoma is a member of Ewing's family tumors (ESFT) and the second most common bone and soft tissue solid tumor arising in children and young adults. The t(11;22)(q24:q12) chromosomal translocation is present in ~85% of tumors and generates a fusion of part of the *EWS* gene with the 3′ portion of *FLI-1*. The resulting fusion protein likely behaves aberrantly as a transcriptional activator that alters target genes in a permissive environment. The EWS protein is a member of a family of RNA-binding proteins that is extensively and asymmetrically dimethylated at arginine residues within RGG consensus sequences (Pahlich et al. 2005). A type I protein arginine methyltransferase (PRMT1) is thought to be responsible for asymmetric dimethylations as it recognizes most, if not all, methylation sites of the EWS protein. Potential interactions between PRMT8 and the EWS protein have also been characterized (Pahlich et al. 2008). Although binding of endogenous and recombinant EWS protein to PRMT8 was observed, in vitro methylation assays revealed a rather poor methyltransferase activity of PRMT8 toward the EWS protein in comparison to PRMT1. Thus, while there has been little reported in terms of gene promoter DNA methylation thus far, the EWS protein seems to be regulated through a different type of epigenetic event that requires further investigation. Given the dire clinical situation with sarcomas, this should be a high-priority area of further research.

Osteosarcoma is the most common solid tumor of childhood, with ~75% of cases occurring in those <20 years of age and the remainder found mainly in the elderly group, with middle-aged individuals mostly being spared this disease. Genetic lesions specific for osteosarcoma thus far have not been identified. In a study of potential epigenetic lesions, 5 candidate gene loci were analyzed for aberrant methylation in 30 pairs of osteosarcoma and corresponding normal tissues (Hou et al. 2006). For the *RASSF1A, TIMP3, MGMT*, and *DAPK1* genes, significant differences were observed in the degree of hypermethylation compared to normal tissues. Measurement of "cumulative multiple promoter hypermethylation" revealed striking differences between malignant and normal tissues. There was also a significant difference in the levels of DNA methylation between the metastatic and nonmetastatic high-grade osteosarcomas. Although only a small number of candidate genes were analyzed, this study suggests the need for more detailed epigenetic studies of this tumor.

Wilms' tumor is a common solid tumor of childhood, occurring mainly in young children aged 2–5 years, and interestingly illustrates a frequent association between congenital malformations and this type of cancer. Chromosomal locations 11p13 (WT1 locus) and 11p15.5 (WT2 locus) are known to harbor genetic and/or epigenetic aberrations in these tumors (Satoh et al. 2006). Mutation of the Wilms' tumor 1 (*WT1*) gene at the WT1 locus has been reported, and the WT2 locus, which comprises the two imprinted domains *IGF2/H19* and *KIP2/LIT1*, can undergo either maternal deletion or alterations of imprinting. In 35 sporadic Wilms' tumors, 29% revealed loss of heterozygosity at 11p15.5 and 40% showed loss of imprinting of *IGF2* as the most frequent genetic and epigenetic alterations in these tumors.

Overall, 83% of the tumors had at least one alteration at 11p15.5 and/or 11p13. One-third of the tumors had alterations at multiple loci. Thus, chromosome 11p appears to be not only genetically but also epigenetically altered in the majority of pediatric Wilms' tumors.

The incidence of lymphoma subtypes differs in an age-related manner. While most adult cases of lymphoma are more indolent, pediatric lymphomas are characterized by their aggressive subtypes (Sandlund et al. 1996). Hodgkin's (HL) and non-Hodgkin's (NHL) lymphoma represent 10–15% of all cancers occurring in children <20 years of age. The most important subtype of HL is nodular sclerosis. Pediatric NHL has four major histologic subtypes: Burkitt lymphoma, diffuse large B-cell lymphoma, anaplastic large cell lymphoma, and lymphoblastic lymphoma. These NHL classes are all considered high-grade, aggressive tumors, but are amenable to successful treatment in many cases.

Rather than cancers involving less-differentiated cells and tissue, as in children, other types of lymphohematopoietic and epithelial tumors, such as chronic leukemias (Guo et al. 2005; Rahmatpanah et al. 2006), mature non-Hodgkin's lymphomas (Baur et al. 1999; Esteller et al. 1999; Katzenellenbogen et al. 1999; Esteller et al. 2002; Esteller 2003; Fung et al. 2003; Kaneko et al. 2003; Nakatsuka et al. 2003; Deligezer et al. 2005; Guo et al. 2005; Reddy et al. 2005; Hayslip and Montero 2006; Martin-Subero et al. 2006; Rahmatpanah et al. 2006; Shi et al. 2007; Taylor et al. 2007a, c), and carcinomas of the breast (Huang et al. 1999; Yan et al. 2000, 2001; Lehmann et al. 2002; Yan et al. 2003; Leu et al. 2004; Szyf et al. 2004), colorectum (Issa et al. 1994; Toyota et al. 1999a, b; Furukawa et al. 2002; Suzuki et al. 2002; Yan et al. 2002; Suzuki et al. 2004; Pufulete et al. 2005; Frigola et al. 2006; Weisenberger et al. 2006), prostate (Usmani et al. 2000; Varambally et al. 2002; Jeronimo et al. 2004; Anway and Skinner 2008b), and lung (Palmisano et al. 2000; Rauch et al. 2007) are more common in adults and show very prominent epigenetic markings. Since DNA methylation in cancer was one of the first areas to be studied, and cancer is much more common in adults than in younger children and adolescents, there are considerably more data published with respect to all the known epigenetic manifestations in these tumors.

In the beginning, most reports involved one or a few candidate genes that demonstrated DNA methylation and/or histone modifications. In some reports, the affected sites were well described and clearly associated with regulatory regions of the specific genes, while in others, the locations and/or densities of methylation were not as well described. This obviously makes direct comparisons problematic. More recent studies involve sub-genomic or genomic studies that, while still early in the process, will likely yield the large-scale data in various tumors and normal tissues that will allow more of a systems biology approach to modeling all the chromatin modifications, gene expression patterns, and other important attributes in order to fully appreciate the epigenomes of various diseases.

Of hematologic malignancies, chronic leukemias, myeloproliferative disorders, and myelodysplastic syndromes occur almost entirely in older adults. In general terms, these are typically more indolent diseases than the acute leukemias, and all have been studied for epigenetic alterations.

The best epigenetic studies of chronic leukemias thus far have come from B chronic lymphocytic leukemia (CLL), once considered a single disease occurring mainly in the elderly, with a variable but generally indolent clinical course. It is now clear that CLL is a heterogeneous disease with at least two subtypes in terms of both biological makeup and prognosis (reviewed in Zent and Kay (2007)). Some patients survive for many years without any therapy, while others progress rapidly within months of diagnosis requiring early initiation of treatment. Biologically, the two subtypes of cells are thought to derive from either a pre-germinal center stage of differentiation or a post-germinal center stage. A number of studies have now shown alterations of DNA methylation (and histone modifications) in CLL (Lipsanen et al. 1988; Wahlfors et al. 1992; Hanada et al. 1993; Kn et al. 2004; Lyko et al. 2004; Rush et al. 2004; Chim et al. 2006; Liu et al. 2006; Rahmatpanah et al. 2006; Yu 2006; Motiwala et al. 2007; Raval et al. 2007; Shi et al. 2007). In some of these studies, only one or a few candidate genes were studied, while in others, larger numbers were examined using high-throughput methods. Myeloproliferative disorders have not been very well represented in epigenetics research thus far.

Myelodysplastic syndromes (MDS) are a group of related clonal disorders that generally produce a functional bone marrow failure through ineffective hematopoiesis. Sometimes, the disease progresses to a form of acute leukemia, most commonly of myeloid lineage (List et al. 2004). Alterations of DNA methylation and histone modifications occur commonly in myeloid malignancies and MDS. In fact, MDS was first studied in the context of clinical treatment with a demethylating agent (Issa 2005). Treatment with azacytidine (Vidaza) or decitabine (Dacogen), inhibitors of DNA methyltransferases, prevents methylation of cytosine residues on DNA and reactivates silenced genes (Issa et al. 2005; Issa 2005). Additionally, vorinostat (suberoylanilide hydroxamic acid, SAHA), a histone deacetylase inhibitor known to be active for treatment of cutaneous T-cell lymphoma, was also tested in a phase 1 study involving patients with relapsed or refractory leukemias or MDS and untreated patients who were not candidates for other chemotherapy (Garcia-Manero et al. 2008b). Of 41 patients, 31 had acute myeloid leukemia (AML), 4 chronic lymphocytic leukemia, 3 MDS, 2 ALL, and 1 chronic myelocytic leukemia. Although there were identified toxicities, there were no drug-related deaths; seven patients had hematologic improvement response, including two complete responses and two complete responses with incomplete blood count recovery (all with AML). Thus, further evaluation of vorinostat in AML/MDS is warranted. While the complete epigenetic profiles of malignancies such as these are not well defined, the fact that there are alterations of DNA methylation and/or histone deacetylations and the clinical responses in early-phase trials support an epigenetic basis for some tumor abnormalities and certainly warrant further clinical and translational research (Garcia-Manero et al. 2002; Issa et al. 2005; Yang et al. 2005; Garcia-Manero et al. 2006; Soriano et al. 2007; Garcia-Manero et al. 2008a, b). However, it is likely that newer second- and third-generation epigenetic modifiers will be needed as there are currently serious toxicities under certain conditions that limit usage, particularly of histone deacetylase inhibitors.

The Spectrum of Acute Lymphoblastic Leukemia Over a Lifetime

As illustrated above, children and adults have very significant differences in the number and type of cancers they may present. One interesting disease, ALL, that has a much higher incidence in children, but also occurs in adults, illustrates a number of age-related differences that may lead to clues into the epigenetic impact(s) on this disease (Fig. 3). Leukemia accounts for approximately 33% of all new cancer diagnoses under the age of 15 years and about 80% of these are ALL; the remaining cases are acute myeloid leukemias (AML). In adults, the reverse is true, with AML being much more common. Approximately 90% of patients with ALL have numeric or structural chromosomal abnormalities that likely contribute to the genetic/epigenetic basis for cancer. Thus, ALL qualifies as an example of an

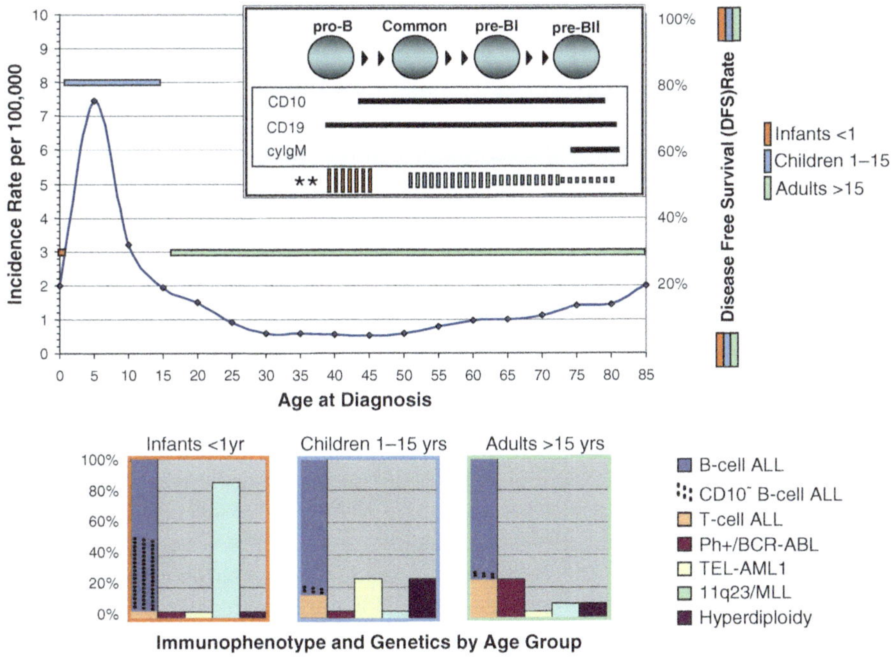

Fig. 3 Characteristics of individuals with ALL by age group. (**A**) Incidence (*left axis, blue curve*) and DFS (*right axis, orange, blue, and green horizontal bars*). A (inset). Characteristic cell surface markers for B-cell developmental stages. ** Represents the stage of B-cell development at which individuals within a particular age group have arrested. The bar height is a rough estimation of the percentages of individuals arrested in each stage. For example, the majority of infants with ALL have blasts similar to the pro-B stage while children and adults have more blasts similar to the common B stage with fewer in the pre-BI and even fewer in the pre-BII. (**B**) Percentage of individuals with B-cell ALL (*blue bar*), CD10 negative B-cell ALL (*dotted blue bar*), T-cell ALL (*light orange bar*), and characteristic genetic anomalies (*magenta, yellow, green, and purple bars*)

age-related cancer that spans the entire lifetime from birth to very elderly and demonstrates a number of age-related differences in clinical behavior, genetics, epigenetics, and biology. Two questions that have not yet been entirely answered are why are these differences age related, and how are they determined?

In Utero Development of ALL

It is likely that at least one form of ALL, and perhaps others such as acute megakaryoblastic leukemia, develops during in utero development. Retrospective identification of leukemia-specific fusion genes *TEL-AML1* and *MLL-AF4* from archived neonatal blood spotted on Guthrie cards and studies of ALL in monozygotic twins suggested that some cases were clearly of a prenatal origin while others likely require at least one more "hit." For *MLL-AF4*, monozygotic twins had nearly a 100% concordance rate when both developed the disease with a very short latency measured in weeks. Thus, the presence of the *MLL-AF4* fusion gene may possibly be leukemogenic alone, require only a few additional short-term alterations, or based on an epigenetically altered progenitor cell model (Feinberg et al. 2006; Feinberg 2008), might be a consequence of the programmed epigenome of a subset of these progenitor cells.

In the case of *TEL-AML1* fusion genes, the twin concordance rate was lower and these fusion genes were found in ~1% of newborns, a frequency ~100 times higher than the prevalence of ALL defined by this fusion gene later in childhood. Thus, while some of these babies with *TEL-AML1* fusion genes went on to develop ALL, most did not, or perhaps did not get exposure to the proper additional hit(s) to develop the disease. The variable incubation period and clinical outcome of such cases, and the 10% concordance rate of leukemia in identical twins with this genotype, support the notion that additional postnatal events are needed for full leukemic transformation involving *TEL-AML1*.

The normal mixed lineage leukemia (*MLL*) gene encodes a DNA-binding protein that specifically methylates histone H3 on lysine residue 4 (H3K4) and thereby positively regulates a number of targets including several *HOX* genes (reviewed in Krivtsov and Armstrong (2007)). However, the leukemogenic fusion forms (there are many *MLL* fusion partners) have lost the H3K4 methyltransferase function, but instead gained the ability to transform normal hematopoietic cells into leukemia creating a strong link between the epigenome and normal hematopoietic stem cell development. Infant ALL (and AML) shows specific translocations involving the *MLL* gene in >70% of cases, but this is much less common in older children and occurs in only ~10% of AML (not ALL) in adults (reviewed in Krivtsov and Armstrong (2007)). Overall, *MLL* mutations occur in only about 5% of cases including all ages. More than 50 fusion partners have been described, but 5 of these encompass 80% of cases. In any case, the presence of an *MLL* translocation portends a poor prognosis compared to those without the translocation. One yet unanswered question relates to how the *MLL* fusion is acquired prenatally, and for those older children and adults, why there is a longer latency compared to infancy.

Childhood and Adult ALL

Non-infant childhood ALL is typically associated with a different set of genetic abnormalities, some of which portend a better or worse prognosis. The t(12;21)(p12;q22) is common in children (~25% of cases) and is associated with a good prognostic group, generally between ages 1 and 10 years, with a CD10+ B-cell immunophenotype. This particular translocation is much less common in adults (4%). Another interesting group of genetic abnormalities in good-risk childhood ALL includes those whose karyotype shows >50 chromosomes (hyperdiploidy). While ~30% of childhood ALL cases are hyperdiploid, this only occurs in 4% of adult cases. In adult ALL, the most frequent translocation is t(9;22), or the Philadelphia chromosome, that involves fusion of the BCR protein to the ABL tyrosine kinase, resulting in constitutive tyrosine kinase activity. This occurs in ~30% of adults and up to 50% in adults >60 years of age, whereas it occurs only in ~5% of children with ALL. Patients with hyperdiploidy, *TEL-AML1* fusion, or t(1;19)/*E2A-PBX1* fusion generally have the most favorable outcome using conventional treatments, whereas those with the t(9;22) or t(4;11) fusion have a dismal prognosis. Age also affects the prognostic importance of genetic abnormalities for unknown reasons. In children with t(9;22)-positive ALL, those aged 1–9 years fared better than adolescents, who in turn responded better than adults. Thus, there is a reciprocal relationship between age and prognosis in this form of ALL. In patients with *MLL-AF4* fusion, infants and adults both have a worse prognosis than children. The t(1;19) with *E2A-PBX1* fusion has no prognostic implications in childhood ALL but is still associated with a poor prognosis in some adult cases.

Cooperating genetic and epigenetic events are not quite as well understood in the spectrum of ALL. Evidence suggests that they must act in concert with several other genetic (or epigenetic) lesions to induce acute leukemia. Early methylation studies in ALL focused on a few or even one candidate gene and revealed that hypermethylation of multiple promoters is common in ALL (Garcia-Manero et al. 2002) and is associated with poor prognosis (Roman-Gomez et al. 2004) and relapse (Matsushita et al. 2004). In addition, methylation of a specific gene was shown to be associated with clinical subtypes of ALL. For example, Zheng et al. (2004) demonstrated that methylation of the *FHIT* gene is associated with hyperdiploid and translocation-negative subtypes of pediatric leukemia and Shteper et al. (2001) established that methylation of *ABL1* is associated with t(9;22)-positive ALL. Finally, studies investigating the relationship between methylation and abnormal Wnt signaling in t(9;22)-positive ALL found that the upregulation of Wnt target genes was correlated with the methylation and subsequent downregulation of the Wnt inhibitors (*sFRP1, sFRP2, sFRP4, sFRP5, WIF1, Dkk3*, and *Hdpr1*) (Martin et al. 2008). Taken together these studies provided evidence that CpG methylation is common in ALL and set the stage for in-depth analyses of the aberrant methylation.

Genome-wide approaches in childhood and adult ALL have identified aberrant methylation in numerous loci which include genes involved in a variety of important cellular processes such as transcription (*ID1, FOXD2, SMAD9*), cell growth (*ICAM1, DLC-1, NES1*), apoptosis (*APAF1, DAPK, DCC*), and cell adhesion (cadherins and metalloproteinases) (Taylor et al. 2007b; Kuang et al. 2008). These studies have not only identified novel targets of aberrant methylation but also provided compelling evidence that the aberrant methylation observed in ALL is targeted, clustering within functional networks and/or within chromosomal neighborhoods. To date the genes identified hold promise as potential epigenetic biomarkers but the true promise lies in identifying those genes that are relevant clinically or functionally to the pathogenesis of ALL that will serve as targets for epigenetic therapy.

Reconciling Development of ALL, Other Cancers, and Aging

In a very thought-provoking 2006 review, Feinberg, Ohlsson, and Henikoff introduced a discussion regarding the potential epigenetic progenitor origin of human cancer (Feinberg et al. 2006). In this model scenario, although cancer is perceived as a heterogeneous group of disorders, they suggest that cancer has a fundamentally common basis grounded in a polyclonal epigenetic patterning in stem and/or progenitor cells, and this is mediated by a group of genes termed "tumor-progenitor genes." It is argued that cancer cell heterogeneity is due, at least in part, to epigenetic variations in progenitor cells and that an epigenetic plasticity along with genetic lesions drives tumor progression and actually addresses many of the known attributes of human cancer. This crucial early role for epigenetic alterations in cancer (and aging) is in addition to epigenetic alterations that can contribute to genetic alterations later in tumor progression. Thus, non-neoplastic but epigenetically altered progenitor cells might be a crucial target for early development of patterns that may then lead to diseases with variable latency. It was proposed that cancer arises in three steps: epigenetic disruption of progenitor cells, an initiating mutation, and maintenance of genetic and epigenetic plasticity.

More recently, this model has been revised into what is termed the common disease genetic and epigenetic (CDGE) model (Bjornsson et al. 2004). This model takes into account more recent evidence from epigenetic studies in a variety of human diseases to further posit that many/most attributes of these disorders with variable latencies can be described using the concepts of an epigenetic overlay on human genetics. We now better understand the importance of cellular plasticity and its potential effects in health and disease through studies that have demonstrated the ability to take differentiated somatic cells and restore them to a pluripotent state or induced progenitor cells (Park et al. 2008). While this process clearly did not alter the DNA sequence, it was an excellent demonstration of epigenetic modifications and cellular plasticity. An important area in need of further work is that related to epigenetic processes involved in tissue-specific differentiation and de-differentiation during development, aging, and in diseases such as cancer. As proposed by Feinberg, with attribution to Longo, aging may be a result of loss of phenotypic plasticity over time, and a unifying theme of epigenetic diseases such as cancer is disruption of normal phenotypic plasticity (Feinberg 2008).

Superimposed on these epigenetic progenitor cell models of disease are pleiotropic effects of environmental exposures as they relate to disease susceptibility (Jirtle and Skinner 2007). As we now know, epigenetic changes can be inherited mitotically in somatic cells and therefore provide a potential mechanism whereby environmental effects on the epigenome can have long-term effects on gene expression. In support of the importance of such a mechanism, increasing evidence from animal studies indicates that prenatal and early postnatal environmental factors including nutritional status, xenobiotic chemicals, reproductive factors, and low-dose radiation can result in epigenetic re-programming and subsequent phenotypic changes that may also be inherited transgenerationally, thereby potentially affecting the health of future generations (reviewed in Jirtle and Skinner (2007)). Recent animal and human studies support the hypothesis that is sometimes referred to as the developmental origins of adult-onset disease but also is consistent with the epigenetic progenitor model. This model includes the idea that the evolution of developmental plasticity which allows an organism to adapt to environmental signals during early life may also increase the risk of developing chronic diseases, effects of aging, and cancer. This might occur as a result of a mismatch between the elevated levels of perceived environmental signals compared to that normally encountered in adulthood.

A very interesting set of observations regarding early development and perturbation of epigenetic states may relate to development of cancers (and other diseases) with variable temporal latencies. The impact(s) of environmental exposures on the mammalian epigenome provides intriguing possibilities to understand how intrauterine exposures to certain types of chemicals might affect heritable changes in DNA methylation, the overall epigenome, and clinical manifestations of these transgenerational phenotypic changes (Anway et al. 2006; Anway and Skinner 2006; Skinner and Anway 2007; Anway et al. 2008; Anway and Skinner 2008a, b). When outbred pregnant female rats were exposed to methoxychlor (a replacement for DDT) or vinclozolin during early pregnancy, epigenetic alterations occurred in the offspring that were then transgenerational in nature. That is, the affected offspring could then be backcrossed to normal mates to create additional pregnancies and this could then be pursued through multiple generations. In one study, epigenetic programming of the germ line occurred during embryonic development in a sex-specific manner (Anway and Skinner 2008a). The male germ line became imprinted following sex determination. Exposure to an endocrine disruptor during the phase of embryonic gonadal sex determination appeared to alter DNA methylation in the germ line and to transmit transgenerational adult-onset diseases, including spermatogenic defects, prostate disease, kidney disease, and cancer. Additional examples of such in utero exposures to various compounds have shown various phenotypic consequences and various adult diseases that seem to be epigenetically predetermined. In some of these studies, animals developed a number of disease states or tissue abnormalities during aging including prostate and kidney disease, immune system abnormalities, testicular abnormalities, and tumor development (Anway et al. 2005; Anway et al. 2006; Anway et al. 2006; Anway and Skinner 2006; Anway et al. 2008; Anway and Skinner 2008a, b).

How then can we reconcile these models of potential epigenetic progenitor cell imprinting, environmental exposures, and aging to better understand the processes that may determine age-related development of certain types of cancer? There is increasingly convincing data to support these models and concepts, and the models discussed above are not inconsistent with each other (Kinlen 2004; Greaves 2005; Feinberg et al. 2006; Jirtle and Skinner 2007; Feinberg 2008). While ALL is certainly a model disease with many genetic and epigenetic lesions identified, there is much more to learn, and it will likely require a systems biology approach to model all the known and yet unknown features of this disease. An even larger hurdle though comes from then attempting to determine which, if any, model can describe commonalities across all human ages and diseases. Cancer is one group, but comprises at least 200 distinct types, some of which show clear age-related distributions, and others that do not. It is true that better biological understanding has led to improved cure rates in some forms of cancer, we still have many more to attack.

The Rapidly Evolving Epigenetic Landscape During Aging and Cancer

Where do we go from here? Research into aging, cancer, and epigenetics is actually at a very early stage for being able to translate findings into clinical use. For instance, we are just now beginning to see that additional epigenetic mechanisms may be involved that are not yet well understood. For instance, what is the role of mitochondrial DNA alterations (mtDNA), microRNAs (miRNAs), or SIRT1 in aging and cancer?

A very recent study reported by Smiraglia et al. (2008) describes an interesting potential role of mtDNA in cancer. A number of tumor types demonstrate decreased mtDNA, and cancers are very commonly affected by defective mitochondria. This group reported that depletion of mtDNA results in significant changes in the DNA methylation patterns of a number of genes within the nucleus and that this can be reversed by restoring mtDNA levels in the cells. This represents an intriguing type of epigenetic modification that deserves further research in both aging and cancer. The miRNAs were discovered more than a decade ago as noncoding, single-stranded small RNAs (~22 nucleotides) that controlled gene expression in the *Caenorhabditis elegans* life cycle. Several of these have been shown to regulate mammalian cell growth, differentiation, and apoptosis. An individual miRNA can modulate the expression of a network of mRNAs. Human cancers commonly exhibit some altered expression of miRNAs with oncogenic (miR-21, miR-106a, and miR-155) or tumor-suppressive (let-7, miR-15a/16, miR-34a, and miR-143/145) activity. Abnormal epigenetic and miRNA regulation of genes is well documented in cancer (Croce and Calin 2005; Calin and Croce 2006, 2007). An increasing number of studies are showing that certain abnormalities of the epigenome and of miRNAs are not independent and could be explained both by an epigenetic regulation of miRNA expression and by miRNA control of components in the epigenetic machinery. This is another exciting area that will contribute to the overall epigenome of aging and cancer.

The sirtuins, already discussed in Chapter 5, represent another category of genes functioning in aging and in cancer. DNA repair, insulin signaling, and sirtuins represent common regulatory mechanisms at the crossroads between cancer and aging. The same defects in these pathways that distinguish normal cells from their malignant counterparts appear to also influence how we age. Based on the role(s) of insulin-signaling and DNA repair pathways in those processes, it may be that dysregulation of these pathways could produce two seemingly disparate outcomes. However, the fact that IIS and DNA repair appear to act in concert suggests that cross talk between these pathways is crucial to maintain homeostasis. In this context, sirtuins may represent bona fide regulators of cellular homeostasis, determining whether cells take a path toward cancer or rather that of aging. A checkpoint arrest response following excessive DNA damage and/or dysregulated IIS might lead to depletion of critical progenitor cells leading to tissue aging, degenerative diseases, and eventually aging of the entire body. In contrast, in those situations where the defects impinge directly on putative gatekeeper genes, uncontrolled proliferation and accumulation of additional DNA damage may eventually result in the appearance of cancer.

When DNA damage continues chronically, it may lead to downregulation of IIS in an attempt to lower ROS levels and prevent further DNA damage. Alternatively, disordered metabolic processes could lead to increased ROS production and increase the amount of DNA damage. As suggested, sirtuins might modulate both the DDR and IIS pathways and modulate these responses in complex ways. In this context, SIRT6 might play a role in reducing IIS in response to genotoxic stress, decreasing ROS levels, and retarding accumulation of DNA damage, a metabolic "rheostat" contributing to cell-fate decisions. A better understanding of the cross talk between these pathways will no doubt provide new insights into the processes of epigenetics, aging, and cancer that may in the long term result in new therapeutic approaches to prevent age-related diseases including cancer.

One closing comment involves our ability, or lack thereof, to integrate these genetic and epigenetic data in a meaningful manner that create new knowledge. Computational epigenetics research relies on large-scale data sets, but recognizes the need for robust hardware and software that can link such data across multiple layers and facilitate both supervised and unsupervised interrogations of the data sets (Das et al. 2006; Dopazo 2006; Jensen et al. 2006; Bock and Lengauer 2008). The future success of such studies is critically dependent on continued development of robust analytical tools to define patterns of existing and newly defined epigenetic alterations and the computational tools to understand and share these valuable data resources. This is true not only of research into epigenetics, aging, and cancer, but all other biomedical fields as well.

Acknowledgements Grant Support: National Cancer Institute grants CA100055 and CA097880 (C.W.C.), CA123018 (H.S.), and the CRC Missouri Chair in Cancer Research (CWC).

References

Alaminos, M., Davalos, V., Cheung, N. K., Gerald, W. L., and Esteller, M. 2004. Clustering of gene hypermethylation associated with clinical risk groups in neuroblastoma. J. Natl. Cancer Inst. 96:1208–1219.

Anway, M. D., Cupp, A. S., Uzumcu, M., and Skinner, M. K. 2005. Epigenetic transgenerational actions of endocrine disruptors and male fertility. Science 308:1466–1469.

Anway, M. D., Leathers, C., and Skinner, M. K. 2006. Endocrine disruptor vinclozolin induced epigenetic transgenerational adult-onset disease. Endocrinology 147:5515–5523.

Anway, M. D., Memon, M. A., Uzumcu, M., and Skinner, M. K. 2006. Transgenerational effect of the endocrine disruptor vinclozolin on male spermatogenesis. J. Androl 27:868–879.

Anway, M. D., Rekow, S. S., and Skinner, M. K. 2008. Transgenerational epigenetic programming of the embryonic testis transcriptome. Genomics 91:30–40.

Anway, M. D. and Skinner, M. K. 2006. Epigenetic transgenerational actions of endocrine disruptors. Endocrinology 147:S43–S49.

Anway, M. D. and Skinner, M. K. 2008a. Epigenetic programming of the germ line: effects of endocrine disruptors on the development of transgenerational disease. Reprod. Biomed. Online. 16:23–25.

Anway, M. D. and Skinner, M. K. 2008b. Transgenerational effects of the endocrine disruptor vinclozolin on the prostate transcriptome and adult onset disease. Prostate 68:517–529.

Baur, A. S., Shaw, P., Burri, N., Delacretaz, F., Bosman, F. T., and Chaubert, P. 1999. Frequent methylation silencing of p15(INK4b) (MTS2) and p16(INK4a) (MTS1) in B-cell and T-cell lymphomas. Blood 94:1773–1781.

Bernstein, B. E., Meissner, A., and Lander, E. S. 2007. The mammalian epigenome. Cell 128: 669–681.

Bjornsson, H. T., Fallin, M. D., and Feinberg, A. P. 2004. An integrated epigenetic and genetic approach to common human disease. Trends Genet. 20:350–358.

Bock, C. and Lengauer, T. 2008. Computational epigenetics. Bioinformatics. 24:1–10.

Bracken, A. P., Pasini, D., Capra, M., Prosperini, E., Colli, E., and Helin, K. 2003. EZH2 is downstream of the pRB-E2F pathway, essential for proliferation and amplified in cancer. EMBO J. 22:5323–5335.

Cai, S., Lee, C. C., and Kohwi-Shigematsu, T. 2006. SATB1 packages densely looped, transcriptionally active chromatin for coordinated expression of cytokine genes. Nat. Genet. 38: 1278–1288.

Calin, G. A. and Croce, C. M. 2006. MicroRNA signatures in human cancers. Nat. Rev. Cancer 6:857–866.

Calin, G. A. and Croce, C. M. 2007. Investigation of microRNA alterations in leukemias and lymphomas. Methods Enzymol. 427:193–213.

Chim, C., Fung, T., Wong, K., Lau, J., Law, M., and Liang, R. 2006. Methylation of INK4 and CIP/KIP families of cyclin-dependent kinase inhibitor (CKI) in Chronic Lymphocytic Leukemia (CLL) in Chinese. J. Clin. Pathol. 59(9):921–926.

Ching, T. T., Maunakea, A. K., Jun, P., Hong, C., Zardo, G., Pinkel, D., Albertson, D. G., Fridlyand, J., Mao, J. H., Shchors, K., Weiss, W. A., and Costello, J. F. 2005. Epigenome analyses using BAC microarrays identify evolutionary conservation of tissue-specific methylation of SHANK3. Nat. Genet. 37:645–651.

Cohen, Y., Merhavi-Shoham, E., Avraham, R. B., Frenkel, S., Pe'er, J., and Goldenberg-Cohen, N. 2008. Hypermethylation of CpG island loci of multiple tumor suppressor genes in retinoblastoma. Exp. Eye Res. 86:201–206.

Croce, C. M. and Calin, G. A. 2005. miRNAs, cancer, and stem cell division. Cell 122:6–7.

Das, R., Dimitrova, N., Xuan, Z., Rollins, R. A., Haghighi, F., Edwards, J. R., Ju, J., Bestor, T. H., and Zhang, M. Q. 2006. Computational prediction of methylation status in human genomic sequences. Proc. Natl. Acad. Sci. USA 103:10713–10716.

Deligezer, U., Erten, N., Akisik, E. E., and Dalay, N. 2005. Circulating fragmented nucleosomal DNA and caspase-3 mRNA in patients with lymphoma and myeloma. Exp. Mol. Pathol. 80(1):72–76.

Deruiter, M. C., Alkemade, F. E., Gittenberger-de Groot, A. C., Poelmann, R. E., Havekes, L. M., and van Dijk, K. W. 2008. Maternal transmission of risk for atherosclerosis. Curr. Opin. Lipidol. 19:333–337.

Dopazo, J. 2006. Bioinformatics and cancer: an essential alliance. Clin. Transl. Oncol. 8:409–415.

Dukers, D. F., van Galen, J. C., Giroth, C., Jansen, P., Sewalt, R. G., Otte, A. P., Kluin-Nelemans, H. C., Meijer, C. J., and Raaphorst, F. M. 2004. Unique polycomb gene expression pattern in Hodgkin's lymphoma and Hodgkin's lymphoma-derived cell lines. Am. J. Pathol. 164: 873–881.

Estecio, M. R., Yan, P. S., Ibrahim, A. E., Tellez, C. S., Shen, L., Huang, T. H., and Issa, J. P. 2007. High-throughput methylation profiling by MCA coupled to CpG island microarray. Genome Res. 17:1529–1536.

Esteller, M. 2003. Profiling aberrant DNA methylation in hematologic neoplasms: a view from the tip of the iceberg. Clin. Immunol. 109:80–88.

Esteller, M. 2007. Cancer epigenomics: DNA methylomes and histone-modification maps. Nat. Rev. Genet.

Esteller, M., Gaidano, G., Goodman, S. N., Zagonel, V., Capello, D., Botto, B., Rossi, D., Gloghini, A., Vitolo, U., Carbone, A., Baylin, S. B., and Herman, J. G. 2002. Hypermethylation of the DNA repair gene O(6)-methylguanine DNA methyltransferase and survival of patients with diffuse large B-cell lymphoma. J. Natl. Cancer Inst. 94:26–32.

Esteller, M., Hamilton, S. R., Burger, P. C., Baylin, S. B., and Herman, J. G. 1999. Inactivation of the DNA repair gene O6-methylguanine-DNA methyltransferase by promoter hypermethylation is a common event in primary human neoplasia. Cancer Res. 59:793–797.

Feinberg, A. P. 2008. Epigenetics at the epicenter of modern medicine. JAMA 299:1345–1350.

Feinberg, A. P., Ohlsson, R., and Henikoff, S. 2006. The epigenetic progenitor origin of human cancer. Nat. Rev. Genet. 7:21–33.

Fraga, M. F., Agrelo, R., and Esteller, M. 2007. Cross-talk between aging and cancer: the epigenetic language. Ann. NY. Acad. Sci. 1100:60–74.

Fraga, M. F. and Esteller, M. 2007. Epigenetics and aging: the targets and the marks. Trends Genet. 23:413–418.

Frigola, J., Song, J., Stirzaker, C., Hinshelwood, R. A., Peinado, M. A., and Clark, S. J. 2006. Epigenetic remodeling in colorectal cancer results in coordinate gene suppression across an entire chromosome band. Nat. Genet. 38:540–549.

Fung, M. K., Au, W. Y., Liang, R., Srivastava, G., and Kwong, Y. L. 2003. Aberrant promoter methylation in gastric lymphoma. Haematologica 88:231–232.

Furukawa, T., Konishi, F., Masubuchi, S., Shitoh, K., Nagai, H., and Tsukamoto, T. 2002. Densely methylated MLH1 promoter correlates with decreased mRNA expression in sporadic colorectal cancers. Genes Chromosomes. Cancer 35:1–10.

Gal-Yam, E. N., Saito, Y., Egger, G., and Jones, P. A. 2008. Cancer epigenetics: modifications, screening, and therapy. Annu. Rev. Med. 59:267–280.

Garcia-Manero, G., Assouline, S., Cortes, J., Estrov, Z., Kantarjian, H., Yang, H., Newsome, W. M., Miller, W. H., Jr., Rousseau, C., Kalita, A., Liu, J., Dubay, M., Patterson, T. A., Li, Z., Besterman, J. M., Reid, G., Laille, E., Martell, R. E., and Minden, M. D. 2008a. Phase

I study of the oral isotype specific histone deacetylase inhibitor MGCD0103 in leukemia. Blood. 112(4):981–989.

Garcia-Manero, G., Daniel, J., Smith, T. L., Kornblau, S. M., Lee, M. S., Kantarjian, H. M., and Issa, J. P. 2002. DNA methylation of multiple promoter-associated CpG islands in adult acute lymphocytic leukemia. Clin. Cancer Res. 8:2217–2224.

Garcia-Manero, G., Kantarjian, H. M., Sanchez-Gonzalez, B., Yang, H., Rosner, G., Verstovsek, S., Rytting, M., Wierda, W. G., Ravandi, F., Koller, C., Xiao, L., Faderl, S., Estrov, Z., Cortes, J., O'Brien, S., Estey, E., Bueso-Ramos, C., Fiorentino, J., Jabbour, E., and Issa, J. P. 2006. Phase 1/2 study of the combination of 5-aza-2'-deoxycytidine with valproic acid in patients with leukemia. Blood 108:3271–3279.

Garcia-Manero, G., Yang, H., Bueso-Ramos, C., Ferrajoli, A., Cortes, J., Wierda, W. G., Faderl, S., Koller, C., Morris, G., Rosner, G., Loboda, A., Fantin, V. R., Randolph, S. S., Hardwick, J. S., Reilly, J. F., Chen, C., Ricker, J. L., Secrist, J. P., Richon, V. M., Frankel, S. R., and Kantarjian, H. M. 2008b. Phase 1 study of the histone deacetylase inhibitor vorinostat (suberoylanilide hydroxamic acid [SAHA]) in patients with advanced leukemias and myelodysplastic syndromes. Blood 111:1060–1066.

Greaves, M. 2005. In utero origins of childhood leukaemia. Early Hum. Dev. 81:123–129.

Guo, J., Burger, M., Nimmrich, I., Maier, S., Becker, E., Genc, B., Duff, D., Rahmatpanah, F., Chitma-Matsiga, R., Shi, H., Berlin, K., Huang, T. H., and Caldwell, C. W. 2005. Differential DNA methylation of gene promoters in small B-cell lymphomas. Am. J. Clin. Pathol. 124: 430–439.

Gyory, I. and Minarovits, J. 2005. Epigenetic regulation of lymphoid specific gene sets. Biochem. Cell Biol. 83:286–295.

Han, H. J., Russo, J., Kohwi, Y., and Kohwi-Shigematsu, T. 2008. SATB1 reprogrammes gene expression to promote breast tumour growth and metastasis. Nature 452:187–193.

Hanada, M., Delia, D., Aiello, A., Stadtmauer, E., and Reed, J. C. 1993. bcl-2 gene hypomethylation and high-level expression in B-cell chronic lymphocytic leukemia. Blood 82:1820–1828.

Harada, K., Toyooka, S., Maitra, A., Maruyama, R., Toyooka, K. O., Timmons, C. F., Tomlinson, G. E., Mastrangelo, D., Hay, R. J., Minna, J. D., and Gazdar, A. F. 2002. Aberrant promoter methylation and silencing of the RASSF1A gene in pediatric tumors and cell lines. Oncogene 21:4345–4349.

Hayslip, J. and Montero, A. 2006. Tumor suppressor gene methylation in follicular lymphoma: a comprehensive review. Mol. Cancer 5:44.

Hitzler, J. and Zipursky, A. 2005. GATA 1 mutations as clonal markers of minimal residual disease in acute megakaryoblastic leukemia of Down syndrome – a new tool with significant potential applications. Leuk. Res. 29:1239–1240.

Hitzler, J. K. 2007. Acute megakaryoblastic leukemia in Down syndrome. Pediatr. Blood Cancer 49:1066–1069.

Hou, P., Ji, M., Yang, B., Chen, Z., Qiu, J., Shi, X., and Lu, Z. 2006. Quantitative analysis of promoter hypermethylation in multiple genes in osteosarcoma. Cancer 106: 1602–1609.

Huang, T. H., Perry, M. R., and Laux, D. E. 1999. Methylation profiling of CpG islands in human breast cancer cells. Hum. Mol. Genet. 8:459–470.

Issa, J. P. 2005. Optimizing therapy with methylation inhibitors in myelodysplastic syndromes: dose, duration, and patient selection. Nat. Clin. Pract. Oncol. 2 Suppl 1:S24–S29.

Issa, J. P. 2003. Age-related epigenetic changes and the immune system. Clin. Immunol. 109: 103–108.

Issa, J. P., Gharibyan, V., Cortes, J., Jelinek, J., Morris, G., Verstovsek, S., Talpaz, M., Garcia-Manero, G., and Kantarjian, H. M. 2005. Phase II study of low-dose decitabine in patients with chronic myelogenous leukemia resistant to imatinib mesylate. J. Clin. Oncol. 23(17):3948–3956.

Issa, J. P., Ottaviano, Y. L., Celano, P., Hamilton, S. R., Davidson, N. E., and Baylin, S. B. 1994. Methylation of the oestrogen receptor CpG island links ageing and neoplasia in human colon. Nat. Genet. 7:536–540.

Jensen, L. J., Saric, J., and Bork, P. 2006. Literature mining for the biologist: from information retrieval to biological discovery. Nat. Rev. Genet. 7:119–129.
Jeronimo, C., Henrique, R., Hoque, M. O., Mambo, E., Ribeiro, F. R., Varzim, G., Oliveira, J., Teixeira, M. R., Lopes, C., and Sidransky, D. 2004. A quantitative promoter methylation profile of prostate cancer. Clin. Cancer Res. 10:8472–8478.
Jirtle, R. L. and Skinner, M. K. 2007. Environmental epigenomics and disease susceptibility. Nat. Rev. Genet. 8:253–262.
Kaneko, Y., Sakurai, S., Hironaka, M., Sato, S., Oguni, S., Sakuma, Y., Sato, K., Sugano, K., and Saito, K. 2003. Distinct methylated profiles in Helicobacter pylori dependent and independent gastric MALT lymphomas. Gut 52:641–646.
Katzenellenbogen, R. A., Baylin, S. B., and Herman, J. G. 1999. Hypermethylation of the DAP-kinase CpG island is a common alteration in B-cell malignancies. Blood 93:4347–4353.
Kinlen, L. 2004. Infections and immune factors in cancer: the role of epidemiology. Oncogene 23:6341–6348.
Kn, H., Bassal, S., Tikellis, C., and El-Osta, A. 2004. Expression analysis of the epigenetic methyltransferases and methyl-CpG binding protein families in the normal B-cell and B-cell chronic lymphocytic leukemia (CLL). Cancer Biol. Ther. 3:989–994.
Krivtsov, A. V. and Armstrong, S. A. 2007. MLL translocations, histone modifications and leukaemia stem-cell development. Nat. Rev. Cancer 7:823–833.
Kuang, S. Q., Tong, W. G., Yang, H., Lin, W., Lee, M. K., Fang, Z. H., Wei, Y., Jelinek, J., Issa, J. P., and Garcia-Manero, G. 2008. Genome-wide identification of aberrantly methylated promoter associated CpG islands in acute lymphocytic leukemia. Leukemia 22(8):1529–1538.
Lehmann, U., Langer, F., Feist, H., Glockner, S., Hasemeier, B., and Kreipe, H. 2002. Quantitative assessment of promoter hypermethylation during breast cancer development. Am. J. Pathol. 160:605–612.
Leu, Y. W., Yan, P. S., Fan, M., Jin, V. X., Liu, J. C., Curran, E. M., Welshons, W. V., Wei, S. H., Davuluri, R. V., Plass, C., Nephew, K. P., and Huang, T. H. 2004. Loss of estrogen receptor signaling triggers epigenetic silencing of downstream targets in breast cancer. Cancer Res. 64:8184–8192.
Li, G., Weyand, C. M., and Goronzy, J. J. 2008. Epigenetic mechanisms of age-dependent KIR2DL4 expression in T cells. J. Leukoc. Biol.
Lindsey, J. C., Anderton, J. A., Lusher, M. E., and Clifford, S. C. 2005. Epigenetic events in medulloblastoma development. Neurosurg. Focus. 19:E10.
Lindsey, J. C., Lusher, M. E., Anderton, J. A., Bailey, S., Gilbertson, R. J., Pearson, A. D., Ellison, D. W., and Clifford, S. C. 2004. Identification of tumour-specific epigenetic events in medulloblastoma development by hypermethylation profiling. Carcinogenesis 25:661–668.
Lindsey, J. C., Lusher, M. E., Anderton, J. A., Gilbertson, R. J., Ellison, D. W., and Clifford, S. C. 2007. Epigenetic deregulation of multiple S100 gene family members by differential hypomethylation and hypermethylation events in medulloblastoma. Br. J. Cancer 97: 267–274.
Lipsanen, V., Leinonen, P., Alhonen, L., and Janne, J. 1988. Hypomethylation of ornithine decarboxylase gene and erb-A1 oncogene in human chronic lymphatic leukemia. Blood 72: 2042–2044.
List, A. F., Vardiman, J., Issa, J. P., and DeWitte, T. M. 2004. Myelodysplastic syndromes. Hematology. Am. Soc. Hematol. Educ. Program. pp. 297–317.
Liu, T. H., Raval, A., Chen, S. S., Matkovic, J. J., Byrd, J. C., and Plass, C. 2006. CpG island methylation and expression of the secreted frizzled-related protein gene family in chronic lymphocytic leukemia. Cancer Res. 66:653–658.
Lyko, F., Stach, D., Brenner, A., Stilgenbauer, S., Dohner, H., Wirtz, M., Wiessler, M., and Schmitz, O. J. 2004. Quantitative analysis of DNA methylation in chronic lymphocytic leukemia patients. Electrophoresis 25:1530–1535.
Martin, V., Agirre, X., Jimenez-Velasco, A., Jose-Eneriz, E. S., Cordeu, L., Garate, L., Vilas-Zornoza, A., Castillejo, J. A., Heiniger, A., Prosper, F., Torres, A., and Roman-Gomez, J. 2008.

Methylation status of Wnt signaling pathway genes affects the clinical outcome of Philadelphia-positive acute lymphoblastic leukemia. Cancer Sci. 99(9):1865–1868.
Martin-Subero, J. I., Ballestar, E., Esteller, M., and Siebert, R. 2006. Towards defining the lymphoma methylome. Leukemia 20:1658–1660.
Matsushita, C., Yang, Y., Takeuchi, S., Matsushita, M., Van Dongen, J. J., Szczepanski, T., Bartram, C. R., Seo, H., Koeffler, H. P., and Taguchi, H. 2004. Aberrant methylation in promoter-associated CpG islands of multiple genes in relapsed childhood acute lymphoblastic leukemia. Oncol. Rep. 12:97–99.
Monk, M. and Holding, C. 2001. Human embryonic genes re-expressed in cancer cells. Oncogene 20:8085–8091.
Mostoslavsky, R. 2008. DNA repair, insulin signaling and sirtuins: at the crossroads between cancer and aging. Front Biosci. 13:6966–6990.
Motiwala, T., Majumder, S., Kutay, H., Smith, D. S., Neuberg, D. S., Lucas, D. M., Byrd, J. C., Grever, M., and Jacob, S. T. 2007. Methylation and Silencing of Protein Tyrosine Phosphatase Receptor Type O in Chronic Lymphocytic Leukemia. Clin. Cancer Res. 13:3174–3181.
Nafee, T. M., Farrell, W. E., Carroll, W. D., Fryer, A. A., and Ismail, K. M. 2008. Epigenetic control of fetal gene expression. BJOG. 115:158–168.
Nakatsuka, S., Takakuwa, T., Tomita, Y., Hoshida, Y., Nishiu, M., Yamaguchi, M., Nishii, K., Yang, W. I., and Aozasa, K. 2003. Hypermethylation of death-associated protein (DAP) kinase CpG island is frequent not only in B-cell but also in T- and natural killer (NK)/T-cell malignancies. Cancer Sci. 94:87–91.
Pahlich, S., Bschir, K., Chiavi, C., Belyanskaya, L., and Gehring, H. 2005. Different methylation characteristics of protein arginine methyltransferase 1 and 3 toward the Ewing Sarcoma protein and a peptide. Proteins 61:164–175.
Pahlich, S., Zakaryan, R. P., and Gehring, H. 2008. Identification of proteins interacting with protein arginine methyltransferase 8: The Ewing sarcoma (EWS) protein binds independent of its methylation state. Proteins. 72(4):1125–1137.
Palmisano, W. A., Divine, K. K., Saccomanno, G., Gilliland, F. D., Baylin, S. B., Herman, J. G., and Belinsky, S. A. 2000. Predicting lung cancer by detecting aberrant promoter methylation in sputum. Cancer Res. 60:5954–5958.
Park, I. H., Zhao, R., West, J. A., Yabuuchi, A., Huo, H., Ince, T. A., Lerou, P. H., Lensch, M. W., and Daley, G. Q. 2008. Reprogramming of human somatic cells to pluripotency with defined factors. Nature 451:141–146.
Pasini, D., Bracken, A. P., and Helin, K. 2004. Polycomb group proteins in cell cycle progression and cancer. Cell Cycle 3:396–400.
Pufulete, M., Al-Ghnaniem, R., Khushal, A., Appleby, P., Harris, N., Gout, S., Emery, P. W., and Sanders, T. A. 2005. Effect of folic acid supplementation on genomic DNA methylation in patients with colorectal adenoma. Gut 54:648–653.
Raaphorst, F. M. 2005. Deregulated expression of Polycomb-group oncogenes in human malignant lymphomas and epithelial tumors. Hum. Mol. Genet. 14 Spec No 1:R93–R100.
Rahmatpanah, F. B., Carstens, S., Guo, J., Sjahputera, O., Taylor, K. H., Duff, D., Shi, H., Davis, J. W., Hooshmand, S. I., Chitma-Matsiga, R., and Caldwell, C. W. 2006. Differential DNA methylation patterns of small B-cell lymphoma subclasses with different clinical behavior. Leukemia 20:1855–1862.
Rauch, T., Wang, Z., Zhang, X., Zhong, X., Wu, X., Lau, S. K., Kernstine, K. H., Riggs, A. D., and Pfeifer, G. P. 2007. Homeobox gene methylation in lung cancer studied by genome-wide analysis with a microarray-based methylated CpG island recovery assay. Proc. Natl. Acad. Sci. USA 104:5527–5532.
Raval, A., Tanner, S. M., Byrd, J. C., Angerman, E. B., Perko, J. D., Chen, S. S., Hackanson, B., Grever, M. R., Lucas, D. M., Matkovic, J. J., Lin, T. S., Kipps, T. J., Murray, F., Weisenburger, D., Sanger, W., Lynch, J., Watson, P., Jansen, M., Yoshinaga, Y., Rosenquist, R., de Jong, P. J., Coggill, P., Beck, S., Lynch, H., de la, C. A., and Plass, C. 2007. Downregula-

tion of death-associated protein kinase 1 (DAPK1) in chronic lymphocytic leukemia. Cell 129: 879–890.

Reddy, J., Shivapurkar, N., Takahashi, T., Parikh, G., Stastny, V., Echebiri, C., Crumrine, K., Zochbauer-Muller, S., Drach, J., Zheng, Y., Feng, Z., Kroft, S. H., McKenna, R. W., and Gazdar, A. F. 2005. Differential methylation of genes that regulate cytokine signaling in lymphoid and hematopoietic tumors. Oncogene 24:732–736.

Riggi, N. and Stamenkovic, I. 2007. The Biology of Ewing sarcoma. Cancer Lett. 254:1–10.

Ringrose, L. 2007. Polycomb comes of age: genome-wide profiling of target sites. Curr. Opin. Cell Biol. 19:290–297.

Roman-Gomez, J., Jimenez-Velasco, A., Castillejo, J. A., Agirre, X., Barrios, M., Navarro, G., Molina, F. J., Calasanz, M. J., Prosper, F., Heiniger, A., and Torres, A. 2004. Promoter hypermethylation of cancer-related genes: a strong independent prognostic factor in acute lymphoblastic leukemia. Blood 104:2492–2498.

Rush, L. J., Raval, A., Funchain, P., Johnson, A. J., Smith, L., Lucas, D. M., Bembea, M., Liu, T. H., Heerema, N. A., Rassenti, L., Liyanarachchi, S., Davuluri, R., Byrd, J. C., and Plass, C. 2004. Epigenetic profiling in chronic lymphocytic leukemia reveals novel methylation targets. Cancer Res. 64:2424–2433.

Sandlund, J. T., Downing, J. R., and Crist, W. M. 1996. Non-Hodgkin's lymphoma in childhood. N. Engl. J. Med. 334:1238–1248.

Satoh, Y., Nakadate, H., Nakagawachi, T., Higashimoto, K., Joh, K., Masaki, Z., Uozumi, J., Kaneko, Y., Mukai, T., and Soejima, H. 2006. Genetic and epigenetic alterations on the short arm of chromosome 11 are involved in a majority of sporadic Wilms' tumours. Br. J. Cancer 95:541–547.

Schlesinger, Y., Straussman, R., Keshet, I., Farkash, S., Hecht, M., Zimmerman, J., Eden, E., Yakhini, Z., Ben-Shushan, E., Reubinoff, B. E., Bergman, Y., Simon, I., and Cedar, H. 2007. Polycomb-mediated methylation on Lys27 of histone H3 pre-marks genes for de novo methylation in cancer. Nat. Genet. 39:232–236.

Shen, L., Kondo, Y., Guo, Y., Zhang, J., Zhang, L., Ahmed, S., Shu, J., Chen, X., Waterland, R. A., and Issa, J. P. 2007. Genome-wide profiling of DNA methylation reveals a class of normally methylated CpG island promoters. PLoS. Genet. 3:e181.

Shi, H., Guo, J., Duff, D. J., Rahmatpanah, F., Chitima-Matsiga, R., Al-Kuhlani, M., Taylor, K. H., Sjahputera, O., Andreski, M., Wooldridge, J. E., and Caldwell, C. W. 2007. Discovery of novel epigenetic markers in non-Hodgkin's lymphoma. Carcinogenesis 28:60–70.

Shteper, P. J., Siegfried, Z., Asimakopoulos, F. A., Palumbo, G. A., Rachmilewitz, E. A., Ben-Neriah, Y., and Ben-Yehuda, D. 2001. ABL1 methylation in Ph-positive ALL is exclusively associated with the P210 form of BCR-ABL. Leukemia 15:575–582.

Skinner, M. K. and Anway, M. D. 2007. Epigenetic transgenerational actions of vinclozolin on the development of disease and cancer. Crit Rev. Oncog. 13:75–82.

Smiraglia, D. J., Kulawiec, M., Bistulfi, G. L., Ghoshal, S., and Singh, K. K. 2008. A novel role for mitochondria in regulating epigenetic modification in the nucleus. Cancer Biol. Ther. 7.

Soriano, A. O., Yang, H., Faderl, S., Estrov, Z., Giles, F., Ravandi, F., Cortes, J., Wierda, W. G., Ouzounian, S., Quezada, A., Pierce, S., Estey, E. H., Issa, J. P., Kantarjian, H. M., and Garcia-Manero, G. 2007. Safety and clinical activity of the combination of 5-azacytidine, valproic acid, and all-trans retinoic acid in acute myeloid leukemia and myelodysplastic syndrome. Blood 110:2302–2308.

Suzuki, H., Gabrielson, E., Chen, W., Anbazhagan, R., Van, E. M., Weijenberg, M. P., Herman, J. G., and Baylin, S. B. 2002. A genomic screen for genes upregulated by demethylation and histone deacetylase inhibition in human colorectal cancer. Nat. Genet. 31:141–149.

Suzuki, H., Watkins, D. N., Jair, K. W., Schuebel, K. E., Markowitz, S. D., Chen, W. D., Pretlow, T. P., Yang, B., Akiyama, Y., Van, E. M., Toyota, M., Tokino, T., Hinoda, Y., Imai, K., Herman, J. G., and Baylin, S. B. 2004. Epigenetic inactivation of SFRP genes allows constitutive WNT signaling in colorectal cancer. Nat. Genet. 36:417–422.

Szyf, M., Pakneshan, P., and Rabbani, S. A. 2004. DNA methylation and breast cancer. Biochem. Pharmacol. 68:1187–1197.

Taylor, K. H., Kramer, R. S., Davis, J. W., Guo, J., Duff, D. J., Xu, D., Caldwell, C. W., and Shi, H. 2007a. Ultradeep bisulfite sequencing analysis of DNA methylation patterns in multiple gene promoters by 454 sequencing. Cancer Res. 67:8511–8518.

Taylor, K. H., Pena-Hernandez, K. E., Davis, J. W., Arthur, G. L., Duff, D. J., Shi, H., Rahmatpanah, F. B., Sjahputera, O., and Caldwell, C. W. 2007b. Large-scale CpG methylation analysis identifies novel candidate genes and reveals methylation hotspots in acute lymphoblastic leukemia. Cancer Res. 67:2617–2625.

Taylor, K. H., Rahmatpanah, F., Davis, J. W., and Caldwell, C. W. 2007c. Chromosomal localization of DNA methylation in small B-cell lymphoma. Leukemia.

Toyota, M., Ahuja, N., Ohe-Toyota, M., Herman, J. G., Baylin, S. B., and Issa, J. P. 1999a. CpG island methylator phenotype in colorectal cancer. Proc. Natl. Acad. Sci. USA 96:8681–8686.

Toyota, M., Ho, C., Ahuja, N., Jair, K. W., Li, Q., Ohe-Toyota, M., Baylin, S. B., and Issa, J. P. 1999b. Identification of differentially methylated sequences in colorectal cancer by methylated CpG island amplification. Cancer Res. 59:2307–2312.

Usmani, B. A., Shen, R., Janeczko, M., Papandreou, C. N., Lee, W. H., Nelson, W. G., Nelson, J. B., and Nanus, D. M. 2000. Methylation of the neutral endopeptidase gene promoter in human prostate cancers. Clin. Cancer Res. 6:1664–1670.

Valinluck, V., Liu, P., Kang, J. I., Jr., Burdzy, A., and Sowers, L. C. 2005. 5-halogenated pyrimidine lesions within a CpG sequence context mimic 5-methylcytosine by enhancing the binding of the methyl-CpG-binding domain of methyl-CpG-binding protein 2 (MeCP2). Nucleic Acids Res. 33:3057–3064.

Valinluck, V. and Sowers, L. C. 2007a. Endogenous cytosine damage products alter the site selectivity of human DNA maintenance methyltransferase DNMT1. Cancer Res. 67:946–950.

Valinluck, V. and Sowers, L. C. 2007b. Inflammation-mediated cytosine damage: a mechanistic link between inflammation and the epigenetic alterations in human cancers. Cancer Res. 67:5583–5586.

Valinluck, V., Tsai, H. H., Rogstad, D. K., Burdzy, A., Bird, A., and Sowers, L. C. 2004. Oxidative damage to methyl-CpG sequences inhibits the binding of the methyl-CpG binding domain (MBD) of methyl-CpG binding protein 2 (MeCP2). Nucleic Acids Res. 32:4100–4108.

van Galen, J. C., Dukers, D. F., Giroth, C., Sewalt, R. G., Otte, A. P., Meijer, C. J., and Raaphorst, F. M. 2004. Distinct expression patterns of polycomb oncoproteins and their binding partners during the germinal center reaction. Eur. J. Immunol. 34:1870–1881.

van, V., Dai, H., van, d., V., He, Y. D., Hart, A. A., Mao, M., Peterse, H. L., van der, K. K., Marton, M. J., Witteveen, A. T., Schreiber, G. J., Kerkhoven, R. M., Roberts, C., Linsley, P. S., Bernards, R., and Friend, S. H. 2002. Gene expression profiling predicts clinical outcome of breast cancer. Nature 415:530–536.

Varambally, S., Dhanasekaran, S. M., Zhou, M., Barrette, T. R., Kumar-Sinha, C., Sanda, M. G., Ghosh, D., Pienta, K. J., Sewalt, R. G., Otte, A. P., Rubin, M. A., and Chinnaiyan, A. M. 2002. The polycomb group protein EZH2 is involved in progression of prostate cancer. Nature 419:624–629.

Visser, H. P., Gunster, M. J., Kluin-Nelemans, H. C., Manders, E. M., Raaphorst, F. M., Meijer, C. J., Willemze, R., and Otte, A. P. 2001. The Polycomb group protein EZH2 is upregulated in proliferating, cultured human mantle cell lymphoma. Br. J. Haematol. 112:950–958.

Wagner, W., Horn, P., Bork, S., and Ho, A. D. 2008. Aging of hematopoietic stem cells is regulated by the stem cell niche. Exp. Gerontol.

Wahlfors, J., Hiltunen, H., Heinonen, K., Hamalainen, E., Alhonen, L., and Janne, J. 1992. Genomic hypomethylation in human chronic lymphocytic leukemia. Blood 80:2074–2080.

Waly, M., Olteanu, H., Banerjee, R., Choi, S. W., Mason, J. B., Parker, B. S., Sukumar, S., Shim, S., Sharma, A., Benzecry, J. M., Power-Charnitsky, V. A., and Deth, R. C. 2004. Activation of methionine synthase by insulin-like growth factor-1 and dopamine: a target for neurodevelopmental toxins and thimerosal. Mol. Psychiatry 9:358–370.

Weisenberger, D. J., Siegmund, K. D., Campan, M., Young, J., Long, T. I., Faasse, M. A., Kang, G. H., Widschwendter, M., Weener, D., Buchanan, D., Koh, H., Simms, L., Barker, M., Leggett, B., Levine, J., Kim, M., French, A. J., Thibodeau, S. N., Jass, J., Haile, R., and Laird, P. W. 2006. CpG island methylator phenotype underlies sporadic microsatellite instability and is tightly associated with BRAF mutation in colorectal cancer. Nat. Genet. 38:787–793.

Wong, I. H., Chan, J., Wong, J., and Tam, P. K. 2004. Ubiquitous aberrant RASSF1A promoter methylation in childhood neoplasia. Clin. Cancer Res. 10:994–1002.

Yan, P. S., Chen, C. M., Shi, H., Rahmatpanah, F., Wei, S. H., Caldwell, C. W., and Huang, T. H. 2001. Dissecting complex epigenetic alterations in breast cancer using CpG island microarrays. Cancer Res. 61:8375–8380.

Yan, P. S., Efferth, T., Chen, H. L., Lin, J., Rodel, F., Fuzesi, L., and Huang, T. H. 2002. Use of CpG island microarrays to identify colorectal tumors with a high degree of concurrent methylation. Methods 27:162–169.

Yan, P. S., Perry, M. R., Laux, D. E., Asare, A. L., Caldwell, C. W., and Huang, T. H. 2000. CpG island arrays: an application toward deciphering epigenetic signatures of breast cancer. Clin. Cancer Res. 6:1432–1438.

Yan, P. S., Shi, H., Rahmatpanah, F., Hsiau, T. H., Hsiau, A. H., Leu, Y. W., Liu, J. C., and Huang, T. H. 2003. Differential distribution of DNA methylation within the RASSF1A CpG island in breast cancer. Cancer Res. 63:6178–6186.

Yang, H., Hoshino, K., Sanchez-Gonzalez, B., Kantarjian, H., and Garcia-Manero, G. 2005. Antileukemia activity of the combination of 5-aza-2′-deoxycytidine with valproic acid. Leuk. Res. 29:739–748.

Yu, M. K. 2006. Epigenetics and chronic lymphocytic leukemia. Am. J. Hematol. 81:864–869.

Zent, C. S. and Kay, N. E. 2007. Chronic lymphocytic leukemia: biology and current treatment. Curr. Oncol. Rep. 9:345–352.

Zhang, Q., Wang, H. Y., Marzec, M., Raghunath, P. N., Nagasawa, T., and Wasik, M. A. 2005. STAT3- and DNA methyltransferase 1-mediated epigenetic silencing of SHP-1 tyrosine phosphatase tumor suppressor gene in malignant T lymphocytes. Proc. Natl. Acad. Sci. USA 102:6948–6953.

Zheng, S., Ma, X., Zhang, L., Gunn, L., Smith, M. T., Wiemels, J. L., Leung, K., Buffler, P. A., and Wiencke, J. K. 2004. Hypermethylation of the 5′ CpG island of the FHIT gene is associated with hyperdiploid and translocation-negative subtypes of pediatric leukemia. Cancer Res. 64: 2000–2006.

DNA Methylation and Alzheimer's Disease

Thomas van Groen

Abstract Epigenetics plays a direct and indirect role in the chances of developing Alzheimer's disease. The decreased DNA methylation status with increasing age of the amyloid precursor protein (*APP*) gene promoter will boost transcription of this gene, leading to higher levels of *APP*. Furthermore, both the *BACE* and *PS1* genes show similar decreased promoter methylation with aging, causing higher levels – and activity – of β- and γ-secretases, increasing APP processing toward Aβ production. Together, this increases the levels of Aβ that will lead to the development of the pathology that is characteristic of sporadic AD. Furthermore, epigenetics plays a role through the nutritional status of the individual, i.e., through low folate and high homocysteine levels the DNA methylation level can be decreased. It is of interest to note that it has been shown that AD patients tend to have low levels of folate and high levels of homocysteine. Finally, parental influences in the inheritance of AD have been demonstrated, likely caused by gene imprinting.

Keywords Alzheimer's disease · Epigenetics · APP · Gamma secretase · Beta secretase

Introduction

Although epigenetic changes have a significant effect on many major physiological processes, such as development, cancer, and aging, surprisingly little is, yet, known about the epigenetic regulation of the age-related decline in cognition. DNA methylation is one of the major epigenetic processes that changes cellular activity by imposing a heritable pattern of gene expression without altering the sequence of DNA. Furthermore, DNA methylation together with histone acetylation is involved in the short- and long-term regulation of gene activity during the lifespan of the individual (Jaenisch and Bird 2003; Robert 2004).

T. van Groen (✉)
Department of Cell Biology, University of Alabama, Birmingham, AL, USA; Department of Neurobiology, University of Alabama, Birmingham, AL, USA
e-mail: vangroen@uab.edu

T.O. Tollefsbol (ed.), *Epigenetics of Aging*, DOI 10.1007/978-1-4419-0639-7_17,

DNA methylation is catalyzed by the DNA methyltransferases (DNMTs), which add a methyl moiety to the cytosine (mostly) located within the CpG dinucleotide (non-CpG cytosine methylation also exists and could play a role in gene regulation). There are three DNA methyltransferases: DNMT1, DNMT2, and DNMT3a and DNMT3b, which are encoded by three independent genes (Bestor 2000; Cheng and Blumenthal 2008). DNMT2 exists, but its role is still not clear (Jeltsch et al. 2006); DNMT3a and DNMT3b primarily function as de novo methyltransferases, whereas DNMT1 is the most abundant enzyme that preferentially methylates hemimethylated DNA and is responsible for stabilizing methylation patterns that have been established early in development. Moreover, MBD2 is the only member of a family of methyl-CpG-binding proteins that has been reported to be both a transcriptional repressor and a DNA demethylase (dMTase) (Detich et al. 2002). It should be noted that it has been shown that the MBD2 protein acts as a methyltransferase (Berger and Bird 2005; Hamm et al. 2008). Mutations in the *DNMT* genes have been shown to cause tumorigenesis (Bestor 2000; Lopatina et al. 2002); furthermore, gene knockout studies have demonstrated the essential role of these proteins in development (e.g., Klose and Bird 2006). For instance, homozygous loss of *DNMT1* in proliferating neuronal precursor cells generates demethylated neurons that are rapidly eliminated postnatally (Li et al. 1992). This suggests that DNA methylation by *DNMT1* is crucial for proper neuronal and CNS functioning during development. In contrast, mice with a conditional neuron-specific ablation of DNMT3a develop normally, but they do develop neuromuscular defects after birth and have a shortened life span (Nguyen et al. 2007).

Aging and Alzheimer's Disease

A genome-wide decrease in DNA methylation has been shown to occur during in vitro aging (Singhal et al. 1987; Wilson et al. 1987). Similarly, it has been demonstrated that a decrease in the general level of DNA methylation is present in the in vivo aging process (Bjornsson et al. 2008). Recently, we have shown that *DNMT1* expression and genomic methylation decline in the aging brain (Liu et al. 2008).

Whereas DNA methylation tends to decrease with age, the risk of developing Alzheimer's disease (AD) increases with age, and it has been demonstrated that age is one of the biggest risk factors for developing Alzheimer's disease (Qiu et al. 2007; Swerdlow 2007). AD is a progressive dementia for which the earliest clinical manifestation is memory dysfunction, and currently, it is the most common dementia in older people (Hebert et al. 2003; Hirtz et al. 2007). Pathologically the disease is characterized by its two defining hallmark lesions: (1) the presence of large numbers of intracellular neurofibrillary tangles and (2) extracellular neuritic plaques in the brain (e.g., Braak and Braak 1993, 1998; Selkoe 2001). Neurofibrillary tangles are predominantly present within neurons; tangles consist of hyperphosphorylated tau filaments (Hyman et al. 2005) – tau is one of the microtubule-associated proteins. The plaques primarily consist of amyloid β protein (Aβ) (Selkoe 2001). Amyloid β

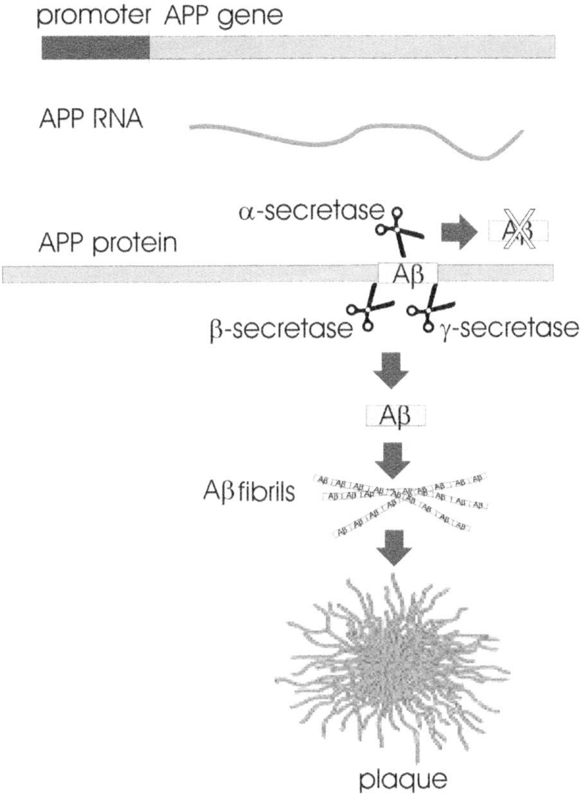

Fig. 1 Schematic showing the sequence of cleavage of the APP by two pathways: the α-secretase cleavage, which produces no Aβ since the cleavage is within the Aβ sequence, and the β- and γ-secretase cleavages, leading to Aβ production and deposition

is a 39–43-amino-acid-long peptide derived through the sequential proteolytic processing of APP by the β-secretase and the γ-secretase (Fig. 1) (Selkoe and Wolfe 2007). The γ-secretase cleaves at position 40 or 42 of the Aβ sequence, thus forming Aβ40 or Aβ42; Aβ42 has been shown to be much more prone to aggregation and plaque forming than Aβ40. Together, this sequence of event has been termed the "amyloid cascade hypothesis" (Hardy and Higgins 1992; Hardy 2006). In contrast, when the APP is first cleaved by the α-secretase (which cleaves within the Aβ sequence; Fig. 1), no Aβ is formed (Van Broeck et al. 2007). The two fragments that are produced are APP_S-α and the C83 peptide; APP_S-α has been suggested to be neuroprotective in contrast to Aβ42, which has been suggested to be neurotoxic (Van Broeck et al. 2007). Genetic causes of AD include genetic mutations of the *presenilin* and *amyloid precursor protein (APP)* genes (e.g., Tanzi and Bertram 2005), which are thought to contribute largely to the familial form of the AD disease (FAD) that usually manifests around the forties. However, FAD accounts for only a small proportion of AD, and the majority of the AD cases are known to occur sporadically.

The demonstration of mutations in the *APP* gene (Tanzi and Bertram 2005) together with cases of *APP* gene duplication (Cabrejo et al. 2006; Rovelet-Lecrux et al. 2006) causing early, inherited forms of AD suggest a central role for aberrant APP processing in the series of pathological changes occurring during the development of Alzheimer's disease. In Down's syndrome (i.e., trisomy 21), it is likely that the *APP* gene triplication causes overproduction of the APP, leading to the Alzheimer pathology in these patients (Beyreuther et al. 1993, Schupf and Sergievsky 2002). Together, these data show further evidence for the role of overexpression of the APP in the development of Alzheimer's disease pathology.

It should be noted, however, that the largest number of mutations that cause early-onset Alzheimer's disease are found in the *presenilin 1 (PS1)* gene [a number of AD-causing mutations are also present on the *presenilin 2 (PS2)* gene; Tanzi and Bertram 2005]. It has been demonstrated that PS1 is an essential component of the γ-secretase complex, which consists of, at least, four proteins (including PS1; Selkoe and Wolfe 2007). The three other major components of the γ-secretase complex are nicastrin, Aph-1, and Pen-2. It has been demonstrated that each is necessary for the functioning of the γ-secretase complex, but each plays a slightly different role in the maturation and activity of the complex (Selkoe and Wolfe 2007).

As indicated above the enzymatic activity of both the β-secretase and the γ-secretase plays a major role in the development of the pathology of Alzheimer's disease, since the combined activity of these two enzymes leads to the production of Aβ from the amyloid precursor protein. It is of interest to note that many of the current drugs that are developed as AD treatments are secretase inhibitors (Pallàs and Camins 2006), both for the β-secretase and the γ-secretase.

It has been suggested that the DNA methylation status has an influence on APP metabolism (Ledoux et al. 1993, 1994), since the promoter region of both the β- and the γ-secretase genes has been shown to have methylation sites. Furthermore, they get hypomethylated during aging, which would lead to increased gene expression levels and, likely, increased enzymatic activity due to increased protein levels. It should be noted that the change in methylation status is most likely related to the *S*-adenosyl methionine (SAM) metabolism (Fuso et al. 2005; Fuso et al. 2007). The combined extra activities of these two secretases are expected to lead to increased Aβ production and thus to earlier and more severe Alzheimer's disease pathology (Fuso et al. 2007; Scarpa et al. 2006; Fig. 2). It should be noted that the PS1 protein, through its activity in the γ-secretase complex, also contributes to proteolysis of other proteins that are directly involved in gene regulation, such as Notch (Fortini et al. 2002, Wolfe 2006). Furthermore, there is strong evidence that the γ-secretase cleavage of the APP leads to the production of the amyloid precursor protein intracellular domain (AICD) fragment that has been suggested to control gene regulation, similar to the Notch intracellular domain fragment (Cao and Sudfhof 2004; Hébert et al. 2006; Kaether et al. 2006; Müller et al. 2007, 2008). Notch signaling mediates many different intercellular communication events that are essential for determining the fates of neural and nonneural cells during development and in adulthood. The Notch receptor acts in a core pathway as a membrane-bound transcription factor that is released to the nucleus by a two-step cleavage mechanism called regulated

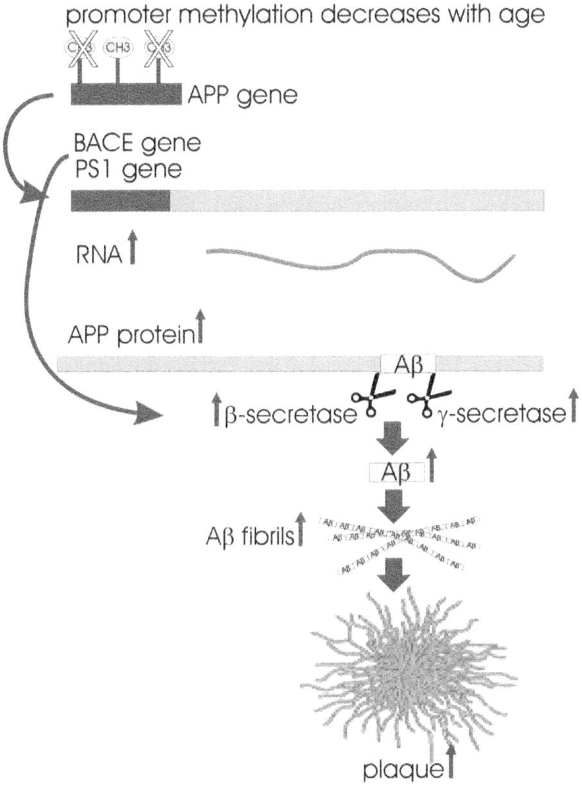

Fig. 2 Schematic showing the effects of decreased promoter methylation on the cleavage of the APP, together leading to increased Aβ production and deposition

intramembrane proteolysis (RIP). The second cleavage is effected by the γ-secretase (i.e., the enzyme complex that includes presenilin 1), an unusual polytopic aspartyl protease that apparently cleaves Notch and numerous other single-transmembrane substrates (e.g., APP) within the lipid bilayer of the membrane (Selkoe and Kopan 2003).

Whereas increases in the amyloidogenic processing of APP lead to Alzheimer's disease, in contrast to changes in the activity of the α-secretase, which would lead to decreases in the levels of Aβ, no age-related changes in the activity of this enzyme have been reported. It has been demonstrated that besides the mutations in the *APP* gene directly causing early-onset AD, several mutations in the *APP* gene promoter (Theuns et al. 2006; Cabrejo et al. 2006), which increase APP levels, also cause early-onset Alzheimer's disease.

Furthermore, the DNA methylation status of the promoter region of the amyloid protein precursor gene has also been shown to play an important role; the upstream promoter of the *beta-amyloid precursor protein* gene has two GC elements (Salbaum et al. 1988; Pollwein et al. 1992; Querfurth et al. 1999). In addition, it has been shown that the *APP* promoter has differential patterns of methylation in the human brain (Rogaev et al. 1994), and the levels of methylation of the promoter region correspond with differences in the levels of APP. Together, these findings indicate that changes in the methylation status of the *APP* promoter region will likely change the chances of developing Alzheimer's disease. Finally, it has been demonstrated that in AD patients the promoter region of the *APP* gene is hypomethylated (West et al. 1995; Tohgi et al. 1999). Together these data indicate that a low methylation status of the promoter region of the *APP* gene likely contributes to early development of AD, by increasing APP levels and, thus, Aβ production.

The relation between nutrition and cognition is widely recognized (e.g., Gordon 1997; Luchsinger et al. 2007c). Dietary components that are known to affect brain development and cognition include alcohol consumption, vitamins (especially A, B6, and B12), cholesterol, folate, and choline (Del Parigi et al. 2006; Gillette Guyonnet et al. 2007). Although the underlying molecular mechanisms remain to be determined, some of these nutrients are thought to interact with biological DNA methylation, for instance, folate and vitamin B_{12} (Davis and Uthus 2003; Wolters et al. 2004, Niculescu et al. 2006). Dietary folate is one of the most extensively investigated substrates that are linked with biological DNA methylation, and it has been suggested that it also plays a role in the pathogenesis of Alzheimer's disease (Snowdon et al. 2000; Luchsinger and Mayeux 2004, Tchantchou et al. 2006; Haan et al. 2007). Most AD patients have been shown to have low folate and high homocysteine levels in blood (Wolters et al. 2004; Haan et al. 2007).

Folate is an important substrate in one-carbon metabolism and is the dietary source of methyl groups for biological methylation (Fig. 3) (Wolters et al. 2004).

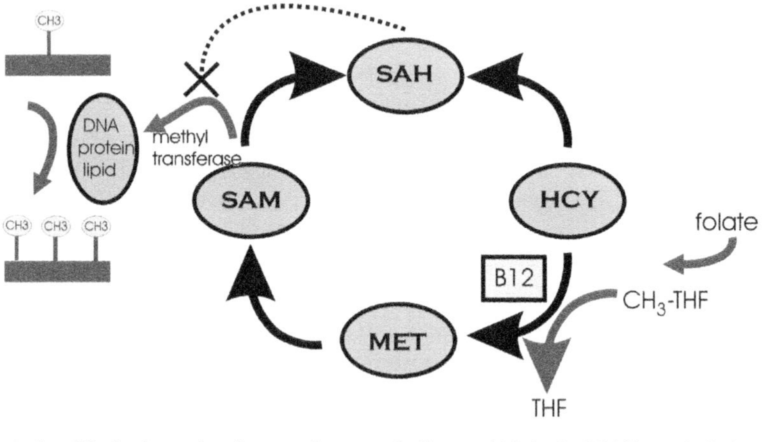

Fig. 3 A simplified schematic of one-carbon metabolism and biological DNA methylation. MET, methionine; SAM, *S*-adenosyl methionine; SAH, *S*-adenosyl homocysteine; HCY, homocysteine; THF, tetrahydrofolate; CH_3-THF, 5-methyltetrahydrofolate. The expected effects of folate/B_{12} deficiency include (1) impairing the conversion of HCY into MET (and resulting in a low level of SAM); (2) increasing the level of HCY; (3) reversing the reaction dynamics between HCY and SAH that favors SAH synthesis. High levels of SAH inhibit the methyltransferase activity; together, these effects will eventually lead to defective genomic methylation, whereas sufficient folate/B_{12} prevents this

It is required for the conversion of homocysteine to methionine and the formation of *S*-adenosyl methionine (SAM). SAM participates in biological methylation reactions, which generate *S*-adenosylhomocysteine (SAH) that subsequently forms homocysteine (Fig. 3). Folate depletion can thus cause genomic DNA hypomethylation, which can be reversed upon dietary folate restoration (Durga et al. 2007). Folate depletion can also lead to cellular accumulation of SAH and dramatically increase blood homocysteine levels. High homocysteine levels have been reported in Alzheimer's disease patients (Haan et al. 2007), but it is not clear if these are a cause or an effect (Wolters et al. 2004). Aging in the adult population is often associated with poor dietary intake, reduced nutrient adsorption, and less efficient utilization of nutrients, all of which could contribute to increased homocysteine levels in aging individuals (Wolters et al. 2004). Furthermore, SAH is a potent inhibitor of DNMT activity through the product inhibition pathway and this can lead to genome hypomethylation (Sibani et al. 2002; Trasler et al. 2003). It has been demonstrated that the disruption of homocysteine/folate metabolism can adversely affect both the developing and the adult brain (e.g., Davis and Uthus 2003). In fact, the levels of folic acid in the diet have been shown to affect CpG island methylation, and folate supplementation increases promoter methylation in an age-dependent manner (Wolters et al. 2004). In addition, decreased folate has been found to impair memory in humans, while folate replenishment increases memory (Haan et al. 2007), and recent exciting studies have shown that folate supplementation in humans for 3 years significantly improved cognitive functions that decline during aging (Durga et al. 2007). Together this will likely lead to impaired DNA methylation in aging individuals and thus increased chances of developing Alzheimer's disease (Wu et al. 2008). Finally, recent studies have indicated that AD patients with a late-onset, i.e., sporadic, Alzheimer's disease have an increased epigenetic drift in several genes that have been shown to be related to the development of AD, compared to age-matched control individuals (Wang et al. 2008). It is of interest to note that the gene with the largest epigenetic drift was the *PS1* gene. Several studies have shown that Alzheimer's disease also shows epigenetic characteristics such that the "parent of origin effect" can be detected (Bassett et al. 2002, 2006; Mosconi et al. 2007). These studies demonstrated that in late-onset Alzheimer's disease there are strong indications for a maternal disease transmission, i.e., maternal origin of the gene(s) that causes or contributes to Alzheimer's disease development. Similarly, studies have shown that several genes that are involved in the development of AD are imprinted (Peters and Beechey 2004; Luedi et al. 2005). Furthermore, Lahiri et al. (2007) have proposed a "Latent Early-Life Associated Regulation" (LEARn) model, which postulates a latent expression of specific genes triggered at the developmental stage. According to this model, environmental agents (e.g., heavy metals) (Basha et al. 2005), intrinsic factors (e.g., cytokines), and dietary factors (e.g., cholesterol) (Hartmann et al. 2007) perturb gene regulation in a long-term fashion, beginning at early developmental stages; however, these perturbations do not have significant pathological results until later in life. Thus, such actions would perturb *APP* gene regulation at a very early stage via its transcriptional machinery, leading to delayed overexpression of APP and subsequently of Aβ deposition. This model operates on the regulatory region (promoter) of the gene and by the effect of methylation at certain sites within the promoter of specific genes such that the promoter activity can be altered by changes in the primary DNA sequence and by epigenetic changes through mechanisms such as DNA methylation at CpG dinucleotides.

Intrinsic epigenetic differences are also present in the APP mRNA expression levels; gene expression is lower in female mouse brain than in male mouse brain (Thakur and Mani 2005), which is inversely correlated with the level of *APP* promoter methylation (Mani and Thakur 2006; Schäfer et al. 2007). It is of interest to note, however, that in both mouse AD models (Wang et al. 2003) and humans the prevalence of AD is higher in females (Gao et al. 1998; Kaminsky et al. 2006). The levels of estrogen have been suggested to play a role in this gender difference in the prevalence of AD (Thakur and Mani 2005); however, other studies do not seem to indicate a predominant role of estrogen in this sex difference (Schäfer et al. 2007).

In conclusion, epigenetics plays a direct and indirect role in the chances of developing Alzheimer's disease; the decreased DNA methylation status with increasing age of the *APP*, *PS1*, and *BACE* gene promoter boosts transcription of these genes. Together, this increases the production of Aβ that will lead to the development of the pathology that is characteristic of AD. Indirectly, epigenetics plays a role through the nutritional status of the individual, i.e., through low folate and high homocysteine levels, the DNA methylation level is decreased. Finally, parental influences in the inheritance of AD have been demonstrated, likely caused by gene imprinting.

References

Basha, M. R., Murali, M., Siddiqi, H. K., Ghosal, K., Siddiqi, O. K., Lashuel, H. A., Ge, Y. W., Lahiri, D. K., and Zawia, N. H. 2005. Lead (Pb) exposure and its effect on APP proteolysis and Abeta aggregation. FASEB J. 19:2083–2084.

Bassett, S. S., Avramopoulos, D., and Fallin, D. 2002. Evidence for parent of origin effect in late-onset Alzheimer disease. Am J Med Genet. 114:679–686.

Bassett, S. S., Avramopoulos, D., Perry, R. T., Wiener, H., Watson, B. Jr., Go, R. C., and Fallin, M. D. 2006. Further evidence of a maternal parent-of-origin effect on chromosome 10 in late-onset Alzheimer's disease. Am J Med Genet. 141B:537–540.

Berger, J., and Bird, A. 2005. Role of MBD2 in gene regulation and tumorigenesis. Biochem Soc Trans. 33(Pt 6):1537–1540.

Bestor, T. H. 2000. The DNA methyltransferases of mammals. Hum Mol Genet. 9:2395–2402

Beyreuther, K., Pollwein, P., Multhaup, G., Mönning, U., König, G., Dyrks, T., Schubert, W., and Masters, C. L. 1993. Regulation and expression of the Alzheimer's beta/A4 amyloid protein precursor in health, disease, and Down's syndrome. Ann NY Acad Sci. 695:91–102.

Bjornsson, H. T., Sigurdsson, M. I., Fallin, M. D., Irizarry, R. A., Aspelund, T., Cui, H., Yu, W., Rongione, M. A., Ekström, T. J., Harris, T. B., Launer, L. J., Eiriksdottir, G., Leppert, M. F., Sapienza, C., Gudnason, V., and Feinberg, A. P. 2008. Intra-individual change over time in DNA methylation with familial clustering. JAMA. 299:2877–2883.

Braak, H., Braak, E. 1993. Neuropathological stageing of Alzheimer-related changes. Acta Neuropathol. 82:239–259.

Braak, H., Braak, E. 1998. Evolution of neuronal changes in the course of Alzheimer's disease. J. Neural. Transm. 53:127–140.

Cabrejo, L., Guyant-Maréchal, L., Laquerrière, A., Vercelletto, M., De la Fournière, F., Thomas-Antérion, C., Verny, C., Letournel, F., Pasquier, F., Vital, A., Checler, F., Frebourg, T., Campion,

D., and Hannequin, D. 2006. Phenotype associated with APP duplication in five families. Brain. 129:2966–2976.
Cao, X., and Südhof, T. C. 2004. Dissection of amyloid-beta precursor protein-dependent transcriptional transactivation. J Biol Chem. 279:24601–24611.
Cheng, X., and Blumenthal, R. M. 2008. Mammalian DNA methyltransferases: a structural perspective. Structure. 16:341–350.
Davis, C. D., and Uthus, E. O. 2003. Dietary folate and selenium affect dimethylhydrazine-induced aberrant crypt formation, global DNA methylation and one-carbon metabolism in rats. J Nutr. 133:2907–2914.
Del Parigi, A., Panza, F., Capurso, C., and Solfrizzi, V. 2006. Nutritional factors, cognitive decline, and dementia. Brain Res Bull. 69:1–19.
Detich, N., Theberge, J., and Szyf M. 2002. Promoter-specific activation and demethylation by MBD2/demethylase. J Biol Chem. 277:35791–35794.
Durga, J., van Boxtel, M. P., Schouten, E. G., Kok,F. J., Jolles, J., Katan, M. B., and Verhoef, P. 2007. Effect of 3-year folic acid supplementation on cognitive function in older adults in the FACIT trial: a randomised, double blind, controlled trial. Lancet. 369:208–216.
Fortini, M. E. 2002. Gamma-secretase-mediated proteolysis in cell-surface-receptor signalling. Nat Rev Mol Cell Biol. 3:673–684.
Fuso, A., Seminara, L., Cavallaro, R. A., D'Anselmi, F., and Scarpa, S. 2005. S-adenosylmethionine/homocysteine cycle alterations modify DNA methylation status with consequent deregulation of PS1 and BACE and beta-amyloid production. Mol Cell Neurosci. 28:195–204.
Fuso, A., Cavallaro, R. A., Zampelli, A., D'Anselmi, F., Piscopo, P., Confaloni, A., and Scarpa, S. 2007. gamma-Secretase is differentially modulated by alterations of homocysteine cycle in neuroblastoma and glioblastoma cells. J Alzheimer's Dis. 11:275–290
Gao, S., Hendrie, H. C., Hall, K. S., and Hui, S. 1998. The relationships between age, sex, and the incidence of dementia and Alzheimer disease: a meta-analysis. Arch Gen Psychiatry. 55: 809–815.
Gillette Guyonnet, S., Abellan Van Kan, G., Andrieu, S., Barberger Gateau, P., Berr, C., Bonnefoy, M., Dartigues, J. F., de Groot, L., Ferry, M., Galan, P., Hercberg, S., Jeandel, C., Morris, M. C., Nourhashemi, F., Payette, H., Poulain, J. P., Portet, F., Roussel, A. M., Ritz, P., Rolland, Y., and Vellas, B. 2007. IANA task force on nutrition and cognitive decline with aging. J Nutr Health Aging. 11:132–52.
Gordon, N. 1997. Nutrition and cognitive function. Brain Dev. 19:165–170.
Haan, M. N., Miller, J. W., Aiello, A. E., Whitmer, R. A., Jagust, W. J., Mungas, D. M., Allen, L. H., and Green, R. 2007. Homocysteine, B vitamins, and the incidence of dementia and cognitive impairment: results from the Sacramento Area Latino Study on Aging. Am J Clin Nutr. 85:511–517.
Hamm, S., Just, G., Lacoste, N., Moitessier, N., Szyf, M., and Mamer O. 2008. On the mechanism of demethylation of 5-methylcytosine in DNA. Bioorg Med Chem Lett. 18:1046–1049.
Hardy, J. 2006. Alzheimer's disease: the amyloid cascade hypothesis: an update and reappraisal. J Alzheimers Dis. 9(3 Suppl):151–153.
Hardy, J. A., and Higgins, G. A. 1992. Alzheimer's disease: the amyloid cascade hypothesis. Science. 256:184–185.
Hartmann, T., Kuchenbecker, J., and Grimm, M. O. 2007. Alzheimer's disease: the lipid connection. J Neurochem. 103(1 Suppl):159–170.
Hebert, L. E., Scherr, P. A., Bienias, J. L., Bennett, D. A., and Evans, D. A. 2003. Alzheimer disease in the US population: prevalence estimates using the 2000 census. Arch Neurol. 60: 1119–1122.
Hébert, S. S., Serneels, L., Tolia, A., Craessaerts, K., Derks, C., Filippov, M. A., Müller, U., and De Strooper, B. 2006. Regulated intramembrane proteolysis of amyloid precursor protein and regulation of expression of putative target genes. EMBO Rep. 7:739–745

Hirtz, D., Thurman, D. J., Gwinn-Hardy, K., Mohamed, M., Chaudhuri, A. R., and Zalutsky, R. 2007. How common are the "common" neurologic disorders? Neurology. 68:326–337.

Hyman, B. T., Augustinack, J. C., and Ingelsson, M. 2005. Transcriptional and conformational changes of the tau molecule in Alzheimer's disease. Biochim Biophys Acta. 1739: 150–157.

Jaenisch, R., and Bird, A. 2003. Epigenetic regulation of gene expression: how the genome integrates intrinsic and environmental signals. Nat Genet. 33 (Suppl):245–254.

Jeltsch, A., Nellen, W., and Lyko, F. 2006. Two substrates are better than one: dual specificities for Dnmt2 methyltransferases. Trends Biochem Sci. 31:306–308.

Kaether, C., Schmitt, S., Willem, M., and Haass, C. 2006. APP and Notch intracellular domains are generated after transport of their precursors to the cell surface. Traffic. 7:408–415.

Kaminsky, Z., Wang, S. C., and Petronis, A. 2006. Complex disease, gender and epigenetics. Ann Med. 38:530–544.

Klose, R. J., and Bird, A. P. 2006. Genomic DNA methylation: the mark and its mediators. Trends Biochem Sci. 31:89–97.

Lahiri, D. K., Maloney, B., Basha, M. R., Ge, Y. W., and Zawia, N. H. 2007. How and when environmental agents and dietary factors affect the course of Alzheimer's disease: the "LEARn" model (latent early-life associated regulation) may explain the triggering of AD. Curr Alzheimer Res. 4:219–228.

Ledoux, S., Rebai, N., Dagenais, A., Shaw, I. T., Nalbantoglu, J., Sekaly, R. P., and Cashman, N. R. 1993. Amyloid precursor protein in peripheral mononuclear cells is up-regulated with cell activation. J Immunol. 150:5566–5575.

Ledoux, S., Nalbantoglu, J., and Cashman, N. R. 1994. Amyloid precursor protein gene expression in neural cell lines: influence of DNA cytosine methylation. Mol Brain Res. 24: 140–144.

Li, E., Bestor, T. H., and Jaenisch, R. 1992. Targeted mutation of the DNA methyltransferase gene results in embryonic lethality. Cell. 69:915–926.

Liu, L., van Groen, T., Kadish, I., and Tollefsbol, T. O. 2007. DNA methylation impacts on learning and memory in aging. Neurobiol Aging. Epub Sep 10

Lopatina, N., Haskell, J. F., Andrews, L. G., Poole, J. C., Saldanha, S., and Tollefsbol, T. 2002. Differential maintenance and de novo methylating activity by three DNA methyltransferases in aging and immortalized fibroblasts. J Cell Biochem. 84:324–334.

Luchsinger, J. A., and Mayeux, R. 2004. Dietary factors and Alzheimer's disease. Lancet Neurol. 3:579–587.

Luchsinger, J. A., Tang, M. X., Miller, J., Green, R., Mehta, P. D., and Mayeux, R. 2007a. Relation of plasma homocysteine to plasma amyloid beta levels. Neurochem Res. 32:775–781

Luchsinger, J. A., Tang, M. X., Miller, J., Green, R., and Mayeux, R. 2007b. Relation of higher folate intake to lower risk of Alzheimer disease in the elderly. Arch Neurol. 64:86–92.

Luchsinger, J. A., Noble, J. M., and Scarmeas, N. 2007c. Diet and Alzheimer's disease. Curr Neurol Neurosci Rep. 7:366–372.

Luedi, P. P., Hartemink, A. J., and Jirtle, R. L. 2005. Genome-wide prediction of imprinted murine genes. Genome Res. 15:875–884.

Mani, S. T., and Thakur, M. K. 2006. In the cerebral cortex of female and male mice, amyloid precursor protein (APP) promoter methylation is higher in females and differentially regulated by sex steroids. Brain Res. 1067:43–47.

Mosconi, L., Brys, M., Switalski, R., Mistur, R., Glodzik, L., Pirraglia, E., Tsui, W., De Santi, S., and de Leon, M.J. 2007. Maternal family history of Alzheimer's disease predisposes to reduced brain glucose metabolism. Proc Natl Acad Sci USA. 104(48):19067–19072.

Müller, T., Concannon, C. G., Ward, M. W., Walsh, C. M., Tirniceriu, A. L., Tribl, F., Kögel, D., Prehn, J.H., and Egensperger, R. 2007. Modulation of gene expression and cytoskeletal dynamics by the amyloid precursor protein intracellular domain (AICD). Mol Biol Cell. 18: 201–210.

Müller, T., Meyer, H. E., Egensperger, R., and Marcus, K. 2008. The amyloid precursor protein intracellular domain (AICD) as modulator of gene expression, apoptosis, and cytoskeletal dynamics-Relevance for Alzheimer's disease. Prog Neurobiol. Epub Jul 5.

Nguyen, S., Meletis, K., Fu, D., Jhaveri, S., and Jaenisch, R. 2007. Ablation of de novo DNA methyltransferase Dnmt3a in the nervous system leads to neuromuscular defects and shortened lifespan. Dev Dyn. 236:1663–1676.

Niculescu, M. D., Craciunescu, C. N., and Zeisel, S. H. 2006. Dietary choline deficiency alters global and gene-specific DNA methylation in the developing hippocampus of mouse fetal brains. FASEB J. 20:43–49.

Pallàs, M., and Camins, A. 2006. Molecular and biochemical features in Alzheimer's disease. Curr Pharm Des. 12:4389–4408.

Peters, J., and Beechey, C. 2004. Identification and characterisation of imprinted genes in the mouse. Brief Funct Genomic Proteomic. 2:320–333.

Pollwein, P., Masters, C. L., and Beyreuther, K. 1992. The expression of the amyloid precursor protein (APP) is regulated by two GC-elements in the promoter. Nucleic Acids Res. 20:63–68.

Querfurth, H. W., Jiang, J., Xia, W., and Selkoe, D. J. 1999. Enhancer function and novel DNA binding protein activity in the near upstream betaAPP gene promoter. Gene. 232:125–141.

Qiu, C., De Ronchi, D., and Fratiglioni, L. 2007. The epidemiology of the dementias: an update. Curr Opin Psychiatry. 20:380–385.

Robert, L. 2004. Epigenetic post-transcriptional mechanisms for regulating physiological functions, and their decline during aging. J Soc Biol. 198:257–262.

Rogaev, E. I., Lukiw, W. J., Lavrushina, O., Rogaeva, E. A., and St George-Hyslop, P. H. 1994. The upstream promoter of the beta-amyloid precursor protein gene (APP) shows differential patterns of methylation in human brain. Genomics. 22:340–347.

Rovelet-Lecrux, A., Hannequin, D., Raux, G., Le Meur, N., Laquerrière, A., Vital, A., Dumanchin, C., Feuillette, S., Brice, A., Vercelletto, M., Dubas, F., Frebourg, T., and Campion, D. 2006. APP locus duplication causes autosomal dominant early-onset Alzheimer disease with cerebral amyloid angiopathy. Nat Genet. 38:24–26.

Salbaum, J. M., Weidemann, A., Lemaire, H. G., Masters, C. L., and Beyreuther, K. 1988. The promoter of Alzheimer's disease amyloid A4 precursor gene. EMBO J. 7:2807–2813.

Scarpa, S., Cavallaro, R. A., D'Anselmi, F., and Fuso, A. 2006. Gene silencing through methylation: an epigenetic intervention on Alzheimer disease. J Alzheimers Dis. 9:407–414.

Schäfer, S., Wirths, O., Multhaup, G., and Bayer, T. A. 2007. Gender dependent APP processing in a transgenic mouse model of Alzheimer's disease. J Neural Transm. 114:387–394.

Schupf, N., and Sergievsky, G. H. 2002. Genetic and host factors for dementia in Down's syndrome. Br J Psychiatry. 180:405–410.

Selkoe, D. J. 2001. Alzheimer's disease: genes, proteins, and therapy. Physiol Rev. 81:741–766.

Selkoe, D., and Kopan, R. 2003. Notch and Presenilin: regulated intramembrane proteolysis links development and degeneration. Annu Rev Neurosci. 26:565–597.

Selkoe, D. J., and Wolfe, M. S. 2007. Presenilin: running with scissors in the membrane. Cell. 131:215–221.

Sibani, S., Melnyk, S., Pogribny, I. P., Wang, W., Hiou-Tim, F., Deng, L., Trasler, J., James, S. J., and Rozen, R. 2002. Studies of methionine cycle intermediates (SAM, SAH), DNA methylation and the impact of folate deficiency on tumor numbers in Min mice. Carcinogenesis. 23:61–65.

Singhal, R. P., Mays-Hoopes, L. L., and Eichhorn, G. L. 1987. DNA methylation in aging of mice. Mech Ageing Dev. 41:199–210.

Snowdon, D. A., Tully, C. L., Smith, C. D., Riley, K. P., and Markesbery, W. R. 2000. Serum folate and the severity of atrophy of the neocortex in Alzheimer disease: findings from the Nun study. Am J Clin Nutr. 71:993–998.

Swerdlow, R. H. 2007. Is aging part of Alzheimer's disease, or is Alzheimer's disease part of aging? Neurobiol Aging. 28:1465–1480.

Tanzi, R. E., and Bertram, L. 2005. Twenty years of the Alzheimer's disease amyloid hypothesis: a genetic perspective. Cell. 120(4):545–555.

Tchantchou, F., Graves, M., Ortiz, D., Chan, A., Rogers, E., and Shea, T. B. 2006. S-adenosyl methionine: A connection between nutritional and genetic risk factors for neurodegeneration in Alzheimer's disease. J Nutr Health Aging. 10:541–544.

Thakur, M. K., and Mani, S. T. 2005. Estradiol regulates APP mRNA alternative splicing in the mice brain cortex. Neurosci Lett. 381:154–157.

Theuns, J., Brouwers, N., Engelborghs, S., Sleegers, K., Bogaerts, V., Corsmit, E., De Pooter, T., van Duijn, C. M., De Deyn, P. P., and Van Broeckhoven, C. 2006. Promoter mutations that increase amyloid precursor-protein expression are associated with Alzheimer disease. Am J Hum Genet. 78:936–946.

Tohgi, H., Utsugisawa, K., Nagane, Y., Yoshimura, M., Genda, Y., and Ukitsu, M. 1999. Reduction with age in methylcytosine in the promoter region -224 approximately -101 of the amyloid precursor protein gene in autopsy human cortex. Mol Brain Res. 70:288–292.

Trasler, J., Deng, L., Melnyk, S., Pogribny, I., Hiou-Tim, F., Sibani, S., Oakes, C., Li, E., James, S. J., and Rozen, R. 2003. Impact of Dnmt1 deficiency, with and without low folate diets, on tumor numbers and DNA methylation in Min mice. Carcinogenesis. 24:39–45.

Van Broeck, B., Van Broeckhoven, C., and Kumar-Singh, S. 2007. Current insights into molecular mechanisms of Alzheimer disease and their implications for therapeutic approaches. Neurodegener. Dis. 4:349–365.

Wang, J., Tanila, H., Puoliväli, J., Kadish, I., and van Groen, T. 2003. Gender differences in the amount and deposition of amyloidbeta in APPswe and PS1 double transgenic mice. Neurobiol Dis. 14:318–327.

Wang, S. C., Oelze, B., and Schumacher, A. 2008. Age-specific epigenetic drift in late-onset Alzheimer's disease. PLoS ONE. 3(7):e2698.

West, R. L., Lee, J. M., and Maroun, L. E. 1995. Hypomethylation of the amyloid precursor protein gene in the brain of an Alzheimer's disease patient. J Mol Neurosci. 6:141–146.

Wilson, V. L., Smith, R. A., Ma, S., and Cutler, R. G. 1987. Genomic 5-methyldeoxycytidine decreases with age. J Biol Chem. 262:9948–9951.

Wolfe, M.S. 2006. The gamma-secretase complex: membrane-embedded proteolytic ensemble. Biochemistry. 45:7931–7939.

Wolters, M., Ströhle, A., and Hahn, A. 2004. Age-associated changes in the metabolism of vitamin B(12) and folic acid: prevalence, aetiopathogenesis and pathophysiological consequences. Z Gerontol Geriatr. 37:109–135.

Wu, J., Basha, M. R., and Zawia, N. H. 2008. The environment, epigenetics and amyloidogenesis. J Mol Neurosci. 34:1–7.

DNA Methylation, Age-Related Immune Defects, and Autoimmunity

Jörg J. Goronzy, Guangjin Li, and Cornelia M. Weyand

Abstract The adaptive immune system is a highly proliferative organ that combines ongoing differentiation with preservation of memory. CpG DNA methylation is central to many of the fate decisions that occur from early lymphoid development to late effector functions of end-differentiated cells. Many of the DNA demethylation-sensitive genes play an important role in autoimmune diseases and in the immune defects that are associated with the aging process, suggesting that failing epigenetic control mechanisms of genes involved in adaptive immune responses contribute to these diseases.

Keywords Epigenetics · Aging · Immune senescence · T cells

Introduction

Continuous self-renewal and differentiation while maintaining memory of previous antigenic encounters are characteristic hallmarks of the adaptive immune system. Differentiation and memory are closely linked to epigenetic imprinting, which plays an important role at all stages of the immune response (Wilson et al. 2005). When B and T lymphocytes develop from hematopoietic stem cells, they make a number of irrevocable, frequently binary fate decisions. Cells commit to either the B-cell or T-cell lineage, and T cells rearrange either their TCR-$\alpha\beta$ or TCR-$\gamma\delta$ loci. They develop into functional lineages, such as CD4 and CD8 T cells, but also NK T cells and regulatory T cells. With the exception of the TCR rearrangements, these lineage choices are not DNA sequence encoded; epigenetically encoded information determines which path is chosen and which genes are expressed.

In addition to the chromatin structure and the various DNA-associated proteins, differential methylation of DNA cytosines is a major mechanism for storing

J.J. Goronzy (✉)
Department of Medicine, Kathleen B. and Mason I. Lowance Center for Human Immunology and Rheumatology, Emory University School of Medicine, Room 1003 Woodruff Memorial Research Building, 101 Woodruff Circle, Atlanta, GA, USA
e-mail: jgoronz@emory.edu

T.O. Tollefsbol (ed.), *Epigenetics of Aging*, DOI 10.1007/978-1-4419-0639-7_18,

epigenetic information. Methylation of CpG islands is associated with the repression or silencing of gene expression through several mechanisms. It can directly inhibit the binding of transcription factors that require a CpG motif for binding; it also recruits a number of methyl-CpG binding proteins that either sterically prevent transcription factor binding or are involved in protein recruitment and protein modification and formation of the chromatin structure. CpG methylation is accomplished by a set of DNA methyltransferases (DNMT). DNMT 3a and 3b mediate de novo methylation, in particular during embryonal development, while DNMT1 is responsible for maintaining methylation patterns. CpG demethylation is frequently passive, occurring during DNA replication. Since the development and the function of the lymphoid system are closely linked to proliferation, passive CpG demethylation and its prevention are highly relevant. DNMT1 is recruited to DNA replication sites and preferentially targets hemi-methylated CpG sites. Insufficient levels, active replacement, or recruitment inhibition of DNMT1 leads to progressive demethylation or infidelity of epigenetic memory. The role of DNMT1 in lymphoid development has been unequivocally shown. Inhibition of DNMT1 at early stages of thymocyte development impairs the survival of TCR/αβ cells and induces the emergence of atypical CD8 TCR/γδ T cells. DNMT1 deletion at later stages does not prevent T-cell development but induces functional aberrations with decreased TCR responsiveness and increased cytokine expression upon stimulation (Lee et al. 2001).

The early stages of T-cell development are restricted to the thymus organ which generates trillions of T cells before it involutes in the adult; thymic output is minimal after the ages of 40–50 years (Goronzy and Weyand 2005). However, T-cell development is not concluded with thymic education. Newly generated T cells are largely undifferentiated and also called naïve. A large pool of such undifferentiated T cells is maintained throughout life until they meet an antigen which they recognize with sufficient affinity. Upon contact with antigen, naïve CD4 and CD8 cells rapidly divide and develop into effector T cells, memory T cells, and regulatory T cells (Akbar et al. 2007). Again, differentiation is associated with fate decisions; in particular CD4 T cells differentiate into Th1, Th2, and Th17 cells biased for the production of certain sets of cytokines (Dong 2008; Mosmann and Coffman 1989). In parallel, stimulated naïve T cells gain the expression of molecules that are necessary for their effector functions such as chemokine or homing receptors. Some of these changes are partially reversible, e.g., some effector cells revert to central memory T cells which share the homing pattern with naïve T cells. However, many of these differentiation steps including the lineage commitments to certain cytokine patterns are irrevocable. This naïve to effector T-cell differentiation is initiated by signaling events that are first translated into specific profiles of transcription factors and eventually epigenetically imprinted (Wilson 2005; Sawalha 2008). This epigenetic imprinting is closely associated and absolutely dependent on T-cell clonal expansion that occurs in the initial days of antigen priming.

Antigen exposure is cumulative, and it is therefore not surprising that T-cell effector populations accumulate at the expense of naïve T cells with age. This is particularly evident for CD8 cells; naïve CD8 T cells dwindle in the second half of life, and differentiated CD45RA-positive effector T cells accumulate. This effect is less pronounced in CD4 T cells, but also here the frequency of naïve T cells declines with age (Czesnikiewicz-Guzik et al. 2008). Homeostatic T-cell proliferation may also be associated with epigenetic changes and T-cell differentiation (Reiner 2005). Even in the absence of exogenous antigenic stimulation, T-cell proliferation occurs throughout life; naïve T cells divide about once to twice per year while memory T cells divide more frequently (Macallan et al. 2004). These findings have given rise to the hypothesis that the immune aging process is in part driven by accumulation of epigenetic imprinting, some of which accumulates as a consequence of homeostatic proliferation and some of which is driven by T-cell differentiation.

As epigenetic imprinting is a major mechanism of T-cell differentiation, it is not surprising that epigenetic changes have been implicated in autoimmunity. Although autoimmunity has germ line-encoded genetic risks, the genetic predisposition for many autoimmune diseases is small with odds ratios of less than 1.5 for many autoimmune genes and odd ratios in the single digits for the total genetic burden, leaving substantial space for environmental causes or chance events. In fact, it is now appreciated that many autoimmune diseases increase in frequency with age (Goronzy et al. 2006), suggestive for a model that causative factors are cumulative. Epigenetic mechanisms could contribute to autoimmune disease in a number of different ways. It is possible that epigenetic changes are actively induced, translating environmental stressors into expression of autoimmune genes. Alternatively, epigenetic changes predisposing for autoimmunity may represent just a consequence of the aging process or the accumulation of T-cell differentiation events.

Epigenetic Regulation of T-Cell Differentiation

Upon encounter with antigen, naïve CD4 T cells differentiate into Th1, Th2, and Th17 cells, characterized by signature cytokines IFN-γ, IL-4, and IL-17 (Reiner 2007). CD8 T cells are already committed to IFN-γ production before they leave the thymus, although this commitment is not absolute. Th1 cells also produce TNF-α and lymphotoxin-α, while Th2 cells produce IL-5, IL-6, IL-13, and, depending on the species, IL-10 (Mosmann and Coffman 1989; Glimcher and Murphy 2000). Th1 cells orchestrate cellular and in particular phagocytic responses, while Th2 cells are more involved in the elimination of extracellular microbes (Mosmann and Coffman 1989). Th17 cells regulate inflammatory pathways (Ouyang et al. 2008). Thus, while the initial antigen-specific cell is multi-potent, it generates distinct progenies (Glimcher and Murphy 2000). The proper choice determines the outcome of the infection or the clinical presentation of chronic inflammatory immune responses. Early studies have shown that Th1 and Th2 cytokines are subject to epigenetic regulation. Treatment with methylation inhibitors induces IFN-γ as well as IL-4 transcription in naïve CD4 T cells; in fact the distinction between Th1 and Th2 cytokine patterns is blurred. IL-4 as well as IFN-γ transcription is amplified in T cells lacking DNMT1 or MBD2 (Lee et al. 2001; Hutchins et al. 2002). Comparison of naïve and memory T cells and of Th1 and Th2 cells yielded differences in the promoter methylation pattern of the IFN-γ promoter. In CD8 T-cell clones, the heritability of a demethylated state directly correlated with IFN-γ expression (Fitzpatrick et al. 1998).

More recent studies have identified enhancer-like regions, designated IFN-γ CNS1 and IFN-γ CNS2, that are important for maximal IFN-γ transcription and that are demethylated in Th1 cells (Lee et al. 2004; Shnyreva et al. 2004). DNA methylation is also important for IL-4 transcription. Th2 cytokines share a locus-control region between the IL-4 and the IL-13 locus within the 3′ region of the ubiquitously expressed RAD50 gene (Lee et al. 2003). Chromatin loops are formed in this region that includes the various promoters and is influenced by the locus-control region (Spilianakis and Flavell 2004). Th2 cytokine transcription appears to be regulated by active demethylation of a CpG motif in this region (Kim et al. 2007). Additional sites of epigenetic control have been mapped to the proximal elements of the IL-4 gene and the enhancer-like elements IL4CNS1 and IL4CNS2 (Lee et al. 2002).

How does DNA methylation relate to Th1 and Th2 differentiation? Th1 and Th2 polarization is a process that requires proliferation, is initially reversible, and is stable only 2–3 weeks after initial stimulation. Initial gene activation correlates with the induction of key transcription factors. Stimulation in the presence of IL-12 leads to STAT4-dependent expression of the IL-12 receptor β-chain and the key transcription factor T-bet and establishes positive feedback loops involving T-bet, IFN-γ, and T-bet-induced transcription factor Hlx (Mullen et al. 2002). Only eventually, a stable Th1 phenotype is established which correlates with extended DNA demethylation of the IFN-γ promoter and enhancer-like regions. At this stage, IFN-γ transcription is independent of T-bet, suggesting that the epigenetic modifications have overcome the requirement for T-bet (Mullen et al. 2002).

Th2 polarization depends on the production of IL-4 by CD4 T cells and mast cells which then induce the transcription factor GATA-3 (Zheng and Flavell 1997). GATA-3 has been directly implicated in inducing chromatin remodeling including competitive displacement of MBD2 from regulatory regions of the IL-4 gene and subsequent progressive DNA demethylation (Hutchins et al. 2002). Similar to T-bet, deletion of GATA-3 in the early stages of Th2 differentiation can inhibit lineage development, while IL-4 transcription in the later stages is relatively GATA-3 independent (Pai et al. 2004; Zhu et al. 2004).

Th17 cells have only been noted in recent years, but appear to play an important role in inflammatory diseases including autoimmune diseases (Bettelli et al. 2008; Korn et al. 2007). Similar to Th1 and Th2 development, differentiation is induced via environmental conditions encountered at the time of T-cell activation and subsequent clonal expansion. Th17-polarizing conditions include the presence of IL-6, IL-23, and TGF-β and the absence of IL-4 (Dong 2008; Bettelli et al. 2006; Veldhoen et al. 2006). Lineage-specific transcription factors for IL-17 include RORα and RORγt (Ivanov et al. 2006; Yang et al. 2008). It is currently unclear whether they act additively or synergistically; however, they appear to be involved in histone modifications that occur during early Th17 development at several CNS elements of the IL-17 gene (Akimzhanov et al. 2007). Whether these histone modifications confer lineage specificity and what the role of DNA methylation may be are currently unclear.

Role of DNA Demethylation in the Expression of T Effector Cell Molecules

In parallel to lineage commitment, naïve T-cell activation induces the transcription of genes that are important for effector cell function. Many of those genes are transiently expressed while cells are activated and return to baseline levels after a resting period of a few days. Such T cells rested after in vitro activation have many similarities to memory T cells in vivo and differ in their gene expression profile from that of naïve T cells. The information on epigenetic control mechanisms that determine the lineages of effector and memory T cells in contrast to naïve T cells is largely unknown; however, indirect conclusions on the role of DNA demethylation can be drawn from comparing the expression patterns of functional T-cell subsets directly obtained ex vivo or activated in vitro and DNMT1-knockout T cells or T cells activated in the presence of a DNMT1 inhibitor such as 5-aza-dC (Wilson et al. 2005; Wilson and Merkenschlager 2006). Key target genes distinguishing functional T-cell subsets are chemokine receptors including CCR7 expressed on naïve and central memory T cells and CCR5 on effector T cells and adhesion molecules including CD44, CD49 isoforms, and CD11a/CD18 also known as LFA-1. Of particular interest is LFA-1, the expression of which is upregulated by DNMT inhibition due to demethylation of the CD11a promoter (Kaplan et al. 2000). The sequences flanking *ITGAL*-encoding CD11a do not contain typical CpG islands but scattered CpG motifs that are progressively demethylated with age (Zhang et al. 2002). In vivo, LFA-1 expression correlates with functional T-cell subsets. A baseline expression is found in naïve T cells that express the CCR7 chemokine receptor in conjunction with the CD45RA molecule. Expression is clearly higher in memory cell populations that express the CD45RO marker and is the highest in end-differentiated effector T cells that are characterized by the loss of the CD28 molecule (Singh et al. 2008). It is of interest that this particular population of effector cells is increased in frequency in autoimmune diseases and accumulates with age (see sections below). LFA-1 is a central molecule in T-cell function, based not only by its ability to direct T-cell homing but also by its involvement in the antigen-recognition process. The LFA-1–ICAM-1 interaction is an integral part of antigen recognition in which LFA-1 is responsible for the attachment of T cells to antigen-presenting cells expressing the relevant MHC/peptide complex. LFA-1–ICAM-1 interaction forms the basis of the T-cell receptor recognition platform or T-cell receptor synapse (Dustin et al. 2004; Sims and Dustin 2002). The mature synapse consists of the T-cell receptor–MHC/peptide pair surrounded by a ring of LFA-1–ICAM-1 pairs that stabilize the complex and allow for the recruitment of additional signaling receptors while restricting the access of negative regulatory receptors (Sims and Dustin 2002). LFA-1 densities, therefore, are likely to have a significant role in lowering T-cell receptor activation thresholds, not only leading to increased responsiveness of memory cells but also potentially increasing the risk for autoimmune recognition (see below).

A second gene that is DNA demethylation sensitive, involved in immune threshold regulation, and differentially expressed in functional T-cell subsets is CD70. CD70 is the ligand for the costimulatory receptor CD27. It is transiently expressed on many antigen-presenting cells including B cells, but can also be expressed on T cells (Borst et al. 2005). The transient expression is important to prevent overshooting T-cell activation. In a transgenic mouse model where CD70 was constitutively expressed on B cells, the overexpression resulted in continuous T-cell activation, eventual T-cell exhaustion, and lymphopenia (Tesselaar et al. 2003). Overexpression of CD70 can also be induced by DNMT inhibition (Oelke et al. 2004). In addition, the regulation of CD70 changes with T-cell differentiation. Naïve and several memory cells do not constitutively express CD70, and T-cell activation induces a very transient expression. In contrast, T-cell effector subpopulations that have lost the expression of the CD28 molecule already have a low level of constitutive expression. Activation of these T cells that accumulate in aging or in autoimmunity induces a lasting expression of CD70. These CD28-negative T cells provide costimulatory function to CD27-expressing naïve T cells which results in the activation of low avidity, potentially autoreactive T-cell receptors contributing to autoimmunity or lymphocyte exhaustion (Lee et al. 2007).

There are several additional parallel findings between CD28-negative T cells and cells treated with DNMT inhibitors. One gene that is DNA methylation sensitive in humans although not in the mouse is perforin. Treatment of human T cells with 5-aza-dC induces perforin expression (Lu et al. 2003; Makar and Wilson 2004). Of interest, perforin expression is activation dependent and confers cytotoxic effector functions in CD8 T cells (Catalfamo and Henkart 2003; Russell and Ley 2002). Again, constitutive expression is seen in CD28-negative T cells including CD4 T cells that normally do not exhibit any perforin/granzyme-mediated cytotoxicity (Warrington et al. 2001).

Epigenetic Control of KIR Expression on T Cells

One of the major transcriptional hallmarks of DNMT inhibition in T cells is the expression of negative and positive regulatory receptors of the killer immunoglobulin-like receptor (KIR) family (Czesnikiewicz-Guzik et al. 2008; Li et al. 2008). Again, the in vitro data parallel the in vivo findings. KIR expression is typical for CD28-negative T cells (Namekawa et al. 2000). KIR comprise a diverse family of regulatory cell surface receptors that are mainly expressed on NK cells and subsets of γδ T cells. With increasing age, KIR expression also emerges on subsets of $CD8^+$ and, to a lesser extent, $CD4^+$ T cells (Snyder et al. 2002; van Bergen et al. 2004). KIR function either as inhibitory or stimulatory receptors depending on the length of the cytoplasmic domain. Inhibitory receptors are long tailed containing one or two immunoreceptor tyrosine-based inhibition motifs (ITIM), while stimulatory receptors are short tailed binding the adapter molecule DAP12 with stimulatory kinase activity (Humphrey et al. 2005; Long et al. 2001). Upon recognition of their HLA class I ligands, KIR can transmit positive and negative signals and modify or even control NK receptor or T-cell receptor-mediated stimulation. A characteristic example is that the loss of MHC class I molecules on virus-infected or tumor cells renders them susceptible to NK cell-mediated killing due to the vanishing of KIR-transmitted negative signals (Martin and Carrington 2005; Parham 2005).

On NK cells, expression of different KIR family members is clonally distributed in a stochastic fashion. Early studies have shown that this clonal distribution pattern is entirely maintained by DNA CpG methylation of the KIR promoters. DNMT inhibition transforms the clonal to a global KIR expression pattern (Chan et al. 2003; Santourlidis et al. 2002). Pre-existing lineage-specific histone signatures appear to be a prerequisite for DNMT inhibition sensitivity. Histone H4 Lys 8 acetylation and histone H3 Lys 9 dimethylation at the KIR promoter are already found in KIR-negative NK cells that do not express KIR, but not in hematopoietic stem cells or B cells that never express KIR (Santourlidis et al. 2008).

KIR expression is absent in T cells of neonates but is gained with age, in particular in the CD8 compartment, where it is mostly found on end-differentiated effector memory T cells that have lost the CD28 molecule. It remains so far unknown how KIR transcription in T cells is activated. Attempts to induce KIR expression in vitro by cytokines have been unsuccessful. In vivo, acquisition of KIR in T cells occurs successively during clonal expansion in a stochastic pattern, but is restricted to a small subset of T cells. However, reporter gene assays using the minimal KIR promoters have shown that all T cells, including naïve and memory CD4 and CD8 T cells, have the transcriptional machinery to support KIR transcription (Xu et al. 2005). As in NK cells, KIR promoters in KIR-negative cells have hypermethylated CpG sites, while the corresponding sites in KIR-positive cells are hypomethylated. DNMT inhibition equally induces potent transcription in KIR-negative CD4 and CD8 T cells, suggesting that CpG demethylation is the ultimate regulator also in the T-cell lineage. Pre-existing histone signatures may again contribute to susceptibility to DNMT inhibition; CD8 T cells have a similar histone and methylation pattern in the KIR promoter as do NK cells (Xu et al. 2005). In CD4 T cells, histone H3 and H4 acetylation is low at the KIR promoter; however, CD4 T cells that are more (HUT78 or normal $CD4^{+}CD28^{-}$ T cells) or less (Jurkat or normal $CD4^{+}CD28^{+}$ T cells) prone to KIR2DL4 expression upon DNMT inhibition can be distinguished by increased dimethylated H3-Lys 4 (Li et al. 2008).

What are the mechanisms that drive increasing KIR expression with age? One possible explanation is that KIR promoter demethylation is a normal differentiation step in effector cell differentiation and that end-differentiated CD8 effector T cells accumulate with age. Indeed, accumulation of CD28-negative effector T cells with age is more common in CD8 than in CD4 T cells which could explain the differential KIR expression pattern. In addition, non-directed promoter demethylation with age appears to play a role. In CD4 as well as CD8 T cells, global CpG methylation decreases with age [(Golbus et al. 1990) and unpublished observation]. CpG sites in KIR promoters are methylated in hematopoietic stem cells and in T cells in cord blood (Santourlidis et al. 2008). As described above, these CpG sites are demethylated in CD8 T cells that express the respective KIR allele. In addition, with age, a random and partial demethylation of the KIR2DL3 promoter was even present in CD8 T cells that were negative for cell surface expression and had only minimal transcription. Correspondingly, the transcription activity increased with age in total CD8 T cells and even in KIR-negative CD8 cells. These age-dependent promoter demethylations are confined to CD8 T cells and exclude CD4 cells. Age-dependent differential recruitment of DNMT1 to the KIR2DL3 promoter may play a central role in maintaining promoter methylation in KIR-negative cells. Chromatin precipitation assays with anti-DNMT1 showed a reduced signal in CD8 compared to CD4 T cells, which further decreased with age. In contrast, precipitations for histone trimethylated H3 Lys 27 were not different (unpublished data).

In summary, KIR expression on T cells appears to be controlled by DNA demethylation during clonal T-cell expansion as well as through random and eventual cumulative demethylation during aging. As described in the following, KIR expression changes the contextual signals of antigen recognition by T cells and contributes to the age-dependent defects in immune competence as well as to breeches in tolerance and autoimmunity.

DNA Demethylation and Age-Associated Immune Defects

With increasing age, the immune system deteriorates and the ability to mount adaptive immune responses declines. The dysfunctional immune system is associated not only with an increased susceptibility to developing tumors or succumbing to infections but also with an increased risk to develop autoimmune diseases or a chronic inflammatory state that predisposes for cardiovascular disease, Alzheimer's disease, as well as frailty among others (Gruver et al. 2007; Pawelec and Larbi 2008). The decline in immune function is complex and multifactorial, with epigenetic control of gene expression only recently emerging as one mechanism.

One major reason for immune dysfunction is that homeostatic mechanisms change with age. The immune system is under constant renewal and proliferative stress and highly depends on the constant influx of immune precursor cells. All immune competent cells derive from hematopoietic stem cells, which decrease with age in numbers and proliferative capacity (Gruver et al. 2007; Aw et al. 2007). B-cell generation in the bone marrow declines with age; in fact, the differentiation from hematopoietic stem cells shifts from predominantly the lymphoid lineage in the young individual to the myeloid lineage in the old individual (Signer et al. 2007). This decline is even more striking for the T-cell lineage which requires the thymus for development. Thymic output exponentially decreases with age and essentially ceases by the age of 40–50 years. This decline in production of new T cells puts an extraordinary burden on the self-renewal capacity of already existing peripheral

T cells. The replicative stress in the hematopoietic stem cell compartment, as well as the peripheral T-cell compartment, results in progressive telomere shortening which is only partially counteracted by telomerase activity. Both hematopoietic stem cells and naïve T cells are able to express telomerase, which is composed of the reverse transcriptase hTERT and the hTR complementary template for the telomeric DNA sequence (Goronzy et al. 2006; Hodes et al. 2002). hTERT transcription is suppressed with differentiation from naïve to memory T cells, and telomerase activity is, therefore, reduced in the elderly where memory cells dominate and even naïve cells acquire partial phenotypic characteristics of memory cells. Telomerase activity in naïve T cells is necessary to elongate telomeres and protect T cells from senescence during homeostatic proliferation or oligoclonal expansion in response to antigen. It is also necessary to protect naïve CD4 T cells from apoptotic death through the intrinsic pathways during antigen-induced clonal burst (Goronzy et al. 2006). The epigenetic control of hTERT is complex. DNA methylation has a dual role on hTERT expression; methylation of CpG sites within the core promoter suppresses hTERT transcription; on the contrary, hTERT expression requires the methylation of CpG sites within the CTCF-binding region to prevent the binding of the inhibitory factor CTCF (Renaud et al. 2007). The impact of aging on the epigenetic control of hTERT transcription in different cell lineages, in particular in hematopoietic stem cells and in naïve T cells, has not been studied.

T-cell homeostatic proliferation and clonal expansion is not only dependent on telomerase activity, but also regulated by the expression of growth factor receptors and the availability of the corresponding growth factors. T-cell homeostatic mechanisms appear to be relatively robust for CD4 T cells up to the eighth decade of life (Naylor et al. 2005); however, they break down for CD8 cells earlier in life. The two compartments differ in their sensitivity to growth factors, in particular IL-6, IL-7, IL-15, and IL-21 (Boyman et al. 2007). How far growth factor and receptor expression change with age, how far loss of expression is responsible for the eventual failure of T-cell homeostatic mechanisms, and how far epigenetic control mechanisms are involved are currently under study. Expression of the IL-7 receptor α-chain again is controlled by DNA methylation (Kim et al. 2007). T cells with a high expression of the IL-7 receptor α-chain exhibit decreased methylation of the IL-7 receptor promoter compared to T cells that have a lower expression. Furthermore, DNMT1 inhibition results in increased receptor expression.

In addition to immune cell homeostasis, gene expression at the individual cell level appears to contribute to defects in immune aging. The best validated examples are the loss of costimulatory molecules and the gain in MHC class I-recognizing negative regulatory receptors (Fig. 1). Loss of CD28 expression is very common in CD8 T cells and has occurred in more than half of all CD8 T cells by the age of 70. The mechanism appears to be related to the loss of DNA-binding factors that bind to the CD28 initiator region and that are negatively regulated by cytokines, such as IL-15 and TNF-α (Bryl et al. 2001; Chiu et al. 2006). In contrast and as already described above, epigenetic control mechanisms, in particular DNA demethylation, are responsible for the KIR expression on CD8 and to a lesser degree CD4 T cells with age. KIR are not the only negative regulatory receptor

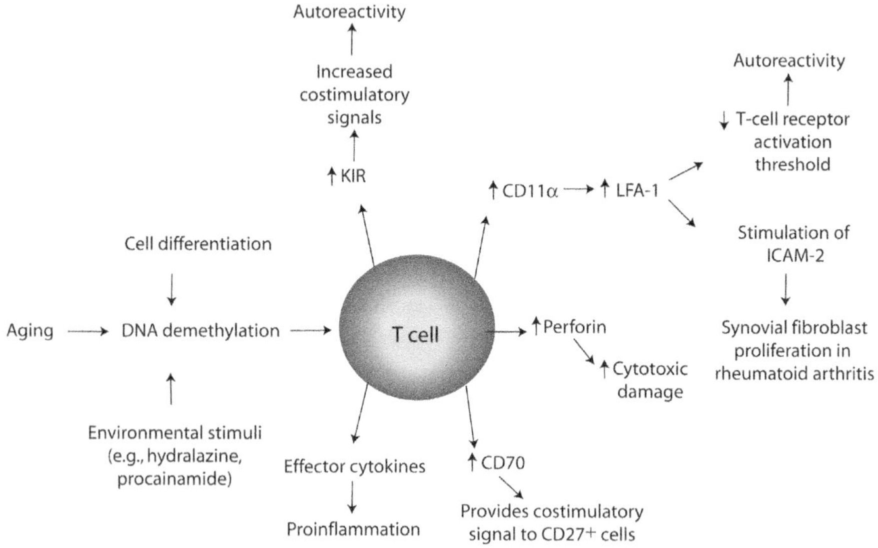

Fig. 1 Age-related expression of negative regulatory receptors on T cells. With increasing age, human T cells lose the main costimulatory molecule CD28 and gain the expression of regulatory receptors such as killer immunoglobulin-like receptors (KIR) that recognize MHC class I molecules. Most of these regulatory receptors deliver a negative signal and block the T-cell receptor-initiated activation events, dampening immune responses with age. The expression of these regulatory receptors is epigenetically controlled

that is found on aging CD8 T cells. Other examples include KLRG-1 and CD85j or ILT-2 (Voehringer et al. 2002; Abedin et al. 2005). In particular, CD85j is highly frequent in the elderly and involves not only the CD8 effector T-cell population but also central memory cells (Czesnikiewicz-Guzik et al. 2008). The genetic control mechanisms are only beginning to be studied; however, preliminary data indicate that the expression of CD85j can also be induced by inhibition of DNMT1 (unpublished data).

All of these negative regulatory receptors bind MHC class I determinants that are ubiquitously expressed on antigen-presenting cells and most somatic cells. Why CD8 and some CD4 T cells express these receptors with age is unknown; however, the net effect is certainly a dampening of the T-cell response. It has been suggested that these receptors are responsible for paralyzing CD8 T-cell responses to tumor antigens. Studies in CD4 T cells have shown that the involvement of these receptors in T-cell signaling is complex, and their negative effect depends on the affinity of T-cell recognition (Henel et al. 2006). In contrast to NK cells, they do not fully suppress activation of T cells, but selectively modify their effector functions. As an example, cytotoxic activity is not affected by these negative regulatory receptors while the induction of IFN-γ transcription that requires prolonged T-cell receptor signaling is clearly suppressed.

A key cytokine that appears to be affected by aging is IL-2. In mice older than 12–18 months, the production of IL-2 is compromised, and this defect is central to their immune defect. IL-2 is produced by non-committed naïve CD4 T cells and also by central memory cells. IL-2 production is central to immune function because it is involved in downstream effector function, in particular, the expression of CD40 ligand necessary for dendritic cell stimulation as well as for providing help in B-cell responses (Haynes and Eaton 2005). Suppression of IL-2 production is characteristic of anergic T cells. IL-2 production is strictly dependent upon costimulatory signals by CD28. T-cell activation through stimulation of the T-cell receptor in conjunction with CD28 induces the demethylation of CpG sites at the IL-2 promoter enhancer region (Bruniquel and Schwartz 2003). Methylation of the site abrogates the binding of the transcription factor OCT-1 and is sufficient to repress IL-2 transcription (Murayama et al. 2006). In addition to DNA demethylation, CD28 costimulation also promotes histone acetylation and gene accessibility (Thomas et al. 2005). Silencing of the IL-2 gene in anergic T cells appears to involve binding of the transcription factor Ikaros and recruitment of histone deacetylases (Bandyopadhyay et al. 2007). These epigenetic mechanisms controlling IL-2 transcription have been well substantiated to play a central role in anergy induction. Whether similar mechanisms act in T-cell aging remains to be studied.

DNA Demethylation and Autoimmunity in Systemic Lupus Erythematosus (SLE) and Rheumatoid Arthritis

Epigenetic mechanisms are essential for normal development and function of the immune system. It is not surprising that failure in epigenetic control can be associated with aberrant gene expression contributing to autoimmunity (Fig. 2). This concept was first explored by Richardson and colleagues in SLE (Richardson et al. 1990; Deng et al. 2001). SLE is a systemic autoimmune disease characterized by wide production of autoantibodies and in particular antibodies to a diverse set of nuclear antigens. It has a genetic predisposition that is incomplete with a low concordance rate in monozygotic twins, stressing exogenous factors or chance events in the development of the disease. SLE occurs predominantly in females during early adulthood; however, the incidence of SLE in men increases with age (Rovensky and Tuchynova 2008). SLE exists as a variant in which the exogenous factors have been clearly defined. This variant of drug-induced lupus can be induced by various medications including procainamide and hydralazine. In patients taking these medications, antinuclear and in particular anti-histone antibodies are induced in the majority of individuals, while overt disease only occurs in a small subset. Both of these medications also inhibit DNA methylation; procainamide is a direct inhibitor of DNMT1, while hydralazine can inhibit the upregulation of DNMT1 and 3a in T cells and B cells, presumably due to an inhibition of the PKCδ and the subsequent ERK signaling pathways (Deng et al. 2003; Lee et al. 2005). This realization raised the possibility that DNA demethylation is involved in the pathogenesis of idiopathic SLE.

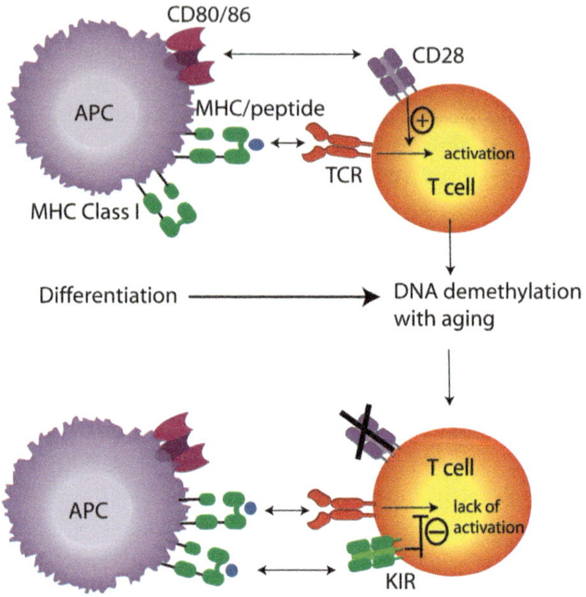

Fig. 2 Epigenetic control of genes involved in autoimmunity. Cell differentiation, aging, and environmental stimuli shape the gene expression profile in T cells by influencing DNA methylation. Many of these expressed genes are involved in autoimmune diseases. The graph lists a selection of genes that have been shown to be demethylation sensitive and their putative mechanisms in autoimmunity

DNMT1 inhibition studies in vitro have characterized several genes whose expression is demethylation sensitive and that are important in autoimmunity; some of these genes have already been discussed. The first prime candidate was the CD11a chain of LFA-1. The overexpression of LFA-1 facilitates low affinity immune responses by stabilizing the immunological synapse. In adoptive transfer studies, T cells treated with the demethylation inhibitor caused an autoimmune syndrome with SLE-like features including the production of autoantibodies. A similar disease was seen when LFA-1 was overexpressed in CD4 T cells that were adoptively transferred, suggesting that the LFA-1 overexpression is a major mechanism for how DNA-demethylated CD4 T cells cause autoimmune disease (Yung et al. 1996). Additional genes that are DNA demethylation sensitive and contribute to the polyclonal B-cell activation seen in SLE are CD70 and CD40 ligands. Co-culture of demethylated CD4 T cells with B cells induces polyclonal antibody production that can be reversed with anti-CD70 antibodies (Oelke et al. 2004). In the case of CD40 ligand, DNMT inhibition increases expression only in women, but not in males, presumably because CD40 ligand is encoded on the X chromosome (Lu et al. 2007).

These experimental findings are paralleled by observations in patients with primary idiopathic SLE. These patients have a decreased global deoxymethylcytosine content which may be caused by a decreased expression of DNMT1 in T cells during active disease. Several of the genes that have been found in the in vitro studies have also been implicated, in particular, CD70, LFA-1, CD40 ligand, and perforin. These genes are generally overexpressed in patients with SLE and examining the methylation of the relevant promoters yielded similar demethylation patterns as were found in the in vitro studies (Balada et al. 2007; Sekigawa et al. 2006).

As already discussed, these genes have also been identified in gene expression studies of immune aging and have been implicated in autoimmune diseases other than SLE that are more frequent in the elderly. LFA-1 is overexpressed in CD28-negative T cells that have been identified as risk factors in a number of autoimmune diseases including rheumatoid arthritis and Wegener's granulomatosis (Komocsi et al. 2002; Martens et al. 1997). In addition to the effect of LFA-1 on T-cell receptor threshold tuning described above, it also has direct effects on the target cells expressing its ligand, ICAM. In rheumatoid arthritis, synovial fibroblast proliferation is one of the clinical hallmarks leading to erosive disease. T cells provide a signal to synovial fibroblasts which protects them from undergoing apoptosis and stimulates their proliferation. This function of the T cells is mediated by LFA-1 that binds to ICAM-2 on synovial fibroblasts. ICAM-2 stimulation by LFA-1 activates the ezrine–Akt pathway in the synovial fibroblast, eventually leading to synovial hyperplasia (Singh et al. 2008). CD28-negative T cells that have higher LFA-1 expression and are more frequent in rheumatoid arthritis have a more pronounced effect on synoviocyte proliferation than normal T cells.

Similarly, CD28-negative T cells also upregulate the expression of the demethylation-sensitive genes perforin and CD70. In addition to typical autoimmune diseases, the increased cytotoxic activity of these cells is important in a number of inflammatory diseases, such as the inflammation of the atherosclerotic plaque that causes plaque instability. Cytotoxic activity of plaque-infiltrating T cells contributes to smooth muscle cell and endothelial cell death and, therefore, initiates plaque instability (Nakajima et al. 2002). CD70, in addition to being a stimulatory molecule for B-cell activation, also functions as a costimulatory ligand for CD27-expressing T cells (Borst et al. 2005). CD70 is overexpressed in the aging host on CD4 and CD8 cells and is also expressed in CD28-negative T cells. The mechanism of this CD70 expression in the elderly involves demethylation of a different region than in SLE. Sustained CD70 expression provides costimulatory signals to CD27-expressing naïve T cells and, therefore, is able to initiate T-cell responses to low affinity or autoantigens (Lee et al. 2007; Arens et al. 2004). In a murine model, overexpression of CD70 eventually completely exhausted the T-cell repertoire and led to lymphopenia and immune defect (Tesselaar et al. 2003).

Finally, expression of stimulatory KIR genes has also been implicated in autoimmune diseases including rheumatoid vasculitis, type I diabetes mellitus, and psoriatic arthritis, as well as in tissue-injurious inflammatory responses such as atherosclerosis (Parham 2005; van der Slik et al. 2003). Because expression of the different KIR variants appears to be stochastic, selective demethylation in the KIR region either favors an overexpression of inhibitory or of stimulatory KIRs. A predominance of inhibitory KIR is seen in normal NK cell responses and leads to an immune defect if expression occurs on T cells during immune aging. In contrast, a predominance of stimulatory receptors favors autoimmunity. An overexpression of stimulatory variants was, indeed, seen in patients with rheumatoid vasculitis, as well as in patients with coronary artery disease and plaque instability (Nakajima et al. 2002; Yen et al. 2001). In further support of this hypothesis, stimulatory variants of the KIR genes have been found as disease-risk genes for several autoimmune syndromes.

Synopsis

DNA demethylation plays a central role in the epigenetic regulation of the developmental stage- and cell-specific gene expression in adaptive immune responses. Genes that have been found to be epigenetically controlled and that are important during T-cell differentiation have also been found to be differentially expressed during the aging process. This overexpression is, in part, due to the accumulation of end-differentiated cells, but also represents a consequence of cumulative DNA demethylation defects with age. DNA demethylation plays a central role in autoimmune diseases, such as SLE, even in the young host. Similar epigenetic changes leading to transcriptional activation are also observed in the elderly where they contribute to the immune defects that are frequently seen in this population. These immune defects not only impair immune competence but also account for the increased frequency of autoimmunity in the elderly ranging from asymptomatic autoantibody production to diseases such as rheumatoid arthritis.

References

Abedin, S., Michel, J.J., Lemster, B., and Vallejo, A.N. 2005. Diversity of NKR expression in aging T cells and in T cells of the aged: the new frontier into the exploration of protective immunity in the elderly. *Exp Gerontol* 40:537–548.

Akbar, A.N., Vukmanovic-Stejic, M., Taams, L.S., and Macallan, D.C. 2007. The dynamic co-evolution of memory and regulatory CD4+ T cells in the periphery. *Nat Rev Immunol* 7: 231–237.

Akimzhanov, A.M., Yang, X.O., and Dong, C. 2007. Chromatin remodeling of interleukin-17 (IL-17)-IL-17F cytokine gene locus during inflammatory helper T cell differentiation. *J Biol Chem* 282:5969–5972.

Arens, R., Schepers, K., Nolte, M.A., van Oosterwijk, M.F., van Lier, R.A., Schumacher, T.N., and van Oers, M.H. 2004. Tumor rejection induced by CD70-mediated quantitative and qualitative effects on effector CD8+ T cell formation. *J Exp Med* 199:1595–1605.

Aw, D., Silva, A.B., and Palmer, D.B. 2007. Immunosenescence: emerging challenges for an ageing population. *Immunology* 120:435–446.

Balada, E., Ordi-Ros, J., and Vilardell-Tarres, M. 2007. DNA methylation and systemic lupus erythematosus. *Ann NY Acad Sci* 1108:127–136.

Bandyopadhyay, S., Dure, M., Paroder, M., Soto-Nieves, N., Puga, I., and Macian, F. 2007. Interleukin 2 gene transcription is regulated by Ikaros-induced changes in histone acetylation in anergic T cells. *Blood* 109:2878–2886.

Bettelli, E., Carrier, Y., Gao, W., Korn, T., Strom, T.B., Oukka, M., Weiner, H.L., and Kuchroo, V.K. 2006. Reciprocal developmental pathways for the generation of pathogenic effector TH17 and regulatory T cells. *Nature* 441:235–238.

Bettelli, E., Korn, T., Oukka, M., and Kuchroo, V.K. 2008. Induction and effector functions of T(H)17 cells. *Nature* 453:1051–1057.

Borst, J., Hendriks, J., and Xiao, Y. 2005. CD27 and CD70 in T cell and B cell activation. *Curr Opin Immunol* 17:275–281.

Boyman, O., Purton, J.F., Surh, C.D., and Sprent, J. 2007. Cytokines and T-cell homeostasis. *Curr Opin Immunol* 19:320–326.
Bruniquel, D., and Schwartz, R.H. 2003. Selective, stable demethylation of the interleukin-2 gene enhances transcription by an active process. *Nat Immunol* 4:235–240.
Bryl, E., Vallejo, A.N., Weyand, C.M., and Goronzy, J.J. 2001. Down-regulation of CD28 expression by TNF-alpha. *J Immunol* 167:3231–3238.
Catalfamo, M., and Henkart, P.A. 2003. Perforin and the granule exocytosis cytotoxicity pathway. *Curr Opin Immunol* 15:522–527.
Chan, H.W., Kurago, Z.B., Stewart, C.A., Wilson, M.J., Martin, M.P., Mace, B.E., Carrington, M., Trowsdale, J., and Lutz, C.T. 2003. DNA methylation maintains allele-specific KIR gene expression in human natural killer cells. *J Exp Med* 197:245–255.
Chiu, W.K., Fann, M., and Weng, N.P. 2006. Generation and growth of CD28nullCD8+ memory T cells mediated by IL-15 and its induced cytokines. *J Immunol* 177:7802–7810.
Czesnikiewicz-Guzik, M., Lee, W.W., Cui, D., Hiruma, Y., Lamar, D.L., Yang, Z.Z., Ouslander, J.G., Weyand, C.M., and Goronzy, J.J. 2008. T cell subset-specific susceptibility to aging. *Clin Immunol* 127:107–118.
Deng, C., Kaplan, M.J., Yang, J., Ray, D., Zhang, Z., McCune, W.J., Hanash, S.M., and Richardson, B.C. 2001. Decreased Ras-mitogen-activated protein kinase signaling may cause DNA hypomethylation in T lymphocytes from lupus patients. *Arthritis Rheum* 44:397–407.
Deng, C., Lu, Q., Zhang, Z., Rao, T., Attwood, J., Yung, R., and Richardson, B. 2003. Hydralazine may induce autoimmunity by inhibiting extracellular signal-regulated kinase pathway signaling. *Arthritis Rheum* 48:746–756.
Dong, C. 2008. TH17 cells in development: an updated view of their molecular identity and genetic programming. *Nat Rev Immunol* 8:337–348.
Dustin, M.L., Bivona, T.G., and Philips, M.R. 2004. Membranes as messengers in T cell adhesion signaling. *Nat Immunol* 5:363–372.
Fitzpatrick, D.R., Shirley, K.M., McDonald, L.E., Bielefeldt-Ohmann, H., Kay, G.F., and Kelso, A. 1998. Distinct methylation of the interferon gamma (IFN-gamma) and interleukin 3 (IL-3) genes in newly activated primary CD8+ T lymphocytes: regional IFN-gamma promoter demethylation and mRNA expression are heritable in CD44(high)CD8+ T cells. *J Exp Med* 188:103–117.
Glimcher, L.H., and Murphy, K.M. 2000. Lineage commitment in the immune system: the T helper lymphocyte grows up. *Genes Dev* 14:1693–1711.
Golbus, J., Palella, T.D., and Richardson, B.C. 1990. Quantitative changes in T cell DNA methylation occur during differentiation and ageing. *Eur J Immunol* 20:1869–1872.
Goronzy, J.J., and Weyand, C.M. 2005. T cell development and receptor diversity during aging. *Curr Opin Immunol* 17:468–475.
Goronzy, J.J., Fujii, H., and Weyand, C.M. 2006. Telomeres, immune aging and autoimmunity. *Exp Gerontol* 41:246–251.
Gruver, A.L., Hudson, L.L., and Sempowski, G.D. 2007. Immunosenescence of ageing. *J Pathol* 211:144–156.
Haynes, L., and Eaton, S.M. 2005. The effect of age on the cognate function of CD4+ T cells. *Immunol Rev* 205:220–228.
Henel, G., Singh, K., Cui, D., Pryshchep, S., Lee, W.W., Weyand, C.M., and Goronzy, J.J. 2006. Uncoupling of T-cell effector functions by inhibitory killer immunoglobulin-like receptors. *Blood* 107:4449–4457.
Hodes, R.J., Hathcock, K.S., and Weng, N.P. 2002. Telomeres in T and B cells. *Nat Rev Immunol* 2:699–706.
Humphrey, M.B., Lanier, L.L., and Nakamura, M.C. 2005. Role of ITAM-containing adapter proteins and their receptors in the immune system and bone. *Immunol Rev* 208:50–65.
Hutchins, A.S., Mullen, A.C., Lee, H.W., Sykes, K.J., High, F.A., Hendrich, B.D., Bird, A.P., and Reiner, S.L. 2002. Gene silencing quantitatively controls the function of a developmental transactivator. *Mol Cell* 10:81–91.

Ivanov, II, McKenzie, B.S., Zhou, L., Tadokoro, C.E., Lepelley, A., Lafaille, J.J., Cua, D.J., and Littman, D.R. 2006. The orphan nuclear receptor RORgammat directs the differentiation program of proinflammatory IL-17+ T helper cells. *Cell* 126:1121–1133.

Kaplan, M.J., Deng, C., Yang, J., and Richardson, B.C. 2000. DNA methylation in the regulation of T cell LFA-1 expression. *Immunol Invest* 29:411–425.

Kim, H.R., Hwang, K.A., Kim, K.C., and Kang, I. 2007. Down-regulation of IL-7Ralpha expression in human T cells via DNA methylation. *J Immunol* 178:5473–5479.

Kim, S.T., Fields, P.E., and Flavell, R.A. 2007. Demethylation of a specific hypersensitive site in the Th2 locus control region. *Proc Natl Acad Sci USA* 104:17052–17057.

Komocsi, A., Lamprecht, P., Csernok, E., Mueller, A., Holl-Ulrich, K., Seitzer, U., Moosig, F., Schnabel, A., and Gross, W.L. 2002. Peripheral blood and granuloma CD4(+)CD28(-) T cells are a major source of interferon-gamma and tumor necrosis factor-alpha in Wegener's granulomatosis. *Am J Pathol* 160:1717–1724.

Korn, T., Oukka, M., Kuchroo, V., and Bettelli, E. 2007. Th17 cells: effector T cells with inflammatory properties. *Semin Immunol* 19:362–371.

Lee, B.H., Yegnasubramanian, S., Lin, X., and Nelson, W.G. 2005. Procainamide is a specific inhibitor of DNA methyltransferase 1. *J Biol Chem* 280:40749–40756.

Lee, D.U., Agarwal, S., and Rao, A. 2002. Th2 lineage commitment and efficient IL-4 production involves extended demethylation of the IL-4 gene. *Immunity* 16:649–660.

Lee, D.U., Avni, O., Chen, L., and Rao, A. 2004. A distal enhancer in the interferon-gamma (IFN-gamma) locus revealed by genome sequence comparison. *J Biol Chem* 279:4802–4810.

Lee, G.R., Fields, P.E., Griffin, T.J., and Flavell, R.A. 2003. Regulation of the Th2 cytokine locus by a locus control region. *Immunity* 19:145–153.

Lee, P.P., Fitzpatrick, D.R., Beard, C., Jessup, H.K., Lehar, S., Makar, K.W., Perez-Melgosa, M., Sweetser, M.T., Schlissel, M.S., Nguyen, S., et al. 2001. A critical role for Dnmt1 and DNA methylation in T cell development, function, and survival. *Immunity* 15:763–774.

Lee, W.W., Yang, Z.Z., Li, G., Weyand, C.M., and Goronzy, J.J. 2007. Unchecked CD70 expression on T cells lowers threshold for T cell activation in rheumatoid arthritis. *J Immunol* 179: 2609–2615.

Li, G., Weyand, C.M., and Goronzy, J.J. 2008. Epigenetic mechanisms of age-dependent KIR2DL4 expression in T cells. *J Leukoc Biol* 84:824–834.

Long, E.O., Barber, D.F., Burshtyn, D.N., Faure, M., Peterson, M., Rajagopalan, S., Renard, V., Sandusky, M., Stebbins, C.C., Wagtmann, N., et al. 2001. Inhibition of natural killer cell activation signals by killer cell immunoglobulin-like receptors (CD158). *Immunol Rev* 181:223–233.

Lu, Q., Wu, A., Ray, D., Deng, C., Attwood, J., Hanash, S., Pipkin, M., Lichtenheld, M., and Richardson, B. 2003. DNA methylation and chromatin structure regulate T cell perforin gene expression. *J Immunol* 170:5124–5132.

Lu, Q., Wu, A., Tesmer, L., Ray, D., Yousif, N., and Richardson, B. 2007. Demethylation of CD40LG on the inactive X in T cells from women with lupus. *J Immunol* 179:6352–6358.

Macallan, D.C., Wallace, D., Zhang, Y., De Lara, C., Worth, A.T., Ghattas, H., Griffin, G.E., Beverley, P.C., and Tough, D.F. 2004. Rapid turnover of effector-memory CD4(+) T cells in healthy humans. *J Exp Med* 200:255–260.

Makar, K.W., and Wilson, C.B. 2004. DNA methylation is a nonredundant repressor of the Th2 effector program. *J Immunol* 173:4402–4406.

Martens, P.B., Goronzy, J.J., Schaid, D., and Weyand, C.M. 1997. Expansion of unusual CD4+ T cells in severe rheumatoid arthritis. *Arthritis Rheum* 40:1106–1114.

Martin, M.P., and Carrington, M. 2005. Immunogenetics of viral infections. *Curr Opin Immunol* 17:510–516.

Mosmann, T.R., and Coffman, R.L. 1989. TH1 and TH2 cells: different patterns of lymphokine secretion lead to different functional properties. *Annu Rev Immunol* 7:145–173.

Mullen, A.C., Hutchins, A.S., High, F.A., Lee, H.W., Sykes, K.J., Chodosh, L.A., and Reiner, S.L. 2002. Hlx is induced by and genetically interacts with T-bet to promote heritable T(H)1 gene induction. *Nat Immunol* 3:652–658.

Murayama, A., Sakura, K., Nakama, M., Yasuzawa-Tanaka, K., Fujita, E., Tateishi, Y., Wang, Y., Ushijima, T., Baba, T., Shibuya, K., et al. 2006. A specific CpG site demethylation in the human interleukin 2 gene promoter is an epigenetic memory. *EMBO J* 25:1081–1092.

Nakajima, T., Schulte, S., Warrington, K.J., Kopecky, S.L., Frye, R.L., Goronzy, J.J., and Weyand, C.M. 2002. T-cell-mediated lysis of endothelial cells in acute coronary syndromes. *Circulation* 105:570–575.

Namekawa, T., Snyder, M.R., Yen, J.H., Goehring, B.E., Leibson, P.J., Weyand, C.M., and Goronzy, J.J. 2000. Killer cell activating receptors function as costimulatory molecules on CD4+CD28null T cells clonally expanded in rheumatoid arthritis. *J Immunol* 165:1138–1145.

Naylor, K., Li, G., Vallejo, A.N., Lee, W.W., Koetz, K., Bryl, E., Witkowski, J., Fulbright, J., Weyand, C.M., and Goronzy, J.J. 2005. The influence of age on T cell generation and TCR diversity. *J Immunol* 174:7446–7452.

Oelke, K., Lu, Q., Richardson, D., Wu, A., Deng, C., Hanash, S., and Richardson, B. 2004. Overexpression of CD70 and overstimulation of IgG synthesis by lupus T cells and T cells treated with DNA methylation inhibitors. *Arthritis Rheum* 50:1850–1860.

Ouyang, W., Kolls, J.K., and Zheng, Y. 2008. The biological functions of T helper 17 cell effector cytokines in inflammation. *Immunity* 28:454–467.

Pai, S.Y., Truitt, M.L., and Ho, I.C. 2004. GATA-3 deficiency abrogates the development and maintenance of T helper type 2 cells. *Proc Natl Acad Sci USA* 101:1993–1998.

Parham, P. 2005. MHC class I molecules and KIRs in human history, health and survival. *Nat Rev Immunol* 5:201–214.

Pawelec, G., and Larbi, A. 2008. Immunity and ageing in man: Annual Review 2006/2007. *Exp Gerontol* 43:34–38.

Reiner, S.L. 2005. Epigenetic control in the immune response. *Hum Mol Genet* 14 Spec No 1: R41–46.

Reiner, S.L. 2007. Development in motion: helper T cells at work. *Cell* 129:33–36.

Renaud, S., Loukinov, D., Abdullaev, Z., Guilleret, I., Bosman, F.T., Lobanenkov, V., and Benhattar, J. 2007. Dual role of DNA methylation inside and outside of CTCF-binding regions in the transcriptional regulation of the telomerase hTERT gene. *Nucleic Acids Res* 35: 1245–1256.

Richardson, B., Scheinbart, L., Strahler, J., Gross, L., Hanash, S., and Johnson, M. 1990. Evidence for impaired T cell DNA methylation in systemic lupus erythematosus and rheumatoid arthritis. *Arthritis Rheum* 33:1665–1673.

Rovensky, J., and Tuchynova, A. 2008. Systemic lupus erythematosus in the elderly. *Autoimmun Rev* 7:235–239.

Russell, J.H., and Ley, T.J. 2002. Lymphocyte-mediated cytotoxicity. *Annu Rev Immunol* 20: 323–370.

Santourlidis, S., Graffmann, N., Christ, J., and Uhrberg, M. 2008. Lineage-specific transition of histone signatures in the killer cell Ig-like receptor locus from hematopoietic progenitor to NK cells. *J Immunol* 180:418–425.

Santourlidis, S., Trompeter, H.I., Weinhold, S., Eisermann, B., Meyer, K.L., Wernet, P., and Uhrberg, M. 2002. Crucial role of DNA methylation in determination of clonally distributed killer cell Ig-like receptor expression patterns in NK cells. *J Immunol* 169:4253–4261.

Sawalha, A.H. 2008. Epigenetics and T-cell immunity. *Autoimmunity* 41:245–252.

Sekigawa, I., Kawasaki, M., Ogasawara, H., Kaneda, K., Kaneko, H., Takasaki, Y., and Ogawa, H. 2006. DNA methylation: its contribution to systemic lupus erythematosus. *Clin Exp Med* 6:99–106.

Shnyreva, M., Weaver, W.M., Blanchette, M., Taylor, S.L., Tompa, M., Fitzpatrick, D.R., and Wilson, C.B. 2004. Evolutionarily conserved sequence elements that positively regulate IFN-gamma expression in T cells. *Proc Natl Acad Sci USA* 101:12622–12627.

Signer, R.A., Montecino-Rodriguez, E., Witte, O.N., McLaughlin, J., and Dorshkind, K. 2007. Age-related defects in B lymphopoiesis underlie the myeloid dominance of adult leukemia. *Blood* 110:1831–1839.

Sims, T.N., and Dustin, M.L. 2002. The immunological synapse: integrins take the stage. *Immunol Rev* 186:100–117.

Singh, K., Colmegna, I., He, X., Weyand, C.M., and Goronzy, J.J. 2008. Synoviocyte stimulation by the LFA-1-intercellular adhesion molecule-2-Ezrin-Akt pathway in rheumatoid arthritis. *J Immunol* 180:1971–1978.

Snyder, M.R., Muegge, L.O., Offord, C., O'Fallon, W.M., Bajzer, Z., Weyand, C.M., and Goronzy, J.J. 2002. Formation of the killer Ig-like receptor repertoire on CD4+CD28null T cells. *J Immunol* 168:3839–3846.

Spilianakis, C.G., and Flavell, R.A. 2004. Long-range intrachromosomal interactions in the T helper type 2 cytokine locus. *Nat Immunol* 5:1017–1027.

Tesselaar, K., Arens, R., van Schijndel, G.M., Baars, P.A., van der Valk, M.A., Borst, J., van Oers, M.H., and van Lier, R.A. 2003. Lethal T cell immunodeficiency induced by chronic costimulation via CD27-CD70 interactions. *Nat Immunol* 4:49–54.

Thomas, R.M., Gao, L., and Wells, A.D. 2005. Signals from CD28 induce stable epigenetic modification of the IL-2 promoter. *J Immunol* 174:4639–4646.

van Bergen, J., Thompson, A., van der Slik, A., Ottenhoff, T.H., Gussekloo, J., and Koning, F. 2004. Phenotypic and functional characterization of CD4 T cells expressing killer Ig-like receptors. *J Immunol* 173:6719–6726.

van der Slik, A.R., Koeleman, B.P., Verduijn, W., Bruining, G.J., Roep, B.O., and Giphart, M.J. 2003. KIR in type 1 diabetes: disparate distribution of activating and inhibitory natural killer cell receptors in patients versus HLA-matched control subjects. *Diabetes* 52:2639–2642.

Veldhoen, M., Hocking, R.J., Atkins, C.J., Locksley, R.M., and Stockinger, B. 2006. TGFbeta in the context of an inflammatory cytokine milieu supports de novo differentiation of IL-17-producing T cells. *Immunity* 24:179–189.

Voehringer, D., Koschella, M., and Pircher, H. 2002. Lack of proliferative capacity of human effector and memory T cells expressing killer cell lectinlike receptor G1 (KLRG1). *Blood* 100: 3698–3702.

Warrington, K.J., Takemura, S., Goronzy, J.J., and Weyand, C.M. 2001. CD4+,CD28- T cells in rheumatoid arthritis patients combine features of the innate and adaptive immune systems. *Arthritis Rheum* 44:13–20.

Wilson, C.B., and Merkenschlager, M. 2006. Chromatin structure and gene regulation in T cell development and function. *Curr Opin Immunol* 18:143–151.

Wilson, C.B., Makar, K.W., Shnyreva, M., and Fitzpatrick, D.R. 2005. DNA methylation and the expanding epigenetics of T cell lineage commitment. *Semin Immunol* 17:105–119.

Xu, J., Vallejo, A.N., Jiang, Y., Weyand, C.M., and Goronzy, J.J. 2005. Distinct transcriptional control mechanisms of killer immunoglobulin-like receptors in natural killer (NK) and in T cells. *J Biol Chem* 280:24277–24285.

Yang, X.O., Pappu, B.P., Nurieva, R., Akimzhanov, A., Kang, H.S., Chung, Y., Ma, L., Shah, B., Panopoulos, A.D., Schluns, K.S., et al. 2008. T helper 17 lineage differentiation is programmed by orphan nuclear receptors ROR alpha and ROR gamma. *Immunity* 28:29–39.

Yen, J.H., Moore, B.E., Nakajima, T., Scholl, D., Schaid, D.J., Weyand, C.M., and Goronzy, J.J. 2001. Major histocompatibility complex class I-recognizing receptors are disease risk genes in rheumatoid arthritis. *J Exp Med* 193:1159–1167.

Yung, R., Powers, D., Johnson, K., Amento, E., Carr, D., Laing, T., Yang, J., Chang, S., Hemati, N., and Richardson, B. 1996. Mechanisms of drug-induced lupus. II. T cells overexpressing lymphocyte function-associated antigen 1 become autoreactive and cause a lupuslike disease in syngeneic mice. *J Clin Invest* 97:2866–2871.

Zhang, Z., Deng, C., Lu, Q., and Richardson, B. 2002. Age-dependent DNA methylation changes in the ITGAL (CD11a) promoter. *Mech Ageing Dev* 123:1257–1268.

Zheng, W., and Flavell, R.A. 1997. The transcription factor GATA-3 is necessary and sufficient for Th2 cytokine gene expression in CD4 T cells. *Cell* 89:587–596.

Zhu, J., Min, B., Hu-Li, J., Watson, C.J., Grinberg, A., Wang, Q., Killeen, N., Urban, J.F., Jr., Guo, L., and Paul, W.E. 2004. Conditional deletion of Gata3 shows its essential function in T(H)1-T(H)2 responses. *Nat Immunol* 5:1157–1165.

Epigenetic Silencing of Progeroid Syndromes

Ruben Agrelo

Abstract The risk of developing cancer increases with age, and the accumulation of epigenetic modifications has been found to be an important factor in tumorigenesis. A global loss of DNA methylation is detected both in aging and in cancer, and promoter hypermethylation is associated with the silencing of genes with roles both in tumor suppression and in progeria, such as the Werner syndrome (*WRN*) and lamin A/C (*LMNA*) genes. The epigenetic inactivation of these two genes in different tumors demonstrated for the first time that genes involved in the aging process could also be of crucial importance in the development of human cancer. In this review, I will explore the physiological significance of these epigenetic modifications during aging and the neoplastic process.

Keywords DNA methylation · Epigenetics · Aging · Cancer · Werner syndrome · Hutchinson Guilford Progeria Syndrome · Lamin A/C

Introduction

Epigenetic changes that take place during the neoplastic process in cells have been a focus of interest for more than a decade (Baylin et al. 2001; Hollyday 2006). Epigenetic changes, which include DNA methylation, chromatin modifications, and methylation, acetylation, ubiquitylation, and phosphorylation of histone tails (Esteller 2007), could also be important determinants of cellular senescence and organismal aging (Richardson 2002).

In this chapter, I will briefly discuss changes in DNA methylation occurring during aging and cancer and the mechanisms whereby aging-related genes could undergo epigenetic changes that contribute to tumorigenesis. I concentrate on recent findings concerning the epigenetic inactivation of two such genes in human cancer: the Werner syndrome gene, *WRN*, and the lamin A/C gene, *LMNA* (Agrelo et al.

R. Agrelo (✉)
Research Institute of Molecular Pathology (IMP), Dr. Bohr-Gasse 7, 1030 Vienna, Austria
e-mail: agrelo@imp.univie.ac.at, ragrelo@adinet.com.uy

T.O. Tollefsbol (ed.), *Epigenetics of Aging*, DOI 10.1007/978-1-4419-0639-7_19,

2005, 2006). I review the epigenetic modifications in cells derived from patients with Hutchinson–Guilford syndrome produced by *LMNA* mutations and present an epigenetic model that could account for the accumulation of progerin in cell normal aging (Scaffidi and Misteli 2005, 2006).

WRN promoter is hypermethylated in a wide variety of human neoplasms (Agrelo et al. 2006), while *LMNA* promoter hypermethylation is restricted to lymphomas and leukemias (Agrelo et al. 2005). This has prognostic and treatment implications for patients with cancer.

DNA Methylation: From Aging to Cancer

In normal cells, CpG methylation plays a role in maintaining the gene silencing that is necessary for a variety of functions including tissue- and development-specific gene expression, silencing the X chromosome, genomic imprinting, and protecting against the expression of intragenomic parasitic elements such as Alu and line sequences (Esteller 2003).

CpG dinucleotides (CpGs) are concentrated in CpG islands in the promoters of the genes and are generally unmethylated, while in areas outside the CpG islands CpGs are sparse and are generally methylated (m^5CpGs).

Cancer and aging are two epigenetically linked processes characterized by global genomic hypomethylation and focal hypermethylation of CpG islands.

The patterns of global DNA methylation are highly specific. After mitosis they are inherited with an adequate degree of fidelity through the action of the DNA methyltransferases (DNMTs). Early studies showed that there was a gradual loss of CpG methylation during the aging process; this was true for normal cells cultured in vitro with increasing passage number (Wilson and Jones 1983) and also during organism aging (Wilson et al. 1987). It has been shown that mouse, hamster, and human fibroblasts cultured to senescence gradually lose methylation (Wilson and Jones 1983).

This reduction in global methylation during aging could be the result of passive demethylation of repetitive DNA (Hornsby et al. 1992) and/or of the coding region and introns of genes (Esteller 2003).

The fact that expression of the maintenance methyltransferase, DNMT1, is down-regulated in aged cultured fibroblasts supports this possibility (Casillas et al. 2003). The question is, does DNA hypomethylation contribute to carcinogenesis and, if so, how?

First, DNA hypomethylation could lead to the loss of normal imprinting patterns, and studies in mice have shown that this loss may in itself be tumorigenic (Sakatani et al. 2005).

DNA hypomethylation might also allow the transcription of parasitic sequences, such as DNA viruses or transposon elements (these are intragenomic parasitic DNA, such as L1 [long interspersed nuclear elements, LINES] and Alu [recombinogenic sequence] repeats). These silent transposons could be transcribed or relocated to other genomic regions, where they could disrupt the function of cellular genes (Jones 1999).

Second, hypomethylation might also contribute to carcinogenesis via its ability to generate genomic instability. Hypomethylation could facilitate illegitimate mitotic recombination leading to chromosome breaks, translocations, or allelic loss (Hoffmann and Schulz 2005). In particular, loss of methylation in the pericentromeric regions of chromosomes may result in breakage of the regions and aneuploidy. As evidence for this, it has been reported that patients with inherited DNMT3b deficiency (ICF syndrome), which is associated with hypomethylation of the pericentromeric regions, have numerous chromosome aberrations (Ehrlich 2003). The old concept that demethylation of oncogenes leads to their activation is no longer widely accepted. Nevertheless, it has been suggested that this phenomenon is responsible for the upregulation of certain genes like *PAX2* (transcription factor) and oncogenic microRNAs (miRNAs) (Esteller 2008).

It has also been observed that expression of de novo DNA methyltransferase DNMT3b is upregulated in aging cultured fibroblasts. This upregulation might be a natural response to loss of methylation in repetitive DNA (Lopatina et al. 2002; Casillas et al. 2003). It is possible that DNMT3b upregulation could cause aberrant hypermethylation of regions such as the promoter CpG islands, which are unmethylated in normal cells. This has been reported for the *MLH1* and $p14^{ARF}$ genes in aged tissues (Nakagawa et al. 2001; Issa 2003; Shen et al. 2003).

Methylation of CpG islands in gene promoter regions is associated with aberrant silencing of transcription and is a mechanism that, together with mutation, inactivates tumor suppressor genes (Fig. 1A) (Esteller 2003). According to the Knudson hypothesis, the abrogation of the function of a tumor suppressor gene requires the loss of function of both copies of the gene (Fig. 1B) (Knudson 2001). The first hit could be due to the inactivation of one copy of the gene, either by germ line mutation (in familial cancers) or by somatic mutation (in sporadic tumors) within the coding region of the gene. The second hit could involve either the loss of the chromosomal region containing the second copy of the gene or loss of heterozygosity (LOH). As a consequence, both alleles of the gene are lost, resulting in neoplastic transformation of the cell. Abnormal promoter hypermethylation can also inactivate one copy of the gene. In familial cancers, abnormal promoter hypermethylation acts only as a second hit, but in sporadic cancers, both copies can be inactivated by this mechanism (Esteller et al. 2001). The importance of promoter hypermethylation has been demonstrated by the ability of demethylating drugs such as 5-azacytidine (5AZA) to reactivate the affected genes and to restore production of the corresponding protein (Herman et al. 1998).

There are many examples of tumor suppressor genes that are lost by epigenetic inactivation, including the *VHL* gene in renal cancer, the cell-cycle-control gene $p16^{Ink4a}$ in many types of cancer, and the mismatch repair (MMR) gene *hMLH1* in colorectal cancer (CRC) and other neoplasms (Esteller 2008).

In certain tumors, gene silencing caused by promoter hypermethylation occurs at a very high frequency. Systematic disruption of multiple suppressor genes has been found in a subset of colorectal cancers in which suppressor genes are targeted by promoter hypermethylation within the tumor (Toyota et al. 1999). This alteration is known as the CpG island methylator phenotype (CIMP). In colonic tissue, it is thought to be associated with a particular genotype that includes mutations of the

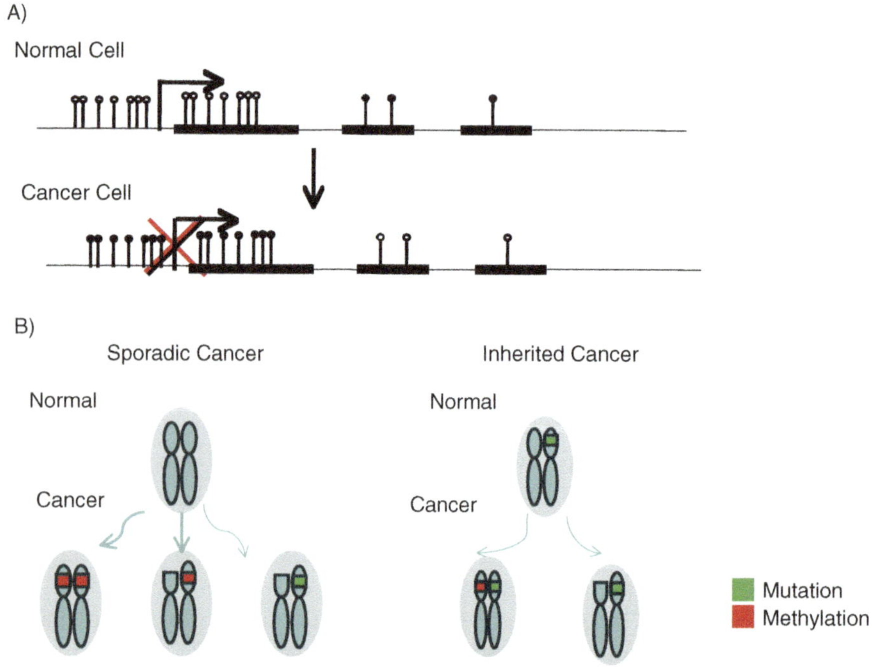

Fig. 1 (**A**) Graphical representation of a typical gene with a CpG island. (**a**) In normal cells, the CpG islands are devoid of methylation, allowing gene expression. (**b**) In cancer cells, the CpG island becomes hypermethylated, preventing gene transcription. (**B**) Genetic and epigenetic events that inactivate tumor suppressor genes, according to the Two-Hit Hypothesis. In sporadic cancers, promoter hypermethylation (*red*) or point mutations (*green*) could account for the first hit, while the second hit could be due to either chromosomal deletions (short arm of the blue chromosome) or promoter hypermethylation. In inherited cancers, with germ line mutations in one copy of the tumor suppressor gene, the second hit could be due to a chromosomal deletion or promoter hypermethylation

BRAF oncogene and microsatellite instability (MSI) due to promoter hypermethylation of *hMLH1* (Weisenberger et al. 2006).

hMLH1 is, therefore, a good example of a tumor suppressor gene that is methylated during the aging process, is involved in DNA repair, and generates MSI, which in a subset of colorectal cancers is the hallmark of CIMP: The DNA MMR system corrects DNA base-pairing errors in newly-replicated DNA (Jascur and Boland 2006). Deficiencies in this system result in high mutation rates (Thomas et al. 1996; Jiricny 2006), particularly in microsatellite sequences (consisting of repeats of 1–4 bp), causing microsatellite instability (MSI). MSI is therefore a hallmark of MMR gene-deficient cancers and has been observed in ~13% of sporadic CRC and in almost all CRC arising in patients with hereditary non-polyposis colorectal cancer (HNPCC) or Lynch syndrome (Aaltonen et al. 1993; Thibodeau et al. 1993). In patients with HNPCC, the defect is due to germ line mutations in DNA MMR genes, mainly *hMLH1* and *hMSH2* (Bocker et al. 1999). These same genes could be inactivated by mutations in MSI+ cancers in patients without HNPCC, but in the subset of these sporadic cancers where no mutation exists, the epigenetic inactivation of *hMLH1* by promoter hypermethylation accounts for MSI+. This association has been observed in colorectal (Kane et al. 1997; Herman et al. 1998), endometrial (Esteller et al. 1998), and gastric tumors (Fleisher et al. 1999), the three tumor types common in HNPCC patients, while *hMLH1* promoter hypermethylation is absent in other tumor types (Esteller et al. 1998). It has been reported that about 90% of MSI+ tumors are methylated at *hMLH1* (the remaining 10% probably have somatic mutations in either *hMLH1* or *hMSH2*) and that MSI– tumors are unmethylated in 95% of cases (Herman et al. 1998).

Because microsatellites are very often present in coding regions, the MSI exhibited by MMR-deficient tumors can be manifested as frameshift mutations that inactivate different genes, including genes that suppress tumor formation (Malkhosyan et al. 1996). As a result, we can consider the inactivation of tumor suppressor genes to be the final step in a malignant progression characterized by increased genomic instability that is triggered by the early epigenetic inactivation of *hMLH1*.

The evidence presented suggests that this accumulation of epigenetic alterations during the aging process could also contribute to neoplastic transformation, as it is well known that global DNA hypomethylation, aberrant promoter hypermethylation associated with CpG islands, and DNMT overexpression are key epigenetic alterations in cancer (Esteller 2003). This hypothesis, however, should be confirmed by future research.

Following the establishment of a link between DNA methylation and cancer, we decided to study the tight connection between aging and cancer from an epigenetic point of view. In order to achieve this objective, we concentrated our efforts on the study of two genes involved in premature aging, *LMNA* and the *WRN*.

Werner Syndrome: The Epigenetic Link Between Aging and Cancer

WS is a rare autosomal recessive disorder characterized by features indicative of accelerated aging (Epstein et al. 1966; Martin 1978). WS patients experience normal development until puberty when no growth spurt occurs, and they consequently have short stature as adults (Huang et al. 2006). Important features include bilateral ocular cataracts, type 2 diabetes mellitus, hypogonadism, and osteoporosis (Tollefsbol and Cohen 1984; Goto 1997; Martin 1999). Some features associated with the prematurely aged appearance of these patients are loss of hair, high-pitched voice, scleroderma-like skin changes, regional atrophy, and abnormal distribution of subcutaneous fat. The two main causes of death are myocardial infarction and neoplasia, which are reported in 44 and 39.5% of patients, respectively (Huang et al. 2006), typically at about 48 years of age. Although considered by some researchers as a segmental progeroid syndrome, as it recapitulates only some aspects of normal aging, the sequential appearance of clinical and biological deterioration in the body systems observed in WS suggests that it is more than this (Goto 1997). This view is supported by the finding that gene expression in WS closely resembles that of normal aging (Kyng et al. 2003). WS fibroblasts exhibit an aging phenotype in culture that is characterized by a reduced replicative life span (Martin et al. 1990). In addition, WS cells exhibit genetic instability, manifested by variegated translocated mosaicism (Salk et al. 1981) and increased mutation rates (Fokuchi et al. 1989). The chromosome 8p12 *WRN* gene was identified in 1996 by positional cloning and encodes a 162-kDa RecQ helicase protein (Yu et al. 1996). All of the disease-associated *WRN* mutations lead to truncations that lack the C-terminus and the nuclear localization signal (Matsumoto et al. 1997). The inability of WRN to be transported to the nucleus in this disease seems to be critical for its pathogenesis. Once in the nucleus, WRN localizes in the nucleoli and the nucleoplasm (von Kobbe and Bohr 2002; Opresko et al. 2003). However, some patients clinically diagnosed with WS do not have mutations in *WRN* and are classified as having atypical WS, and in a subset of these patients, novel mutations of *LMNA* gene have been described (Chen et al. 2003).

The WRN protein is a member of the RecQ helicase family that also includes BLM, RecQ4, RecQ5, and RecQ1 (Bachrati and Hickson 2003). Mutations in the first two are responsible for Bloom syndrome (Ellis and German 1996) and Rothmund–Thompson/RAPADILINO/Baller–Gerold syndrome, respectively (Kitao et al. 1998; Siitonen et al. 2003; Van Maldergem 2006), both of which are associated with chromosomal instability and cancer. No syndromes have yet been associated with mutations in RecQ5 and RecQ1.

WRN protein has DNA-dependent ATPase and 3′–5′ helicase and 3′–5′ exonuclease activities (Gray et al. 1997; Huang et al. 1998; Kamath Loeb et al. 1998). Remarkably, it is the only family member that possesses exonuclease activity (Shen and Loeb 2000). The substrate specificity of the WRN helicase in vitro includes several DNA replication, recombination, and repair intermediates such as Holliday junctions, forked duplexes, and also RNA–DNA structures (Brosh and Bohr 2002). Some interactions are functional, as the Ku heterodimer can stimulate WRN exonuclease (Cooper et al. 2000) and the telomere-binding protein, TRF2, stimulates WRN helicase (Opresko et al. 2002). p53 and BLM can also inhibit WRN exonuclease activity (Brosh et al. 2001).

The WRN protein is ubiquitously expressed. Individuals heterozygous for *WRN* show genetic instability in vivo and cell lines derived from these individuals have intermediate sensitivities to DNA-damaging agents such as DNA-crosslinking drugs and topoisomerase I inhibitors that are lethal to WS cells (Ogburn et al. 1997; Moser et al. 2000; Poot et al. 2001).

WRN has roles in several different pathways, such as DNA replication, recombination, DNA repair, p53-mediated pathways, and telomere metabolism (Opresko et al. 2003). An important in vivo function for WRN is in homology-dependent recombination repair (HDR), which can be used to repair DNA damage while suppressing gene loss or rearrangements (Prince et al. 2001; Saintigny et al. 2001). Another function during recovery from replication arrest has also been postulated: In mammalian cells, WRN seems to repair DNA strand breaks that arise from replication arrest, limiting cell death and genome instability (Pichierri et al. 2001; Prince et al. 2001; Saintigny et al. 2001). WRN also has an important role in the maintenance of telomere length and the suppression of telomere sister chromatid exchanges (Chang et al. 2004). Telomeres are G-rich sequences that need to be disassembled in order to be replicated and repaired. Failure of this process leads to the generation of telomeric DNA ends that can be detected as DNA double-strand breaks (DSB), initiating a damage response that can trigger cellular senescence. Accurate telomere metabolism is also important for the suppression of genetic instability and chromosome rearrangements (Crabbe et al. 2007). WRN could also be required to make telomeres accessible for replication and/or repair. Moreover, WRN has multiple roles in DNA repair, since it is also important for the repair of chromosomal double-strand break(s) (DSB) by homologous recombination (HR), non-homologous end-joining (NHEJ) and single-strand annealing (SSA), and the repair of single-strand breaks (SSB) by base excision repair (BER) (Bohr 2005).

These roles of WRN in different aspects of DNA metabolism highlight its importance in aging and cancer, through prevention of genome instability and senescence. A model has been proposed (Kudlow et al. 2007) whereby WS pathogenesis is due to a defect in DNA metabolism that produces genetic instability, mutagenesis, and cell loss that may in turn produce cell lineage- or tissue-specific defects. The outcomes could be as different as senescence or mutation-dependent neoplastic proliferation.

Epigenetic Silencing of Werner Syndrome in Human Cancer

Tumor suppressor genes can be divided into two types, caretakers and gatekeepers (Kinzler and Vogelstein 1997); the former control genomic stability, while the latter can prevent the growth of tumor cells. WRN could have dual functions as both a caretaker and a gatekeeper protein. This is common among proteins involved in DNA repair, e.g., the mismatch repair proteins MSH2, MSH6, or MLH1 that is often inactivated by promoter hypermethylation. *WRN* qualifies as a caretaker as it prevents excessive recombination from occurring. Loss of this function in homozygous mutant cells confers the hyper-recombination phenotype (Salk et al. 1981) and generates large deletions (Fokuchi et al. 1989) and other kinds of chromosomal aberrations via mitotic recombination that can eventually induce LOH at various loci. The causal relationship between LOH in tumor suppressor genes and the development of human cancers is well established.

The role of *WRN* as a gatekeeper is suggested by the attenuation of p53-mediated apoptosis associated with WRN deficiencies. Since p53-mediated apoptosis is supposed to be critical for tumor suppression, its attenuation could contribute to cancer predisposition (Spillare et al. 1999; Wang et al. 2001). The telomere connection may account, at least to some extent, for the caretaker function of WRN family helicases in maintaining genomic integrity and stability.

Because WS patients develop epithelial and mesenchymal tumors, which are one of the two major causes of death before the age of 48 (Oshima 2000), a tumor suppressor function for WRN protein has been proposed. This tumor suppressor gene role is also supported by a very high rate of LOH at the chromosomal *WRN* loci at 8p11.2–p12 in many tumor types, including colorectal and breast cancer (Chughtai et al. 1999; Armes et al. 2004). However, somatic mutations in WRN have not been described in sporadic neoplasms.

Recently, it has been reported that the *WRN* gene underwent epigenetic inactivation by CpG island promoter hypermethylation in a wide variety of tumors of both mesenchymal and epithelial origin, including those that are observed in WS patients (such as osteosarcomas, thyroid, and gastric tumors) (Yamamoto et al. 2003; Agrelo et al. 2006).

WS patients have a high incidence of malignant neoplasms (Epstein et al. 1966). The kind of neoplasms appearing in WS patients is different from that observed in people who do not have the syndrome. The ratio of mesenchymal:epithelial cancers is 1:1, as compared with 1:10 in the normal population (Chen and Oshima 2002; Hickson 2003). Thus, it is possible that the accelerated aging process in WS patients contributes to the higher incidence of tumors, but that the specific loss of the *WRN* gene confers susceptibility to a particular tumor type, as observed with other tumor suppressor genes, such as *hMLH1* or *BRCA1* (Fig. 2A).

Treating cancer cells in which the *WRN* promoter is hypermethylated by administering the demethylating drug 5-aza-2′-deoxycytidine restored the expression of WRN RNA transcript and protein. WRN is the only RecQ member that exhibits exonuclease activity, and treatment with 5-aza-2′-deoxycytidine leads to a significant increase in WRN-associated exonuclease activity (Agrelo et al. 2006).

Interestingly, the reintroduction of *WRN* into transformed cell lines lacking *WRN* as a consequence of hypermethylation results in a reduction in colony formation and a decrease in growth of tumor xenografts, giving support to the hypothesis of *WRN* as a tumor suppressor gene (Agrelo et al. 2006). Moreover, the analysis of a large panel of cancer cell lines and primary tumors shows that loss of *WRN* expression by CpG island promoter hypermethylation gene silencing is a general mechanism of genome instability, as WRN seems to maintain genomic stability by functioning at the interface between DNA replication and DNA repair (Fig. 2B).

Because of *WRN* hypermethylation, these cancer cells were highly sensitive to the actions of inhibitors of topoisomerase and to the actions of DNA-damaging agents, as is the case with WS cells. This has been shown for camptothecin (a topoisomerase I inhibitor) and mitomycin (an interstrand crosslinker) when the apoptotic rate was measured by flow cytometry and chromosomal breakage by cytogenetic analysis. Regarding apoptosis, both camptothecin and mitomycin C were optimal inductors of apoptosis in cells with a hypermethylated *WRN* promoter. The same phenomenon was observed for the cell line from a WS patient which was used as a positive control, and this is in agreement with previous reports. In contrast, cells unmethylated at the *WRN* promoter were more resistant to camptothecin and

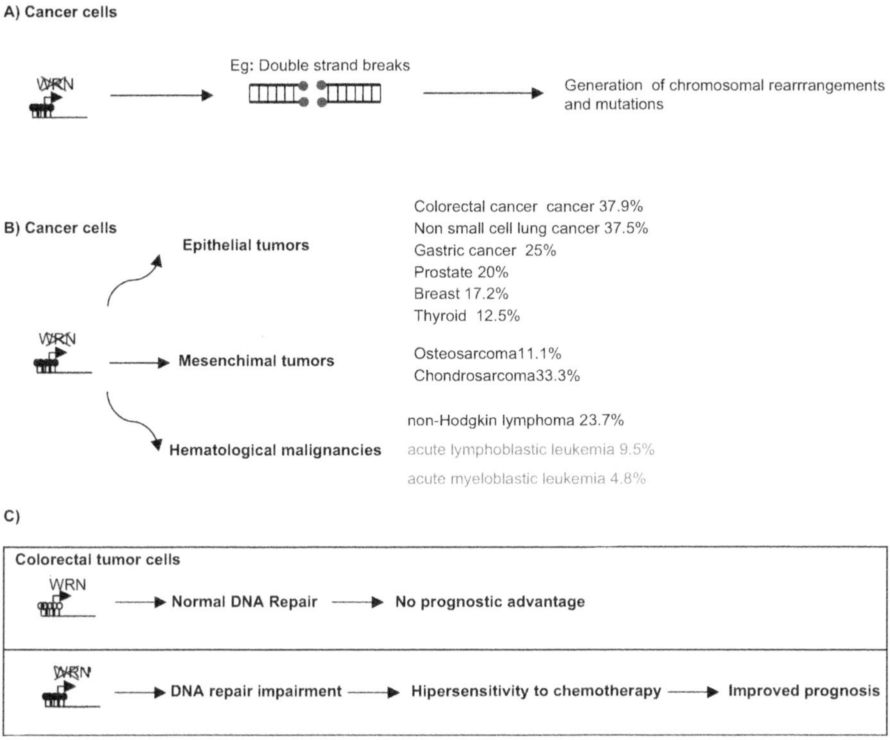

Fig. 2 (**A**) Epigenetic inactivation of *WRN* in human cancer. The epigenetic inactivation of the WRN protein in various tumor types of both mesenchymal and epithelial origin causes the accumulation of chromosomal rearrangements and somatic mutations, as the protein plays an important role in the repair of DSBs. (**B**) WRN hypermethylation in primary tumors. Epithelial tumors were most prevalent in CRC (37.9%) followed by non-small-cell lung cancer (37.5%), gastric (25%), prostate (20%), breast (17.2%), and thyroid (12.5%) tumors. In hematological tumors, promoter hypermethylation of the *WRN* gene was often found in non-Hodgkin's lymphomas (23,7%), acute lymphoblastic leukemia (9.5%), but was much less common in acute myeloblastic leukemia (4,8%). In mesenchymal tumor types, *WRN* hypermethylation was present in chondrosarcomas (33.3%) and osteosarcomas (11.1%). (**C**) Epigenetic changes associated with *WRN*. When *WRN* is silenced by methylation at its CpG island promoter in CRC, DNA repair processes are impaired due to the lack of WRN protein. These neoplastic cells become susceptible to the effects of chemotherapy, resulting in markedly improved patient prognoses

mitomycin C-induced apoptosis (Agrelo et al. 2006). Moreover, introduction of the exogenous *WRN* gene into hypermethylated cells made them more resistant to the apoptosis mediated by both drugs.

Human cancer cell lines with an unmethylated WRN promoter also demonstrated minimal chromosomal breakage following exposure to mitomycin C. In contrast, cells from a WS patient and cells with aberrant *WRN* methylation were extremely sensitive to the drug. Furthermore, when WRN-transfected cells were exposed to mitomycin C, they acquired resistance to chromosomal breakage.

Interestingly, downregulation of WRN by RNA interference in cells with an unmethylated promoter of WRN leads to an increase in the hypersensitivity ratio of chromosomal breakage following exposure to mitomycin C. These findings led us to hypothesize that WRN methylation may be a good predictor of response to camptothecin-based chemotherapy. Irinotecan (CPT-11), a camptothecin analog, is used in the treatment of colon cancers, in which CpG island hypermethylation frequently silences *WRN*. Following treatment with irinotecan, the median survival time of patients with *WRN*-hypermethylated colorectal CRC was significantly longer ($n = 45$, 39.2 months) than that of patients with unmethylated cancers ($n = 43$, 20.7 months). This shows that low levels of WRN in tumor cells predict a good response to irinotecan-based therapy (Fig. 2C).

Importantly, WRN might have a role as a second-line defense mechanism against methyl adducts (N3-methyladenine and O6-methylguanine) that can lead to blocked DNA replication and increased mutation rates (Blank et al. 2004). It has also been suggested that low levels of WRN expression might contribute to the genesis of sporadic tumors by increasing the risk of methylation-induced mutagenesis. O6-methylguanine residues are recombinogenic, cytotoxic, and had the potential to generate point mutations (Blank et al. 2004).

It has been demonstrated that temozolomide (TMZ) cytotoxicity is increased by WRN suppression in O6-methylguanine–DNA methyltransferase (MGMT)-deficient but not MGMT-proficient glioblastoma cell lines (Blank et al. 2004).

MGMT protein activity is the primary cell defense against O6-methylguanine. It removes mutagenic and cytotoxic adducts from O6-guanine in DNA.

Lack of MGMT could induce mutations that lead to cancer, given the tendency of O6-methylguanine to pair with thymine during DNA replication, resulting in the conversion of guanine–cytosine to adenine–thymine pairs in DNA (Blank et al. 2004). It is remarkable that, in a wide variety of human cancers, MGMT expression is suppressed by promoter hypermethylation-mediated silencing and generates G:C to A:T transitions in *KRAS* and *TP53* genes (Esteller 2008). Moreover, in glioma cells, the presence of a hypermethylated *MGMT* promoter leads to enhanced sensitivity to the action of alkylating agents compared to the sensitivity of cells with unmethylated *MGMT* promoters and better responses in patients treated with these types of drugs (Esteller 2008). In consequence, tumor cells in which *WRN* and *MGMT* promoter hypermethylation-mediated silencing occurs simultaneously might constitute a particular group of tumors with high genomic instability and a better response to alkylating drugs such as TMZ (frequently used in chemotherapy) compared to tumors with *MGMT* promoter hypermethylation only.

An association between *WRN* promoter methylation in CRC and mucinous differentiation, independent of the MSI and CpG island methylator phenotype (CIMP) status, has recently been reported (Kawasaki et al. 2008). They also found a correlation between *WRN* methylation and MSI/CIMP. The authors proposed that *WRN* methylation may provide the link between mucinous differentiation and MSI and/or CIMP. As WS patients have elevated serum and urine levels of hyaluronic acid and because WS cells exhibit abnormal glycosaminoglycan metabolism (Salk 1982; Kieras et al. 1986; Tanabe and Goto 2001), it has been hypothesized that *WRN* promoter hypermethylation and functional loss of the protein in CRC cells may explain the excessive mucin overproduction and secretion in a subset of colon cancers with CIMP and/or MSI (Kawasaki et al. 2008). It has been argued that this hypothesis could account for the well-known association between mucinous differentiation and MSI/CIMP in colorectal cancer, as positive correlations between *WRN* methylation and MSI/CIMP and between *WRN* methylation and mucinous differentiation have been demonstrated (Kawasaki et al. 2008). Moreover, another study showed that mucinous colon tumors overexpressed markers of resistance to fluorouracil (5-FU) and oxaliplatin and also that patients treated with 5-FU had reduced disease-free survival. However, no differences in markers of resistance to irinotecan were identified in the same study (Glasgow et al. 2005). From this study, we may conclude that mucinous tumors associated with *WRN* methylation could benefit from irinotecan treatment for two reasons: first, because of the sensitivity conferred by the lack of WRN and second because they have no resistance to the drug.

Finally, it would be interesting to determine whether some members of the RecQ helicase family other than WRN are repressed by promoter hypermethylation in cancerous cells. If so, it would be useful to explore what happens to the expression levels of other members of the RecQ family in cancer cells where the *WRN* promoter is unmethylated and WRN is expressed. Genome instability could be a consequence, at least in some cases, of the repression of other RecQ helicases (for example, BLM). This might explain why hypermethylation of *WRN* occurs in only a subset of tumors.

Epigenetic Connection Between Lamin A/C and Cancer

The nuclear lamina is a network of lamin filaments and proteins and is located in the inner side of the nuclear membrane. It is composed of B-type lamins, A-type lamins, and lamin-associated proteins (Gruenbaum et al. 2000). It has recently been demonstrated that nuclear lamin is highly dynamic and has many possible roles, such as overall organization of chromatin, non-random positioning of subchromosome domains, and probably also in the regulation of gene expression (Broers et al. 1999; Taddei et al. 2004). It is also involved in the maintenance of nuclear structure, cell cycle regulation, DNA replication, differentiation, and apoptosis. Lamins can be classified as type A or type B, the first being expressed in most differentiated somatic cells and the latter being present in all cells and essential for cell viability (Hutchison and Worman 2004).

The lamin A/C gene, *LMNA*, encodes the A-type lamins A and C, isoforms that arise as a result of alternative RNA splicing. Important roles for lamin A/C are the maintenance of lamina stability and also in regulating transcription (Hutchison and Worman 2004). Mutations in the human *LMNA* gene cause a wide range of inherited diseases called laminopathies. These pathologies can be tissue specific and include muscle (as in Emery Dreifuss muscular dystrophies, limb girdle muscular dystrophies, and cardiomyopathy), adipose tissue (as in Dunnigan-type familial partial lipodystrophy), peripheral neuronal tissue (as in Charcot–Marie–Tooth disorder type 2), bone and adipose tissues (as in mandibuloacral dysplasia) or can be more generalized (as in Hutchinson–Guilford progeria and atypical Werner syndrome) (Broers et al. 2004). The fact that atypical WS has a mutation in *LMNA* prompted us to study possible epigenetic alterations of its promoter in human cancer. It has been well known for over a decade that lamin A/C was not present in hematological malignancies, but the molecular mechanism remained unknown (Stadelmann et al. 1990; Lyn and Worman 1997). We have demonstrated for the first time that promoter CpG island hypermethylation of *LMNA* is associated with loss of RNA and protein expression in leukemias and lymphomas (Agrelo et al. 2005) (Fig. 3A). It is

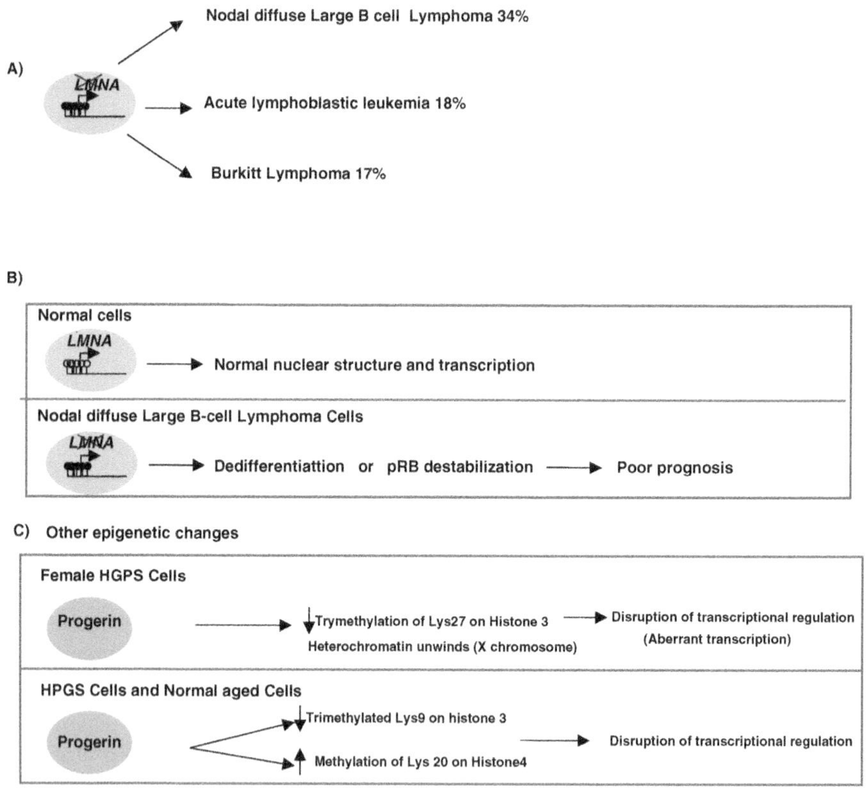

Fig. 3 (**A**) *LMNA* hypermethylation in hematological malignancies. Lamin A/C is hypermethylated in primary lymphoblastic leukemias (18%) and in non-Hodgkin's lymphomas, including diffuse large B-cell lymphoma (DLBCL) (34%) and Burkitt's lymphomas (17%). This epigenetic state is associated with poorer prognoses. (**B**) Epigenetic changes in HGPS cells. In female HGPS cells, progerin expression reduces trimethylation of Lys 27 in histone H3, causing unraveling of condensed heterochromatin on the inactive X chromosome. Moreover, HGPS cells and cells from aged individuals have reduced levels of trimethylated Lys 9 in histone H3 (which is reversible upon inhibition of the aberrant splicing of progerin), as well as increased levels of methylation of Lys 20 in histone H4. These altered patterns of histone modification lead to the disruption of transcriptional regulation

also well known that normal T and B lymphocytes express lamin A/C (Guilly et al. 1990), and loss of its expression in malignant B and T lymphocytes could reflect a block in the differentiation pathways of these cells. Thus the link between reduced expression of *LMNA* and malignancy might rely on the fact that lamin A/C is normally expressed in differentiated cells involved in the chromatin reorganization and reprogramming necessary for terminal differentiation and growth arrest (Prokocimer et al. 2006). We can therefore predict that A-type lamins would facilitate an undifferentiated phenotype that could be an advantage for the aggressive progression of a tumor. However, another attractive model might also account for the advantage that a lymphoma or leukemia cell could gain by the loss of expression of lamin A/C following promoter hypermethylation. It has been demonstrated that A-type lamins interact with retinoblastoma protein (pRB). During the G1 phase of the cell cycle, pRB is anchored to the nuclear matrix by direct binding of lamin A/C and it is also localized in lamin A/C internal nuclear foci (Kennedy et al. 2000). In the absence of lamin A/C, pRB is degraded in a proteasome-dependent fashion, although the mechanism by which lamin A/C stabilizes pRB remains unknown. pRB regulates cell cycle progression (Johnson et al. 2004).

During the cell cycle, pRB is hyperphosphorylated by the cyclin D/cyclin-dependent kinases 4 and 6 (cyclin D/Cdk4/6) complex (Sharpless and de Pinho 1999). Hyperphosphorylated pRB is unable to bind to and inhibit E2F transcription factor, allowing transcription of E2F target genes that are important for the G1/S transition (such as DNA polymerase II, cyclin E, p19 Myb, and dihydrofolate reductase) (Vernell et al. 2003). This allows cell cycle progression. The Ink4a locus encodes $p16^{Ink4a}$ and $p14^{ARF}$ using different promoters. $p16^{Ink4a}$ prevents binding of Cdk4/6 to cyclin D, inhibiting kinase activity. This results in hypophosphorylated pRB, which binds to E2F and inhibits E2F-mediated transcription, leading to cell cycle arrest and senescence (Quelle et al. 1995) (Fig. 4A). Hypermethylation of *LMNA* promoter may cause cells to become refractory to $p16^{Ink4a}$, as has been shown for lamin A/C–/– mouse embryonic fibroblasts (MEFS) (Nitta et al. 2006). This is because hypophosphorylated pRB is destabilized, allowing E2F to trigger cell cycle progression (Fig. 4B). The *LMNA* gene could then be regarded as a tumor suppressor gene.

Moreover, $p16^{Inka}$ is an upstream player in this pathway and is often inactivated by promoter hypermethylation in lymphomas such as diffuse large B-cell lymphomas (DLBCL) (Amara et al. 2008). Further studies should address the issue of whether both genes are methylated simultaneously, strengthening inactivation of this pathway.

Interestingly, lamin A/C–/– cells are refractory to $p14^{ARF}$-induced cell cycle arrest (Nitta, Smith and Kennedy 2007). $p14^{ARF}$ sequesters MDM2 protein and inhibits p53 degradation, resulting in p53-mediated cell cycle arrest and apoptosis (Weber et al. 1999). In fact, $p14^{ARF}$ expression is often lost in hematological neoplasias (Taniguchi 1997).

To conclude, a mouse line that does not express lamin A or C proteins (lamin A/C–/–) dies at 6–7 weeks of age from muscular dystrophy and cardiomyopathy (Sullivan et al. 1999), raising the possibility that mice die before hematological neoplasia appears.

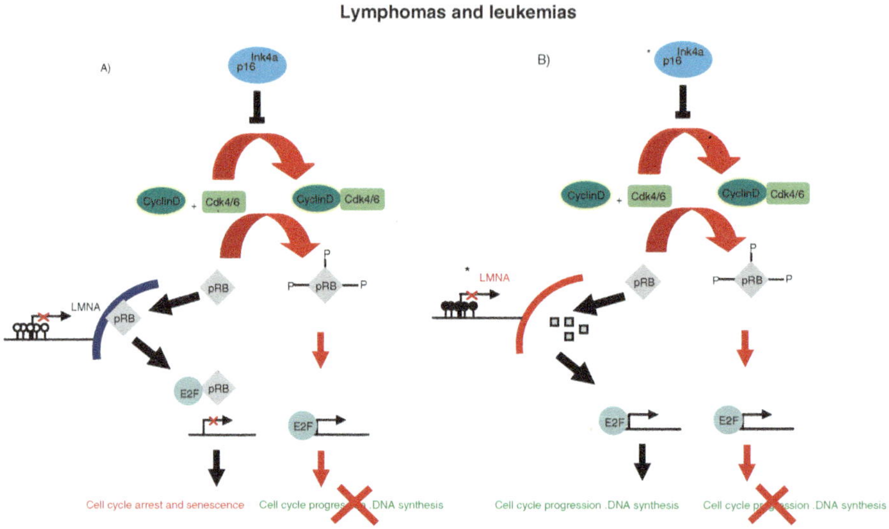

Fig. 4 Author's proposed model for the role of lamin A/C suppression in hematological neoplasia. (**A**) In lymphoma and leukemia cells during the G1 phase of the cell cycle, pRB is anchored to the nuclear matrix by direct binding of lamin A/C and is stabilized, making possible its binding to E2F and therefore promoting cell cycle arrest and senescence. (**B**) If lamin A/C undergoes promoter hypermethylation-mediated silencing, pRB is destabilized and degraded in a proteasome-dependent fashion, allowing E2F to trigger cell cycle progression and creating cells that are refractory to the effects $p16^{Ink4a}$. *Genes that are usually silenced by promoter hypermethylation (see text for details)

We have also demonstrated that *LMNA* promoter CpG island hypermethylation is a good predictor of shorter failure-free and overall survival in nodal diffuse large B-cell lymphomas (Agrelo et al. 2005) (Fig. 3B). This finding suggests that this epigenetic modification is a clinically relevant parameter for the management of these patients. These lymphomas account for approximately 30–40% of adult non-Hodgkin's lymphomas, for which the most commonly used prognostic indicator is the International Prognostic Index (IPI). The IPI takes into account factors that are mostly linked to patient characteristics (age, performance status) and to disease progression (disease stage, lactate dehydrogenase levels, and extent of extranodal involvement). Importantly, lamin A/C hypermethylation is independent of the IPI value, confirming its potential role as a clinical marker.

The outcome of patients with hematologic malignancies has been improved as a result of current chemotherapeutic regimens, but many patients do not respond to or benefit from their use. Interestingly, this methylation could be susceptible to reversion by demethylating agents, such as 5AZA, which is currently approved by the FDA for the treatment of myelodysplastic syndrome, based on its capacity to reactivate CpG island-hypermethylated genes (Esteller 2005). *LMNA* is an excellent target for demethylating agents used in treating hematological malignancies (Agrelo et al. 2005).

The findings concerning the epigenetic mechanism responsible for silencing *LMNA* could help to provide a better understanding of how lamins contribute to the neoplastic process (Prokocimer et al. 2006).

Epigenetic Modifications in Hutchinson–Guilford Cells

HGPS is characterized by alopecia, atherosclerosis, prominent scalp veins, deficiencies in adipose storage tissue, and a high-pitched voice, with death occurring at approximately 13 years of age, usually due to atherosclerosis of the coronary and cerebrovascular arteries (Hennekam 2006). In approximately 80% of HGPS cases, a de novo silent mutation (G608G: GGC→GGT) in *LMNA* activates a cryptic donor splice site and eliminates 150 bp of exon 11, while other mutations have also been described including E145K, R471C, R527C, G608S, T623S, and 1824C>T. To produce mature lamin A, prelamin A undergoes post-translational modification of its CaaX motif; farnesylation of the cysteine residue by farnesyltransferase cleavage of the aaX amino acids, probably by the FACE-1/ZMPSTE24 metalloproteinase, carboxymethylation of the farnesylated cysteine by prenylcysteine carboxymethyltransferase, and, finally, a second cleavage of the 15 terminal residues, also by FACE-1/ZMPSTE24. While the C-terminal CaaX motif required for farnesylation is retained, the signal directing the second cleavage by FACE-1/ZMPSTE24 is not, resulting in a truncated isoform of prelamin A called progerin, which cannot be cleaved by this metalloprotease and consequently remains farnesylated (De Sandre-Giovannoli et al. 2003; Eriksson et al. 2003). The accumulation of progerin at the nuclear envelope is toxic for cells in HGPS and is considered to be the primary cause of nuclear abnormalities such as nuclear herniations, blebbing, and chromatin stress that leads to premature cell death (Glynn and Glover 2005; Mallampalli et al. 2005; Capell et al. 2005; Toth et al. 2005). HGPS fibroblasts accumulate progerin when they age in culture and display changes in nuclear shape and architecture and also loss of peripheral heterochromatin. To account for this loss of heterochromatin from the epigenetic point of view, it has been demonstrated that cultured cells from both HGPS patients and normal, aged individuals display reduced levels of trimethylation of lysine 9 in histone H3 (H3K9me3) (Fig. 3C) and an altered association of this marker with heterochromatin protein 1 (Hp1) and the CREST antigen (Scaffidi and Misteli 2006). This modification of histone H3, which is associated with active transcription and was previously thought to be permanent, has recently been shown to be reversible through the action of JHDM3A, a histone demethylase (Klose et al. 2006). If JHDM3A is shown to be involved in the reduction of H3K9me3 levels in HPGS and aged cells, a therapeutic approach based on inhibition of its demethylase activity could be explored.

It has also been shown that low levels of progerin can disrupt trimethylation of lysine 27 in histone H3 (H3K27me3) (Fig. 3C), which is a marker of facultative chromatin in the inactive X chromosome, with or without the characteristic nuclear deformations (Shumaker et al. 2006). The methyltransferase responsible for this marker was also downregulated. Moreover, the upregulation of H4K20me3 (Fig. 3C), a marker for constitutive heterochromatin, in both HGPS cells and old rats (Sarg et al. 2002; Shumaker et al. 2006) indicated that similar mechanisms may be at work in aging individuals and HGPS patients. Further evidence connecting HGPS and normal aging comes from the finding that cell nuclei from old individuals show similar alterations to those of HGPS patients, including increased DNA damage and morphological abnormalities, as well as changes in histone modifications (Scaffidi and Misteli 2005, 2006). It is a remarkable finding that cells from normal aged individuals exhibit low levels of progerin. These data suggest that progerin accumulation leads to perturbations in the epigenetic control of chromatin structure and subsequently to altered gene expression patterns (Scaffidi and Misteli 2005, 2006).

An Epigenetic Molecular Hypothesis That Accounts for Progerin Accumulation in Normal Aging Cells

How could this small amount of progerin be produced in normal cells of aged individuals?

m^5C residues facilitate the generation of point mutations in several ways. In particular, methylated cytosine can suffer spontaneous deamination to thymine, and so could be considered an endogenous mutagen (Fig. 5A). If this mutation remains uncorrected and occurs, for example, in the coding regions of genes, it could be responsible for disease.

In the neoplastic process, these mutations can modify the functions of genes that are important for survival and cell growth regulation. For instance, many of the *TP*53 gene mutations that take place in somatic cells (including the 248, 273, and 282 hotspots) are caused by C–T transitions (Rideout et al. 1990). It has also been shown that CpGs in these regions are methylated in normal tissues and that half of the *TP*53 mutations occur at methylated CpG (m^5CpG). The risk of mutations at m^5C relative to C is estimated to be tenfold (Rideout et al. 1990). In more than 80% of HGPS cases, a de novo silent mutation (G608G: GGC→GGT) in *LMNA* activates a cryptic donor splice site and eliminates 150 bp of exon 11 (De Sandre-Giovannoli et al. 2003; Eriksson et al. 2003). In the model I propose here, accumulation of progerin in somatic cells is a result of deamination of the predicted methylated m^5CpG (GGCG → GGTG) that generates the use of the cryptic splice site because it is converted to TG (Fig. 5B–C).

Remarkably, the R644C mutation that causes atypical progeria is affected at the cleavage site by FACE-1/ZMPSTE24, generating a Δ35 lamin (Csoka et al. 2004) (Fig. 5B–C). Deamination of the mCpG might also cause accumulation of this lamin precursor. A recent study in which several cancer-associated genes from colon and breast cancers were sequenced emphasizes the importance of mutations at CpG

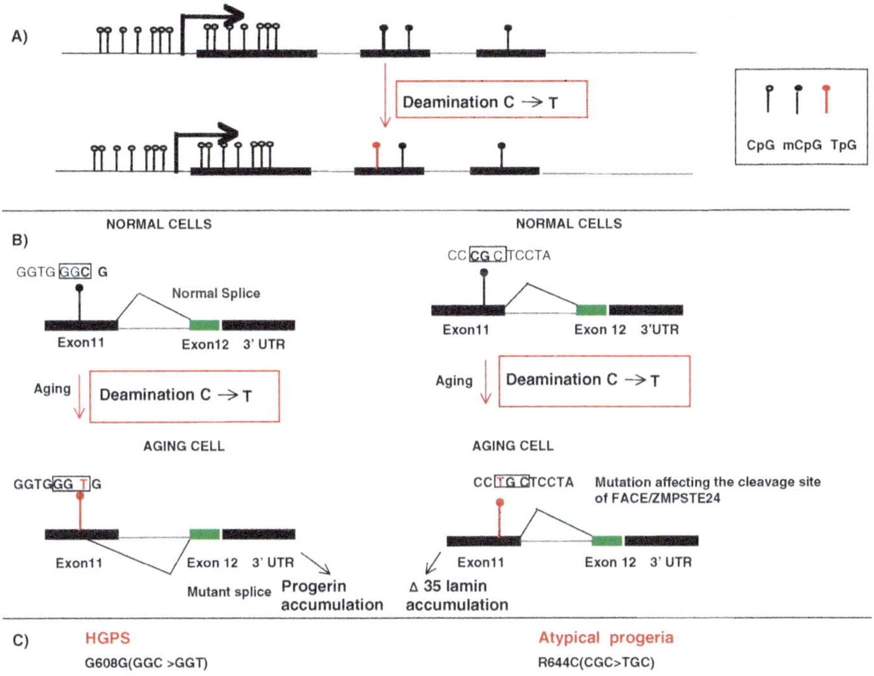

Fig. 5 Author's proposed model for the accumulation of progerin (LΔ50) and mutant lamin in aging cells. (**A**) Methylated cytosine within the coding regions of genes may promote mutations by spontaneous hydrolytic deamination to thymine ($m^5C \rightarrow T$). (**B**) A model in which deamination of $m^5C \rightarrow T$ activates the cryptic donor splice site in aging cells, leading to the accumulation of progerin (*left*). (**B**) Alternatively, this could lead to a mutation affecting the cleavage site of FACE1/ZMPSTE24, as is seen in atypical progeria [R644C(CGC>TGC)] (*right*). (**C**) In HGSP, a de novo silent mutation (G608G: GGC→GGT) in *LMNA* activates a cryptic donor splice site and eliminates 150 bp of exon 11 generating progerin (Δ50 lamin); in atypical progeria, the R644C (CGC>TGC) mutation affects the cleavage site of FACE1/ZMPSTE24, generating a Δ35 lamin

sites. In colon cancer, 44% of mutations were C–T transitions; in breast cancer, the figure was 17%.

This shows that the frequency of methylation-induced mutations varies in different tissues, and this is probably a result of different abilities to repair T:G mispairs in individual cell types. Consequently, this model predicts that the amount of progerin produced during the aging process or replicative senescence would vary in different tissues and organs (Sjöblom et al. 2006). Very recently, a case of Hutchinson–Guilford syndrome was reported which was associated with osteosarcoma (Shalev et al. 2007). It would therefore be interesting to test whether progerin is also accumulated in cancer cells and, if so, whether m^5C deamination is the underlying mechanism.

Discussion and Conclusions

In this chapter I have summarized several of the epigenetic alterations with direct roles in cell transformation that accumulate during aging. The role of methylation in the aging and neoplastic processes, the global loss of DNA methylation associated with aging and cancer attributed to downregulation of DNMT1, and DNMT3b upregulation has been reviewed. Importantly, the recent discovery that promoter hypermethylation occurs in genes with roles in tumor suppression and progeria, such as *WRN* and *LMNA*, has also been reviewed, as have some of the epigenetic modifications affecting the cells in HGPS patients and normal, aged individuals, where low amounts of progerin were found.

Regarding these two genes, it could be important to test if the *LMNA* promoter is progressively hypermethylated in normal, aged tissues, as seen in lymphoma and leukemia cells, as the nuclei of older individuals show alterations comparable to those seen in HGPS (Mehta et al. 2007, Haithcock et al. 2005). Progressive methylation of the promoter might inactivate *LMNA* expression, making the cells cancer-prone because of pRB destabilization and proteasome-dependent degradation (Johnson et al. 2004) that promotes elevated ploidy and genome instability (Srinivasan et al. 2007). However, *LMNA* is also involved in DNA replication: It has been shown to be associated with sites of early replication in primary fibroblasts from early G1 until early S phase (Bridger et al. 1993; Kennedy et al. 2000; Barbie et al. 2004). pRB and E2F are also localized in these sites (Kennedy et al. 2000). Additionally, it is well known that E2F is released to initiate the S phase, activating the transcription of genes whose products regulate the initiation of replication (Ohtani 1999). Its activity is also needed for a normal rate of replication, because E2F activates the transcription of components of the DNA replication machinery, such as DNA polymerase α (Nevins 1992). Moreover, it has also been postulated that E2F might directly regulate the initiation of replication that is stimulated by histone acetyltransferase (HAT) activity (which allows assembly of the replication complex at the origin) (Cayirlioglu and Duronio 2001). Where the balance between defects of replication (because a lack of lamin A/C to inhibit the formation of sites of early replication) and replication stimulation (following release of E2F) lies could lead to different results – senescence or cancer – and this would depend on the cellular context (Fig. 6).

It will also be important to test whether the splicing site is subjected to epigenetic regulation, in agreement with the model presented, as this could account for the small amounts of progerin found in normal aged cells.

The *WRN* gene is epigenetically silenced in a wide variety of tumor types of both mesenchymal and epithelial origin, including those commonly observed in WS patients (such as osteosarcoma, thyroid, and gastric tumors), in a similar fashion to that observed with other familial tumor suppressor genes with DNA repair functions, such as *hMLH1*.

Since WS cells accumulate chromosomal rearrangements and somatic mutations at an increased rate in an age-dependent manner, we can speculate that inactivation of the *WRN* gene creates a new mutator epigenetic pathway in human cancer

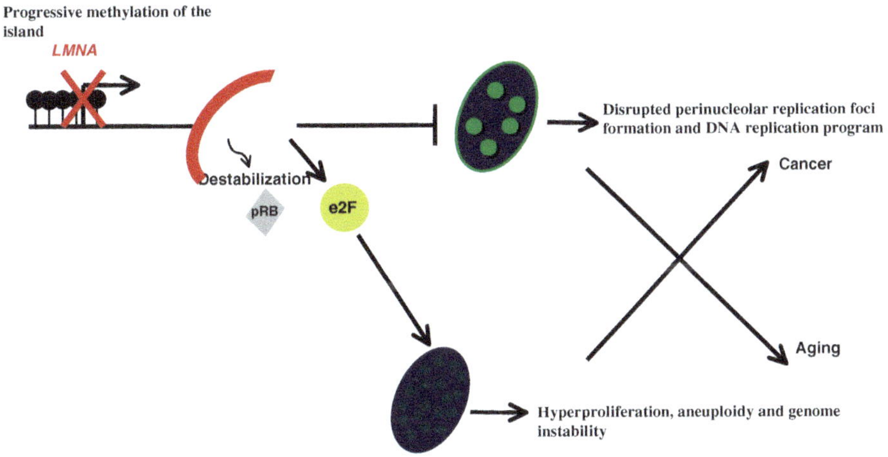

Fig. 6 *LMNA* methylation during aging. Progressive methylation of the lamin A/C promoter in aging, if it occurs, might disrupt replication, given that lamin has been associated with early sites of DNA replication (*green*). On the other hand, E2F would be released, activating transcription of genes that control the initiation of replication and components of the replication machinery. Loss of pRB could promote ploidy and genome instability. The balance between the two processes could promote cell senescence or cancer, depending on the cellular context

(Fig. 2A). This hypothesis is supported by the fact that cancer cells in which *WRN* is silenced by promoter hypermethylation are extremely sensitive to the action of DNA-damaging agents. Future studies will determine if this is the case. It will also be interesting to test if promoter hypermethylation of *WRN* in human cancers is partly due to the progressive methylation of CpG islands, and as in the aging process, as is the case in other genes such as *MLH1* and $p14^{ARF}$ in CRC, even though no mechanism for the association of these two genes and aging has been demonstrated. In this context a new model of cellular aging suggests that WS cells mimic the epigenetic changes that occur during normal cellular aging (Agrelo 2007).

Acknowledgments I carried out the research involving the epigenetic silencing of progeroid genes in the Cancer Epigenetics Laboratory, directed by Dr. Manel Esteller at the CNIO (National Spanish Cancer Centre, Madrid). I would like to address special thanks to Manel Esteller for his friendship and for sharing days of exciting discoveries during my post-doctoral stage in his laboratory. I would also like to thank two anonymous reviewers for useful suggestions.

References

Aaltonen, L.A., Peltomäki, P., Leach, F.S., Sistonen, P., Pylkkänen, L. et al. 1993. Clues to the pathogenesis of familial colorectal cancer. Science. 260:812–816.

Agrelo, R., Setien, F., Espada, J., Artiga, M.J., Rodriguez, M. et al. 2005. Inactivation of the lamin A/C gene by CpG island promoter hypermethylation in hematologic malignancies, and its association with poor survival in nodal diffuse large B-cell lymphoma. J Clin Oncol. 23:3940–3947.

Agrelo, R., Cheng, W.H., Setien, F., Ropero, S., Espada, J. et al. 2006. Epigenetic inactivation of the premature aging Werner syndrome gene in human cancer. Proc Natl Acad Sci USA. 103:8822–8827.

Agrelo R. 2007. A new molecular model of cellular aging based on Werner syndrome. Med Hypotheses. 68:770–80.

Amara, K., Trimeche, M., Ziadi., S., Laatiri, A., Hachana, M., Korbi, S. 2008. Prognostic significance of aberrant promoter hypermethylation of CpG islands in patients with diffuse large B-cell lymphomas. Ann Oncol. 19:1774–1786.

Armes, J.E., Hammet, F., de Silva, M., Ciciulla, J., Ramus, S.J. et al. 2004. Candidate tumor suppressor genes on chromosome arm 8p in early-onset and high-grade breast cancers. Oncogene. 23:5697–5702.

Bachrati, C.Z., Hickson, I.D. 2003. RecQ helicases: suppressors of tumorigenesis and premature aging. Biochem J. 374:577–606.

Barbie, D.A., Kudlow, B.A., Frock, R., Zhao, J., Johnson, B.R., et al. 2004. Nuclear reorganization of mammalian DNA synthesis prior to cell cycle exit Mol Cell Biol. 24:595–607.

Baylin, S.B., Esteller, M., Rountree, M.R., Bachman, K.E., Schuebel, K., Herman, J.G. 2001. Aberrant patterns of DNA methylation, chromatin formation and gene expression in cancer. Hum Mol Genet. 10:687–692.

Blank, A., Bobola, M.S., Gold, B., Varadarajan, S.D., Kolstoe, D. et al. 2004. The Werner syndrome protein confers resistance to the DNA lesions N3-methyladenine and O6-methylguanine: implications for WRN function. DNA Repair 3:629–638.

Bocker, T., Ruschoff, J., Fishel, R. 1999. Molecular diagnostics of cancer predisposition: hereditary non-polyposis colorectal carcinoma and mismatch repair defects. Biochim Biophys Acta. 1423:O1–O10.

Bohr, V.A. 2005. Deficient DNA repair in the human progeroid disorder, Werner syndrome. Mutat Res. 577:252–259.

Bridger, J.M., Kill, I.R., O'Farrell, M., Hutchison, C,J. 1993. Internal lamin structures within G1 nuclei of human dermal fibroblasts. J Cell Sci. 104 :297–306

Broers, J.L., Machiels, B.M., van Eys, G.J., Kuijpers, H.J., Manders, E.M. et al. 1999. Dynamics of the nuclear lamina as monitored by GFP-tagged A-type lamins. J Cell Sci. 112:3463–3475.

Broers, J.L., Hutchison, C.J., Ramaekers F.C. 2004. Laminopathies. J Pathol. 204:478–488.

Brosh, R.M., Jr., Karmakar, P., Sommers, J.A., Yang, Q., Wang, X.W. et al. 2001. p53 Modulates the exonuclease activity of Werner syndrome protein. J Biol Chem. 276:35093–35102.

Brosh, R.M. Jr., Bohr, V.A. 2002. Roles of the Werner syndrome protein in pathways required for maintenance of genome stability. Exp Gerontol. 37:491–506.

Capell, B.C., Erdos, M.R., Madigan, J.P., Fiordalisi, J.J., Varga, R. et al. 2005. Inhibiting farnesylation of progerin prevents the characteristic nuclear blebbing of Hutchinson-Gilford progeria syndrome. Proc Natl Acad Sci USA. 102:12879–12884.

Casillas, M.A., Jr Lopatina, N., Andrews, L.G., Tollefsbol, T.O. 2003. Transcriptional control of the DNA methyltransferases is altered in aging and neoplastically-transformed human fibroblasts. Mol Cell Biochem. 252:33–43.

Cayirlioglu P; Duronio, R. 2001. Cell cycle: Flies teach and old dogma new tricks Current Biology 11:R178–R181.

Cooper, M.P., Machwe, A., Orren, D.K., Brosh, R.M., Ramsden, D., Bohr, V.A. 2000. Ku complex interacts with and stimulates the Werner protein. Genes Dev. 14:907–912.

Crabbe, L., Jauch, A., Naeger, C.M., Holtgreve-Grez, H., Karlseder, J. 2007. Telomere dysfunction as a cause of genomic instability in Werner syndrome. Proc Natl Acad Sci USA. 104: 2205–2210.

Chang, S., Multani, A.S., Cabrera, N.G., Naylor, M.L., Laud, P. et al. 2004. Essential role of limiting telomeres in the pathogenesis of Werner syndrome. Nature Genet. 36:877–882.

Chen, L., Oshima, J.J. 2002. Werner syndrome. Biomed Biotechnol. 2:46–54.

Chen, L., Lee, L., Kudlow, B.A., Dos Santos, H.G., Sletvold, O. et al. 2003. LMNA mutations in atypical Werner's syndrome. Lancet. 362:440–445.

Chughtai, S.A., Crundwell, M.C., Cruickshank, N.R., Affie, E., Armstrong, S. et al. 1999. Two novel regions of interstitial deletion on chromosome 8p in colorectal cancer. Oncogene. 18:657–665.

Csoka, A.B., Cao, H., Sammak, P.J., Constantinescu, D., Schatten, G.P., Hegele, R.A. 2004. Novel lamin A/C gene (LMNA) mutations in a typical progeroid syndromes. J Med Genet: 41: 304–308.

De Sandre-Giovannoli, A., Bernard, R., Cau, P., Navarro, C., Amiel, J. et al. 2003. Lamin a truncation in Hutchinson-Gilford progeria. Science. 300:2055.

Ehrlich, M. 2003. The ICF syndrome, a DNA methyltransferase 3B deficiency and immunodeficiency disease. Clin Immunol. 109:17–28.

Ellis, N.A., German, J. 1996. Molecular genetics of Bloom's syndrome. Hum Mol Genet. 5 (Spec. No.):1457–1463.

Epstein, C.J., Martin, G.M., Schultz, A.L., Motulsky, A.G. 1966. Werner's syndrome: A review of its symptomatology, natural history, pathologic features, genetics and relationship to the natural aging process. Medicine (Baltimore). 45:177–221.

Eriksson, M., Brown, W.T., Gordon, L.B., Glynn, M.W., Singer, J. et al. 2003. Recurrent de novo point mutations in lamin A cause Hutchinson- Gilford progeria syndrome. Nature. 423: 293–298.

Esteller, M., Levine, R., Baylin, S.B., Ellenson, L.H., Herman, J.G. 1998. MLH1 promoter hypermethylation is associated with the microsatellite instability phenotype in sporadic endometrial carcinomas. Oncogene. 17:2413–2417.

Esteller, M., Fraga, M.F., Guo, M., Garcia-Foncillas, J., Hedenfalk, I., et al. 2001. DNA methylation patterns in hereditary human cancer s mimics sporadic tumorigenesis. Hum Mol Genet. 10:3001–3007.

Esteller, M. 2003. Relevance of DNA methylation in the management of cancer. Lancet Oncol. 4:351–358.

Esteller, M. 2005. DNA methylation and cancer therapy. New developments and expectations. Curr Opin Oncol. 17:55–60.

Esteller, M. 2007. Cancer epigenomics: DNA methylomes and histone-modification maps. Nat Rev Genet. 4:286–298.

Esteller M Epigenetics in cancer. 2008. N Engl J Med. 358:1148–1159.

Fleisher, A.S., Esteller, M., Wang, S., Tamura, G., Suzuki, H. et al. 1999. Hypermethylation of the hMLH1 gene promoter in human gastric cancers with microsatellite instability. Cancer Res. 59:1090–1095.

Fokuchi, K., Martin G.M., Monnat, J.R. 1989. Mutator phenotype of Werner syndrome is characterized by extensive deletions. Proc Natl Acad Sci USA. 86:5893–5897.

Glasgow, S.C., Yu, J., Carvalho, L.P., Shannon, W.D., Fleshman, J.W., McLeod, H.L. 2005. Unfavourable expression of pharmacologic markers in mucinous colorectal cancer. Br J Cancer. 92:259–264.

Glynn, M.W., Glover, T.W. 2005. Incomplete processing of mutant lamin A in Hutchinson-Gilford progeria leads to nuclear abnormalities, which are reversed by farnesyltransferase inhibition. Hum Mol Genet. 14:2959–2969.

Goto, M. 1997. Hierarchical deterioration of body systems in Werner's syndrome: Implications for normal ageing. Mech Ageing Dev. 98:239–254.

Gray, M.D., Shen, J.C., Kamath-Loeb, A.S., Blank, A., Sopher, B.L. et al. 1997. The Werner syndrome protein is a DNA helicase. Nature Genet. 17:100–103.

Gruenbaum,Y., Wilson, K.L., Harel, A., Goldberg, M., Cohen, M. 2000. Review: Nuclear lamins–structural proteins with fundamental functions. J Struct Biol 129:313–323.

Guilly, M.N., Kolb, J.P., Gosti, F., Godeau, F., Courvalin, J.C. 1990. Lamins A and C are not expressed at early stages of human lymphocyte differentiation. Exp Cell Res. 189: 145–147.

Haithcock, E., Dayani, Y., Neufeld, E., Zahand, A.J., Feinstein, N. et al. 2005. Age-related changes of nuclear architecture in *Caenorhabditis elegans*. Proc Natl Acad Sci USA 102: 16690–16695.

Hennekam, R.C. 2006. Hutchinson-Gilford progeria syndrome: Review of the phenotype. Am J Med Genet A. 140:2603–2624.

Herman, J.G., Umar, A., Polyak, K., Graff, J.R., Ahuja, N. et al. 1998. Incidence and functional consequences of hMLH1 promoter hypermethylation in colorectal carcinoma. Proc Natl Acad Sci USA 95:6870–6875.
Hickson, I.D. 2003. RecQ helicases: caretakers of the genome. Nat Rev Cancer. 3:169–178.
Hoffmann, M.J., Schltz, W.A. 2005. Causes and consequences of DNA hypomethylation in human cancer. Biochem Cell Biol 83:296–321.
Hollyday,R. 2006. Epigenetics: a historical overview. Epigenetics. 1:76–80.
Hornsby, P.J., Yang, L., Gunter, L.E. 1992. Demethylation of satellite I DNA during senescence of bovine adrenocortical cells in culture. Mutat Res. 275:13–19.
Huang, S., Li, B., Gray, M.D., Oshima, J., Mian, I.S., Campisi, J. 1998. The premature ageing syndrome protein, WRN, is a 3′→5′ exonuclease. Nature Genet. 20:114–116.
Huang, S., Lee, L., Hanson, N.B., Lenaerts, C., Hoehn, H. et al. 2006. The spectrum of WRN mutations in Werner syndrome patients. Hum Mutat. 27:558–567.
Hutchison, C.J., Worman, H.J. 2004. A-type lamins: Guardians of the soma? Nat. Cell. Biol 6:1062–1067.
Issa, J.P. 2003. Age-related epigenetic changes and the immune system. Clin Immunol. 109: 103–108.
Jascur, T., Boland, C.R. 2006. Structure and function of the components of the human DNA mismatch repair system. Int J Cancer. 119:2030–2035.
Jiricny, J. 2006. The multifaceted mismatch-repair system. Nat Rev Mol Cell Biol. 7:335–346.
Johnson, B.R., Nitta, R.T., Frock, R.L., Mounkes, L., Barbie, D.A et al. 2004. A-type lamins regulate retinoblastoma protein function by promoting subnuclear localization and preventing proteasomal degradation. Proc Natl Acad USA. 101: 9677–9682.
Jones, P.L. 1999. The methylation paradox. Trends Genet 15:34–37.
Kamath-Loeb, A.S., Shen, J.C., Loeb, L.A., Fry, M. 1998. Werner syndrome protein. II. Characterization of the integral 3′→5′ DNA exonuclease. J Biol Chem 273:34145–34150.
Kane, M.F., Loda, M., Gaida, G.M., Lipman, J., Mishra, R. et al. 1997. Methylation of the hMLH1 promoter correlates with lack of expression of hMLH1 in sporadic colon tumors and mismatch repair-defective human tumor cell lines. Cancer Res. 57:808–811.
Kawasaki, T., Ohnishi, M., Suemoto, Y., Kirkner, G.J., Liu, Z. et al. 2008. WRN promoter methylation possibly connects mucinous cancer, differentiation microsatellite instability and CpG island methylator phenotype in colorectal cancer. Mod Pathol. 21:150–158.
Kennedy, B.K., Barbie, D.A., Classon, M., Dyson, N., Harlow, E. 2000. Nuclear Organization of DNA replication in primary mammalian cells. Genes Dev. 14:2855–2868.
Kieras, F.J., Brown, W.T., Houck, G.E. Jr., Zebrower, M. et al. 1986. Elevation of urinary hyaluronic acid in Werner's syndrome and progeria. Biochem Med Metab Biol. 36:276–282.
Kinzler, K.W., Vogelstein, B. 1997. Cancer-susceptibility genes. Gatekeepers and caretakers. Nature. 386:761–763.
Kitao, S., Ohsugi, I., Ichikawa, K., Goto, M., Furuichi, Y., Shimamoto, A. 1998. Cloning of two new human helicase genes of the RecQ family: biological significance of multiple species in higher eukaryotes. Genomics 54:443–452.
Klose, R.J., Yamane, K., Bae, Y., Zhang, D., Erdjument-Bromage, H. et al. 2006. The transcriptional repressor JHDM3A demethylates trimethyl histone H3 lysine 9 and lysine 36. Nature. 442:312–316.
Knudson, A.G. 2001. Two genetic hits (more or less) to cancer. Nat Rev Cancer. 1:157–162.
Kyng, J., May, A., Kølvraa. S., Bohr, V.A. 2003. Gene expression profiling in Werner syndrome closely resembles that of normal aging. Proc Natl Acad Sci USA. **100**:12259–12264.
Kudlow, B.A., Kennedy, B.K., Monnat R.J. Jr. 2007. Werner and Hutchinson-Gilford progeria syndromes: mechanistic basis of human progeroid diseases. Nat Rev Mol Cell Biol. 5: 394–404.
Lyn, F., Worman, H.J. 1997. Expression of nuclear lamins in human tissues and cancer cell lines and transcription from the promoters of the lamin A/C and B1 genes. Exp Cell Res. 236: 378–384.

Lopatina, N., Haskell, J.F., Andrews, L.G., Poole, J.C., Saldanha, S., Tollefsbol, T. 2002. Differential maintenance and de novo methylating activity by three DNA methyltransferases in aging and immortalized fibroblasts. J Cell Biochem. 84:324–334.

Malkhosyan, S., Rampino, N., Yamamoto, H., Perucho, M. 1996. Frameshift mutator mutations. Nature. 382:499–500.

Mallampalli, M.P., Huyer, G., Bendale, P., Gelb, M.H., Michaelis, S. 2005. Inhibiting farnesylation reverses the nuclear morphology defect in a HeLa cell model for Hutchinson- Gilford progeria syndrome. Proc Natl Acad Sci USA. 102:14416–14421.

Martin, G.M. 1978. Genetic syndromes in man with potential relevance to the pathobiology of aging. Birth Defects Orig Artic Ser. 14:5–39.

Martin, G.M., Sprague C.A., Epstein, C.J. 1990. Replicative life-span of cultivated human cells. Lab Invest. 23:86–92.

Martin, G.M., Oshima, J., Gray, M.D., Poot, M. 1999. What geriatricians should know about the Werner syndrome. J Am Geriatr Soc. 47:1136–1144.

Matsumoto, T., Shimamoto, A., Goto, M., Furuichi, Y. 1997. Impaired nuclear localization of defective DNA helicases in Werner's syndrome. Nature Genetics. 16:335–336.

Mehta, I.S., Figgitt, M., Clements, C.S., Kill, I.R., Bridger, J.M. 2007. Alterations to nuclear architecture and genome behavior in senescent cells. Ann NY Acad Sci. 1100:250–263.

Moser, M.J., Bigbee, W.L., Grant, S.G., Emond, M.J., Langlois, R.G. et al. 2000. Genetic instability and hematologic disease risk in Werner syndrome patients and heterozygotes. Cancer Res. 60:2492–2496.

Nakagawa, H., Nuovo, G.J., Zervos, E.E., Martin, E.W. Jr., Salovaara, R. et al. 2001. Age-related hypermethylation of the 5′ region of MLH1 in normal colonic mucosa is associated with microsatellite-unstable colorectal cancer development. Cancer Res. 61:6991–6995.

Nevins, J.R. 1992. Transcriptional regulation. A closer look at E2F. Nature. 358:375–376.

Nitta, R.T., Jameson, S.A., Kudlow, B.A., Conlan, L.A., Kennedy, B.K. 2006. Stabilization of the retinoblastoma protein by A-type nuclear lamins is required for INK4A-mediated cell cycle arrest. Mol Cell Biol. 265360–265372.

Nitta, R.T., Smith, C.L., Kennedy, B.K. 2007. Evidence that proteasome-dependent degradation of the retinoblastoma protein in cells lacking A-type lamins occurs independently of gankyrin and MDM2. PLoS One. 2:1–9.

Ogburn, C.E., Oshima, J., Poot, M., Chen, R., Hunt, K.E et al. 1997. An apoptosis-inducing genotoxin differentiates heterozygotic carriers for Werner helicase mutations from wild-type and homozygous mutants. Hum Genet. 101:121–125.

Ohtani,K. 1999. Implication of transcription factor E2F in regulation of DNA replication Front Biosci. 4:D793–804.

Opresko, P.L., von Kobbe, C., Laine, JP., Harrigan, J., Hickson, I.D., Bohr, V.A. 2002. Telomere-binding protein TRF2 binds to and stimulates the Werner and Bloom syndrome helicases. J Biol Chem. 277:41110–41119.

Opresko, P.L., Cheng, W.H., von Kobbe, C., Harrigan, J.A., Bohr, V.A. 2003. Werner syndrome and the function of the Werner protein; what they can teach us about the molecular aging process. Carcinogenesis. 24:791–802.

Oshima, J. 2000. The Werner syndrome protein: an update. Bioassays. 22:894–901.

Pichierri, P., Franchitto, A., Mosesso, P., Palitti, F. 2001. Werner's syndrome protein is required for correct recovery after replication arrest and DNA damage induced in S-phase of cell cycle. Mol Biol Cell 12:2412–2421.

Poot, M., Yom, J.S., Whang, S.H., Kato, J.T., Gollahon, K.A., Rabinovitch, P.S. 2001. Werner syndrome cells are sensitive to DNA cross-linking drugs. FASEB J. 15:1224–1226

Prince, P.R., Emond, M.J., Monnat, R.J. Jr. 2001. Loss of Werner syndrome protein function promotes aberrant mitotic recombination. Genes Dev. 15:933–938.

Prokocimer, M., Margalit, A., Gruenbaum Y. 2006. The nuclear lamina and its proposed roles in tumorigenesis: projection on the hematologic malignancies and future targeted therapy. J Struct Biol. 155:351–360.

Quelle, D.E., Zindy, F., Ashmun, R.A., Sherr, C.J. 1995. Alternative reading frames of the INK4a tumor suppressor gene encode two unrelated proteins capable of inducing cell cycle arrest. Cell. 84:993–1000.

Richardson, B.C. 2002. Role of DNA Methylation in the Regulation of Cell Function: Autoimmunity. Aging and Cancer. J Nutr. 132(8 Suppl):2401S–2405S.

Rideout, W.M 3rd., Coetzee, G.A., Olumi, A.F., Jones, P.A. 1990. 5-Methylcytosine as an endogenous mutagen in the human LDL receptor and p53 genes. 249:1288–1290.

Saintigny, Y., Makienko, K., Swanson, C., Emond, M.J., Monnat, R.J. Jr. et al. 2002. Homologous recombination resolution defect in Werner syndrome. Mol Cell Biol. 22:6971–6978.

Sakatani, T., Kaneda, A., Iacobuzio-Donahue, C.A., Carter, M.G., de Boom, W.S. et al. 2005. Loss of imprinting of Igf2 alters intestinal maturation and tumorigenesis in mice. Science 307: 1976–1978.

Salk, D., Au, K., Hoehn, H., Martin, G.M. 1981. Cytogenetics of Werner's syndrome cultured skin fibroblast: variegated translocation mosaicism. Cytogenet Cell Genet. 30:92–107.

Salk D. 1982. Werner's syndrome: a review of recent research with an analysis of connective tissue metabolism, growth control of cultured cells, and chromosomal aberrations. Hum Genet. 62: 1–5.

Sarg, B., Koutzamani, E., Helliger, W., Rundquist, I., Lindner, H.H. 2002. Postsynthetic trimethylation of histone H4 at lysine 20 in mammalian tissues is associated with aging. J Biol Chem. 277:39195–39201.

Scaffidi, P., Misteli, T. 2005. Reversal of the cellular phenotype in the premature aging disease Hutchinson-Gilford progeria syndrome. Nat Med. 11:440–445.

Scaffidi, P., Misteli, T. 2006. Lamin A-dependent nuclear defects in human aging. Science. 312:1059–1063.

Shalev, S.A., De Sandre Giovannoli, A., Shani, A.A. Levi, N. 2007. An association of Hutchinson Guilford Progeria and malignancy. Am J Med Genet A. 143A: 1821–1826.

Sharpless, N., de Pinho,R. 1999. The INK4a/ARF locus and its two gene products. Curr Opin Genet Dev. 9:22–30.

Shen, J.C., Loeb, L.A. 2000. Werner syndrome exonuclease catalyzes structure dependent degradation of DNA. Nucleic Acids Res. 28:3260–3268.

Shen, L., Kondo, Y., Hamilton, S.R., Rashid, A., Issa, J.P. 2003. P14 methylation in human colon cancer is associated with microsatellite instability and wild-type p53. Gastroenterology. 124:626–633.

Shumaker, D.K., Dechat, T., Kohlmaier, A., Adam, S.A., Bozovsky, M.R. et al. 2006. Mutant nuclear lamin A leads to progressive alterations of epigenetic control in premature aging. Proc Natl Acad Sci USA. 103:8703–8708.

Siitonen, H.A., Kopra, O., Kääriäinen, H., Haravuori, H., Winter, R.M. et al. 2003. Molecular defect of RAPADILINO syndrome expands the phenotype spectrum of RECQL diseases. Hum Mol Genet. 12:2837–2844.

Sjöblom, T., Jones, S., Wood, L.D., Parsons, D.W., Lin, J. et al. 2006. The consensus coding sequences of human breast and colorectal cancers. Science. 314:268–274.

Spillare, EA., Robles, A.I., Wang, X.W., Shen, J.C., Yu, C.E. et al. 1999. p53-mediated apoptosis is attenuated in Werner syndrome cells. Genes Dev. 13:1355–1360.

Srinivasan, S.V., Mayhew, C.N., Schwemberger, S., Zagorski, W., Knudsen, E.S. 2007. RB loss promotes aberrant ploidy by deregulating levels and activity of DNA replication factors. J Biol Chem. 282:23867–23877

Stadelmann, B., Khandjian, E., Hirt, A., Lüthy, A., Weil, R., Wagner, H.P. 1990. Repression of nuclear lamin A and C gene expression in human acute lymphoblastic leukemia and non-Hodgkin's lymphoma cells. Leuk Res. 14:815–821,

Sullivan, T., Escalante-Alcalde, D., Bhatt, H., Anver, M., Bhat, N et al. 1999. Loss of A-type lamin expression compromises nuclear envelope integrity leading to muscular dystrophy. J Cell Biol. 147:913–920.

Taddei., A., Hediger, F., Neumann, F.R., Gasser, S.M. 2004. The function of nuclear architecture: A genetic approach. Annu Rev Genet. 38:305–345
Tanabe, M., Goto, M. 2001. Elevation of serum hyaluronan level in Werner's syndrome. Gerontology. 47:77–81.
Taniguchi, T., Chikatsu, N., Takahashi S., Fujita, A., Uchimaru, K., Asano, S. et al. 1999. Expression of p16INK4A and p14ARF in hematological malignancies. Leukemia. 13:1760–1769
Thibodeau, S.N., Bren, G., Schaid, D. 1993. Microsatellite instability in cancer of the proximal colon. Science. 260:816–819.
Thomas, D.C., Umar, A., Kunkel, T.A. 1996. Microsatellite instability and mismatch repair defects in cancer. Mutat Res. 350:201–205.
Tollefsbol, T.O., Cohen, H.J. 1984. Werner 's syndrome: an underdiagnosed disorder resembling premature aging. Age. 7:75–88.
Toth, J.I., Yang, S.H., Qiao, X., Beigneux, A.P., Gelb, M.H. et al. 2005. Blocking protein farnesyltransferase improves nuclear shape in fibroblasts from humans with progeroid syndromes. Proc Natl Acad Sci USA. 102:12873–12878.
Toyota, M., Ahuja, N., Ohe-Toyota, M., Herman, J.G., Baylin, S.B., Issa, J.P. 1999. CpG island methylator phenotype in colorectal cancer. Proc Natl Acad Sci USA. 96:8681–8686
Van Maldergem, L., Siitonen, H.A., Jalkh, N., Chouery, E., De Roy, M. et al. 2006. Revisiting the craniosynostosisradial ray hypoplasia association: Baller-Gerold syndrome caused by mutations in the RECQL4 gene. J Med Genet. 43:148–152.
Vernell, R., Helin, K., Muller, H. 2003. Identification of target genes of the p16INK4A-pRB-E2F pathway J Biol Chem. 278:46124–46137
von Kobbe, C., Bohr V.A. 2002. A nucleolar targeting sequence in the Werner syndrome protein resides within residues 949–1092. J Cell Sci. 115:3901–3907.
Wang, X.W., Tseng, A., Ellis, N.A., Spillare, E.A., Linke, S.P. et al. 2001. Functional interaction of p53 and BLM DNA helicase in apoptosis. J Biol Chem. 276:32948–32955.
Weber, J.D., Taylor, L.J., Roussel, M.F., Sherr, C.J., Bar-Sagi, D. 1999. Nucleolar Arf sequesters Mdm2 and activates p53. Nat Cell Biol. 1:20–26
Weisenberger, D.J., Siegmund, K.D., Campan, M., Young, J., Long, T.I. et al. 2006. CpG island methylator phenotype underlies sporadic microsatellite instability and is tightly associated with BRAF mutation in colorectal cancer. Nat Genet. 38:787–793
Wilson, V.L., Jones, P.A. 1983. DNA methylation decreases in aging but not in immortal cells. Science. 220:1055–1057.
Wilson, V.L., Smith, R.A., Ma, S., Cutler, R.G. 1987. Genomic 5-methyldeoxycytidine decreases with age. J Biol Chem. 262:9948–9951
Yamamoto, K., Imakiire, A., Miyagawa, N., Kasahara, T. 2003. A report of two cases of Werner's syndrome and review of the literature. J Orthop Surg. **11**:224–23
Yu, C.E., Oshima, J., Fu, Y.H., Wijsman, E.M., Hisama, F. et al. 1996. Positional cloning of the Werner's syndrome gene. Science. 272:258–262.

DNA Methylation and Osteoarthritis

Helmtrud I. Roach

Abstract Osteoarthritis is a debilitating and progressive disease that affects around two-thirds of people of retirement age. A key feature is the degradation of articular cartilage, leading to loss of shock-absorbing capacity, pain, and difficulties in articulation of the joints. The enzymes involved in the cartilage degradation are, paradoxically, produced by the chondrocytes, which undergo a phenotypic change from normal chondrocytes, which express the typical chondrocytic genes (collagens type II, IX, and XI, aggrecan, Sox-9, etc.), to cells that aberrantly express cartilage matrix-degrading proteases and other genes that are not part of the normal repertoire of chondrocytes. This chapter evaluates the evidence that DNA de-methylation at critical CpG sites in the relevant promoters underlies the aberrant expression of non-chondrocytic genes. Although definitive data are still limited, evidence is presented that the aberrant expression of MMP-3, MMP-9, MMP-13, ADAMTS-4, and leptin is associated with loss of CpG methylation. Another feature of osteoarthritis is the silencing of many genes that are expressed by normal chondrocytes. This does not seem to be connected to hyper-methylation of the CpG island promoters of type II collagen, aggrecan or p21WAF1/CIP1 (an inhibitor of proliferation), although there is some evidence that hyper-methylation might contribute to the silencing of osteogenic protein-1, an anabolic factor for cartilage. Further work is urgently needed to investigate not only the CpG methylation status of other genes that are induced or silenced in osteoarthritis but also the mechanisms involved in the loss of methylation and the factors that might initiate this loss.

Keywords osteoarthritis · DNA methylation · epigenetics · cytokines · cartilage degradation · matrix metalloprotease

Introduction

One of the most frequent diseases associated with aging is osteoarthritis (OA), which affects ~60% of men and ~70% of women over the age of65 (Sarzi-Puttini

H.I. Roach (✉)
Bone & Joint Research Group, Southampton General Hospital, University of Southampton, Southampton, SO16 6YD, UK
e-mail: h.roach@southampton.ac.uk

T.O. Tollefsbol (ed.), *Epigenetics of Aging*, DOI 10.1007/978-1-4419-0639-7_20,

et al. 2005). The lifetime risk of symptomatic knee OA has been estimated at 44% with the risk increasing to 66% among obese people (Murphy et al. 2008). OA does not cause death, but severely affects the quality of life by significantly limiting everyday activities because movement becomes difficult and painful. This is a consequence of failure of the articular cartilage that covers the ends of long bones and, in a healthy individual, acts as a shock absorber and facilitates articulation (Aigner et al. 2006b; Goldring 2006; Roach and Tilley 2007). However, OA not only affects the cartilage, but is a disease of the joint. Swelling and sporadic inflammation of the synovial capsule and membrane and stiffening of the subchondral bone account for many of the symptoms of the disease (Roach et al. 2007). Given that OA is such a widespread disease, it is perhaps surprising that we still do not fully understand the cause and have so far no disease-modifying drugs available (Fajardo and Di Cesare 2005).

OA used to be considered an inevitable result of 'wear and tear' as a consequence of increasing age, often exacerbated by obesity. However, not all elderly persons suffer from OA (Carrington 2005). Moreover, the disease also arises in young people, for example, as secondary arthritis following failed treatment for developmental dysplasia of the hip, a condition that becomes clinically apparent when the patient is 20–30 years of age (da Silva et al. 2008). Alterations in joint anatomy, e.g., impingement of a non-spherical head into the acetabulum, also lead to osteoarthritic changes in young adults (Wagner et al. 2003). More commonly, joint injury, as sustained in skiing accidents during the second to fourth decade, increases the risk of developing OA after the age of 60 (Lohmander et al. 2007).

While these factors may explain secondary OA in some cases, this does not apply to the majority of patients with primary or idiopathic OA. In these patients, OA can be precipitated by many factors, either extrinsic or intrinsic. External factors include an imbalance between the load exerted and the capacity of the cartilage to absorb it, for example, the skiing injuries mentioned above. Intrinsic factors include genetic susceptibility. Recent years have seen extensive research directed at identifying susceptibility genes for OA. The disease is clearly polygenetic, i.e., the result from the interaction of a number of genes. Linkage studies have implicated interleukin-1 (IL-1), matrilin 3, IL-4 receptor alpha, secreted frizzled-related protein 3 (sfrp3), ADAM12, and asporin (ASPN) (Loughlin 2005). Of these IL-1, IL-4Rα, sfrp3, and ASPN are involved in chondrocyte differentiation. Genome-wide association studies additionally implicated the vitamin D receptor (VDR), estrogen receptor alpha (ERα), transforming growth factor-beta (TGF-β), insulin-like growth factor-1 (IGF-1), cartilage oligomeric matrix protein (COMP), collagen type IX α1 chain (COL9A1), collagen type X1α1 chain (COL11A1), and the ankylosing gene (ANK) (Spector and Macgregor 2004). The effects of individual genes are relatively modest, but the combination of several gene variations may result in additive effects and thereby increase the risk of OA (Valdes et al. 2008).

Twin studies have shown that the effect of genetic factors varies at different target joints: the combined effect is 39–65% for hand and knee OA in women, 60% for OA of the hip, and up to 70% for OA of the spine (Spector and Macgregor 2004). This incomplete concordance in monozygotic twins points to the importance of environmental factors. Moreover, genetic factors determine predisposition or risk, but not the actual incidence of disease in a given individual which likely results from a combination of genetic predisposition and environmental factors. Chondrocytes seem to be particularly prone to changes in their microenvironment, as indicated by the well-known loss of phenotype in monolayer culture (Dessau et al. 1978). This raises the question of how environmental factors could cause the disease in genetically susceptible individuals. At the molecular level, the integration of environmental signals with the genome is thought to involve changes in the epigenome (Jaenisch and Bird 2003). Epigenetic changes result in long-term altered gene expression, either through inappropriate activation of genes that are normally suppressed or through silencing of genes that are essential for normal cellular function. Epigenetics could thus provide a molecular explanation for changed gene expression patterns. To understand how and why epigenetic changes might be involved in the initiation and/or progression of OA, one must understand the function of articular cartilage and how disease-associated changes compromise that function in OA. The cartilage undergoes two kinds of changes: (i) degradation of the extracellular cartilage matrix and (ii) an abnormal and radical change in the gene expression repertoire of the cartilage cells (chondrocytes).

Function and Characteristics of Normal Articular Cartilage

Articular cartilage covers the ends of long bones. This cartilage is avascular and aneural, acts as shock absorber and, together with synovial fluid, is essential for the pain-free movement of bones during articulation. As a result of cartilage damage, the characteristic smooth surface protecting the bone ends is lost, and friction during joint movement becomes abnormally high, causing pain and making movement difficult. It is the structure of the extracellular matrix that gives articular cartilage the capacity to absorb shock. Cartilage is a highly hydrated gel. Its organic matrix consists of a loose mesh of type II collagen along with collagen molecules of type IX and XI and other minor types (Aigner and Stove 2003). These provide tensile strength and compressive stiffness. Highly negatively charged aggrecan molecules are interspersed within the collagen mesh. They absorb water, causing the matrix to swell, but the swelling is limited by the meshwork of collagen fibers. The aggrecan/water structure thus provides cartilage with its elasticity and shock-absorbing capacity.

The cartilage matrix is normally maintained by the articular chondrocytes, which make up about 3% of the cartilage matrix. These chondrocytes divide rarely, have a long life, and, their low metabolic activity notwithstanding, typically express cartilage matrix genes (Aigner et al. 2003). These include among many others the genes that give rise to aggrecan, the typical cartilage collagens (types II, IX, XI as well as many minor types), and cartilage oligomeric protein (COMP). On the other hand, matrix-degrading enzymes, such as the matrix metalloproteinases MMP-1,

MMP-2, MMP-9, MMP-13 or the aggrecanase ADAMTS-4, are *not* part of the gene expression repertoire of normal articular chondrocytes (Aigner et al. 2004). Normal chondrocytes also do not express inflammatory cytokines, such as IL-1β or TNF-α (Ayache et al. 2002; Dozin et al. 2002; Fan et al. 2007; Goldring and Goldring 2004), or the inducible nitric oxide synthase (iNOS) (Hauselmann et al. 1998) or cyclo-oxygenase-2 (Cox-2) (Lianxu et al. 2005), enzymes involved in the synthesis of nitric oxide or prostaglandins, respectively. Yet all of these genes are expressed by some OA chondrocytes.

Changes in Extracellular Matrix During Progression of OA

Unlike animal models of OA, where the disease develops within days or weeks, the human disease often takes decades before OA becomes clinically apparent. This suggests that the changes are gradual and progressive, with clinical symptoms becoming apparent only after a threshold has been overcome. Another feature is that OA progression cannot be halted once cartilage erosion has started.

The availability of human articular cartilage as a consequence of joint replacements has made it possible to map the stages of the disease (for recent reviews, see Aigner et al. (2006b; Goldring and Goldring (2007, 2006); Roach and Tilley (2007); Wu and Kalunian (2005)). Articular cartilage can be divided into superficial, intermediate, and deep zones. The surface of the superficial zone of healthy cartilage is smooth, the chondrocytes are flattened cells, aligned parallel to the surface (Fig. 1A), as are the collagen fibers. In the intermediate zone, chondrocytes are rounded and the main orientation of collagen fibers is vertical. This zone merges into the deep zone and calcified cartilage, which abuts the subchondral bone. In OA, aggrecan is lost first from the superficial zone in weight-bearing regions (Roach and Tilley 2007). With time, the region of aggrecan loss progresses to the intermediate zone. The concurrent loss of collagen from the superficial zone leads to an overall thinning of cartilage matrix, first in the weight-bearing regions, but eventually from other regions as well.

The enzymes involved in the progressive erosion of the articular cartilage have been well defined. The aggrecanases ADAMTS-4 and ADAMTS-5 (Malfait et al. 2002; Sandy and Verscharen 2001; Stanton et al. 2005; Tortorella et al. 2001) as well as MMP-3 (stromelysin) (Freemont et al. 1997; Mehraban et al. 1998; Okada et al. 1992) degrade aggrecan. MMP-13 (collagenase-3) is the major collagenase involved (Neuhold et al. 2001; Shlopov et al. 2000), followed by the gelatinases MMP-2 and MMP-9 (Tsuchiya et al. 1996) (Duerr et al. 2004; Freemont et al. 1997; Masuhara et al. 2000; Soder et al. 2006), which cleave the denatured collagen fibrils, while MMP-7 (Ohta et al. 1998), MMP-8 (Chubinskaya et al. 1999), and MMP-14, a membrane type-I MMP (Imai et al. 1997), may also be involved, possibly as activators of the other MMPs. Initially, some of these enzymes are probably produced by synovial fibroblasts, especially if the synovium is inflamed (Okada et al. 1992; Wolfe et al. 1993). Synthetic MMP inhibitors were found to be chondro-protective

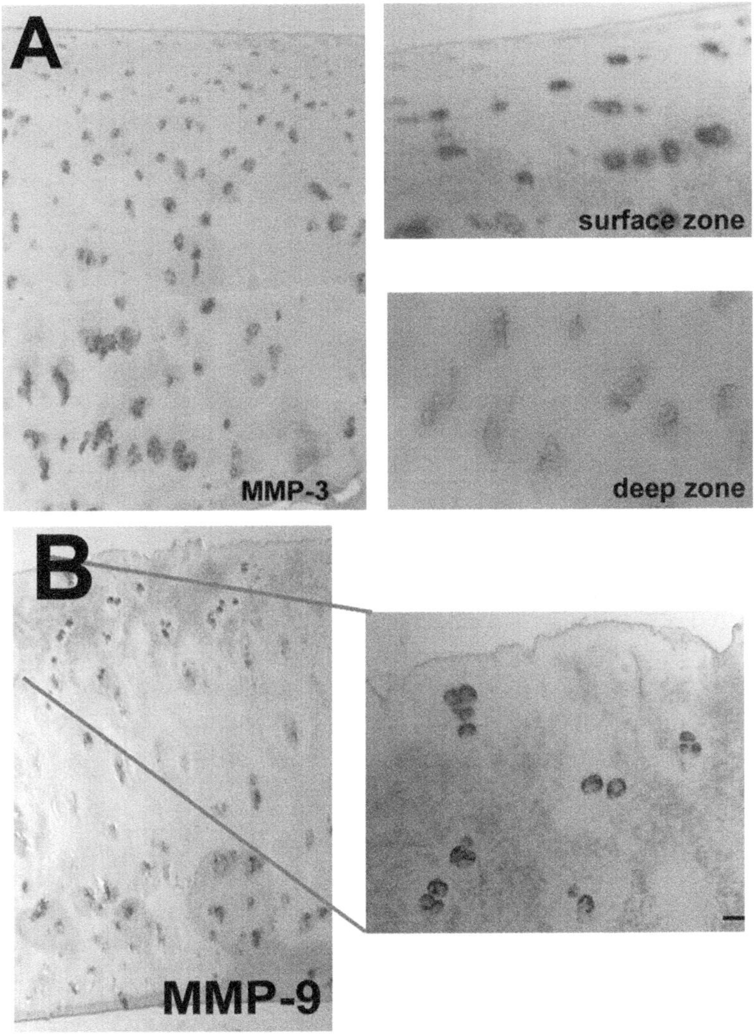

Fig. 1 Distribution of 'degradative' chondrocytes in (**A**) control cartilage from a patient with a femoral neck fracture (#NOF) and (**B**) early-stage OA cartilage. 'Degradative' chondrocytes are characterized by production of MMPs and aggrecanases, as illustrated by immunocytochemistry for MMP-3 (**A**) and MMP-9 (**B**). The same localization is found for MMP-13 and ADAMTS-4. Very few 'degradative' chondrocytes are present in the surface zone of control cartilage, but their numbers and location have increased in early OA cartilage. Figure 20.1A reprinted with permission from Roach et al. (2005b)

in animal models of OA (Milner and Cawston 2005), which are trauma induced and accompanied by inflammation. However, results for the use of MMP inhibitors in clinical trials have been equivocal (Milner and Cawston 2005). This suggests that synovium-derived enzymes may be less important in human OA and articular chondrocytes could, paradoxically, be the main source of the enzymes that degrade the cartilage matrix (Roach and Tilley 2007; Tetlow et al. 2001).

The 'Degradative' Chondrocyte Phenotype

The central role of chondrocytes in the disease process has been recognized for some time (Goldring 2000; Goldring 2006; Goldring and Goldring 2007; Roach and Tilley 2007; Sandell and Aigner 2001) and many studies have compared the characteristics of chondrocytes from non-OA patients with those from OA patients. Immunohistochemical studies have been particularly informative in that they showed that not all chondrocytes of an OA patient have the same characteristics, but that the cell phenotype depends on location and disease severity. As already mentioned, normal articular chondrocytes, on the whole, do not produce the degradative enzymes. However, even in non-OA cartilage, there are always a few cells near the surface that are immunopositive for the degradative enzymes (shown for MMP-3 in 20.1A). In low-grade OA, the number of chondrocytes located in the upper zone of the cartilage that produce degradative enzymes has increased, as illustrated in Fig. 1B for MMP-9, whereas the cells in the deep zone are still normal chondrocytes, at least as far as the expression of degradative enzymes is concerned. These findings have been confirmed by other immunolocalization studies (Okada et al. 1992; Roach et al. 2005b; Tetlow et al. 2001; Walter et al. 1998; Wu et al. 2002). To distinguish those OA chondrocytes that express degradative enzymes from those that do not, we have termed the former 'degradative' chondrocytes (Roach et al. 2007; Roach et al. 2005b). Recognizing this heterogeneity is important for understanding the cellular changes leading to OA, because it indicates that not all chondrocytes in an OA patient differ from normal chondrocytes. Yet this heterogeneity is often overlooked in gene expression studies, where the need to obtain enough cells for RNA extraction leads to the pooling of all chondrocytes from a given OA patient. As a result, the differences between chondrocytes located near the surface and those of the deep zone will not be recognized. This difference is illustrated in Fig. 2, where the expression of ADAMTS-4 was examined by quantitative real-time PCR in normal chondrocytes and then separately for the surface and deep zones of OA chondrocytes. As expected, there was negligible expression of ADAMTS-4 in normal chondrocytes, but expression of ADAMTS-4 was increased 70- to 700-fold in OA chondrocytes located in the surface zone. This increase was generally not found in deep zone chondrocytes, although some increase was observed in two samples. This suggests that some 'degradative' chondrocytes were already present in the deep zone of these two samples. The considerable variability between the samples might have been partly due to inter-patient variability, but may also have resulted from variations in the zonal dissection of the articular cartilage.

As OA progresses, the enzyme-expressing cells divide to form doublets (Fig. 3B), quadruplets (Fig. 3C), and finally the typical clones of severe OA (Fig. 3D), in which all cells express the matrix-degrading enzymes (shown for

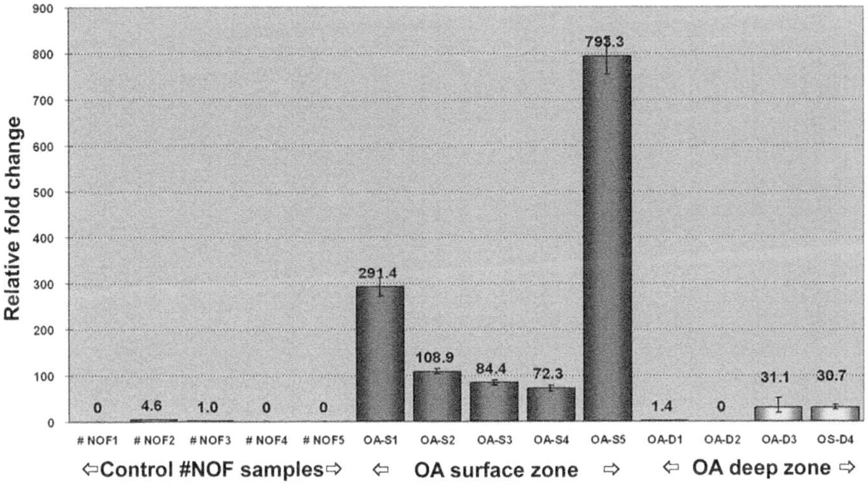

Fig. 2 Real-time RT-PCR of ADAMTS-4 expression in control chondrocytes, the surface zone of OA patients (containing 'degradative' chondrocytes), and the deep zone of OA patients. Relative quantification of gene expression was performed using the Applied Biosystems ABI Prism 7500 sequence detection system. PCR reactions for all samples were performed in triplicate in 96-well optical plates. Glyceraldehyde-3-phosphate dehydrogenase (GAPDH) was used as internal control gene for normalization and the expression of ADAMTS-4 was relatively quantified by $2^{-\Delta\Delta Ct}$ method. The highest expression levels are in the surface zone of OA patients

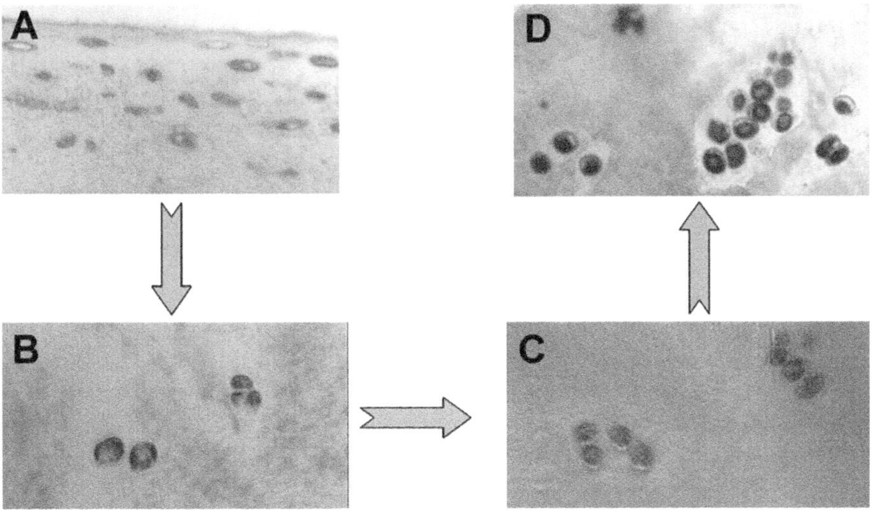

Fig. 3 Proliferation of 'degradative' chondrocytes with increasing severity of OA. Only the occasional protease-expressing cell is seen near the surface in control cartilage (**A**). In low-grade OA cartilage, doublets are frequently found (**B**), which become quadruplets in medium-grade OA cartilage (**C**) and finally the typical clones of high-grade OA cartilage (**D**). Immunocytochemistry for MMP-9. Reprinted with permission from Roach et al. (2005b)

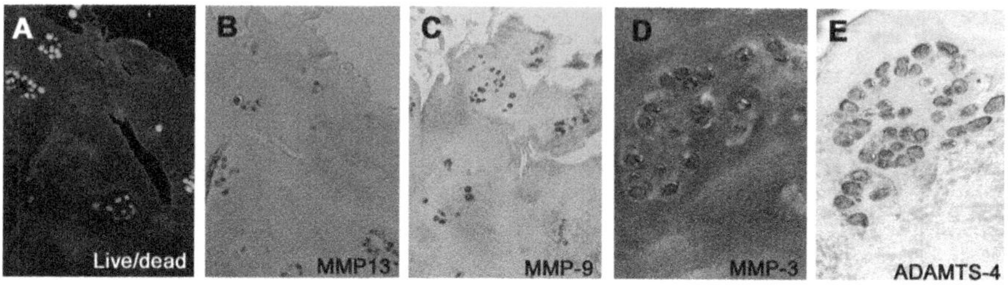

Fig. 4 Chondrocytes in the articular cartilage of severe OA. (**A**) Bright CellTracker green staining. CellTracker green reacts with house-keeping esterases to produce a green fluorescent product and thus labels viable, metabolically active cells (Roach and Clarke 2000). All the cells within a clone produce the degradative enzymes MMP-13, MMP-9, MMP-3, and ADAMTS-4 (**B–E**). The enzymes are released into the matrix, as shown in B–D, which is better visible when the sections had been treated with hyaluronidase. The proteoglycans present in the cartilage matrix tend to cover up the epitope-binding sites for the antibodies in the matrix. Treatment with hyaluronidase prior to staining digests the proteoglycans and thus liberates the epitopes. Absence of matrix staining for ADAMTS-4 is due to lack of hyaluronidase treatment in this particular section. Reprinted with permission from Roach (2008)

MMP-9 in Fig. 3). This clearly illustrates that the aberrant gene expression is transmitted to daughter cells. Interestingly, the cells that are present as clones or clusters in severe OA are metabolically active, as shown by the intense CellTracker green stain (Fig. 4A). These viable cells express several matrix-degrading enzymes, such as MMP-13 (Fig. 4B), MMP-9 (Fig. 4C), MMP-3 (Fig. 4D), and ADAMTS-4 (Fig. 4E). As these enzymes are released into the matrix, they cause degradation and ultimately loss of the cartilage matrix. It is the author's hypothesis that these phenotypically modulated 'degradative' chondrocytes lie at the heart of disease progression in OA and that the severity of the disease increases in parallel with the number of chondrocytes that change to the 'degradative' phenotype (Roach 2008; Roach and Tilley 2007).

Reasons for Suspecting Epigenetic Changes in OA

As already pointed out, OA progression involves a phenotypic modulation of an increasing number of articular chondrocytes (Aigner and Dudhia 1997) to cells of the 'degradative' phenotype. In addition to the matrix-degrading enzymes already mentioned, the 'degradative' cells also express other non-chondrocytic genes, e.g., VEGF (Enomoto et al. 2003), pleiotrophin (Pufe et al. 2003), and leptin (Iliopoulos et al. 2007). This suggests that a fundamental change in gene expression has taken place, characterized by inappropriate activation of many genes. At the same time, expression of typical chondrocytic genes may be suppressed, although some activation as part of an attempted repair response also takes place (Aigner et al. 2006a; Ijiri et al. 2008). The fact that abnormal 'degradative' cells appear first as individual chondrocytes, later as doublets, then as quadruplets and clones suggests that, in contrast to the normal chondrocytes, the aberrant cells have proliferated and transmitted the pattern of aberrant gene expression to their daughter cells. In other words, all cluster cells are of the 'degradative' phenotype and *the aberrant phenotype is both stable and heritable*. With increasing severity of OA, 'degradative' chondrocytes increase in number, first as a result of cell division of the abnormal chondrocytes, and second because cartilage cells in the middle and deep zones change phenotype to 'degradative' chondrocytes (Roach and Tilley 2007). Moreover, OA chondrocytes express not only one, but many abnormal genes. This suggests that *an overall permanent switch in the phenotype* of the chondrocytes has occurred rather than short-term upregulation of one or more factors. The permanent switch in phenotype, as well as the heritability of the gene expression changes, suggests that destabilization of the chondrocytic phenotype in osteoarthritis may well be the result of changes in epigenetic status. Pathological epigenetic disruption may either activate normally silent genes or silence normally active genes (Feinberg 2007). This is precisely the situation in arthritic OA chondrocytes, where cartilage-degrading proteases are permanently activated, whereas the normal chondrocytic genes tend to be silenced.

Based on our knowledge of epigenetics, one would predict that those genes that are not part of the repertoire of normal chondrocytes would be silenced. Silencing would be mediated by a high level of DNA methylation combined with histone de-acetylation on histones 3 and 4, methylation at histone 3 lysine 9 (H3K9) and H3K27, interaction with heterochromatin proteins, and a closed chromatin structure, all of which are the hallmarks of transcriptionally silent genes (Feinberg 2007; Gan et al. 2007). When those genes are induced in 'degradative' OA chondrocytes, one would predict a change in epigenetic status to a low level of DNA methylation, acetylated histones 3 and 4 and methylated lysine 4 on H3 (H3K4), interaction with euchromatin proteins, and an open chromatin structure. Since both normal and aberrant DNA methylation patterns are stable in somatic cells (Bird 2002; Reik 2007), whereas histone modifications are more readily reversible by specific enzymes (Berger 2007), DNA methylation status constitutes a good index of epigenetic status. Moreover, the DNA methylation pattern is rapidly reproduced during cell division on the nascent strand by the maintenance methyl transferase DNMT1 (Attwood et al. 2002; Martin and Zhang 2007). The histone code can be re-established after cell division by involving methyl-binding domains (MBDs) and DNMTs, which interact with histone methyltransferases and histone deacetylases (Fujita et al. 2003; Fuks et al. 2003), so that both the DNA methylation pattern and the histone code are normally faithfully reproduced during mitosis.

Loss of DNA Methylation and Aberrant Expression of Proteases in 'Degradative' OA Chondrocytes

Data on DNA methylation of specific CpG sites in specific genes in relation to osteoarthritis are still sparse. Kim et al. (1996) found that inflammatory arthritis was associated with overall DNA hypomethylation in peripheral blood mononuclear cells, but did not investigate specific genes or chondrocytes. Preliminary studies suggest that epigenetics may be important in rheumatoid arthritis (RA) (Brooks 2005; Sanchez-Pernaute et al. 2008). For example, one feature of RA is the chronic infiltration of peripheral blood mononuclear cells (PBMCs) into the synovium, a process stimulated by the multifunctional cytokine interleukin-6 (IL-6). Nile et al. (2008) found that a loss of methylation at a single CpG site at –1099 bp in the IL-6 promoter, as found in PBMCs of RA patients, correlated with increased IL-6 expression in RA. Surprisingly, little change was found between healthy and RA patients in the methylation status of 16 CpG sites in the region –666 to +27 bp, which were mostly un-methylated, or in the 7 CpG sites between –1096 and –1001, which were highly methylated. The observation that methylation status at a single CpG site can be crucial for gene expression is indeed remarkable.

To demonstrate changes in DNA methylation in osteoarthritic cartilage, it is important to ensure that samples of control cartilage contain predominantly normal cells, whereas samples of OA cartilage contain mostly 'degradative' chondrocytes. By discarding the surface layer of control cartilage, the sample will contain only normal chondrocytes (Fig. 5A). On the other hand, by selecting only the surface zone of OA cartilage (Fig. 5B) and the thin, partially eroded cartilage from the weight-bearing regions in OA patients (Fig. 4), one obtains a sample rich in degradative OA chondrocytes. These carefully selected samples make it possible to identify changes in DNA methylation of the major cartilage-degrading enzymes. The first

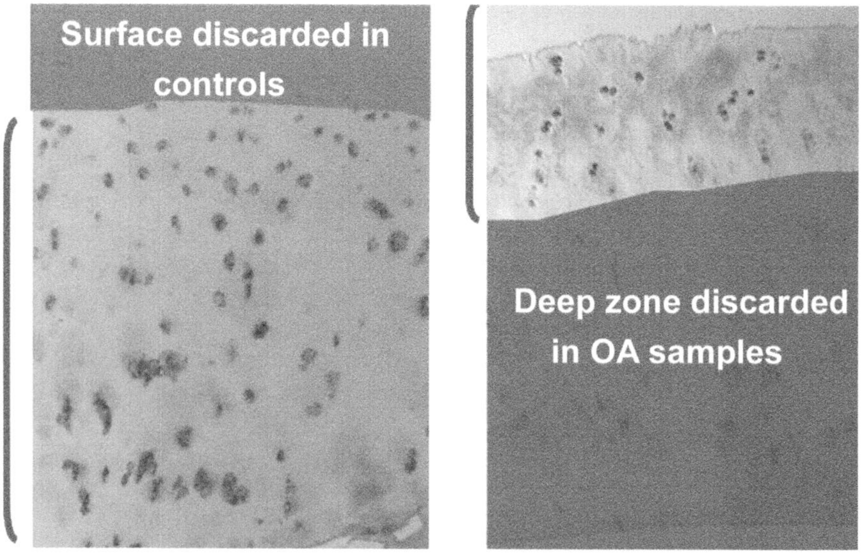

Fig. 5 Micro-dissection prior to extracting RNA or DNA to ensure that control chondrocytes are contrasted with 'degradative' OA chondrocytes. For controls (*left*), only the middle and deep zones should be analyzed, while for OA cartilage (*right*) only the surface zones should be sampled

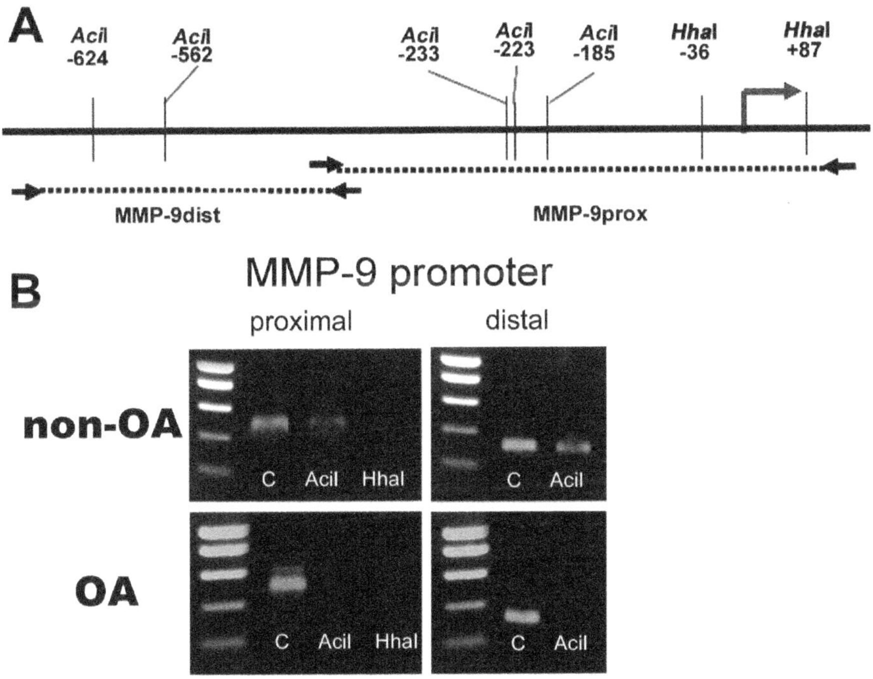

Fig. 6 Loss of DNA methylation in the MMP-9 promoter in osteoarthritis. (**A**) The MMP-9 promoter: Location of CpG sites (vertical bars) with methylation-sensitive transcription enzymes and regions spanned by primers. Another HhaI site, also covered by the proximal primers, is present at +87 bp. (**B**) DNA methylation status as determined by treatment with methylation-sensitive restriction enzymes, followed by PCR. Controls (**C**) are non-enzyme treated. Presence of a band in the samples treated with either *Aci*I or *Hha*I indicates presence of methylation, whereas absence of a band indicates loss of DNA methylation at one or more of the sites cut by the enzyme. (**B**) Reprinted with permission from Roach et al. (2005a)

gene examined was MMP-9, which contains just six CpGs in the 670 bp promoter sequence (Huhtala et al. 1991), see Fig. 6A. DNA methylation status was examined using a technique that depends on methylation-sensitive restriction enzymes followed by PCR (Pogribny et al. 2000; Roach and Hashimoto 2007). This technique takes advantage of the fact that certain restriction enzymes, for example, *Aci*I and *Hha*I, will only cleave the DNA at their recognition site if the DNA is not methylated. If cleavage does occur, no PCR amplification with primers bracketing this site is possible, whereas presence of DNA methylation allows PCR amplification. Hence presence of a PCR band indicates presence of CpG methylation and vice versa. Roach et al. (2005a) demonstrated that the three *Aci*I digestible sites located in the proximal promoter at –185, –223, and –233 bp relative to the transcription start site were methylated in normal chondrocytes, which did not express MMP-9, but at least one of these sites was un-methylated in OA chondrocytes that expressed MMP-9. The same applied to the two *Aci*I sites at –562 and –624 bp in the distal promoter. However, at least one of the two *Hha*I sites at –36 bp or at +87 bp was un-methylated in both non-OA and OA samples, see Fig. 6B.

Further studies (Cheung et al. 2008; Roach et al. 2005b) also investigated the methylation status of the promoters of MMP-3, MMP-13, and of ADAMTS-4, which are other degradative enzymes typically expressed de novo in OA chondrocytes. All of these promoters contain relatively few CpG sites (7 in 2000 bp for MMP-3, 8 in 600 bp for MMP-13, and 13 in 900 bp for ADAMTS-4), a situation that seems to favor pathological de-methylation. Figure 7 shows typical results. The methylation status of individual CpG sites varied widely, from sites that were methylated in both control and OA samples (most of the CpG sites in the MMP-13 promoter) to sites that were largely un-methylated in either group (the *Ava*I site at –721 in the MMP-9 promoter, see Fig. 7). However, each enzyme had specific

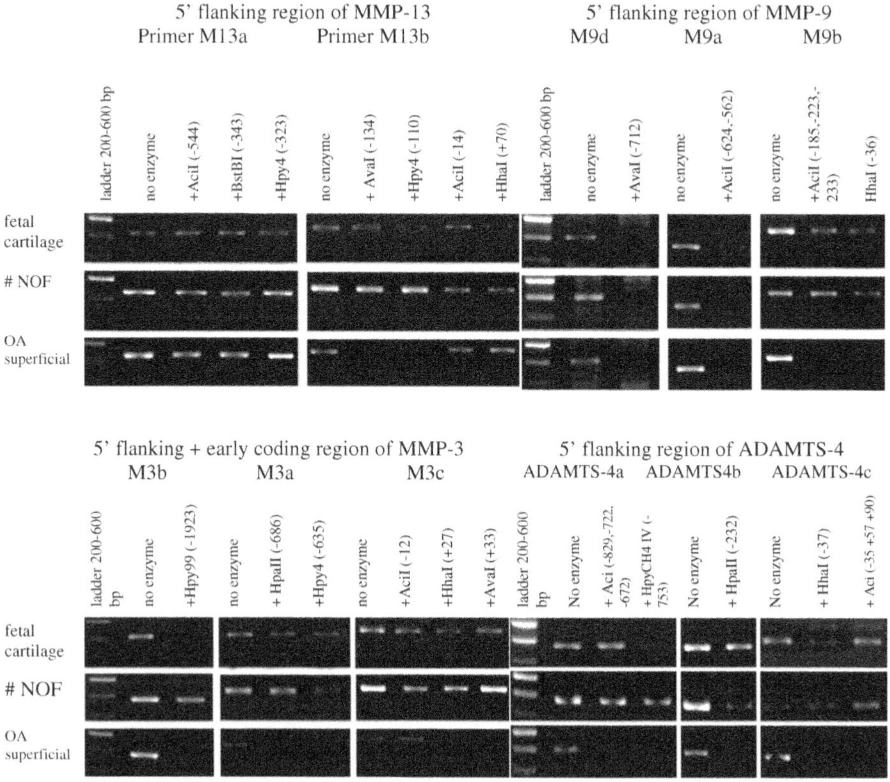

Fig. 7 Representative results of PCR amplifications of the promoter regions with or without digestion by methylation-sensitive restriction enzymes. PCR reactions were performed for each sample without or after enzyme digestions as indicated. A typical pattern of reactions is shown for DNA extracted from fetal control cartilage, control cartilage from a patient with femoral neck fracture (#NOF), and from the surface zone or around the weight-bearing area of an OA patient. The ladder shows 200 bp intervals. Hpy4, HpyCH4IV. Reprinted with permission from Roach et al. (2005b)

CpG sites that were methylated in control chondrocytes, but had lost methylation in most OA chondrocytes. Overall, the percentage of methylated CpG sites decreased from 80 to 52% in OA samples (Roach et al. 2005b). These data, together with the heritable nature of abnormal enzyme expression, are consistent with the concept that de-methylation is responsible for the focal gene activation pattern typical of 'degradative' cartilage cells in vivo.

Epigenetic Regulation of Leptin in Osteoarthritis

Leptin is normally secreted by white adipose tissue. Although mainly associated with food intake and energy consumption, leptin has pleiotropic actions. Expression in OA has detrimental effects by inducing the synthesis of nitric oxide synthesis, several pro-inflammatory cytokines, and MMP-13 (Dumond et al. 2003; Iliopoulos et al. 2007). Normal articular chondrocytes do not express leptin, but the gene is abnormally expressed by OA chondrocytes. Iliopoulos et al. (2007) investigated whether this changed gene expression pattern was due to altered epigenetic regulation. They divided cartilage samples into three groups, normal, minimal, and maximal OA. DNA methylation status of 32 CpG sites in the leptin promoter was determined by bisulfite modification. In normal cartilage, 22/32 CpG sites were methylated and expression of leptin was negligible (Fig. 8B). In minimally damaged

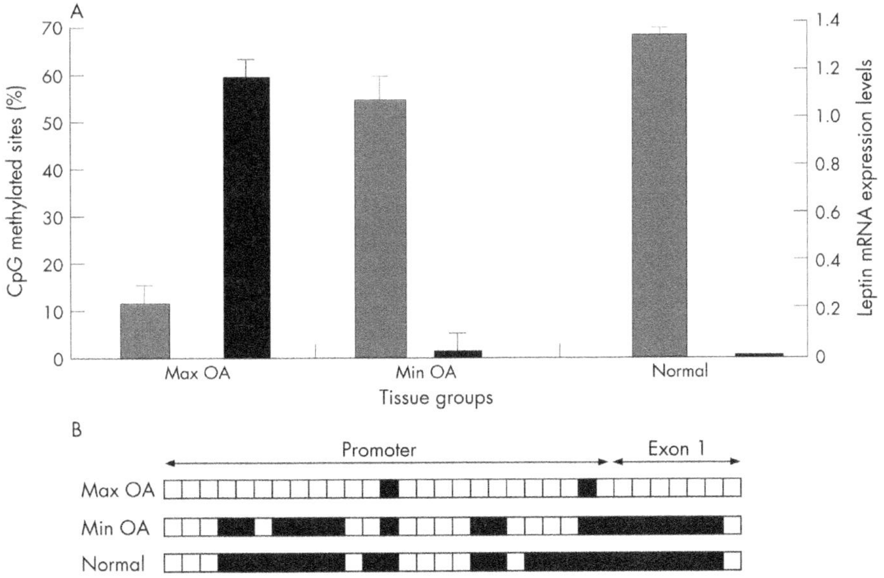

Fig. 8 (**A**) Loss of DNA methylation in the leptin promoter (*gray*) with increasing OA severity correlates with increased expression (*black bars*) of leptin. (**B**) Methylation status of individual CpG sites in the leptin promoter and exon 1. Normal chondrocytes show a high level of methylation, which had decreased in minimal OA, while in maximal (high-grade OA) only two CpG sites remained methylated. Reprinted with permission from Iliopoulos et al. (2007)

cartilage, 5 more CpG sites had lost DNA methylation, with a modest increase in leptin expression, while in maximal OA only 2/32 CpG sites remained methylated and leptin expression had increased ~60-fold. Treatment of normal chondrocytes with the DNA methylation inhibitor 5-aza-deoxycytidine resulted in loss of DNA methylation and an increase in leptin expression, thereby confirming the correlation between DNA methylation status and leptin expression.

Is Downregulation of Chondrocytic Genes Associated with Silencing by DNA Hyper-methylation?

Many chondrocytic genes are downregulated in aged articular cartilage. This may be due to the generally low metabolic activity of aged chondrocytes, downregulation in response to cytokines or transcription factors, or epigenetic silencing by DNA hyper-methylation or by silencing histone modifications. To date very few studies have examined which of these possibilities applies. The main chondrocytic genes are type II collagen (*COL2A1*) and aggrecan. Pöschl et al. (2005) examined the DNA methylation status at 33 CpG sites located in a 340 bp CpG island in the proximal promoter of aggrecan. All CpG sites were un-methylated with no difference between normal aged and OA cartilage. This suggests that silencing by increased DNA methylation was not involved in aggrecan downregulation in OA. Similarly, the CpG island of the *COL2A1* promoter was un-methylated in both non-expressing (marrow stromal cells) and COL2A1-expressing (stimulated MSCs; chondrocytes) cells (Zimmermann et al. 2008). However, expression of COL2A1 also depends on Sox-9, Sox-5, and Sox-6 binding to an enhancer region located in the first intron (de Crombrugghe et al. 2000; de Crombrugghe et al. 2001; Leung et al. 1998) and it is possible that differential DNA methylation patterns could be found in this 309 bp enhancer region, which contains 20 CpG sites. A further example of lack of association between decreased expression and DNA methylation is *p21WAF1/CIP1*, an inhibitor of proliferation. This gene is expressed by normal chondrocytes, which have a very low proliferation rate. In OA, p21WAF1/CIP1 is downregulated, which may explain the increased cell division observed in OA. However, Sesselmann et al. (2008) could not demonstrate any differential DNA methylation between normal and OA chondrocytes.

In view of these negative findings that DNA hyper-methylation is unlikely to be responsible for the downregulation of chondrocytic genes, one might be tempted to conclude that this applies to all. However, Loeser et al. (2008) provided evidence that the age-related decrease in the expression of osteogenic protein-1 (OP-1) by articular chondrocytes was associated with increased methylation of the promoter. OP-1, also known as bone morphogenetic protein-7, is an anabolic factor that stimulates cartilage matrix synthesis. Hence loss of expression would be detrimental to the maintenance of articular cartilage and may contribute to OA. Using methylation-specific PCR, Loeser et al. (2008) were able to show that some CpG sites were unmethylated in young people, but methylated in old individuals with methylation increasing with age. Treatment with 5-azacytidine, an inhibitor of DNA methylation, increased expression of chondrocytic genes ~2-fold, supporting the notion that DNA methylation had resulted in partial silencing of some chondrocytic genes.

Promoters vary considerably with regard to the number of CpG sites from those containing only sparse CpGs to CpG island promoters. In the past epigenetic regulation was thought to be particularly applicable to genes with CpG island promoters. However, as shown in the Human Epigenome project (www.epigenome.org), the vast majority of CpG island promoters are un-methylated in normal cells, irrespective of expression (Eckhardt et al. 2006). Because the *COL2A1*, *aggrecan*, and *p21WAF1/CIP1* promoters contain CpG islands, DNA methylation may not be the major factor that determines silencing in non-expressing cells. The OP-1 promoter also contains an island near the transcription start site, yet this seems to be methylated with age. There are other examples of genes with CpG island promoters that are subject to epigenetic regulation. One is the human secretin receptor gene, regulated by both Sp1/Sp3 and hyper-methylation of a CpG island (Pang et al. 2004). Silencing of the bone-specific collagen type I gene in non-expressing cell types seems to depend on DNA methylation of the CpG island (Guenette et al. 1992; Yamane et al. 2005). Pathological hyper-methylation of the CpG island promoters occurs frequently in tumor suppressor genes (Esteller 2007). Silencing appears to be initiated by removal of Sp1 from specific Sp1/Sp3-binding sites (Li et al. 2008; Stirzaker et al. 2004), followed by histone modifications and de novo CpG methylation (Strunnikova et al. 2005). Once CpG methylation has occurred, Sp1 and other transcription factors, such as CREB, can no longer bind (Mancini et al. 1999; Pang et al. 2004) and the gene is silenced long term (Aoyama et al. 2004). Hence the possibility that pathological hyper-methylation occurs in at least some of those chondrocytic genes that are silenced in OA cannot be excluded.

Summary: Is Osteoarthritis an Epigenetic Disease?

Methylation of genomic DNA clearly represents a very important mechanism to determine tissue and cell differentiation. Aberrant methylation patterns are involved in important pathologies such as tumorigenesis and probably late-onset, complex non-Mendelian diseases (Petronis 2001), including osteoarthritis. The evidence presented in this chapter suggests that changes in the DNA methylation pattern could explain the fact that the gene expression pattern of osteoarthritic chondrocytes is altered in a stable and heritable manner, i.e., this alteration is transmitted to daughter cells. So far, this has only been shown for MMP-3, MMP-9, MMP-13, ADAMTS-4, and leptin. It is not yet known whether loss of DNA methylation also underlies the aberrant expression of other genes in OA cartilage or whether transcription factors, growth factors, and cytokines can induce expression of aberrant genes independent of DNA de-methylation. The interactions between CpG methylation and changes in histone modifications will also be of interest. No doubt future studies will investigate these issues.

The individual molecular steps by which normal chondrocytes are converted to the 'degradative' phenotype remain enigmatic. Stochastic changes as a result of accumulated errors in replicating the DNA methylation pattern during cell division may contribute, but are unlikely to be the only explanation because normal human articular chondrocytes divide only rarely and OA can occur in young adults. Further studies directed at identifying the mechanism(s) by which DNA methylation is lost are urgently needed as are studies investigating which factors initiate the phenotypic change to 'degradative' chondrocytes.

The pathological alterations that result from changes in epigenetic regulation cannot readily be reversed and are transmitted to daughter cells. This being the case, the consequences of epigenetic changes are more disastrous than the short-term up- or downregulation of expression that occurs in response to transcription factors or cytokines. If it proved possible to prevent or reverse epigenetic changes, progression of OA would be slowed, maybe even halted. That would constitute a major therapeutic benefit (Roach 2008) to OA patients whose numbers will increase in the future as a result of the longer life span in the developed and developing worlds. However, at present there are no therapies to treat OA that are based on epigenetics.

References

Aigner, T., and Dudhia, J. 1997. Phenotypic modulation of chondrocytes as a potential therapeutic target in osteoarthritis: a hypothesis. Ann. Rheum. Dis. 56:287–291.

Aigner, T., Fundel, K., Saas, J., Gebhard, P. M., Haag, J., Weiss, T., Zien, A., Obermayr, F., Zimmer, R., and Bartnik, E. 2006a. Large-scale gene expression profiling reveals major pathogenetic pathways of cartilage degeneration in osteoarthritis. Arthritis Rheum. 54:3533–3544.

Aigner, T., Saas, J., Zien, A., Zimmer, R., Gebhard, P. M., and Knorr, T. 2004. Analysis of differential gene expression in healthy and osteoarthritic cartilage and isolated chondrocytes by microarray analysis. Methods Mol. Med. 100:109–128.

Aigner, T., Sachse, A., Gebhard, P. M., and Roach, H. I. 2006b. Osteoarthritis: Pathobiology-targets and ways for therapeutic intervention. Adv. Drug Deliv. Rev. 58:128–149.

Aigner, T., and Stove, J. 2003. Collagens – major component of the physiological cartilage matrix, major target of cartilage degeneration, major tool in cartilage repair. Adv. Drug Deliv. Rev. 55:1569–1593.

Aigner, T., Zien, A., Hanisch, D., and Zimmer, R. 2003. Gene expression in chondrocytes assessed with use of microarrays. J. Bone Joint Surg. Am. 85-A Suppl 2:117–123.

Aoyama, T., Okamoto, T., Nagayama, S., Nishijo, K., Ishibe, T., Yasura, K., Nakayama, T., Nakamura, T., and Toguchida, J. 2004. Methylation in the core-promoter region of the chondromodulin-I gene determines the cell-specific expression by regulating the binding of transcriptional activator Sp3. J. Biol. Chem. 279:28789–28797.

Attwood, J. T., Yung, R. L., and Richardson, B. C. 2002. DNA methylation and the regulation of gene transcription. Cell Mol. Life Sci. 59:241–257.

Ayache, N., Boumediene, K., Mathy-Hartert, M., Reginster, J. Y., Henrotin, Y., and Pujol, J. P. 2002. Expression of TGF-betas and their receptors is differentially modulated by reactive oxygen species and nitric oxide in human articular chondrocytes. Osteoarthritis Cartilage. 10: 344–352.

Berger, S. L. 2007. The complex language of chromatin regulation during transcription. Nature. 447:407–412.

Bird, A. 2002. DNA methylation patterns and epigenetic memory. Genes Dev. 16:6–21.

Brooks, W. H. 2005. Autoimmune disorders result from loss of epigenetic control following chromosome damage. Med. Hypotheses. 64:590–598.

Carrington, J. L. 2005. Aging bone and cartilage: cross-cutting issues. Biochem. Biophys. Res. Commun. 328:700–708.

Cheung, K. S., Hashimoto, K., Yamada, N., and Roach, H. I. 2008. Expression of ADAMTS-4 by chondrocytes in the surface zone of human osteoarthritic cartilage is regulated by epigenetic DNA de-methylation. Rheumatol. Int. [Epub ahead of print]:DOI 10.1007/s00296-008-0744

Chubinskaya, S., Kuettner, K. E., and Cole, A. A. 1999. Expression of matrix metalloproteinases in normal and damaged articular cartilage from human knee and ankle joints. Lab Invest. 79: 1669–1677.

da Silva, M. A., Yamada, N., Clarke, N. M., and Roach, H. I. 2008. Cellular and epigenetic features of a young healthy and a young osteoarthritic cartilage compared with aged control and OA cartilage. J. Orthop. Res. [E-pub ahead of print]:DOI 10.1002/jor.20799

de Crombrugghe, B., Lefebvre, V., Behringer, R. R., Bi, W., Murakami, S., and Huang, W. 2000. Transcriptional mechanisms of chondrocyte differentiation. Matrix Biol. 19:389–394.

de Crombrugghe, B., Lefebvre, V., and Nakashima, K. 2001. Regulatory mechanisms in the pathways of cartilage and bone formation. Curr Opin Cell Biol. 13:721–727.

Dessau, W., Sasse, J., Timpl, R., Jilek, F., and Von der, M. K. 1978. Synthesis and extracellular deposition of fibronectin in chondrocyte cultures. Response to the removal of extracellular cartilage matrix. J. Cell Biol. 79:342–355.

Dozin, B., Malpeli, M., Camardella, L., Cancedda, R., and Pietrangelo, A. 2002. Response of young, aged and osteoarthritic human articular chondrocytes to inflammatory cytokines: molecular and cellular aspects. Matrix Biol. 21:449–459.

Duerr, S., Stremme, S., Soeder, S., Bau, B., and Aigner, T. 2004. MMP-2/gelatinase A is a gene product of human adult articular chondrocytes and is increased in osteoarthritic cartilage. Clin. Exp. Rheumatol. 22:603–608.

Dumond, H., Presle, N., Terlain, B., Mainard, D., Loeuille, D., Netter, P., and Pottie, P. 2003. Evidence for a key role of leptin in osteoarthritis. Arthritis Rheum. 48:3118–3129.

Eckhardt, F., Lewin, J., Cortese, R., Rakyan, V. K., Attwood, J., Burger, M., Burton, J., Cox, T. V., Davies, R., Down, T. A., Haefliger, C., Horton, R., Howe, K., Jackson, D. K., Kunde, J., Koenig, C., Liddle, J., Niblett, D., Otto, T., Pettett, R., Seemann, S., Thompson, C., West, T., Rogers, J., Olek, A., Berlin, K., and Beck, S. 2006. DNA methylation profiling of human chromosomes 6, 20 and 22. Nat. Genet. 38:1378–1385.

Enomoto, H., Inoki, I., Komiya, K., Shiomi, T., Ikeda, E., Obata, K., Matsumoto, H., Toyama, Y., and Okada, Y. 2003. Vascular endothelial growth factor isoforms and their receptors are expressed in human osteoarthritic cartilage. Am J Pathol. 162:171–181.

Esteller, M. 2007. Epigenetic gene silencing in cancer: the DNA hypermethylome. Hum. Mol. Genet. 16 Spec No 1:R50–R59.

Fajardo, M., and Di Cesare, P. E. 2005. Disease-modifying therapies for osteoarthritis: current status. Drugs Aging. 22:141–161.

Fan, Z., Soder, S., Oehler, S., Fundel, K., and Aigner, T. 2007. Activation of interleukin-1 signaling cascades in normal and osteoarthritic articular cartilage. Am. J. Pathol. 171:938–946.

Feinberg, A. P. 2007. Phenotypic plasticity and the epigenetics of human disease. Nature. 447: 433–440.

Freemont, A. J., Hampson, V., Tilman, R., Goupille, P., Taiwo, Y., and Hoyland, J. A. 1997. Gene expression of matrix metalloproteinases 1, 3, and 9 by chondrocytes in osteoarthritic human knee articular cartilage is zone and grade specific. Ann. Rheum. Dis. 56:542–549.

Fujita, N., Watanabe, S., Ichimura, T., Tsuruzoe, S., Shinkai, Y., Tachibana, M., Chiba, T., and Nakao, M. 2003. Methyl-CpG binding domain 1 (MBD1) interacts with the Suv39h1-HP1

heterochromatic complex for DNA methylation-based transcriptional repression. J. Biol. Chem. 278:24132–24138.
Fuks, F., Hurd, P. J., Wolf, D., Nan, X., Bird, A. P., and Kouzarides, T. 2003. The methyl-CpG-binding protein MeCP2 links DNA methylation to histone methylation. J. Biol. Chem. 278:4035–4040.
Gan, Q., Yoshida, T., McDonald, O. G., and Owens, G. K. 2007. Epigenetic Mechanisms Contribute to Pluripotency and Cell Lineage Determination of Embryonic Stem Cells. Stem Cells. 25:2–9.
Goldring, M. B. 2006. Update on the biology of the chondrocyte and new approaches to treating cartilage diseases. Best. Pract. Res. Clin. Rheumatol. 20:1003–1025.
Goldring, M. B. 2000. The role of the chondrocyte in osteoarthritis. Arthritis Rheum. 43: 1916–1926.
Goldring, M. B., and Goldring, S. R. 2007. Osteoarthritis. J. Cell Physiol. 213:626–634.
Goldring, S. R., and Goldring, M. B. 2004. The role of cytokines in cartilage matrix degeneration in osteoarthritis. Clin. Orthop. Relat Res. S27–S36.
Goldring, S. R., and Goldring, M. B. 2006. Clinical aspects, pathology and pathophysiology of osteoarthritis. J. Musculoskelet. Neuronal. Interact. 6:376–378.
Guenette, D. K., Ritzenthaler, J. D., Foley, J., Jackson, J. D., and Smith, B. D. 1992. DNA methylation inhibits transcription of procollagen alpha 2(I) promoters. Biochem. J. 283:699–703.
Hauselmann, H. J., Stefanovic-Racic, M., Michel, B. A., and Evans, C. H. 1998. Differences in nitric oxide production by superficial and deep human articular chondrocytes: implications for proteoglycan turnover in inflammatory joint diseases. J. Immunol. 160:1444–1448.
Huhtala, P., Tuuttila, A., Chow, L. T., Lohi, J., Keski-Oja, J., and Tryggvason, K. 1991. Complete structure of the human gene for 92-kDa type IV collagenase. Divergent regulation of expression for the 92- and 72-kilodalton enzyme genes in HT-1080 cells. J. Biol. Chem. 266: 16485–16490.
Ijiri, K., Zerbini, L. F., Peng, H., Otu, H. H., Tsuchimochi, K., Otero, M., Dragomir, C., Walsh, N., Bierbaum, B. E., Mattingly, D., van, F. G., Komiya, S., Aigner, T., Libermann, T. A., and Goldring, M. B. 2008. Differential expression of GADD45beta in normal and osteoarthritic cartilage: potential role in homeostasis of articular chondrocytes. Arthritis Rheum. 58: 2075–2087.
Iliopoulos, D., Malizos, K. N., and Tsezou, A. 2007. Epigenetic regulation of leptin affects MMP-13 expression in osteoarthritic chondrocytes: possible molecular target for osteoarthritis therapeutic intervention. Ann. Rheum. Dis. 66:1616–1621.
Imai, K., Ohta, S., Matsumoto, T., Fujimoto, N., Sato, H., Seiki, M., and Okada, Y. 1997. Expression of membrane-type 1 matrix metalloproteinase and activation of progelatinase A in human osteoarthritic cartilage. Am. J. Pathol. 151:245–256.
Jaenisch, R., and Bird, A. 2003. Epigenetic regulation of gene expression: how the genome integrates intrinsic and environmental signals. Nat. Genet. 33 Suppl:245–254.
Kim, Y. I., Logan, J. W., Mason, J. B., and Roubenoff, R. 1996. DNA hypomethylation in inflammatory arthritis: reversal with methotrexate. J. Lab Clin. Med. 128:165–172.
Leung, K. K., Ng, L. J., Ho, K. K., Tam, P. P., and Cheah, K. S. 1998. Different cis-regulatory DNA elements mediate developmental stage- and tissue-specific expression of the human COL2A1 gene in transgenic mice. J. Cell Biol. 141:1291–1300.
Li, D., Da, L., Tang, H., Li, T., and Zhao, M. 2008. CpG methylation plays a vital role in determining tissue- and cell-specific expression of the human cell-death-inducing DFF45-like effector A gene through the regulation of Sp1/Sp3 binding. Nucleic Acids Res. 36:330–341.
Lianxu, C., Hongti, J., and Changlong, Y. 2005. NF-kappaBp65-specific siRNA inhibits expression of genes of COX-2, NOS-2 and MMP-9 in rat IL-1beta-induced and TNF-alpha-induced chondrocytes. Osteoarthritis Cartilage 14:367–376.
Loeser, R. F., Im, H. J., Richardson, B., Lu, Q., and Chubinskaya, S. 2008. Methylation of the OP-1 promoter: potential role in the age-related decline in OP-1 expression in cartilage. Osteoarthritis Cartilage [Epub ahead of publication] DOI:10.1016/j.joca.2008.08.003.

Lohmander, L. S., Englund, P. M., Dahl, L. L., and Roos, E. M. 2007. The long-term consequence of anterior cruciate ligament and meniscus injuries: osteoarthritis. Am. J. Sports Med. 35: 1756–1769.

Loughlin, J. 2005. The genetic epidemiology of human primary osteoarthritis: current status. Expert. Rev. Mol. Med. 7:1–12.

Malfait, A. M., Liu, R. Q., Ijiri, K., Komiya, S., and Tortorella, M. D. 2002. Inhibition of ADAM-TS4 and ADAM-TS5 prevents aggrecan degradation in osteoarthritic cartilage. J Biol. Chem. 277:22201–22208.

Mancini, D. N., Singh, S. M., Archer, T. K., and Rodenhiser, D. I. 1999. Site-specific DNA methylation in the neurofibromatosis (NF1) promoter interferes with binding of CREB and SP1 transcription factors. Oncogene. 18:4108–4119.

Martin, C., and Zhang, Y. 2007. Mechanisms of epigenetic inheritance. Curr. Opin. Cell Biol. 19:266–272.

Masuhara, K., Bak, L. S., Nakai, T., Sugano, N., Ochi, T., and Sasaguri, Y. 2000. Matrix metalloproteinases in patients with osteoarthritis of the hip. Int. Orthop. 24:92–96.

Mehraban, F., Lark, M. W., Ahmed, F. N., Xu, F., and Moskowitz, R. W. 1998. Increased secretion and activity of matrix metalloproteinase-3 in synovial tissues and chondrocytes from experimental osteoarthritis. Osteoarthritis Cartilage 6:286–294.

Milner, J. M., and Cawston, T. E. 2005. Matrix metalloproteinase knockout studies and the potential use of matrix metalloproteinase inhibitors in the rheumatic diseases. Curr. Drug Targets. Inflamm. Allergy. 4:363–375.

Murphy, L., Schwartz, T. A., Helmick, C. G., Renner, J. B., Tudor, G., Koch, G., Dragomir, A., Kalsbeek, W. D., Luta, G., and Jordan, J. M. 2008. Lifetime risk of symptomatic knee osteoarthritis. Arthritis Rheum. 59:1207–1213.

Neuhold, L. A., Killar, L., Zhao, W., Sung, M. L., Warner, L., Kulik, J., Turner, J., Wu, W., Billinghurst, C., Meijers, T., Poole, A. R., Babij, P., and DeGennaro, L. J. 2001. Postnatal expression in hyaline cartilage of constitutively active human collagenase-3 (MMP-13) induces osteoarthritis in mice. J. Clin. Invest. 107:35–44.

Nile, C. J., Read, R. C., Akil, M., Duff, G. W., and Wilson, A. G. 2008. Methylation status of a single CpG site in the IL6 promoter is related to IL6 messenger RNA levels and rheumatoid arthritis. Arthritis Rheum. 58:2686–2693.

Ohta, S., Imai, K., Yamashita, K., Matsumoto, T., Azumano, I., and Okada, Y. 1998. Expression of matrix metalloproteinase 7 (matrilysin) in human osteoarthritic cartilage. Lab Invest. 78: 79–87.

Okada, Y., Shinmei, M., Tanaka, O., Naka, K., Kimura, A., Nakanishi, I., Bayliss, M. T., Iwata, K., and Nagase, H. 1992. Localization of matrix metalloproteinase 3 (stromelysin) in osteoarthritic cartilage and synovium. Lab Invest. 66:680–690.

Pang, R. T., Lee, L. T., Ng, S. S., Yung, W. H., and Chow, B. K. 2004. CpG methylation and transcription factors Sp1 and Sp3 regulate the expression of the human secretin receptor gene. Mol. Endocrinol. 18:471–483.

Petronis, A. 2001. Human morbid genetics revisited: relevance of epigenetics. Trends Genet. 17:142–146.

Pogribny, I. P., Pogribna, M., Christman, J. K., and James, S. J. 2000. Single-site methylation within the p53 promoter region reduces gene expression in a reporter gene construct: possible in vivo relevance during tumorigenesis. Cancer Res. 60:588–594.

Pöschl, E., Fidler, A., Schmidt, B., Kallipolitou, A., Schmid, E., and Aigner, T. 2005. DNA methylation is not likely to be responsible for aggrecan down regulation in aged or osteoarthritic cartilage. Ann. Rheum. Dis. 64:477–480.

Pufe, T., Bartscher, M., Petersen, W., Tillmann, B., and Mentlein, R. 2003. Pleiotrophin, an embryonic differentiation and growth factor, is expressed in osteoarthritis. Osteoarthritis. Cartilage. 11:260–264.

Reik, W. 2007. Stability and flexibility of epigenetic gene regulation in mammalian development. Nature. 447:425–432.

Roach, H. I. 2008. Potential directions for drug development for osteoarthritis. Expert. Opin. Drug Discovery. 3:475–487.

Roach, H. I., Aigner, T., Soder, S., Haag, J., and Welkerling, H. 2007. Pathobiology of osteoarthritis: pathomechanisms and potential therapeutic targets. Curr. Drug Targets. 8:271–282.

Roach, H. I., and Clarke, N. M. P. 2000. Physiological cell death of chondrocytes in vivo is not confined to apoptosis: New observations on the mammalian growth plate. J. Bone Joint Surg. [Br]. 82-B:601–613.

Roach, H. I., and Hashimoto, K. 2007. PCR-based methods to determine DNA methylation status at specific CpG sites using methylation-sensitive restriction enzymes. In: Methods Express: PCR, ed. S. Hughes and A. Moody, pp. 279–292. Bloxham, Oxfordshire, UK: Scion Publishing Ltd.

Roach, H. I., Inglis, S., Partridge K. A., Clarke N. M. P., Oreffo R. O. C., and Bronner F. 2005a. Can changes in the DNA methylation pattern explain the clonally inherited altered phenotype of osteoarthritic chondrocytes? In: Proceedings of the 8th International Conference on the Chemistry and Biology of Mineralized Tissues, eds. W. Landis and J. Sodek, pp. 192–195. Toronto: University of Toronto.

Roach, H. I., and Tilley, S. 2007. The pathogenesis of osteoarthritis. In: Bone and Osteoarthritis, Volume 4, Topics in Bone Biology, eds. F. Bronner and M. C. Farach-Carson, pp. 1–18. London: Springer Verlag

Roach, H. I., Yamada, N., Cheung, K. S., Tilley, S., Clarke, N. M., Oreffo, R. O., Kokubun, S., and Bronner, F. 2005b. Association between the abnormal expression of matrix-degrading enzymes by human osteoarthritic chondrocytes and demethylation of specific CpG sites in the promoter regions. Arthritis Rheum. 52:3110–3124.

Sanchez-Pernaute, O., Ospelt, C., Neidhart, M., and Gay, S. 2008. Epigenetic clues to rheumatoid arthritis. J. Autoimmun. 30:12–20.

Sandell, L. J., and Aigner, T. 2001. Articular cartilage and changes in arthritis. An introduction: cell biology of osteoarthritis. Arthritis Res. 3:107–113.

Sandy, J. D., and Verscharen, C. 2001. Analysis of aggrecan in human knee cartilage and synovial fluid indicates that aggrecanase (ADAMTS) activity is responsible for the catabolic turnover and loss of whole aggrecan whereas other protease activity is required for C-terminal processing in vivo. Biochem. J. 358:615–626.

Sarzi-Puttini, P., Cimmino, M. A., Scarpa, R., Caporali, R., Parazzini, F., Zaninelli, A., Atzeni, F., and Canesi, B. 2005. Osteoarthritis: an overview of the disease and its treatment strategies. Semin. Arthritis Rheum. 35:1–10.

Sesselmann, S., Soder, S., Voigt, R., Haag, J., Grogan, S. P., and Aigner, T. 2008. DNA methylation is not responsible for p21WAF1/CIP1 down-regulation in osteoarthritic chondrocytes. Osteoarthritis. Cartilage [Epub ahead of print] DOI:10.1016/j.joca.2008.09.006.

Shlopov, B. V., Gumanovskaya, M. L., and Hasty, K. A. 2000. Autocrine regulation of collagenase 3 (matrix metalloproteinase 13) during osteoarthritis. Arthritis Rheum. 43:195–205.

Soder, S., Roach, H. I., Oehler, S., Bau, B., Haag, J., and Aigner, T. 2006. MMP-9/gelatinase B is a gene product of human adult articular chondrocytes and increased in osteoarthritic cartilage. Clin. Exp. Rheumatol. 24:302–304.

Spector, T. D., and Macgregor, A. J. 2004. Risk factors for osteoarthritis: genetics. Osteoarthritis Cartilage. 12 Suppl A:S39–S44.

Stanton, H., Rogerson, F. M., East, C. J., Golub, S. B., Lawlor, K. E., Meeker, C. T., Little, C. B., Last, K., Farmer, P. J., Campbell, I. K., Fourie, A. M., and Fosang, A. J. 2005. ADAMTS5 is the major aggrecanase in mouse cartilage in vivo and in vitro. Nature. 434:648–652.

Stirzaker, C., Song, J. Z., Davidson, B., and Clark, S. J. 2004. Transcriptional gene silencing promotes DNA hypermethylation through a sequential change in chromatin modifications in cancer cells. Cancer Res. 64:3871–3877.

Strunnikova, M., Schagdarsurengin, U., Kehlen, A., Garbe, J. C., Stampfer, M. R., and Dammann, R. 2005. Chromatin inactivation precedes de novo DNA methylation during the progressive epigenetic silencing of the RASSF1A promoter. Mol. Cell Biol. 25:3923–3933.

Tetlow, L. C., Adlam, D. J., and Woolley, D. E. 2001. Matrix metalloproteinase and proinflammatory cytokine production by chondrocytes of human osteoarthritic cartilage: associations with degenerative changes. Arthritis Rheum. 44:585–594.

Tortorella, M. D., Malfait, A. M., Deccico, C., and Arner, E. 2001. The role of ADAM-TS4 (aggrecanase-1) and ADAM-TS5 (aggrecanase-2) in a model of cartilage degradation. Osteoarthritis Cartilage. 9:539–552.

Tsuchiya, K., Maloney, W. J., Vu, T., Hoffman, A. R., Schurman, D. J., and Smith, R. L. 1996. RT-PCR analysis of MMP-9 expression in human articular cartilage chondrocytes and synovial fluid cells. Biotech. Histochem. 71:208–213.

Valdes, A. M., Doherty, M., and Spector, T. D. 2008. The additive effect of individual genes in predicting risk of knee osteoarthritis. Ann. Rheum. Dis. 67:124–127.

Wagner, S., Hofstetter, W., Chiquet, M., Mainil-Varlet, P., Stauffer, E., Ganz, R., and Siebenrock, K. A. 2003. Early osteoarthritic changes of human femoral head cartilage subsequent to femoro-acetabular impingement. Osteoarthritis Cartilage. 11:508–518.

Walter, H., Kawashima, A., Nebelung, W., Neumann, W., and Roessner, A. 1998. Immunohistochemical analysis of several proteolytic enzymes as parameters of cartilage degradation. Pathol. Res. Pract. 194:73–81.

Wolfe, G. C., MacNaul, K. L., Buechel, F. F., McDonnell, J., Hoerrner, L. A., Lark, M. W., Moore, V. L., and Hutchinson, N. I. 1993. Differential in vivo expression of collagenase messenger RNA in synovium and cartilage. Quantitative comparison with stromelysin messenger RNA levels in human rheumatoid arthritis and osteoarthritis patients and in two animal models of acute inflammatory arthritis. Arthritis Rheum. 36:1540–1547.

Wu, C. W., and Kalunian, K. C. 2005. New developments in osteoarthritis. Clin. Geriatr. Med. 21:589–601.

Wu, W., Billinghurst, R. C., Pidoux, I., Antoniou, J., Zukor, D., Tanzer, M., and Poole, A. R. 2002. Sites of collagenase cleavage and denaturation of type II collagen in aging and osteoarthritic articular cartilage and their relationship to the distribution of matrix metalloproteinase 1 and matrix metalloproteinase 13. Arthritis Rheum. 46:2087–2094.

Yamane, K., Suzuki, H., Ihn, H., Kato, M., Yoshikawa, H., and Tamaki, K. 2005. Cell type-specific regulation of the TGF-beta-responsive alpha2(I) collagen gene by CpG methylation. J. Cell Physiol. 202:822–830.

Zimmermann, P., Boeuf, S., Dickhut, A., Boehmer, S., Olek, S., and Richter, W. 2008. Correlation of COL10A1 induction during chondrogenesis of mesenchymal stem cells with demethylation of two CpG sites in the COL10A1 promoter. Arthritis Rheum. 58:2743–2753.

Part V
Epigenetic Interventions and Aging

Histone-Modifying Drugs in Aging

Ulrich Mahlknecht and Barbara Zschoernig

Abstract Healthy aging remains one of the ideals of modern society. Therefore, interventions regarding the aging process are of considerable interest. Since calorie restriction constitutes the only reproducible and evolutionarily conserved way of extending life span so far, calorie restriction mimetics increasingly attract attention to the field. Among the molecules and strategies currently being considered, the sirtuin family of histone deacetylases emerged as promising potential therapeutic targets, mostly due to their requirement of NAD^+ as a cofactor for enzymatic activity, which determines a crucial link between sirtuins and the energy-dependent regulation of gene transcription. Especially SIRT1 has been implicated to play a crucial role during the aging process. Consequently, recent research focussed on the development of highly specific modulators of SIRT1, which represent the most advanced compounds in terms of potency, selectivity and demonstrated activity with regard to a promising therapeutic potential as calorie restriction mimetics in aging and age-related disease models.

Keywords Sirtuins · Aging · Calorie restriction

Aging, Age-Associated Diseases and Calorie Restriction (CR)

Cellular senescence and organism aging are characterized by progressive loss of physiological functions and metabolic processes which are often accompanied by age-associated diseases, such as neuronal degeneration, e.g. Alzheimer's or Parkinson's disease or metabolic disorders such as type 2 diabetes (Love, 2005). Due to biological complexities, we still lack a complete picture of the molecular

U. Mahlknecht (✉)
Department of Internal Medicine, Saarland University Medical Center, Division of Immunotherapy and Gene Therapy, José Carreras Center for Immunotherapy and Gene Therapy, D-66421 Homburg/Saar, Germany
e-mail: ulrich.mahlknecht@uks.eu

T.O. Tollefsbol (ed.), *Epigenetics of Aging*, DOI 10.1007/978-1-4419-0639-7_21,

mechanisms related to aging, cellular senescence and longevity, particularly with regard to humans. There are several factors associated with the rate of aging, among them the metabolic control, changes of gene expression patterns and the production of reactive oxygen species (ROS) by mitochondria (Gruber et al., 2008). As healthy aging remains one of the ideals of modern society, interventions regarding the aging process are of considerable interest.

Up to now, calorie restriction (CR), a phenomenon first described in the 1930s by McCay and co-workers where an organism is provided with at least 20% fewer calories below ad libitum level, constitutes the most robust and reproducible way of extending health and longevity (McCay, 1935). CR has not only been shown to increase the median and maximum life span of a variety of organisms (Masoro, 2000; Masoro and Austad, 1996; Weindruch and Sohal, 1997; Weindruch et al., 1986; Yu, 1994) but is also associated with a decreased incidence or delayed rate of age-related diseases as demonstrated in several rodent studies (Weindruch et al., 1986; Yu, 1994). Furthermore, the beneficial effects of CR are related to lower circulating insulin levels and an increased insulin sensitivity, thereby reducing the predisposition to diabetes as well as other metabolic disorders which is also associated with life span extension based on experiments in animal models (Katic and Kahn, 2005; Lane et al., 1995). Another key feature of CR is a lowered core body temperature that results in a reduced and more efficient energy expenditure which is related to an increased life span as well (Lane et al., 1996; Roth et al., 1995). Despite the lack of long-term studies, there is also emerging evidence that CR might constitute a life span extending mechanism for humans. Studies on dogs (Kealy et al., 2002), cows (Pinney et al., 1972) and non-human primates (Mattison et al., 2003; Roth et al., 2004) revealed that many of the physiological responses in these organisms resemble those observed in rodents on CR. Importantly, the National Institute of Aging initiated short-term human CR studies (6–12 months) at Washington University, Tufts University and the Pennington Center at Louisiana State which already confirm, albeit on a preliminary basis, reduced plasma insulin levels and body temperature that are key features of the CR response repeatedly observed in animal studies (Hadley et al., 2005; Heilbronn and Ravussin, 2003). In consideration of the already demonstrated positive effect on human health (Roth, 2005), these data are the first evidence that humans might indeed benefit from CR not only due to disease protection but also in terms of increased survival.

Histone Modifications and Histone-Modifying Enzymes in Aging

Aging is accompanied by genomic instability and change of gene expression (Oberdoerffer and Sinclair, 2007). Most recently, changes in epigenetic modifications have been considered as a major determinant of aging and cellular senescence (Fraga et al., 2005; Fraga and Esteller, 2007; Sedivy et al., 2008), due to their ability to regulate gene expression by remodelling of the chromatin structure. DNA methylation and post-transcriptional histone modifications such as methylation, phosphorylation or acetylation are the best-studied epigenetic modifications so far (Jenuwein and Allis, 2001; Strahl and Allis, 2000). Whereas both hypomethylation and hypermethylation of specific DNA sequences, depending on the gene and tissue examined, have been associated with aging in a number of studies (Richardson, 2003), there is only limited information available regarding changes of histone modification patterns in aging. For example, the trimethylation of mammalian histone H4 on lysine 20 (H4K20), an epigenetic mark of constitutive heterochromatin, has been demonstrated to increase with aging in rat kidney and liver cells (Sarg et al., 2002) probably due to mutations in lamin A and associated defects within the nuclear lamina (Shumaker et al., 2006). Additionally, age-related changes in the phosphorylation state of the linker histone H1 subtype profile have been observed. Aging of human peripheral blood lymphocytes is accompanied by a significant dephosphorylation of histone H1.4 and H1.5 arguing for an increased senescence-associated heterochromatin formation (Happel et al., 2008). Similarly, reversible protein acetylation has been demonstrated to be an evolutionary conserved mechanism of modulating life span (Chang and Min, 2002). Originally thought to occur exclusively on ϵ-lysine groups of histones, it has been shown in recent years that the function of several non-histone proteins relies on reversible acetylation mediated by two competing enzyme classes, histone acetylases (HATs) and histone deacetylases (HDACs). So far, only HDACs have been linked to regulation of life span in several model organisms. Based on their homology to their yeast counterparts, HDACs are categorized in four classes (de Ruijter et al., 2003), among them the class III HDACs have been demonstrated to be involved in longevity. The class III histone deacetylases (HDACs), also referred to as the "sirtuins", received their name on the basis of their homology with the yeast silent information regulator 2 (SIR2) protein, which is an NAD^+-dependent HDAC (Landry et al., 2000; Tanner et al., 2000) in contrast to class I, II and IV HDACs, which are Zn^{2+}-dependent hydrolases. The requirement of NAD^+ as a cofactor for enzymatic activity suggested the sirtuins to be a crucial link in the energy-dependent regulation of gene transcription. Indeed, SIRT1, which is the closest mammalian homolog among the seven human sirtuins that have been identified so far (Frye, 1999, 2000), has been reported to be a key element in a variety of physiological processes such as metabolism, neurogenesis and cell survival due to its ability to deacetylate both histone and numerous non-histone substrates (Michan and Sinclair, 2007; Sauve et al., 2006). Studies in a variety of organisms such as *Saccharomyces cerevisiae*, *Drosophila melanogaster*, *Caenorhabditis elegans* and mammalians demonstrated that overexpression or hyperactivity of yeast Sir2 and its orthologs is connected to a prolonged life span (Longo and Kennedy, 2006). Especially SIRT1, the closest human homolog to yeast Sir2, has been implicated to play a crucial role during the aging process for several reasons (Guarente and Picard, 2005; Saunders and Verdin, 2007; Westphal et al., 2007). In summary, SIRT1 is downregulated in senescent cells (Sasaki et al., 2006) and during aging (Sommer et al., 2006). Furthermore, calorie restriction induces SIRT1 expression in mammalian cells and humans thereby promoting cell survival (Cohen et al., 2004), whereas SIRT1 knockout mice fail to display a phenotype of CR (Chen et al., 2005). In addition, SIRT1 is involved in the upregulation of mitochondrial biogenesis due to its capability to deacetylate and thus activate the peroxisome proliferator-activating receptor (PPAR)-gamma co-activator-1α (PGC-1α) (Rodgers et al., 2005), which stimulates mitochondrial activity and subsequently increases glucose metabolism, which in turn improves insulin sensitivity (Lagouge et al., 2006). The regulation of the mitochondrial biogenesis and metabolism is widely accepted as a key component in the regulation of life span and aging (Lopez-Lluch et al., 2008). Interestingly, SIRT1 has been demonstrated not only to mimic calorie restriction but also to possess a neuroprotective function. The resveratrol-mediated activation promotes the SIRT1-induced resistance to axonal degeneration (Araki et al., 2004), and there is emerging evidence that SIRT1 protects neurons from apoptosis (Brunet et al., 2004) and is involved in preventing neurodegeneration in models of Alzheimer's disease (Kim et al., 2007).

Sirtuin Modulating Substances in Aging

Calorie restriction (CR) has been demonstrated to be the most reliable way of extending the life span so far. Due to the fact that the degree and length of a calorie restriction is not applicable for everyone, a key focus of the current research is the development of drugs that mimic the beneficial effects of CR without the necessity of a prolonged diet (Roth et al., 2005). The glycolytic inhibitor 2-deoxyglucose was the first calorie restriction mimetic identified by limiting the glucose level within the cell (Ingram et al., 2006). Due to an apparent narrow window between efficacy and toxicity, this compound is not applicable as a human therapeutic. Among other molecules and strategies currently being considered, activators of sirtuin activity and/or abundance have been of considerable interest. The discovery that sirtuins, especially SIRT1, are involved in the regulation of life span in different organisms due to their ability to mimic the process of calorie restriction outlines their therapeutic potential as anti-aging drugs. There is emerging evidence that besides SIRT1 several other sirtuins such as SIRT2, SIRT3, SIRT4 and SIRT6 participate in the molecular mechanisms of aging as well. Currently, SIRT1-activating compounds as well as inducers of SIRT1 expression are the most advanced compounds in terms of potency, selectivity and demonstrated activity with regard to a promising therapeutic potential as calorie restriction mimetics and in age-related disease models.

Activators of SIRT1 Enzymatic Activity

Resveratrol

The hallmark activator of SIRT1 is resveratrol (3,4,5-trihydroxystilbene, Fig. 1A), a polyphenol that has been identified by high-throughput screening using a fluorescent-based sirtuin enzymatic assay (Howitz et al., 2003). Resveratrol was initially recognized by the *French paradox*, a phenomenon where individuals despite high-fat diets exhibit a low incidence of cardiovascular disease due to regular consumption of red wine. Analysis in non-mammalian organisms revealed that a treatment with resveratrol extends the life span through a direct activation of SIRT1 (Howitz et al., 2003; Wood et al., 2004) by increasing its substrate-binding affinity (Borra et al., 2005). Furthermore, it retards cellular senescence in human diploid fibroblasts (Huang et al., 2008). Several lines of evidence indicate that resveratrol possesses a SIRT1-mediated life extending capacity in mammals as well. In a study of Baur and co-workers resveratrol treatment has been demonstrated to improve the health and longevity of mice on a high-calorie diet (Baur et al., 2006). Although both high-calorie-fed mice suffered from obesity, the group receiving resveratrol lived significantly longer and exhibited the typical molecular changes associated with longer life span including increased insulin sensitivity, reduced insulin-like growth factor-1 (IGF-I) levels, increased peroxisome proliferator-activated receptor-gamma co-activator-1α (PGC-1α) activity and increased mitochondrial number. Consistent with this observation, phenotypes of SIRT1 transgenic mice partially resembled calorie restriction (Bordone et al., 2007). However, treatment of normal-fed mice with resveratrol given the compound beginning at midlife did not reveal a life span extension compared to the control group (Pearson et al., 2008). Remarkably, these mice maintained a greater healthiness during aging compared with mice fed without resveratrol. As a member of the stilbene family, resveratrol exists as *cis*- and *trans*-isomers in a variety of dietary sources such as grapes, red wine, plums or peanuts (Burns et al., 2002). Due to steric reasons, *trans*-resveratrol exhibits a higher abundance than *cis*-resveratrol and is therefore considered the most active form. Upon exposure to UV irradiation, the *trans* form can undergo isomerization to the *cis* form (Lamuela-Raventos, 1995). Although most of the studies connecting resveratrol to life span extension have been performed with *trans*-resveratrol, there is emerging evidence that the *cis* form activates various genes related to nuclear factor kappa B (NF-κB) regulation and whose transcriptional activation is critical in a number of pathologies and may therefore generally possess more beneficial properties than first realized (Leiro et al., 2005). Although *trans*-resveratrol is absorbed efficiently by humans, the stimulation of SIRT1 is strongly affected by polyphenol stability and metabolism (de Boer et al., 2006). It has been demonstrated that resveratrol exhibits a poor compound stability in vivo as well as a poor bioavailability due to extensive gut and liver metabolism (Walle et al., 2004). To overcome these liabilities, SRT501, a stabilized, more bioavailable formulation of resveratrol, has been developed by Sirtis Pharmaceuticals Inc. Hence, the development of specific modulators increasingly gained attention. Most recently several more potent SIRT1-specific activators that are structurally unrelated to resveratrol have been described by two groups (Milne et al., 2007; Nayagam et al., 2006).

SRT1460, SRT1720 and SRT2183

Lately, three novel SIRT1-activating compounds (SRT1460, SRT1720 and SRT2183, Fig. 1A) have been identified by a high-throughput fluorescence

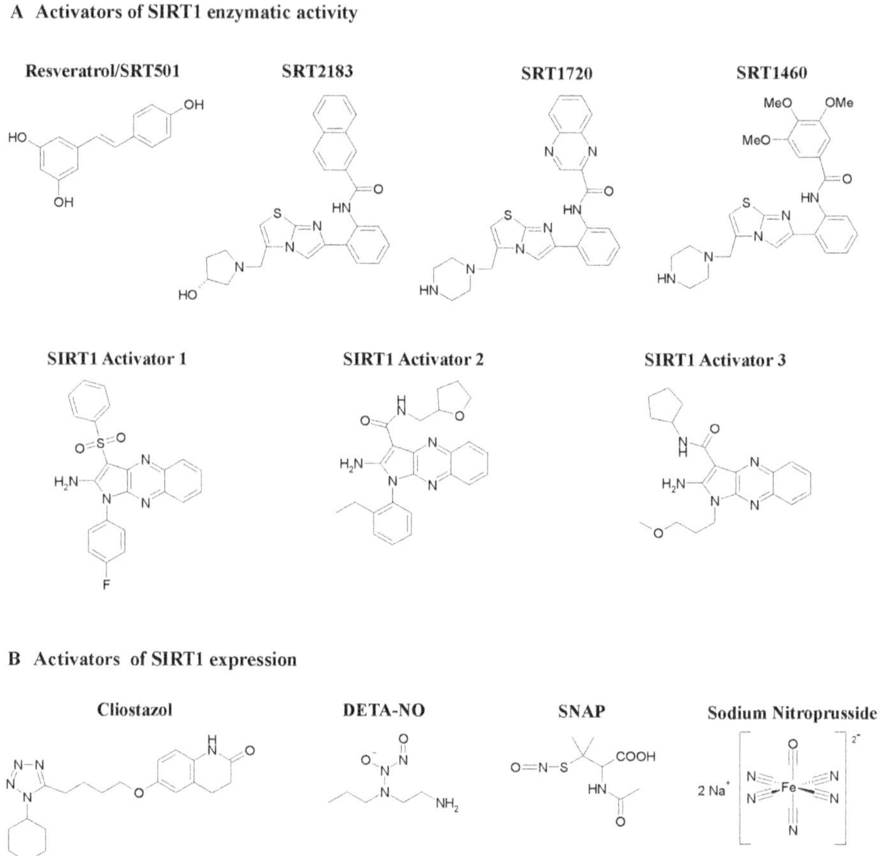

Fig. 1 SIRT1 modulating substances. (**A**) Activators of SIRT1 enzymatic activity. SRT501 constitutes a more bioavailable formulation of resveratrol (3,4,5-trihydroxystilbene). SIRT1 activators developed by Sirtris Pharmaceuticals Inc. (Milne et al., 2007): SRT2183: (R)-*N*-(2-(3-((3-hydroxypyrrolidin-1-yl)methyl)imidazo[2,1-b]thiazol-6-yl)phenyl)-2-naphthamide, SRT1720: *N*-(2-(3-(piperazin-1-ylmethyl)imidazol[2,1-b]thiazol-6-yl)phenyl)quinoxaline-2-carboxamide, SRT1460: 3,4,5-trimethoxy-*N*-(2-(3-(piperazin-1-ylmethyl)imidazo[2,1-b]thiazol-6-yl)phenyl)benzamide. Quinoxaline derivates (Nayagam et al., 2006): SIRT1 Activator 1: 3-Benezenesulfonyl-1-(4-fluorophenyl)-1H-pyrrolol[2,3-b]quinoxaline-2-ylamine, SIRT1 Activator 2: 2-Amino-1-(2-ethyl-phenyl)-1H-pyrrolo[2,3-b]quinoxaline-3-carboxylic acid (tetrahydro-furan-2-ylmehtyl)-amine, SIRT1 Activator 3: 2-Amino-1-(3-methoxy-propyl)-1H-pyrrolo[2,3,b]quinoxaline-3-carboxyclic acid cyclopentylamide (**B**) Activators of SIRT1 expression. Deta-NO: Diethylenetriamine-NO, SNAP: *S*-Nitroso-*N*-acetyl-D,L-penicillamine

polarization assay followed by a high-throughput mass spectrometry assay (Milne et al., 2007). These activators are structurally unrelated to resveratrol and their selectivity and potency were analysed in vitro by determining the concentration of compound required to increase enzyme activity by 50% (EC_{50}) as well as the maximal activation achieved at the highest compound doses tested (%). Being highly selective for SIRT1 as demonstrated in cell-based fluorescence deacetylation assays with SIRT2 and SIRT3, the SRT activators exhibit nanomolar to low micromolar potency towards SIRT1 in vitro (resveratrol $EC_{50} = 46.2\ \mu M$, maximal activation 201%; SRT1720 $EC_{50} = 0.16\ \mu M$, maximal activation 780%; SRT2183 $EC_{50} = 0.36\ \mu M$, maximal activation 296%; SRT1460 $EC_{50} = 2.9\ \mu M$, maximal activation 447%). The mechanism of SIRT1 activation has been demonstrated to rely on an increase of the substrate-binding affinity (lowered K_m) by binding of the activators to an allosteric site located within the amino-terminal domain of SIRT1. Importantly, the compound SRT1720 has been proven to be useful not only in activating SIRT1 in vitro but also in three different in vivo models which displayed the characteristic changes of calorie restriction. In diet-induced obese (DIO) as well as genetically obese mice ($Lep^{ob/ob}$), the treatment with SRT1720 significantly improved the insulin sensitivity, decreased the plasma glucose levels and increased the mitochondrial biogenesis. Consistent with these results, the glucose homeostasis and insulin sensitivity in adipose tissue, skeletal muscle and liver were markedly improved in Zucker fa/fa rats, a genetically obese rodent model. Taken together SIRT1 activation by SRT1720 seems to mimic the effects of calorie restriction on the metabolic and mitochondrial function and therefore constitutes a promising drug for the treatment of age-related diseases such as type 2 diabetes.

Quinoxaline Derivates (Sirtuin Activator 1, 2 and 3)

Nayagam and co-workers identified three novel sirtuin activators (sirtuin activator 1, 2 and 3, Fig. 1A) based on quinoxaline scaffolds using a high-throughput screening approach (Nayagam et al., 2006). These compounds are capable of activating SIRT1 with a micromolar potency in vitro as demonstrated in cell-based adipogenesis differentiation assay in mouse adipocytes where a dose-dependent fat mobilization was observed upon treatment with the discovered activators. Therefore, these compounds are anticipated to possess calorie mimetic modulating capacities, despite the lack of in vivo data from rodent models.

Consistent with the observations that different activators of SIRT1 successfully mimic the beneficial effects of calorie restriction, the specific inhibition of SIRT1 activity by sirtinol, a cell-permeable 2-hydroxy-1-napthaldehyde derivate, induced senescence-like growth arrest in human endothelial and cancer cells as demonstrated by increased histone H3 lysine 14 (H3K14) and histone H4 lysine 16 (H4K16) as well as p53 acetylation levels, accompanied by an attenuated DNA synthesis, an increased SA-β-gal activity as well as senescence-like morphological changes (Ota et al., 2007, 2006). This is further supporting the idea that SIRT1-activating compounds might be useful as promising treatment strategy for aging or age-related diseases.

Activators of SIRT1 Expression

The calorie restriction mimetic capabilities of SIRT1 may be achieved not only by a direct activation of its enzyme activity as demonstrated above but also by a regulation of the protein abundance on a transcriptional level. Interestingly, calorie restriction has been demonstrated to induce the expression of endothelial nitric oxide synthase (eNOS) leading to an increase in NO production in murine white adipocytes, which results in an increase of SIRT1 mRNA and protein levels as well as mitochondrial biogenesis (Nisoli et al., 2005). Therefore, substances capable of modulating intracellular nitric oxide (NO) levels either by an upregulation of eNOS abundance and/or activity or the treatment with NO donors constitute a promising approach to induce the SIRT1-mediated regulation of life span, especially with regard to the fact that NO production and eNOS expression have been demonstrated to be limited in senescent endothelial cells (Sato et al., 1993). Indeed, cliostazol (Fig. 1B), an inhibitor of cellular phosphodiesterase III (PDE3), prevents oxidative stress-induced endothelial premature senescence by an NO-dependent upregulation of SIRT1 in human endothelial cells (HUVEC) both in vitro and in vivo (Ota et al., 2008). Overexpression of SIRT1 prevented oxidative stress-induced endothelial senescence whereas inhibition of SIRT1 activity or expression induced premature senescence-like phenotypes (Ota et al., 2007). Treatment of HUVECs with cliostazol led to an increase in NO production through eNOS activation by phosphorylation at serine 1177 via cAMP/PKA- and PI3K/Akt-dependent mechanisms which results in increasing SIRT1 mRNA and protein levels by an yet unknown molecular mechanism (Ota et al., 2008). Likewise, treatment with the NO donors diethylenetriamine-NO (DETA-NO) and *S*-Nitroso-*N*-acetyl-D,L-penicillamine (SNAP) induced elevated SIRT1 mRNA and protein levels in cultured white adipocytes (Nisoli et al., 2005). Consistent with this observation, treatment of human PBMCs with the NO donor sodium nitroprusside (Fig. 1B) has been demonstrated to extend the life span of human PBMCs about 20% after 14 days which correlates with an increased SIRT1 protein level and a decreased histone H4 lysine 16 (H4K16) acetylation status pointing to an NO-mediated activation of SIRT1 activity (Engel and Mahlknecht, 2008).

Concluding Remarks

Healthy aging remains one of the ideals of modern society. Therefore, interventions regarding the aging process are of considerable interest. Since calorie restriction constitutes the only way of reproducibly extending life span so far, calorie restriction mimetics increasingly attracted attention to the field. Among the potential therapeutic targets currently investigated, the modulation of the enzymatic activity and/or protein abundance of SIRT1, a member of the sirtuin family of histone deacetylases, has been demonstrated to successfully mimic calorie restriction, therefore constituting a promising therapeutic target for the treatment of aging and age-related diseases. Consequently, the development of highly specific SIRT1 modulators represented the most active field of research so far. However, there is emerging evidence for the involvement of other sirtuin family members such as SIRT2, -3, -4, -6 in the regulation of aging and age-related diseases as well. Future investigations regarding specific inhibitors of these remaining sirtuins will not only contribute to a more detailed understanding of the aging process but also emerge as potent calorie mimetic drugs.

References

Araki T., Sasaki Y., Milbrandt J. 2004. Increased nuclear NAD biosynthesis and SIRT1 activation prevent axonal degeneration. Science 305: 1010–1013.

Baur J. A., Pearson K. J., Price N. L., Jamieson H. A., Lerin C., Kalra A. et al. 2006. Resveratrol improves health and survival of mice on a high-calorie diet. Nature 444: 337–342.

Bordone L., Cohen D., Robinson A., Motta M. C., van Veen E., Czopik A. et al. 2007. SIRT1 transgenic mice show phenotypes resembling calorie restriction. Aging Cell 6: 759–767.

Borra M. T., Smith B. C., Denu J. M. 2005. Mechanism of human SIRT1 activation by resveratrol. J. Biol. Chem. 280: 17187–17195.

Brunet A., Sweeney L. B., Sturgill J. F., Chua K. F., Greer P. L., Lin Y. et al. 2004. Stress-dependent regulation of FOXO transcription factors by the SIRT1 deacetylase. Science 303: 2011–2015.

Burns J., Yokota T., Ashihara H., Lean M. E., Crozier A. 2002. Plant foods and herbal sources of resveratrol. J. Agric. Food Chem. 50: 3337–3340.

Chang K. T., Min K. T. 2002. Regulation of lifespan by histone deacetylase. Ageing Res. Rev. 1: 313–326.

Chen D., Steele A. D., Lindquist S., Guarente L. 2005. Increase in activity during calorie restriction requires Sirt1. Science 310: 1641.

Cohen H. Y., Miller C., Bitterman K. J., Wall N. R., Hekking B., Kessler B. et al. 2004. Calorie restriction promotes mammalian cell survival by inducing the SIRT1 deacetylase. Science 305: 390–392.

de Boer V. C., de Goffau M. C., Arts I. C., Hollman P. C., Keijer J. 2006. SIRT1 stimulation by polyphenols is affected by their stability and metabolism. Mech. Ageing Dev. 127: 618–627.

de Ruijter A. J., van Gennip A. H., Caron H. N., Kemp S., van Kuilenburg A. B. 2003. Histone deacetylases (HDACs): characterization of the classical HDAC family. Biochem. J. 370: 737–749.

Engel N., Mahlknecht U. 2008. Aging and anti-aging: unexpected side effects of everyday medication through sirtuin1 modulation. Int. J. Mol. Med. 21: 223–232.

Fraga M. F., Ballestar E., Paz M. F., Ropero S., Setien F., Ballestar M. L. et al. 2005. Epigenetic differences arise during the lifetime of monozygotic twins. Proc. Natl. Acad. Sci. USA. 102: 10604–10609.

Fraga M. F., Esteller M. 2007. Epigenetics and aging: the targets and the marks. Trends Genet. 23: 413–418.

Frye R. A. 1999. Characterization of five human cDNAs with homology to the yeast SIR2 gene: Sir2-like proteins (sirtuins) metabolize NAD and may have protein ADP-ribosyltransferase activity. Biochem. Biophys. Res. Commun. 260: 273–279.

Frye R. A. 2000. Phylogenetic classification of prokaryotic and eukaryotic Sir2-like proteins. Biochem. Biophys. Res. Commun. 273: 793–798.

Gruber J., Schaffer S., Halliwell B. 2008. The mitochondrial free radical theory of ageing – where do we stand? Front. Biosci. 13: 6554–6579.

Guarente L., Picard F. 2005. Calorie restriction – the SIR2 connection. Cell 120: 473–482.

Hadley E. C., Lakatta E. G., Morrison-Bogorad M., Warner H. R., Hodes R. J. 2005. The future of aging therapies. Cell 120: 557–567.

Happel N., Doenecke D., Sekeri-Pataryas K. E., Sourlingas T. G. 2008. H1 histone subtype constitution and phosphorylation state of the ageing cell system of human peripheral blood lymphocytes. Exp. Gerontol. 43: 184–199.

Heilbronn L. K., Ravussin E. 2003. Calorie restriction and aging: review of the literature and implications for studies in humans. Am. J. Clin. Nutr. 78: 361–369.
Howitz K. T., Bitterman K. J., Cohen H. Y., Lamming D. W., Lavu S., Wood J. G. et al. 2003. Small molecule activators of sirtuins extend Saccharomyces cerevisiae lifespan. Nature 425: 191–196.
Huang J., Gan Q., Han L., Li J., Zhang H., Sun Y. et al. 2008. SIRT1 overexpression antagonizes cellular senescence with activated ERK/S6k1 signaling in human diploid fibroblasts. PLoS ONE 3: e1710.
Ingram D. K., Zhu M., Mamczarz J., Zou S., Lane M. A., Roth G. S. et al. 2006. Calorie restriction mimetics: an emerging research field. Aging Cell 5: 97–108.
Jenuwein T., Allis C. D. 2001. Translating the histone code. Science 293: 1074–1080.
Katic M., Kahn C. R. 2005. The role of insulin and IGF-1 signaling in longevity. Cell. Mol. Life Sci. 62: 320–343.
Kealy R. D., Lawler D. F., Ballam J. M., Mantz S. L., Biery D. N., Greeley E. H. et al. 2002. Effects of diet restriction on life span and age-related changes in dogs. J. Am. Vet. Med. Assoc. 220: 1315–1320.
Kim D., Nguyen M. D., Dobbin M. M., Fischer A., Sananbenesi F., Rodgers J. T. et al. 2007. SIRT1 deacetylase protects against neurodegeneration in models for Alzheimer's disease and amyotrophic lateral sclerosis. Embo J. 26: 3169–3179.
Lagouge M., Argmann C., Gerhart-Hines Z., Meziane H., Lerin C., Daussin F. et al. 2006. Resveratrol improves mitochondrial function and protects against metabolic disease by activating SIRT1 and PGC-1alpha. Cell 127: 1109–1122.
Lamuela-Raventos R. M. 1995. Direct HPLC Analysis of cis- and trans-Resveratrol and Piceid Isomers in Spanish Red Vitis vinifera Wines. J. Agric. Food Chem. 43: 281–283.
Landry J., Sutton A., Tafrov S. T., Heller R. C., Stebbins J., Pillus L. et al. 2000. The silencing protein SIR2 and its homologs are NAD-dependent protein deacetylases. Proc. Natl. Acad. Sci. USA. 97: 5807–5811.
Lane M. A., Baer D. J., Rumpler W. V., Weindruch R., Ingram D. K., Tilmont E. M. et al. 1996. Calorie restriction lowers body temperature in rhesus monkeys, consistent with a postulated anti-aging mechanism in rodents. Proc. Natl. Acad. Sci. USA. 93: 4159–4164.
Lane M. A., Ball S. S., Ingram D. K., Cutler R. G., Engel J., Read V. et al. 1995. Diet restriction in rhesus monkeys lowers fasting and glucose-stimulated glucoregulatory end points. Am. J. Physiol. 268: E941–948.
Leiro J., Arranz J. A., Fraiz N., Sanmartin M. L., Quezada E., Orallo F. 2005. Effect of cis-resveratrol on genes involved in nuclear factor kappa B signaling. Int. Immunopharmacol. 5: 393–406.
Longo V. D., Kennedy B. K. 2006. Sirtuins in aging and age-related disease. Cell 126: 257–268.
Lopez-Lluch G., Irusta P. M., Navas P., de Cabo R. 2008. Mitochondrial biogenesis and healthy aging. Exp. Gerontol. 43: 813–819.
Love R. 2005. Calorie restriction may be neuroprotective in AD and PD. Lancet Neurol. 4: 84.
Masoro E. J. 2000. Caloric restriction and aging: an update. Exp. Gerontol. 35: 299–305.
Masoro E. J., Austad S. N. 1996. The evolution of the antiaging action of dietary restriction: a hypothesis. J. Gerontol. A Biol. Sci. Med. Sci. 51: B387–391.
Mattison J. A., Lane M. A., Roth G. S., Ingram D. K. 2003. Calorie restriction in rhesus monkeys. Exp. Gerontol. 38: 35–46.
McCay C. M. 1935. Iodized salt a hundred years ago. Science 82: 350–351.
Michan S., Sinclair D. 2007. Sirtuins in mammals: insights into their biological function. Biochem. J. 404: 1–13.
Milne J. C., Lambert P. D., Schenk S., Carney D. P., Smith J. J., Gagne D. J. et al. 2007. Small molecule activators of SIRT1 as therapeutics for the treatment of type 2 diabetes. Nature 450: 712–716.
Nayagam V. M., Wang X., Tan Y. C., Poulsen A., Goh K. C., Ng T. et al. 2006. SIRT1 modulating compounds from high-throughput screening as anti-inflammatory and insulin-sensitizing agents. J. Biomol. Screen. 11: 959–967.

Nisoli E., Tonello C., Cardile A., Cozzi V., Bracale R., Tedesco L. et al. 2005. Calorie restriction promotes mitochondrial biogenesis by inducing the expression of eNOS. Science 310: 314–317.

Oberdoerffer P., Sinclair D. A. 2007. The role of nuclear architecture in genomic instability and ageing. Nat. Rev. Mol. Cell. Biol. 8: 692–702.

Ota H., Akishita M., Eto M., Iijima K., Kaneki M., Ouchi Y. 2007. Sirt1 modulates premature senescence-like phenotype in human endothelial cells. J. Mol. Cell. Cardiol. 43: 571–579.

Ota H., Eto M., Kano M. R., Ogawa S., Iijima K., Akishita M. et al. 2008. Cilostazol inhibits oxidative stress-induced premature senescence via upregulation of Sirt1 in human endothelial cells. Arterioscler. Thromb. Vasc. Biol. 28: 1634–1639.

Ota H., Tokunaga E., Chang K., Hikasa M., Iijima K., Eto M. et al. 2006. Sirt1 inhibitor, Sirtinol, induces senescence-like growth arrest with attenuated Ras-MAPK signaling in human cancer cells. Oncogene 25: 176–185.

Pearson K. J., Baur J. A., Lewis K. N., Peshkin L., Price N. L., Labinskyy N. et al. 2008. Resveratrol delays age-related deterioration and mimics transcriptional aspects of dietary restriction without extending life span. Cell. Metab. 8: 157–168.

Pinney D. O., Stephens D. F., Pope L. S. 1972. Lifetime effects of winter supplemental feed level and age at first parturition on range beef cows. J. Anim. Sci. 34: 1067–1074.

Richardson B. 2003. Impact of aging on DNA methylation. Ageing Res. Rev. 2: 245–261.

Rodgers J. T., Lerin C., Haas W., Gygi S. P., Spiegelman B. M., Puigserver P. 2005. Nutrient control of glucose homeostasis through a complex of PGC-1alpha and SIRT1. Nature 434: 113–118.

Roth G. S. 2005. Caloric restriction and caloric restriction mimetics: current status and promise for the future. J. Am. Geriatr. Soc. 53: S280–283.

Roth G. S., Ingram D. K., Lane M. A. 1995. Slowing ageing by caloric restriction. Nat. Med. 1: 414–415.

Roth G. S., Lane M. A., Ingram D. K. 2005. Caloric restriction mimetics: the next phase. Ann. NY. Acad. Sci. 1057: 365–371.

Roth G. S., Mattison J. A., Ottinger M. A., Chachich M. E., Lane M. A., Ingram D. K. 2004. Aging in rhesus monkeys: relevance to human health interventions. Science 305: 1423–1426.

Sarg B., Koutzamani E., Helliger W., Rundquist I., Lindner H. H. 2002. Postsynthetic trimethylation of histone H4 at lysine 20 in mammalian tissues is associated with aging. J. Biol. Chem. 277: 39195–39201.

Sasaki T., Maier B., Bartke A., Scrable H. 2006. Progressive loss of SIRT1 with cell cycle withdrawal. Aging Cell 5: 413–422.

Sato I., Morita I., Kaji K., Ikeda M., Nagao M., Murota S. 1993. Reduction of nitric oxide producing activity associated with in vitro aging in cultured human umbilical vein endothelial cell. Biochem. Biophys. Res. Commun. 195: 1070–1076.

Saunders L. R., Verdin E. 2007. Sirtuins: critical regulators at the crossroads between cancer and aging. Oncogene 26: 5489–5504.

Sauve A. A., Wolberger C., Schramm V. L., Boeke J. D. 2006. The biochemistry of sirtuins. Annu. Rev. Biochem. 75: 435–465.

Sedivy J. M., Banumathy G., Adams P. D. 2008. Aging by epigenetics – a consequence of chromatin damage? Exp. Cell. Res. 314: 1909–1917.

Shumaker D. K., Dechat T., Kohlmaier A., Adam S. A., Bozovsky M. R., Erdos M. R. et al. 2006. Mutant nuclear lamin A leads to progressive alterations of epigenetic control in premature aging. Proc. Natl. Acad. Sci. USA. 103: 8703–8708.

Sommer M., Poliak N., Upadhyay S., Ratovitski E., Nelkin B. D., Donehower L. A. et al. 2006. DeltaNp63alpha overexpression induces downregulation of Sirt1 and an accelerated aging phenotype in the mouse. Cell Cycle 5: 2005–2011.

Strahl B. D., Allis C. D. 2000. The language of covalent histone modifications. Nature 403: 41–45.

Tanner K. G., Landry J., Sternglanz R., Denu J. M. 2000. Silent information regulator 2 family of NAD- dependent histone/protein deacetylases generates a unique product, 1-O-acetyl-ADP-ribose. Proc. Natl. Acad. Sci. USA. 97: 14178–14182.

Walle T., Hsieh F., DeLegge M. H., Oatis J. E., Jr., Walle U. K. 2004. High absorption but very low bioavailability of oral resveratrol in humans. Drug Metab. Dispos. 32: 1377–1382.
Weindruch R., Sohal R. S. 1997. Seminars in medicine of the Beth Israel Deaconess Medical Center. Caloric intake and aging. N. Engl. J. Med. 337: 986–994.
Weindruch R., Walford R. L., Fligiel S., Guthrie D. 1986. The retardation of aging in mice by dietary restriction: longevity, cancer, immunity and lifetime energy intake. J. Nutr. 116: 641–654.
Westphal C. H., Dipp M. A., Guarente L. 2007. A therapeutic role for sirtuins in diseases of aging? Trends Biochem. Sci. 32: 555–560.
Wood J. G., Rogina B., Lavu S., Howitz K., Helfand S. L., Tatar M. et al. 2004. Sirtuin activators mimic caloric restriction and delay ageing in metazoans. Nature 430: 686–689.
Yu B. P. 1994. How diet influences the aging process of the rat. Proc. Soc. Exp. Biol. Med. 205: 97–105.

Dietary Effect on Epigenetics During the Aging Process

Yuanyuan Li and Trygve O. Tollefsbol

Abstract It has been proposed that epigenetic modulation plays a critical role in cellular senescence and organismal aging. Two major epigenetic codes are DNA methylation and histone modifications, including acetylation, methylation, phosphorylation and ubiquitylation. Environmental factors such as dietary components can regulate gene expression by altering epigenetic modifications and may thus have long lasting effects on the aging process. In this review, we will focus on the advances in various dietary factors on epigenetic modulation during aging, which will offer exciting new opportunities to explore the role of diet in influencing the biology of aging.

Keywords Epigenetic · Aging · Diet · DNA methylation · Histone modification

Introduction

Epigenetic processes, which literally mean outside conventional genetics and do not involve mutations of DNA itself, have been described to influence gene expression by at least two pathways: DNA methylation and histone acetylation. It has been extensively proposed that epigenetic modulators play an important role in maintenance of gene stability during the process of cell development and proliferation. During the aging process, there is a progressively reduced capability for homeostasis predominantly due to aberrant gene expressions. Nutrition is believed to be a chief contributor to the gene regulation in aging process by impacting epigenetic pathways. Therefore, epigenetic-mediated changes in gene expression in response to dietary and other lifestyle exposures appear to be a major molecular mechanism linking environmental factors with the genome and consequences for cell function and health throughout the life course. Here, we will focus on the advances in various

T.O. Tollefsbol (✉)
Center for Aging and Department of Biology, 175 Campbell Hall, 1300 University of Alabama at Birmingham, AL 35294-1170, USA
e-mail: trygve@uab.edu

T.O. Tollefsbol (ed.), *Epigenetics of Aging*, DOI 10.1007/978-1-4419-0639-7_22,

dietary effects on epigenetic modulation during aging, which will offer exciting new opportunities to explore the role of diet in influencing the biology of aging.

Caloric Restriction (CR) and Aging

In animal models, CR is by far the best established environmental manipulation to extend longevity and curtail the onset of age-related chronic diseases (McCay et al. 1989; Forster et al. 2003; Cooper et al. 2004). In humans, short-term CR trials indicate that 20% CR acting over periods of 2–6 years is associated with reduced body weight, blood pressure, blood cholesterol, and blood glucose – risk factors for the major killer diseases of cardiovascular disease and diabetes (Weindruch et al. 1986; Masoro 1988; Everitt and Le Couteur 2007). Later studies also demonstrated that even relatively short periods of CR (2–8 weeks) produced a rapid and progressive shift in gene expression in the liver toward that seen in animals exposed to long-term CR (Dhahbi et al. 2004). Conversely, shifting from long-term CR to a control diet for as little as 2 months reversed 90% of the gene expression changes of the animals exposed to long-term CR. However, the precise molecular mechanisms by which CR mediates its effects are not clear.

Recently, extensive studies have focused on SIRT1, the ortholog of Sir2 (silence information regulator 2), in yeast (Frye 2000; Imai et al. 2000). SIRT1, encoding a nicotinamide adenine dinucleotide (NAD^+)-dependent histone deacetylase, has been identified as a member of chromatin modifying factors of class III histone deacetylases with the capability of protecting cells from oxidative and genotoxic stress in stress management and metabolism (Vaquero et al. 2004; Yamamoto et al. 2007). The levels of SIRT1 have been reported to increase in rodent and human tissues in response to CR (Cohen et al. 2004), and this increase is proposed to cause favorable changes in metabolism and stress tolerance triggered by this diet. SIRT1 seems to act as a nutrient sensor by decoding fluctuations in cellular NAD^+ levels. This is especially exciting from the perspective that SIRT1 also modulates aging and may consequently serve as the key decoder that translates caloric restriction to increase life span. Future research will establish to what extent interference with SIRT1-regulatory pathways that regulate glucose homeostasis may lead to new therapeutic approaches to metabolic and age-related diseases and syndromes.

Diet and DNA Methylation During Aging

DNA methylation in cytosine–guanine (CpG)-dinucleotide sequences is a heritable, tissue-, and species-specific modification of mammalian DNA (Razin and Riggs 1980). DNA methylation is an important epigenetic determinant in gene expression, maintenance of DNA integrity, and stability. During cell division, methylation patterns in the parental strand of DNA are maintained in the daughter strand by the action of DNA methyltransferase 1 (DNMT1) which catalyses the transfer of a methyl group from *S*-adenosyl methionine (SAM), the methyl donor, to the cytosine

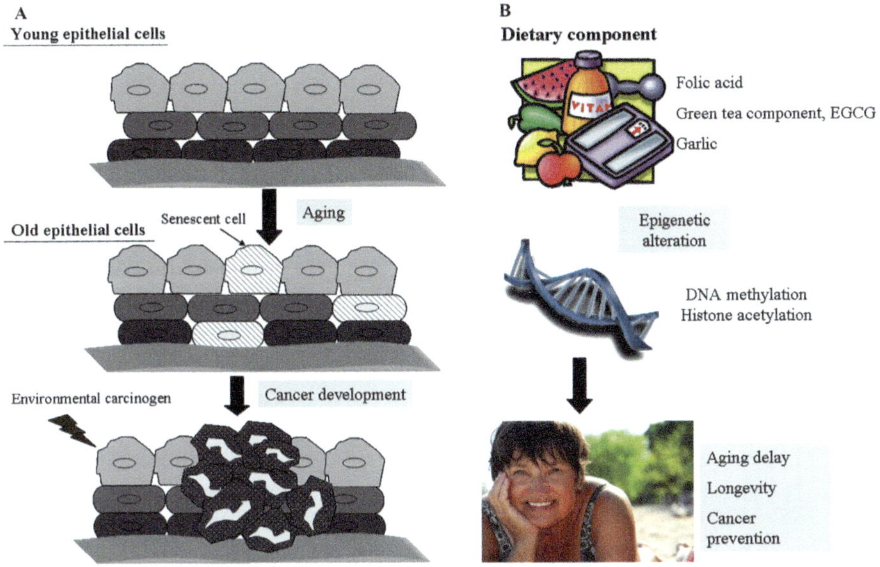

Fig. 1 Dietary components contribute to aging delay and cancer prevention by affecting epigenetic factors. **A**. Senescent-induced genomic instability and cancer development model in epithelial cells. Shown is a typical epithelium layer, in contact with a basement membrane. Continuous epithelial turnover during aging is thought to lead to genomic instability in senescent cells. This process may culminate in massive and rapid abnormal changes in critical gene expression, resulting in cancer development. **B**. Dietary effects on epigenetic modulation during aging. Dietary components, such as folate (Shane and Stokstad, 1983), green tea component (EGCG) (Waladkhani and Clemens, 1998), and garlic compound (DADS) (Myzak and Dashwood, 2006), are believed to prevent aging and aging-related disease by influencing epigenetic factors, including DNA methylation and histone acetylation processes

residues by a relatively complex mechanism (Bestor 2000). Aberrant patterns and dysregulation of DNA methylation mechanistically cause stable, heritable transcriptional silencing of the associated gene and are seen during both tumorigenesis and aging (Jones and Baylin 2002; Egger et al. 2004). Epigenetic variability at specific transcription regulation sites appears to be susceptible to modulation by nutritional changes (Waterland and Jirtle 2004) (Fig. 1). Therefore dietary components, which can affect the process of DNA methylation, may influence the aging process by regulation of the expression of certain key genes (Table 1).

Methyl Donor Diet and Aging

Methyl donor diet, referring to a series of dietary components, can be used for synthesis of SAM, including folate, vitamin B12, and many other compounds (Shane and Stokstad 1983; Brunaud et al. 2003). Folate, a water-soluble B vitamin, has been the focus of intense interest because of its role in the pathogenesis of several chronic diseases, including cancer, and its potential ability to modulate DNA methylation (Brunaud et al. 2003; Pogribny et al. 2004; Kim 2005). Two different

Table 1 Dietary components affecting epigenetic modulation during aging process

Epigenetic process	Component	Component function	Derivation
DNA methylation	Folate, vitamin B12	Methyl donor	Most vegetables (Shane and Stokstad 1983)
	EGCG	DNMT1 inhibition	Green tea (Fang et al. 2003)
	Genistein	DNMTase inhibition	Soybean (Barnes 1995)
Histone acetylation	Butyrate	Histone deacetylase (HDAC) inhibitor	Fermentation of dietary fiber
	Diallyl disulfide (DADS)		Garlic (Myzak and Dashwood 2006)
	Sulforaphane (SFN)		Broccoli sprouts

phenotypes in the Agouti and AxinFused inbred mice models were used in folate studies, which provided evidence for DNA methylation modulation (Morgan et al. 1999; Rakyan et al. 2003). These studies indicated that maternal dietary methyl group supplementation containing folic acid permanently altered the phenotypic coat color of the offspring via increased CpG methylation by using these specific mice models (Wolff et al. 1998; Waterland and Jirtle 2003). Similarly, a methyl group-rich diet (folate, vitamin B12, choline, and betaine) has been shown to significantly reduce the proportion of progeny with a kinked tail in AxinFused mice by one-half via increased CpG methylation in the promoter of the AxinFu gene (Waterland et al. 2006a). Conversely, when mice were fed a methyl-donor-deficient diet (lacking folic acid, vitamin B12, and choline), this led to downregulation of the imprinted *Igf2* gene (Waterland et al. 2006b). These studies represent a clear demonstration of how nutrition can influence the epigenetic organization of genes, and as a consequence, can have long-term effects on gene expression and phenotype.

It has been apparent that the genome-wide DNA methylation content decreases with age both in vitro and in vivo (Mays-Hoopes et al. 1986; Liu et al. 2008) and that this altered epigenetic marking may be responsible for abnormalities in gene expression in both oncogenesis and aging (Mutter 1997). However, the response of DNA methylation to aging is quite tissue specific. Thus, these alterations in cellular biology with aging may reduce folate availability in certain tissues, resulting in impairment of nucleotide synthesis and biological DNA methylation.

Green Tea Component: (-)-epigallocatechin-3-gallate

Surveys indicate that the elderly are particularly at risk for marginal deficiency of vitamins and trace elements. Changes in bodily functions, together with the malnutrition associated with advancing age, increase the risk of developing a number of age-related diseases. Abnormal DNA methylation patterns can cause genomic instability, which have been recognized as important factors in the biology of aging and of many age-associated degenerative diseases. Therefore, dietary components with epigenetic modulation activity have received particular attention because of their potential role in modulating DNA methylation status of certain key genes associated with aging and chronic conditions.

Epigallocatechin-3-gallate (EGCG), the most abundant and active polyphenol in green tea, has been well studied for its anticancer properties. Vegetables, beverages, fruits, and other components of the human diet commonly contain polyphenols which have been shown in numerous studies to have chemoprevention properties (Waladkhani and Clemens 1998; Yang et al. 2001; Adhami et al. 2003). EGCG has a variety of beneficial effects, such as preventing cancer, mutagenesis, diabetes, cardiovascular diseases, and ischemic damage, while promoting energy expenditure (Higdon and Frei 2003).

Various mechanisms have been suggested to explain the chemopreventive and anticancer effects of green tea polyphenols in different experimental models, but it is still not clear how EGCG affects the aging process. One mechanism could be that EGCG is an antioxidant well known for its radical and oxidant scavenging activities (Meng et al. 2008). Another possible mechanism includes the inhibition of DNMT1 leading to demethylation and reactivation of methylation-silenced genes, in which EGCG fits into the catalytic pocket of DNMT1 (Fang et al. 2003) and in doing so acts as a competitive inhibitor for the normal cytosine substrate. It has been reported that EGCG can inhibit DNMT1 activity and lead to hypomethylation and re-expression of the tumor suppressor genes including p16, RARβ, and hMLH1 in human esophageal KYSE 510 and 150 cells (Fang et al. 2003). In our lab, we found EGCG can downregulate the expression of hTERT (human telomerase reverse transcriptase), the catalytic subunit of the human telomerase enzyme, in MCF-7 breast cancer cells in a time- and dose-dependent manner due to epigenetic modulations. Treatment of MCF-7 cells with EGCG resulted in a time-dependent decrease in hTERT promoter methylation and ablated histone H3 Lys9 acetylation (Berletch et al. 2008). Recent studies showed that EGCG can stimulate the deacetylase activity of recombinant SIRT1 protein and could therefore be a potential regulator of aging-associated processes (de Boer et al. 2006). Consistent with our studies, this may provide a new approach to epigenetic mechanisms regulated by EGCG influencing chromatin modification, such as histone acetylation. However, there have few reports on senescence-related gene methylation alterations with long-term green tea or EGCG intake during the aging process either in vivo or in vitro. Further studies will be applied to this particular area and will help to solve the problems for the nutritional epigenetic modulation in aging.

Soybean Component: Genistein

Genistein, a natural isoflavonoid found in soybean products, has been proposed to be associated with a lower incidence and mortality of various cancers and is believed to be a chemopreventive agent (Barnes 1995; Shon et al. 2006). Genistein is the most abundant phytoestrogen in soybeans and shares the similar chemical properties with estradiol (Bradlow and Sepkovic 2002). It is well accepted that females live longer than males, which is partly due to the reason that the higher levels of estrogens in females protect them against aging by upregulating the expression of antioxidants on longevity-related genes. More recently, intense studies have focused on genistein-induced epigenetic changes both in vivo and in vitro. Genistein can also upregulate the expressions of various tumor suppressor genes, such as p16, p21, RARβ, and MGMT by regulating epigenetic events in cancer cells (Fang et al. 2005). However, as an analog of estrogen, it is a controversial topic in terms of long-term dietary genistein intake for modulation of aging and/or aging-related diseases, such as cancer. Moreover, questions related to the safety of early-life exposure to phytoestrogens remain unanswered.

Dietary Component and Histone Modifications During Aging

The histone tails, which protrude from the bundle of eight proteins making up the globular nucleosome core around which DNA is wrapped in the nucleus, are post-translationally modified by methylation, acetylation, phosphorylation, and ubiquitination (Jenuwein and Allis 2001). Current hypotheses suggest that the pattern of histone modifications in any region of the genome alters chromatin structure and thus controls access to the associated DNA by the proteins which make up the transcriptional machinery. In general, histone acetylation is the most prevalent pattern in histone decorations, and acetylated histones lead to a more open chromatin structure and gene expression, while deacetylated histone produces an inaccessible heterochromatin and gene silencing (Burgess-Beusse et al. 2002). Although such epigenetic states can be stably inherited from one cell generation to the next through mitosis, it is becoming apparent that histone modifications may be a "sensor" of the metabolic state of a cell (Jenuwein and Allis 2001). In addition, the proteins responsible for DNA methylation and those which modify histone decoration appear to cooperate to alter the organization of chromatin and thus gene expression (Burgess-Beusse et al. 2002).

Histone Deacetylase (HDAC) Inhibitors

Recent interest in HDAC inhibitors has expanded into the realm of cancer chemoprevention (Fig. 1). The working hypothesis of dietary HDAC inhibitor agents is that DNA/chromatin interactions are kept in a constrained state in the presence of HDAC/corepressor complexes, but HDAC inhibitors enable histone acetyltransferase/co-activator (HAT/CoA) complexes to transfer acetyl groups to lysine "tails" in histones, thereby loosening the interactions with DNA and facilitating transcription factor access and gene activation. Extensive evidence above shows that dietary compounds such as butyrate, diallyl disulfide (DADS), and sulforaphane (SFN) act as weak ligands for HDACs and exhibit HDAC inhibitory activity for aging chemoprevention, due to a lifetime of subtle modifications to the histone code (Myzak and Dashwood 2006) (Table 1).

Dietary SFN enriched in broccoli sprouts has been well studied for its anticancer properties through HDAC inhibition by activating genes such as p21 and Bax to facilitate cell-cycle arrest and/or apoptosis both in vitro and in vivo (Dashwood and Ho 2007). It is well known that epigenetic modulations can impact multiple areas in gene expression, by which a new hypothesis is proposed that the dietary HDAC inhibitor could ameliorate other chronic conditions and degeneration diseases in aging process. Given such widespread implications, it is interesting to speculate further about SFN and other dietary HDAC inhibitors and their impact on development and chronic disease susceptibility.

In addition to SFN, there are many other known or putative diet-derived HDAC inhibitors. This compound is derived from the fermentation of dietary fiber and represents the primary metabolic fuel for the colonocytes in the large bowel. Previous studies have reported that the addition of 4-phenylbutyrate (PBA) to the diet of fruit flies (*Drosophila*) increased longevity without any loss of resistance to stress or reproductive ability (Kang et al. 2002). These effects are probably due to the well-established action of butyrate as a histone deacetylase (HDAC) inhibitor with wide-ranging effects on gene expression and cell function. A further example is diallyl disulphide (DADS), a garlic compound, which has been shown to re-express previously epigenetically silenced genes by suppressing HDAC activity (Druesne et al. 2004).

Concluding Remarks

The concept of diminished homeostasis during aging gives rise to a number of research concerns. Epigenetic modulations are drawing more attention toward aging research due to their subtle and comprehensive regulations of gene expression involving two main pathways, DNA methylation and histone acetylation. Dietary factors are believed to have profound effects on many aspects of health including aging. Therefore, investigations on nutritional factors, which can interact with the epigenetic mechanisms and result in altered gene expression, will lead to extensive interest in the aging and nutrition research areas. Areas which appear to be particularly promising are (i) caloric restriction (CR) and aging; (ii) nutritional modulation of DNA methylation which involves the green tea component EGCG and soybean product genistein; and (iii) nutritional modulation of histone modification which involves butyrate, diallyl disulfide (DADS), and sulforaphane (SFN), acting as a series of HDAC inhibitors. Epigenetic-mediated changes in gene expression in response to dietary exposure appear to be a major molecular mechanism linking environmental factors with the genome with consequences for cell function and health throughout the life course.

Acknowledgment This work was supported by grant from National Cancer Institute (RO1 CA 129415) and Susan G. Komen for the Cure.

References

Adhami, V. M., Ahmad, N., and Mukhtar, H. 2003. Molecular targets for green tea in prostate cancer prevention. J. Nutr. 133: 2417S–2424S.

Barnes, S. 1995. Effect of genistein on in vitro and in vivo models of cancer. J. Nutr. 125: 777S–783S.

Berletch, J., Liu, C., Love, W. K., Andrews, L. G., Katiyar, S. K., and Tollefsbol, T. O. 2008. Epigenetic and genetic mechanisms contribute to telomerase inhibition by EGCG. J. Cell. Biochem. 103: 509–519.

Bestor, T. H. 2000. The DNA methyltransferases of mammals. Hum. Mol. Genet. 9: 2395–2402.

Bradlow, H. L., and Sepkovic, D. W. 2002. Diet and breast cancer. Ann. NY. Acad. Sci. 963: 247–267.

Brunaud, L., Alberto, J. M., Ayav, A., Gérard, P., Namour, F., Antunes, L., Braun, M., Bronowicki, J. P., Bresler, L., and Guéant, J. L. 2003. Effects of vitamin B12 and folate deficiencies on DNA methylation and carcinogenesis in rat liver. Clin. Chem. Lab Med. 41: 1012–1019.

Burgess-Beusse, B., Farrell, C., Gaszner, M., Litt, M., Mutskov, V., Recillas-Targa, F., Simpson, M., West, A., and Felsenfeld, G. 2002. The insulation of genes from external enhancers and silencing chromatin. Proc. Natl. Acad. Sci. USA. 99 Suppl 4: 16433–16437.

Cohen, H. Y., Miller, C., Bitterman, K. J., Wall, N. R., Hekking, B., Kessler, B., Howitz, K. T., Gorospe, M., de Cabo, R., and Sinclair, D. A. 2004. Calorie restriction promotes mammalian cell survival by inducing the SIRT1 deacetylase. Science 305: 390–392.

Cooper, T. M., Mockett, R. J., Sohal, B. H., Sohal, R. S., and Orr, W. C. 2004. Effect of caloric restriction on life span of the housefly, Musca domestica. FASEB. J. 18: 1591–1593.

Dashwood, R. H., and Ho, E. 2007. Dietary histone deacetylase inhibitors: From cells to mice to man. Semin. Cancer Biol. 17: 363–369.

de Boer, V. C., de Goffau, M. C., Arts, I. C., Hollman, P. C., and Keijer, J. 2006. SIRT1 stimulation by polyphenols is affected by their stability and metabolism. Mech. Ageing Dev. 127: 618–627.

Dhahbi, J. M., Kim, H. J., Mote, P. L., Beaver, R. J., and Spindler, S. R. 2004. Temporal linkage between the phenotypic and genomic responses to caloric restriction. Proc. Natl. Acad. Sci. USA. 101: 5524–5529.

Druesne, N., Pagniez, A., Mayeur, C., Thomas, M., Cherbuy, C., Duée, P., Martel, P. H., and Chaumontet, C. 2004. Diallyl disulfide (DADS) increases histone acetylation and p21(waf1/cip1) expression in human colon tumor cell lines. Carcinogenesis 25: 1227–1236.

Egger, G., Liang, G., Aparicio, A., and Jones, P. A. 2004. Epigenetics in human disease and prospects for epigenetic therapy. Nature 429: 457–463.

Everitt, A. V., and Le Couteur, D. G. 2007. Life extension by calorie restriction in humans. Ann. NY. Acad. Sci. 1114: 428–433.

Fang, M. Z., Chen, D., Sun, Y., Jin, Z., Christman, J. K., and Yang, C. S. 2005. Reversal of hypermethylation and reactivation of p16INK4a, RARbeta, and MGMT genes by genistein and other isoflavones from soy. Clin. Cancer Res. 11: 7033–7041.

Fang, M. Z., Wang, Y., Ai, N., Hou, Z., Sun, Y., Lu, H., Welsh, W., and Yang, C. S. 2003. Tea polyphenol (-)-epigallocatechin-3-gallate inhibits DNA methyltransferase and reactivates methylation-silenced genes in cancer cell lines. Cancer Res. 63: 7563–7570.

Forster, M. J., Morris, P., and Sohal, R. S. 2003. Genotype and age influence the effect of caloric intake on mortality in mice. FASEB. J. 17: 690–692.

Frye, R. A. 2000. Phylogenetic classification of prokaryotic and eukaryotic Sir2-like proteins. Biochem. Biophys. Res. Commun. 273: 793–798.

Higdon, J. V., and Frei, B. 2003. Tea catechins and polyphenols: Health effects, metabolism, and antioxidant functions. Crit. Rev. Food Sci. Nutr. 43: 89–143.
Imai, S., Armstrong, C., Kaeberlein, M., and Guarente, L. 2000. Transcriptional silencing and longevity protein Sir2 is an NDA-dependent histone deacetylase. Nature 403: 795–800.
Jenuwein, T., and Allis, C. D. 2001. Translating the histone code. Science 293: 1074–1080.
Jones, P. A., and Baylin, S. B. 2002. The fundamental role of epigenetic events in cancer. Nat. Rev. Genet. 3: 415–428.
Kang, H. L., Benzer, S., and Min, K. T. 2002. Life extension in Drosophila by feeding a drug. Proc. Natl. Acad. Sci. USA. 99: 838–843.
Kim, Y. I. 2005. Nutritional epigenetics: Impact of folate deficiency on DNA methylation and colon cancer susceptibility. J. Nutr. 135: 2703–2709.
Liu, L., Li, Y., and Tollefsbol, T. O. 2008. Gene-environment interactions and epigenetic basis of human diseases. Curr. Issues Mol. Biol. 10: 25–36.
Masoro, E. J. 1988. Food restriction in rodents: An evaluation of its role in the study of aging. J. Gerontol. 43: B59–64.
Mays-Hoopes, L., Chao, W., Butcher, H. C., and Huang, R. C. 1986. Decreased methylation of the major mouse long interspersed repeated DNA during aging and in myeloma cells. Dev. Genet. 7: 65–73.
McCay, C. M., Crowell, M. F., and Maynard, L. A. 1989. The effect of retarded growth upon the length of life span and upon the ultimate body size. 1935. Nutrition 5: 155–172.
Meng, Q., Velalar, C. N., and Ruan, R. 2008. Effects of epigallocatechin-3-gallate on mitochondrial integrity and antioxidative enzyme activity in the aging process of human fibroblast. Free Radic. Biol. Med. 44: 1032–1041.
Morgan, H. D., Sutherland, H. G., Martin, D. I., and Whitelaw, E. 1999. Epigenetic inheritance at the agouti locus in the mouse. Nat. Genet. 23: 314–318.
Mutter, G. L. 1997. Role of imprinting in abnormal human development. Mutat. Res. 396: 141–147.
Myzak, M. C., and Dashwood, R. H. 2006. Histone deacetylases as targets for dietary cancer preventive agents: Lessons learned with butyrate, diallyl disulfide, and sulforaphane. Curr. Drug Targets 7: 443–452.
Pogribny, I. P., James, S. J., Jernigan, S., and Pogribna, M. 2004. Genomic hypomethylation is specific for preneoplastic liver in folate/methyl deficient rats and does not occur in non-target tissues. Mutat. Res. 548: 53–59.
Rakyan, V. K., Chong, S., Champ, M. E., Cuthbert, P. C., Morgan, H. D., Luu, K. V., and Whitelaw, E. 2003. Transgenerational inheritance of epigenetic states at the murine Axin(Fu) allele occurs after maternal and paternal transmission. Proc. Natl. Acad. Sci USA. 100: 2538–2543.
Razin, A., and Riggs, A. D. 1980. DNA methylation and gene function. Science 210: 604–610.
Shane, B., and Stokstad, E. L. 1983. The interrelationships among folate, vitamin B12, and methionine metabolism. Adv. Nutr. Res. 5: 133–170.
Shon, Y. H., Park, S. D., and Nam, K. S. 2006. Effective chemopreventive activity of genistein against human breast cancer cells. J. Biochem. Mol. Biol. 39: 448–451.
Vaquero, A., Scher, M., Lee, D., Erdjument-Bromage, H., Tempst, P., and Reinberg, D. 2004. Human SIRT1 interacts with histone H1 and promotes formation of facultative heterochromatin. Mol. Cell 16: 93–105.
Waladkhani, A. R., and Clemens, M. R. 1998. Effect of dietary phytochemicals on cancer development (review). Int. J. Mol. Med. 1: 747–753.
Waterland, R. A., Dolinoy, D. C., Lin, J. R., Smith, C. A., Shi, X., and Tahiliani, K. G. 2006a. Maternal methyl supplements increase offspring DNA methylation at Axin Fused. Genesis 44: 401–406.
Waterland, R. A., and Jirtle, R. 2003. Transposable elements: Targets for early nutritional effects on epigenetic gene regulation. Mol. Cell Biol 23: 5293–5300.

Waterland, R., and Jirtle, R. L. 2004. Early nutrition, epigenetic changes at transposons and imprinted genes, and enhanced susceptibility to adult chronic diseases. Nutrition 20: 63–68.

Waterland, R. A., Lin, J. R., Smith, C. A., and Jirtle, R. L. 2006b. Post-weaning diet affects genomic imprinting at the insulin-like growth factor 2 (Igf2) locus. Hum. Mol. Genet. 15: 705–716.

Weindruch, R., Walford, R. L., Fligiel, S., and Guthrie, D. 1986. The retardation of aging in mice by dietary restriction: Longevity, cancer, immunity and lifetime energy intake. J. Nutr. 116: 641–654.

Wolff, G. L., Kodell, R. L., Moore, S. R., and Cooney, C. A. 1998. Maternal epigenetics and methyl supplements affect agouti gene expression in Avy/a mice. FASEB. J. 12: 949–957.

Yamamoto, H., Schoonjans, K., and Auwerx, J. 2007. Sirtuin functions in health and disease. Mol. Endocrinol. 21: 1745–1755.

Yang, C. S., Landau, J. M., Huang, M. T., and Newmark, H. L. 2001. Inhibition of carcinogenesis by dietary polyphenolic compounds. Annu. Rev. Nutr. 21: 381–406.

Environmental Effects on Age-Associated Epigenetics

Sabita N. Saldanha, Ashley McCollum, and Trygve O. Tollefsbol

Abstract Aging of an organism appears to be a precursor for most, if not all disease outcomes. However, some diseases associated with aging are due to genetic mutations, while others occur even without altering the genetic sequence, through a phenomenon known as epigenetics. DNA methylation and histone modifications are classic examples of epigenetic modifications. These modifications appear to be in a constant flux and, therefore, can be altered by exposure to environmental insults over the life time of the organism. This chapter focuses on how environmental conditions such as heat stress, exposure to metals, and nurturing can affect the dynamic epigenetic modifications in aging. In addition, how these modifications can either directly or indirectly result in age-associated pathologies are also addressed.

Keywords Epigenetics · Aging · Environment and Epigenetic modifications

Introduction

Organismal systems are comprised of three types of cell populations. The stem cell populations that primarily give rise to new cells, the differentiated cell populations that are tissue and function specific, and apoptotic populations composed of dead cells. Differentiated cell populations are of particular interest as these populations are responsible for maintaining the functional integrity of the organs. Differentiated somatic cells, however, have a finite life span and follow Hayflick's limit (Hayflick 1979, 1980, 1985), that is, after a programmed number of cell divisions, the cells are unable to divide and thereby enter the senescent phase. Senescent cell populations are the hallmark of the aging process and may induce the aging phenotype and

T.O. Tollefsbol (✉)
Center for Aging and Department of Biology, 175 Campbell Hall, 1300 University of Alabama at Birmingham, AL 35294–1170, USA
e-mail: trygve@uab.edu

T.O. Tollefsbol (ed.), *Epigenetics of Aging*, DOI 10.1007/978-1-4419-0639-7_23,

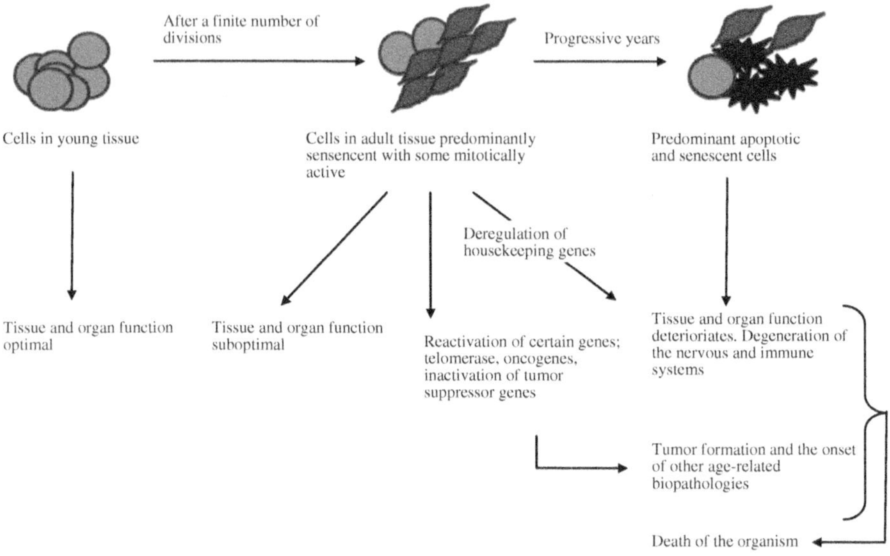

Fig. 1 Cellular and tissue changes occurring with the onset of aging. The circles (*light gray*) symbolize mitotically active and proliferating cells as seen in young tissues. However, an internal genetic clock causes the cells to switch from mitotically active to slow proliferating cells as seen in adult tissue, with predominant silent but metabolically active senescent cells (*middle gray, spindle shaped*). This phase deteriorates further, and genes responsible for the healthy sustenance of cells may be silenced (e.g., housekeeping genes). Eventually cell death (*dark gray, star shaped*) ensues which may trigger many age-related diseases

in some cases age-associated pathologies (Price et al. 2002; Wiemann et al. 2002; Minamino and Komuro 2007) (Fig. 1). Several other determinants contribute to the aging process which are discussed in other reviews (Sedivy et al. 2008). It is interesting to note that although several molecular changes within a cell contribute to the aging phenomenon, of particular interest is the change in gene expression as these fundamental units control almost all cellular processes. The previously held belief that mutations solely affect gene expression has been refuted. Recent investigations have shown otherwise (Sedivy et al. 2008). Altered gene expression can be mediated by an epigenetic process where the sequence of the gene remains unaltered, and the patterns of these changes are stably inherited through cellular divisions. Stable inheritance of these patterns is controlled by mechanisms such as DNA methylation and post-translational modifications of key histone residues (Fraga and Esteller 2007). The resultant accumulation of certain epigenetic marks and the interplay of these events may influence how an organism ages (Fraga and Esteller 2007). Breaking this code can help to understand what marks trigger a cell to age or proliferate. The most predominant factors known to influence such epigenetic events are the environment and nutrition (Fraga et al. 2005; Feil 2006; Fraga and Esteller 2007).

DNA Methylation

DNA methylation is a predominant epigenetic mechanism that controls gene expression. Methylation occurs at CpG dinucleotide residues controlled by the DNA methyltransferase enzymes (Feil 2006). The methylation patterns that are inherited are substrate specific, and maintenance methylation by DNA methyltransferase 1 (DNMT1) will occur only if a premethylated template exists. This allows for a faithful reproduction of the previous methylation pattern, without altering the epigenetic mark. Methylation of CpG dinucleotides is primarily associated with gene repression. Observations of global hypomethylation have been documented in some studies (Wilson and Jones 1983; Mays-Hoopes 1989; Richardson 2003; Liu et al. 2007). However, hypermethylation of locus-specific genes or specific promoter elements can induce gene expression (Feil 2006). Epigenesis is not a simple process but has some degree of complexity, as seen in higher order organisms. The epigenetic process of DNA methylation in gene control has been well established in X-chromosome inactivation and genomic imprinting (Feil 2006).

X-Chromosome Inactivation

Although very few phenotypic genes are present on the X chromosome, the genes on this chromosome control many developmental processes, and thus, maintaining the dosage is of utmost importance. X-chromosome inactivation occurs at the X inactivating center (XIC) (Avner and Heard 2001; Boumil and Lee 2001; Heard et al. 2001; Sado et al. 2001). Inactivation of X (Xi) chromosome is mediated through epigenetic intervention (Wang et al. 2001). The inactivation of X takes place randomly in the somatic cells, and only in the germ line of the females are both X chromosomes active. Inactivation of the X chromosome is initiated at the X-chromosome inactivating center and is mediated by a sole gene *Xist* (X inactive signal transcript) (Avner and Heard 2001; Salstrom 2007; Vincent-Salomon et al. 2007). An RNA molecule transcribed by this gene covers the X chromosomes and initiates the wave of inactivation across the whole chromosome with the exception of a few regions. In addition to the specific RNA transcript that is not translated, heterochromatization of the X chromosome is mediated by DNA methylation and chromatin remodeling.

Modification of histone H3 tails accounts for some of the early events that take place once the X chromosome that is marked for inactivation by the Xist RNA (Heard et al. 2001). Lysine 9 and 4 of histone H3 becomes globally methylated and hypoacetylated, and hypomethylated, respectively (Heard et al. 2001). Methylation of lysine 9 may take part in the chromatin reorganization induced by Xist transcript that stimulates the repression of X chromosome. HP1 heterochromatin proteins have the ability to bind to H3 methylated K9 and to bind to histone methyltransferases that enhance the spread of methylation (Nakayama et al. 2001). This allows for the spread of the inactivation state. Histone H4 hypoacetylation on the other hand is a modification that occurs as a later event (Kaneshiro et al.

2007). This later event suggests that early H3 histone modifications are primarily responsible for chromosome-wide inactivation (Heard et al. 2001; Jenuwein and Allis 2001).

Studies have shown that the Xi chromosome has interspersed regions of active and inactive regions. This indicates that the inactivation is not signaled by individual histone marks but rather the cumulative effects of various histone modifications acting together (Table 1). What has been observed is that there are several genes on the inactive X that escape inactivation (Kaneshiro et al. 2007). A few that are silenced are housekeeping genes such as Hypoxanthine phosphoribosyltransferase 1 *(HPRT1)* and phosphoglycerate kinase1 *(PGK1)*. The genes are silenced through epigenetic mechanisms of DNA methylation as the promoters of these genes are CpG rich. Many of these genes also control various cellular processes. The inactivation and silencing of genes occur early in embryonic development (Kaneshiro

Table 1 Histone modifications involved in X chromosome and gene activation/inactivation

Histone modified	Residue modified	Type of modification	Effect on X chromosome	References
H3	Lysine 27	Trimethylation	Silences certain regions on X chromosome and may associate with some coding regions	Brinkman et al. (2006)
H3	–	Acetylation	Associated with promoters of active genes on X chromosome (but may not be transcribed)	Brinkman et al. (2006)
H4	–	Acetylation	Associated with promoters of active genes on X chromosome (but may not be transcribed)	Brinkman et al. (2006)
H3	Lysine 4	Trimethylation	Associated with promoters of active genes on X chromosome (but may not be transcribed)	Brinkman et al. (2006)
H3	Lysine 9	Trimethylation	Associated with transcription	Brinkman et al. (2006)
H3 H3 H4	Lysine 27 Lysine 9 Lysine 20	Trimethylation Dimethylation Monomethylation	Facultative X chromatin states	Heard et al. (2001; Kohlmaier et al. (2004); Plath et al. (2002); Silva et al. (2008)
H3 H3 H4	Lysine 27 Lysine 9 Lysine 20	Monomethylation Trimethylation Trimethylation	Constitutive X chromatin states	Ebert et al. (2004); Kohlmaier et al. (2004); Schotta et al. (2004a, b)

et al. 2007). However, reprogramming of these genes can be achieved through environmental influences as the epigenome is quite dynamic. The exact relationship between environmental factors and X reactivation via epigenetic mechanisms yet needs to be ascertained. The speculation is that with age, probably this stability can be reprogrammed or erased resulting in reactivation that could possibly trigger or amplify phenotypic defects.

Histone Modifications

Histones are basic proteins that are evolutionary well conserved and are essential to the proper packaging of DNA in the nucleus. About ~146 bp of DNA are wrapped around the histone octamer molecule which is comprised of two subunits each of H2A, H2B, H3, and H4 (Luger et al. 1997; Jenuwein and Allis 2001). This structure is called the nucleosome and forms the repeating unit of chromatin.

The activation and inactivation of gene function has been established by the reversible chemical modification of histone amino 'tails' that initiate euchromatic or heterochromatic states. Commonly, acetylation, phosphorylation, methylation, ubiquitination, and sumoylation are involved in modifying the exposed histone sequences (Jenuwein and Allis 2001; Ke et al. 2006; Iniguez-Lluhi 2006; Sims et al. 2006).

Amino acid residues such as lysine and arginine are prone to such modifications. However, the degree of chemical modification, the number of residues modified, and the type of modification greatly influence the control on gene expression. For example, two processes such as acetylation and methylation of histone residues can act simultaneously either on identical or different subunits, but the overall pattern or code dictates if a gene is turned on or off.

The modified histone states are initiated by enzymes such as histone methyltransferases (add methyl groups), histone acetylases [HATs] (add acetyl groups), or histone deacetylases [HDACs] (remove acetyl groups). Investigations have demonstrated that a cross talk can exist between epigenetics states (Vaissiere et al. 2008). Histone acetylation allows the unwinding of DNA associated with the nucleosome exposing it to regulatory elements. Thus, acetylation not only alters the structure of chromatin but also can dictate whether a gene is expressed or repressed based on the regulatory factors that can bind to exposed promoter elements. Although histone acetylation and its role in carcinogenesis is intensively studied, relatively little is known about the molecular mechanisms by which this epigenetic process influences changes in the histone code in aging. One can speculate that in young healthy cells, loss of methylation of CpG islands with acetylation maintains the expression of crucial genes. However, with age and exposure to environmental insults, a shift in epigenetic profile may occur whereby the spread of methylation to CpG residues with deacetylation of local histones can induce progressive loss of gene expression (Vaissiere et al. 2008). This shift in profiles may enhance age-related etiologies (Cairns 2001; Rowley 1998; Wolffe 2001; Vaissiere et al. 2008).

Environmental Effects and Epigenetics

Organisms are constantly exposed to environmental insults throughout their lifetime that can inadvertently change the landscape of molecular processes. Environmentally induced stresses contributed by an individual's occupation, lifestyle, and external surroundings are considered to be extremely crucial to these changes. Exposure to metals such as nickel, lead, cadmium, arsenic are known to initiate tumorigensis (Herceg 2007; Vaissiere et al. 2008). These factors therefore have a potential effect on chromatin organization, mediated by DNA methylation and histone alterations that code specific epigenetic information. These metals therefore can significantly affect changes in normal epigenetic patterns, initiate aberrant epigenetic signaling, and induce neoplasms. There is a well-established link between ionizing radiations and chromosomal instability. This instability most probably arises from abnormal methylation of unmethylated regions. Chronic exposure to UV radiation has been shown to induce global hypomethylation as seen in studies carried out in mice (Herceg 2007). The end result of these abnormalities early on in life culminates in age-associated malignancies.

Nickel

Epigenetic damage by nickel particles occurs in heterochromatic regions of chromosomes. Studies have shown that nickel tends to increase DNA methylation near heterochromatin (Herceg 2007). Interestingly, tumor suppressor genes and senescence genes may be located in these regions. Therefore, critical cellular genes can be turned off by de novo methylation of the promoters. Sometimes more than a single epigenetic event can transcriptionally silence important genes. This has been seen in a gene such as guanine-hypoxanthine phosphoribosyltransferase *(gpt)*. This gene is transcriptionally silenced not only by DNA methylation but also by a decrease in global acetylation of histones H3 and H4 and increased histone H3K9 dimethylation (Herceg 2007). Studies have shown in the presence of nickel that specific lysine residues are hypoacetylated, primarily at position 12 and 16 of histone H4 (Broday et al. 1999, 2000; Zoroddu et al. 2000. In addition, nickel preferentially binds to a site at histidine 18 (Broday et al. 2000). This binding of nickel to histidine at position 18 prevents or hinders the accessibility of neighboring lysine residues to the HAT complex. Thus histone code patterns are significantly modified or changed by chronic exposure to the nickel molecule (Table 2).

Arsenic and Cadmium

Drinking water and food contain low but permissible levels of arsenic. However, chronic exposure of this element can exert perturbations in epigenetic patterns. Arsenic-related hypomethylation is due to the depletion of the cofactor

Table 2 Changes in histone modification by exposure to nickel

Histone subunit	Epigenetic modification	References
H2A	Cleaves C-terminal tail; increases ubiquitination	Bal et al. (2000a, b); Karaczyn et al. (2003); Ke et al. (2006)
H2B	Truncates, deaminates, and oxidizes; increases ubiquitination	Kaneshiro et al. (2007)
H3	Increase in lysine 9 dimethylation, decrease in acetylation	Chen et al. (2006)
H4	Decrease in acetylation of lysines 12 and 16, respectively	Broday et al. (2000)

S-adenosylmethionine, required for methylation as well as arsenic metabolism. The sodium form of arsenic has been shown to increase genomic hypomethylation (Zhao et al. 1997) of certain crucial genes such as Harvey rat sarcoma viral ocogene *(Ha-Ras)*, and these perturbations can allow for neoplastic development later on in an aged individual. Cadmium inhibits the epigenetic modifying DNA methyltransferases leading to hypomethylation. However, prolonged exposure to cadmium stimulates regional hypermethylation and global hypomethylation which is a commonly observed phenomenon in aged tissues.

Lead (Pb)

The brain undergoes numerous changes throughout life in humans. The set of genes that are passed along from each generation has much to do with the direction of aging in the brain, but environmental influences can contribute to the process as well. For example, the structure of chromatin can be influenced by the environment which can cause a lifetime of phenotypical changes and possibly be passed on to future generations. Diseases such as Alzheimer's and Parkinsons are prime examples of environmental effects on phenotype (Miller and O'Callaghan 2008). Since some of the diseases have a later onset of the phenotype, one can speculate that the genes are initially silenced until affected by an environmental factor, which may reverse the inactivation through epigenetic mechanisms. Methylation is one such epigenetic element that partakes in altering the epigenome. DNA methylation can be affected by environmental factors, such as drugs, maternal care, and social situations, thus confirming the association of DNA methylation to the long-term changes of gene expressions in the brain (Champagne and Meaney 2007; Champagne 2008; Champagne and Curley 2008).

Exposure to lead early in development is known to have effects on brain development (Zawia and Basha 2005, Wu et al. 2008). Brain development normally occurs during early pre- and postnatal periods followed by further development until early adulthood. Lead-mediated changes in brain development have been associated with specific genes such as amyloid precursor proteins (Wu et al. 2008). In normal aging individuals, the accumulation of amyloid plaques and associated proteins has been observed, and this process if accelerated can induce neurodegenerative diseases such as Alzheimer's disease. Exposure to lead decreases methylation of a crucial few genes involved in brain development such as the amyloid beta (A4) precursor protein (*APP*) gene (Wu et al. 2008). Hypomethylation of the *APP* promoter leads to decrease in the methylation burden, thereby inducing *APP* mRNA production. Consequently the increase in APP production is associated with an increase in the amyloidgenic β cleavage product during senescence. This cleavage product forms aggregates and generates free radicals that attack biomolecules such as DNA. The presence of a methylated cytosine impairs the repair of adjacent oxidized guanine bases, thereby increasing the susceptibility of neurons to further damage. Events such as these may collectively enhance the neurodegenerative process.

Maternal behavior appears to play a significant role in altering behavior and physiology. One study (Meaney and Szyf 2005) demonstrated that maternal nurture influenced epigenomic patterns, especially of the brain. This group analyzed the epigenetic patterns of a specific gene, the glucocorticoid receptor. Differential methylation patterns of the promoter of the glucocorticoid receptor were observed in the hippocampus region, dependent on the type of nurture provided (grooming, licking, arched back feeding) (Meaney and Szyf 2005). In addition, the expression of nerve growth factor-induced clone A (NGFIA) that binds to the promoter region of the glucocorticoid receptor is altered by histone acetylation (Weaver et al. 2004a, b). Nevertheless, these marks and consequently the effect of the maternal influence can be reversed through pharmacological interventions indicating that the marks are dynamic rather than stable.

Heat Stress

A few studies have shown that heat stress may induce changes in methylation imprinting patterns (Zhu et al. 2008. However, these changes may be specific to only a few genes involved in embryo implantation and development. Aberrant imprinting patterns have been shown to cause drastic developmental failures in embryos exposed to heat stress (Ealy et al. 1995, Ozawa et al. 2002). Imprinting of genes is primarily controlled by a well-known epigenetic mechanism, DNA methylation. Genomic demethylation and remethylation occur in germ cell development and after fertilization (Reik et al. 1990, Howlett and Reik 1991). However, the methylation patterns acquired are maintained after fertilization and preimplantation development. Thus, these patterns which are otherwise maintained in normal methylation imprinting may be lost due to heat stress. To study the effect of heat stress and its effects on methylation patterns, some groups have assessed four imprinted genes, two paternal and two maternal genes (Lau et al. 1994; Lefebvre et al. 1998; Zhu et al. 2008). These studies showed that the paternal imprinted genes H19 and imprinted maternally expressed transcript (non protein coding) (*H19*) insulin-like growth factor 2 receptor (*Igf2r*) and the maternal imprinted genes parternally expressed gene 1 (*Peg1*) and paternally expressed gene 2 (*Peg2*) reveal a possible link between heat stress and methylation imprinting (Lau et al. 1994; Lefebvre et al. 1998; Zhu et al. 2008). Surprisingly, heat stress-related abnormal methylation imprinting has been shown to affect paternal imprinting rather than maternal (Zhu et al. 2008). But this evidence is limited to only a few genes. For a more concrete conclusion several imprinted genes need to be analyzed. One can speculate that under heat stress, changes may occur in the maternal body. These changes may induce intracellular oxidative damage that can affect sensitive biomolecules such as DNA (Zhu et al. 2008). However, a clear-cut relationship between heat stress and epigenetic changes in the embryo is still unclear. The paternally imprinted genes H19 and Igf2r tend to be hypomethylated under heat stress, which has been supported by several studies (Doherty et al. 2000; Khosla et al. 2001a, b; Zhu et al. 2008).

Abnormal methylation imprinting may affect phenotypic outcomes of the genes resulting in developmental errors. Maternally imprinted genes, at least those that have been studied, have been shown to be unaffected by heat stress (Zhu et al. 2008). However, a large body of evidence is required to support these observations. Therefore, extensive investigations need to be carried out to discern the exact changes in epigenetic patterns induced by heat stress. These patterns may differ from what are seen *in vitro* versus *in vivo*.

Environment and Epigenetic Drift in Populations

In the approach of understanding how environment affects the epigenetic mechanisms of populations as a whole, identical genomes are analyzed for any subtle or drastic changes based on geographical location, lifestyle, or environment. This assessment has been investigated in monozygotic twin population studies (Petronis 2006). Thus any changes in the phenotype of a given trait in identical genomes can be attributed as an environmental contribution. Although biases may exist, in large-scale epigenetic studies of twin populations, these studies can provide a means of understanding the impact of environment on gene expression, the genome, and organism as a whole (Petronis 2006). Studies and observations in different organisms have shown the effect of environment on chromatin modifications. For example, plants exposed to cold temperatures fail to flower and this phenomenon is mediated by histone modifications to genes that encode repressors of flowering (Amasino 2004; Sung and Amasino 2006). Environmental stress can bring about expression of previously silenced genes, which has been observed in *Drosophila* studies. During stress, a specific heat shock chaperon protein Hsp90 increases the activity of histone H3 lysine 4 methyltransferase, inducing the expression of specific target genes (Ruden et al. 2005). Extensive grooming which includes licking and arched back nursing have shown to alter the epigenetic patterns in the promoter of the glucocorticoid receptor in the hippocampus of the pups (Weaver et al. 2004a, b). All these studies indicate that epigenetic modifications serve as a substrate for the environmental insults. Thus, through these studies, assessment of molecular epigenetic changes mediated by environmental factors can be assessed.

Comparison of identical twins for testing environmental epigenetics is highly suitable as their DNA sequences are identical and any change or discordance seen in a given phenotype may be attributed to the environmental factor. Biases and multi-environmental factors can change the overall scenario; however, initial studies using these cohorts can provide a wealth of information. For example, a study which analyzed the Beckwith–Wiedemann syndrome demonstrated that imperfect imprinting of a gene at KCNQ1OT1 (KCNQ1 overlapping transcript 1) due to DNA methylation differences in monozygotic twins (MZ) affected by the syndrome was the cause for this disorder (Weksberg et al. 2002). Therefore, lack of maintenance methylation at a key stage of embryogenesis is a factor for the observed phenomenon. Similar studies that used MZ twins to assess global and locus-specific epigenetic differences showed that in a cohort ranging from 3 to 74 years of age, an age-specific epigenetic drift was observed (Fraga et al. 2005). However, the discordance in the onset of disease does not always correlate with the age of onset (e.g., Alzheimer's versus diabetes mellitus type 1). Twin studies revealed that the concordance of Alzheimer's disease was 83% but that of diabetes type 1 was as low as 23% (Petronis 2006). Further studies are required to analyze the implication of DNA sequence differences as compared to epigenetic differences in the contribution to the synchronous/asynchronous nature of the onset of the disease.

Conclusion

Nurture and nature tend to influence the physiological and behavioral outcomes of individuals. The observation may be attributed to the effects of environment, both extrinsic and intrinsic, on differential developmental patterns. Epigenomic marks are relatively stable and occur early in the developmental program. These stably inherited patterns, however, can be changed or reversed by environmental influences such as stress, toxic metal exposure, or even maternal nurture. Understanding these epigenetic marks is crucial to knowing the patterns associated with aging cells or if particular codes are associated with diseased phenotypes. Since aging epigenetics is a newly explored field, well-structured studies are required to understand the interplay of individual or coordinated codes on gene expression mediated by environmental cues.

References

Amasino, R. 2004. Vernalization, competence, and the epigenetic memory of winter. Plant Cell. 16: 2553–2559.

Avner, P., and Heard, E. 2001. X-chromosome inactivation: counting, choice and initiation. Nat Rev Genet. 2: 59–67.

Bal, W. et al. 2000a. Ni(II) specifically cleaves the C-terminal tail of the major variant of histone H2A and forms an oxidative damage-mediating complex with the cleaved-off octapeptide. Chem Res Toxicol. 13: 616–624.

Bal, W., Kozlowski, H., and Kasprzak, K. S. 2000b. Molecular models in nickel carcinogenesis. J Inorg Biochem. 79: 213–218.

Boumil, R. M., and Lee, J. T. 2001. Forty years of decoding the silence in X-chromosome inactivation. Hum Mol Genet. 10: 2225–2232.

Broday, L., Cai, J., and Costa, M. 1999. Nickel enhances telomeric silencing in Saccharomyces cerevisiae. Mutat Res. 440: 121–130.

Broday, L. et al. 2000. Nickel compounds are novel inhibitors of histone H4 acetylation. Cancer Res. 60: 238–241.

Brinkman, A. B. et al. 2006. Histone modification patterns associated with the human X chromosome. EMBO Rep. 7: 628–634.

Cairns, B. R. 2001. Emerging roles for chromatin remodeling in cancer biology. Trends Cell Biol. 11: S15–S21.

Champagne, F. A. 2008. Epigenetic mechanisms and the transgenerational effects of maternal care. Front Neuroendocrinol. 29: 386–397.

Champagne, F. A., and Meaney, M. J. 2007. Transgenerational effects of social environment on variations in maternal care and behavioral response to novelty. Behav Neurosci. 121: 1353–1363.

Champagne, F. A., and Curley, J. P. 2008, Jan 18. Epigenetic mechanisms mediating the long-term effects of maternal care on development. Neurosci Biobehav Rev. [Epub ahead of print] PMID: 18430469

Chen, H. et al. 2006. Nickel ions increase histone H3 lysine 9 dimethylation and induce transgene silencing. Mol Cell Biol. 26: 3728–3737.

Doherty, A. S. et al. 2000. Differential effects of culture on imprinted H19 expression in the preimplantation mouse embryo. Biol Reprod. 62: 1526–1535.

Ealy, A. D. et al. 1995. Developmental changes in sensitivity of bovine embryos to heat shock and use of antioxidants as thermoprotectants. J Anim Sci. 73: 1401–1407.

Ebert, A. et al. 2004. Su(var) genes regulate the balance between euchromatin and heterochromatin in Drosophila. Genes Dev. 18: 2973–2983.

Feil, R. 2006. Environmental and nutritional effects on the epigenetic regulation of genes. Mutat Res. 600: 46–57.

Fraga, M. F. and Esteller, M. 2007. Epigenetics and aging: the targets and the marks. Trends Genet. 23: 413–418.

Fraga, M. F. et al. 2005. Epigenetic differences arise during the lifetime of monozygotic twins. Proc Natl Acad Sci USA. 102: 10604–10609.

Hayflick, L. 1979. Cell biology of aging. Fed Proc. 38: 1847–1850.

Hayflick, L. 1980. The cell biology of human aging. Sci Am. 242: 58–65.

Hayflick, L. 1985. The cell biology of aging. Clin Geriatr Med. 1: 15–27.

Heard, E. et al. 2001. Methylation of histone H3 at Lys-9 is an early mark on the X chromosome during X inactivation. Cell. 107: 727–378.

Herceg, Z. 2007. Epigenetics and cancer: towards an evaluation of the impact of environmental and dietary factors. Mutagenesis. 22: 91–103.

Howlett, S. K., and Reik, W. 1991. Methylation levels of maternal and paternal genomes during preimplantation development. Development. 113: 119–127.

Iniguez-Lluhi, J. A. 2006. For a healthy histone code, a little SUMO in the tail keeps the acetyl away. ACS Chem Biol. 1: 204–206.

Jenuwein, T., and Allis, C. D. 2001. Translating the histone code. Science. 293: 1074–1080.

Kaneshiro, K. et al. 2007. An integrated map of p53-binding sites and histone modification in the human ENCODE regions. Genomics. 89: 178–188.

Karaczyn, A. A. et al. 2003. The octapeptidic end of the C-terminal tail of histone H2A is cleaved off in cells exposed to carcinogenic nickel (II). Chem Res Toxicol. 16: 1555–1559.

Ke, Q. et al. 2006. Alterations of histone modifications and transgene silencing by nickel chloride. Carcinogenesis. 27: 1481–1488.
Khosla, S. et al. 2001a. Culture of preimplantation embryos and its long-term effects on gene expression and phenotype. Hum Reprod Update. 7: 419–427.
Khosla, S. et al. 2001b. Culture of preimplantation mouse embryos affects fetal development and the expression of imprinted genes. Biol Reprod. 64: 918–926.
Kohlmaier, A. et al. 2004. A chromosomal memory triggered by Xist regulates histone methylation in X inactivation. PLoS Biol. 2: 0991–1003.
Lau, M. M. et al. 1994. Loss of the imprinted IGF2/cation-independent mannose 6-phosphate receptor results in fetal overgrowth and perinatal lethality. Genes Dev. 8: 2953–2963.
Lefebvre, L. et al. 1998. Abnormal maternal behaviour and growth retardation associated with loss of the imprinted gene Mest. Nat Genet. 20: 163–169.
Luger, K. et al. 1997. Crystal structure of the nucleosome core particle at 2.8 A resolution. Nature. 389: 251–260.
Liu, L. et al. 2007, Sep 10. DNA methylation impacts on learning and memory in aging. Neurobiol Aging. [Epub ahead of print] PMID: 17850924
Mays-Hoopes, L. L. 1989. DNA methylation in aging and cancer. J Gerontol. 44: 35–36.
Meaney, M. J., and Szyf, M. 2005. Environmental programming of stress responses through DNA methylation: life at the interface between a dynamic environment and a fixed genome. Dialogues Clin Neurosci. 7: 103–123.
Miller, D. B., and O'Callaghan, J. P. 2008. Do early-life insults contribute to the late-life development of Parkinson and Alzheimer diseases? Metabolism. 57 Suppl 2: S44–49.
Minamino, T., and Komuro, I. 2007. Vascular cell senescence: contribution to atherosclerosis. Circ Res. 100: 15–26.
Nakayama, J. et al. 2001. Role of histone H3 lysine 9 methylation in epigenetic control of heterochromatin assembly. Science. 292: 110–113.
Ozawa, M., Hirabayashi, M., and Kanai, Y. 2002. Developmental competence and oxidative state of mouse zygotes heat-stressed maternally or in vitro. Reproduction. 124: 683–689.
Petronis, A. 2006. Epigenetics and twins: three variations on the theme. Trends Genet. 22: 347–350.
Plath, K. et al. 2002. Xist RNA and the mechanism of X chromosome inactivation. Annu Rev Genet. 36: 233–278.
Price, J. S. et al. 2002. The role of chondrocyte senescence in osteoarthritis. Aging Cell. 1: 57–65.
Reik, W., Howlett, S. K., and Surani, M. A. 1990. Imprinting by DNA methylation: from transgenes to endogenous gene sequences. Dev Suppl. 99–106.
Richardson, B. 2003. Impact of aging on DNA methylation. Ageing Res Rev. 2: 245–61.
Rowley, J. D. 1998. The critical role of chromosome translocations in human leukemias. Annu Rev Genet. 32: 495–519.
Ruden, D. M. et al. 2005. Hsp90 and environmental impacts on epigenetic states: a model for the trans-generational effects of diethylstibesterol on uterine development and cancer. Hum Mol Genet. 14 Spec No 1: R149–15.
Sado, T. et al. 2001. Regulation of imprinted X-chromosome inactivation in mice by Tsix. Development. 128: 1275–1286.
Salstrom, J. L. 2007. X-inactivation and the dynamic maintenance of gene silencing. Mol Genet Metab. 92: 56–62.
Schotta, G. et al. 2004a. A silencing pathway to induce H3-K9 and H4-K20 trimethylation at constitutive heterochromatin. Genes Dev. 18: 1251–1262.
Schotta, G. et al. 2004b. The indexing potential of histone lysine methylation. Novartis Found Symp. 259: 22–37; discussion 37–47, 163–169.
Sedivy, J. M., Banumathy, G., and Adams, P. D. 2008. Aging by epigenetics – a consequence of chromatin damage? Exp Cell Res. 314: 1909–1917.
Silva, S. S. et al. 2008. X-chromosome inactivation and epigenetic fluidity in human embryonic stem cells. Proc Natl Acad Sci USA. 105: 4820–4825.

Sims, J. K. et al. 2006. A trans-tail histone code defined by monomethylated H4 Lys-20 and H3 Lys-9 demarcates distinct regions of silent chromatin. J Biol Chem. 281: 12760–12766.
Sung, S., and Amasino, R. M. 2006. Molecular genetic studies of the memory of winter. J Exp Bot. 57: 3369–3377.
Vaissiere, T., Sawan, C., and Herceg, Z. 2008. Epigenetic interplay between histone modifications and DNA methylation in gene silencing. Mutat Res. 659: 40–48.
van Doorn, R. et al. 2005. Aberrant DNA methylation in cutaneous malignancies. Semin Oncol. 32: 479–487.
Vincent-Salomon, A. et al. 2007. X inactive-specific transcript RNA coating and genetic instability of the X chromosome in BRCA1 breast tumors. Cancer Res. 67: 5134–5140.
Wang, J. et al. 2001. Imprinted X inactivation maintained by a mouse Polycomb group gene. Nat Genet. 28: 371–375.
Weaver, I. C. et al. 2004a. Early environmental regulation of hippocampal glucocorticoid receptor gene expression: characterization of intracellular mediators and potential genomic target sites. Ann NY Acad Sci. 1024: 182–212.
Weaver, I. C. et al. 2004b. Epigenetic programming by maternal behavior. Nat Neurosci. 7: 847–854.
Weksberg, R. et al. 2002. Discordant KCNQ1OT1 imprinting in sets of monozygotic twins discordant for Beckwith-Wiedemann syndrome. Hum Mol Genet. 11: 1317–1325.
Wiemann, S. U. et al. 2002. Hepatocyte telomere shortening and senescence are general markers of human liver cirrhosis. FASEB J. 16: 935–942.
Wilson, V. L. and Jones, P. A. 1983. DNA methylation decreases in aging but not in immortal cells. Science. 220: 1055–1057.
Wolffe, A. P. 2001. Chromatin remodeling: why it is important in cancer. Oncogene. 20: 2988–2990.
Wu, J., Basha, M. R., and Zawia, N. H. 2008. The environment, epigenetics and amyloidogenesis. J Mol Neurosci. 34: 1–7.
Zawia, N. H., and Basha, M. R. 2005. Environmental risk factors and the developmental basis for Alzheimer's disease. Rev Neurosci. 16: 325–337.
Zhao, C. Q. et al. 1997. Association of arsenic-induced malignant transformation with DNA hypomethylation and aberrant gene expression. Proc Natl Acad Sci USA. 94: 10907–10912.
Zoroddu, M. A. et al. 2000. Interaction of Ni(II) and Cu(II) with a metal binding sequence of histone H4: AKRHRK, a model of the H4 tail. Biochim Biophys Acta. 1475: 163–168.
Zhu, J. Q. et al. 2008. Heat stress causes aberrant DNA methylation of H19 and Igf–2r in mouse blastocysts. Mol Cells. 25: 211–215.

Part VI
Future Directions/Perspectives

Future Directions in Research on the Epigenetics of Aging

Huidong Shi and Charles W. Caldwell

Abstract Epigenetics of aging is a new research direction that brings exciting and in-depth revelations in the near future. As reviewed in chapters throughout this book, aging is likely to be influenced by many complex interacting epigenetic factors. These include DNA methylation, histone modifications, chromatin remodeling, and noncoding RNAs. With the help of new technologies, particularly the knowledge derived from the genome-wide epigenomic studies, the upcoming years will be an exciting time to make breakthroughs in research on epigenetics of aging. Here we focus on several new directions in epigenetic research overall and their potential influence on aging research and the potential of anti-aging intervention using personalized, epigenetic-targeted pharmacologic agents.

Keywords Epigenome · Aging · Epigenetic-targeted therapy

Introduction

Epigenetics is a rapidly evolving frontier of biomedical science as evidenced by the exponential increase in literature citations over the last two decades. Epigenetic mechanisms known to modulate activation or silencing of gene transcription include DNA methylation of CpG islands (CGI) in regulatory regions of the genome; chromatin remodeling and higher order chromatin structural alterations; post-translational modifications of histone proteins which include methylation, acetylation, ubiquitination, phosphorylation, and gene regulation through noncoding RNAs (reviewed in Momparler (2003), Laird (2005), Rodenhiser and Mann (2006), Fraga and Esteller (2007), Gal-Yam et al. (2008), Feinberg (2008)). There are likely additional epigenetic modifications and regulatory mechanisms

C.W. Caldwell (✉)
Department of Pathology and Anatomical Sciences, Ellis Fischel Cancer Center, University of Missouri, Columbia, MO 65203, USA
e-mail: caldwellc@health.missouri.edu

T.O. Tollefsbol (ed.), *Epigenetics of Aging*, DOI 10.1007/978-1-4419-0639-7_24,

that have not yet been discovered or elucidated. Increasing evidence demonstrates that alterations in normal gene regulation adversely affect phenotypic plasticity and result in a broad spectrum of tissue dysfunction and disease outcomes. These include, but are not limited to, multiple cancers, autoimmune diseases, neurodegenerative disorders, liver diseases, respiratory disorders, developmental disorders, cardiovascular diseases, and behavioral disorders. Understanding temporal and tissue-specific epigenetic regulation is therefore fundamental to human health and diseases, not only in aging.

Most recent studies show that epigenetic regulation of gene transcription plays a pivotal role in the governance of normal and disease development through dynamic transcriptional activities from gametogenesis through embryonic and neonatal stages, and continuing throughout adolescence, adulthood, and elderly stages. The integration of epigenetics with genetics and environmental influences will be necessary to fully understand mechanisms of complex human diseases associated with normal aging process. Normal development and aging are directed by epigenetic processes and can be affected by various environmental exposures resulting in alterations in the transcriptional potential of a cell or tissue type. This altered gene expression in many cases is stable and may persist leading to increased susceptibility to disease later in life and even across generations.

Epigenetic research in human health is rapidly evolving and has reached a critical point where opportunities exist to make significant inroads toward understanding how epigenetically regulated transcription directs functional processes during development and across the life span as well as in disease states. Thus, discovery and cataloging of epigenetic marks and profiles in normal differentiated tissues/cells and in progenitor stem cells will provide a more comprehensive understanding of differentiation, maintenance, or plasticity of cell-type identity and potentially transform the development of molecular/epigenetic/cell-based strategies for disease intervention.

There are at least five areas of future research directions in epigenetics that will provide amazing opportunities to improve health quality, longevity, and human medicine in the future. First, while many technologies and methods have been developed for epigenetics research (Laird 2005), none so far can provide all the data necessary, and therefore even better high-throughput technologies are needed. Second, in order to fully appreciate the information embedded in high-dimensional data sets, further improvements in bioinformatics and computational science are needed to infer new knowledge from studies of the many layers of the epigenome (Bock and Lengauer 2008). Third, genome-wide studies of normal, tissue-specific epigenetic patterns that occur in individuals should be studied within a person over time, as well as differences that may occur between the epigenomes of different individuals and different ethnic and racial groups. Fourth, the upcoming years are an exciting time to pose questions regarding the most important issues in human health and disease in hopes of defining prevention and treatment strategies based on sound, mechanistic science. Finally, development of disease is a failure of current medical practice. We do not yet know enough to prevent all disease, and therefore newer pharmacologic targets may include new epigenetic modifiers that may target or slow the manifestations of aging itself, or may be targeted to cure diseases without harmful side effects, thus maximizing the quality of human life. The lessons we learn in these areas may also be extended into veterinary medicine and into further basic research including aging-related research.

Technological Advances in Epigenetic Research

The Human Genome Project (HGP) is one of the greatest scientific achievements of our time. Thanks to the firm commitment of NHGRI in developing DNA sequencing technologies toward the goal of the "$1,000 genome," advances of DNA sequencing technology have transformed the landscape of genomic analysis. Several cancer genome sequencing and genome-wide association studies are currently ongoing and a personal genome project is also on the way (http://www.personalgenomes.org/). However, the sequence itself does not fully predict how the genome is packaged in chromosomes and chromatin to provide for the differential expression of genes. Over the last 3 years, an international consensus has emerged in the epigenetics research community for the need of an organized Human Epigenome Project aimed at generating a high-resolution DNA methylation map of the human genome in all major tissues (Jones and Martienssen 2005; Brena et al. 2006; Esteller 2006). Indeed, efforts have already been initiated in Europe and the United States. The Alliance for the Human Epigenome and Disease (AHEAD) has urged the research community to join in a coordinated effort to decode the human epigenome (2008), Just as the Human Genome Project provided a reference "normal" sequence for studying human disease, the goal of the AHEAD project is to provide high-resolution reference epigenome maps. A long-term goal of 1 bp resolution of DNA methylation and a comprehensive analysis of chromatin, including modifications of histones, chromatin factor binding, and structural changes, have been proposed. As a pilot project, Eckhardt and colleagues (2006) used bisulfite sequencing to quantitatively assess DNA methylation at 1.8 M CpGs along 3 chromosomes in 12 different tissues, covering 873 genes in 2524 amplicons. This study has established a benchmark for a possible future epigenome project.

One of the bottlenecks for the human epigenome project is the development of high-throughput, genome-wide technologies for epigenomic profiling. Fortunately, in the past decade, technology development has moved rapidly. Since Huang et al. (Huang, Perry and Laux 1999) developed the first microarray-based approach called differential methylation hybridization (DMH) to study large-scale methylation changes, many other microarray-based methods have been developed (Yan et al. 2000; Shi et al. 2002; Chen et al. 2003; Shi et al. 2003; Gebhard et al. 2006; Rollins et al. 2006; Saxonov et al. 2006; Schumacher et al. 2006; Kuang et al. 2008). The basic idea behind the array-based approach is to hybridize the pool of differentially methylated genomic DNA onto a microarray panel that contains genomic DNA regions that are potentially differentially methylated. The methylated DNA can be enriched by restriction enzyme, methyl-binding proteins, or anti-methyl-cytosine antibodies. The improvements in commercial oligonucleotide microarray manufacturing technologies have made it possible now to screen the non-repetitive regions in the whole genome using tiling arrays from several vendors such as Affymetrix,

Agilent, and Nimblegen. Small-scale arrays that focus on CGI or annotated promoters only are also available from these vendors. In addition, due to highly flexible array manufacture processes, all of these (and other) vendors provide custom array services enabling researchers to design and then chose their own sets of target regions for interrogation.

The power of next-generation DNA sequencing technology, such as Roche/454 GS FLX, Illumina/Solexa 1G Genetic Analyzer, and ABI SOLiD, will further transform the landscape of epigenomic research. For example, by coupling chromatin immunoprecipitation with next-generation sequencing (ChIP-seq), genome-wide distribution of transcription factor-binding sites or chromatin modifications have been characterized (Barski et al. 2007; Mikkelsen et al. 2007; Robertson et al. 2007). This method can also be used to identify DNA-binding sites of other regulatory proteins, or by "re-ChIP," combinations of proteins that may interact in complexes. We recently have performed ultra-deep sequencing analysis of bisulfite-modified DNA with 454 pyrosequencing and demonstrated the robustness and utility of the next-generation sequencing technologies for epigenomic analysis (Taylor et al. 2007). Meissner et al. (2008) described a large-scale random approach termed reduced representation bisulfite sequencing (RRBS) for analyzing and comparing genomic methylation patterns using next-generation sequencing. This approach will sequence a defined fraction of a large genome. In addition, several groups have recently developed methods for targeted genome sequencing (Albert et al. 2007; Dahl et al. 2007; Fredriksson et al. 2007; Porreca et al. 2007; Okou et al. 2007). The underlying concept is very simple. If a method is available to extract only a highly informative subset of sequences from the samples, then the throughput of an assay can be improved dramatically by sequencing only such subsets. Although the studies are designed for the resequencing of all exons in the human genome, the same concepts can be applied to DNA methylation and regulatory protein analyses. In this way, the sequencing cost can be reduced to approximately 1% compared with full-genome sequencing.

One of the primary goals of the proposed Epigenome project is to generate reference epigenomes for normal tissues. There is not one single epigenome, but rather multiple epigenomes or epigenetic profiles, each one defining specific gene expression profiles and therefore the phenotype of a specific cell type (e.g., neuron, T lymphocyte, type II lung cells, or pancreatic beta cells). Priorities have been given to develop reference epigenomes of government-approved human embryonic stem cells, differentiating cells, and selected differentiated cell lines and human primary cells that are relevant to complex human disease. This will be enormously important, but needs to be done not only in a single reference individual, but eventually across multiple individuals, as discussed below.

Methods to investigate the association of specific proteins, such as modified histones or transcription factors, with specific DNA sequences are quite valuable and generally fall into two categories: chromatin immunoprecipitation followed by microarray hybridization (ChIP-chip) or followed by DNA sequencing (ChIP-seq). Both methods have been used with great success to produce genome-wide maps of where specific histone proteins or PcG proteins bind to DNA in the genome

(Weinmann et al. 2002; Kirmizis and Farnham 2004; Oberley et al. 2004; Bernstein et al. 2006; Squazzo et al. 2006; Barski et al. 2007; Mikkelsen et al. 2007). The challenges now are to bring together all these data to infer new meaning in specific situations. These will require bioinformatics and computational science approaches, as well as teams of multidisciplinary scientists.

Computational Science in Future Epigenetic Research

In order to maximally derive useful information from large-scale data sets that will ultimately be produced in biomedical research, including epigenetics, new bioinformatics and computational science solutions will be needed. Systems thinking is used in a variety of scientific and technological fields and can certainly be applied to studies of epigenomics. Biomedical scientists in the post-genomic era are challenged with huge volumes of data (e.g., genome sequences, expression data), originating from heterogeneous technologies (e.g., microarray, ChIP-chip) and representing multiple biological processes. This massive influx of data and the desire to turn it into biologically coherent information have forced us to think not in terms of single molecules but in terms of networks and systems.

An exciting prospect of the post-genomic era is to be able to integrate knowledge across different levels of biological organization and to anchor this at the molecular level. A systems approach can be described as an iterative process that includes (1) data collection and integration of all available information, (2) system modeling, (3) experimentation at a global level, and (4) generation of new hypotheses. The systems approach should help us make predictions, for example, about the effect of an epigenetic perturbation (e.g., CGI DNA methylation) in biological behaviors of interest, such as gene expression, or to determine the conditions that would describe cell biology. Such predictions should then lead to testable hypotheses and further public deposits of related data. Such a systems approach might be used to dissect the epigenome of healthy and diseased cells and then relate this to publicly available gene expression data or other biological attributes. Over time, it is expected that other investigators would add additional layers of epigenetic (and other) information such as histone modifications, microRNA profiles, and proteomic profiles in future studies and contribute this to a central repository of such data.

Controlled vocabularies and ontologies are designed to capture the precise meaning and semantic relationships of terminology utilized in a particular knowledge domain. The creation of a usable ontology is a very demanding, labor-intensive process requiring the cooperation of persons with expertise in the relevant subject matter and/or knowledge engineering. A number of well-constructed ontologies such as gene ontology (GO; http://www.geneontology.org/), gene regulation ontology (GRO; http://obofoundry.org/), and resources through microarray and gene expression data (MGED; http://www.mged.org/) are publicly available. GO, in particular, has been widely employed as a means of analyzing and classifying data through their annotation scheme. There is no doubt that full utilization of high-dimensional data in biomedicine will require further standardization and linkage to infer the greatest meaning and provide useful information.

There are currently multiple sites located around the world that host data and computational tools to analyze the increasingly complicated data that are being produced and will increase dramatically in the future. Established in the United States in 1988 as a national resource for molecular biology information, the National Center for Biotechnology Information (NCBI) creates and hosts public databases, conducts intramural and collaborative research in computational biology, develops software tools for analyzing large-scale data, and disseminates biomedical information for the purpose of better understanding the molecular processes affecting human health and disease (http://www.ncbi.nlm.nih.gov/). One function that will become even more important as large public data sets are produced in the future is that of hosting the data in a sharable manner. The Gene Expression Omnibus (GEO) is a gene expression/molecular abundance repository supporting data submissions, and a curated, online resource for gene expression data browsing, query, and retrieval. Similar functions will support future epigenomics projects, as discussed below. A European counterpart located in the Wellcome Trust Genome Campus south of Cambridge, UK, also was formed for the public sharing of data and resources and continues to be a worldwide resource, the European Bioinformatics Institute (http://www.ebi.ac.uk/) which is part of the European Molecular Biology Institute (EMBL). The J. Craig Venter Institute (http://www.tigr.org/) was formed in October 2006 through the merger of several affiliated and legacy organizations – The Institute for Genomic Research (TIGR), The Center for the Advancement of Genomics (TCAG), The J. Craig Venter Science Foundation, The Joint Technology Center, and the Institute for Biological Energy Alternatives (IBEA). These organizations have now become one large multidisciplinary, genomic-focused, organization with locations in Rockville, Maryland, and La Jolla, California.

Genome-Wide Epigenetic Studies

Several genome- or subgenome-wide studies have recently been published that address important questions regarding epigenetic states of normal, abnormal, and differentiating mammalian cells (Bernstein et al. 2006; Barski et al. 2007; Bernstein et al. 2007; Mikkelsen et al. 2007; Meissner et al. 2008).

Squazzo, et al. (2006) reported genome-wide promoter binding studies of the PRC2 member SUZ12 in five different mouse and human cell lines and demonstrated differential binding across large regions between cell types and demonstrated that SUZ12 targets were bound also by EZH2 and H3K27me3, but not by PolII or H3K9me3. They found that a specific set of genes bound SUZ12 in embryonic stem cells and germ-cell tumors that did not bind in more differentiated adult tumors. In their experiments, SUZ12 binding strictly correlated with recruitment of the EZH2 histone methyltransferase and resulted in silenced chromatin. The SUZ12 targets in embryonic cells were highly enriched for developmental genes such as transcription factors, clustered and unclustered homeobox genes, but in the adult tumors the largest category of SUZ12 targets was glycoprotein genes, and very few transcription factor and homeobox-domain genes. Further, there is evidence to suggest that the transcription factor OCT4 may in part recruit SUZ12 to specific gene promoters.

In human brain, genome-scale cytosine methylation analysis by DNA sequencing identified a number of important functional relationships between site specificities and densities of methylation related to genomic structures including promoters, exons, CGIs, and intergenic regions (Rollins et al. 2006). Another sequencing-based study of methylation across three human chromosomes yielded important insights into the relationships with PcG proteins and transcription in T cells and other tissues (Eckhardt et al. 2006). These, and other studies, have set the stage for analysis and interpretation of data to infer mechanistic processes in epigenetic functions across the spectrum of mouse and human embryonic and adult cells, and in various cancers (Hashimshony et al. 2003; Keshet et al. 2006; Schlesinger et al. 2007).

The major histocompatibility complex (MHC) was recently examined using a genomic tiling array of 2 Kb resolution covering the entire 4 Mb MHC region (Tomazou et al. 2008). The array has been designed to be compatible with chromatin immunoprecipitation (ChIP), methylated DNA immunoprecipitation (MeDIP), and thus has facilitated a very detailed mapping of this region. More, and larger, studies examining regions and genomes will be needed in the future and may include tiling arrays at various resolutions, as well as DNA sequencing approaches.

Biomedical and Epigenetic Studies of the Future

The upcoming years will be an exciting time to pose questions and solve problems regarding the most important issues in human health and disease and perhaps define prevention and treatment strategies based on sound, mechanistic science, including that of epigenetics. Some of the greatest health challenges face us in the near future with an aging population larger than ever before. One of the most concerning problems in this population is that of Alzheimer's disease and other forms of dementia, related to the impact(s) of epigenetics, as discussed in the chapter "DNA Methylation, Age-Related Immune Defects, and Autoimmunity." Further, there also seems to be a progressive increase in autism-spectrum disorders occurring in young children for reasons yet unclear. An epigenetic overlay is now being addressed in this area of research that could have major ramifications into prevention and treatment of this target disorder in our children. There is also a very strong epigenetic impact on cancers, as discussed in the chapter "DNA Methylation and Alzheimer's Disease." This is a very difficult area of study, just based on the shear number of types of cancers and the concomitant contributions of genetics and epigenetics in these tumors. From a historical perspective, it is still much more desirable, and in some instances efficacious, to prevent diseases such as cancer rather than to try to treat this group of diseases. Infectious diseases comprise the largest health threat worldwide, and whatever new information we can learn from epigenetic studies of the normal and disordered immune system, as discussed in the chapter "Epigenetic Silencing of Progeroid Syndromes," and its potential pharmacologic manipulation could have profound health benefits. Autoimmune diseases, including rheumatoid arthritis, are very debilitating and currently we have limited options for care of such patients. The age-related development of osteoarthritis will likely also increase with our aging population, particularly in a society that is increasingly physically active, along with attendant stress on knees, hips, and other joints. Finally, perhaps the greatest challenge is determining how, or if, it is possible to better control the aging process itself, which goes beyond cosmetic issues, but also to head of the currently inevitable health deteriorations associated with aging.

Certainly an area that needs further investigations involve the hypotheses that epigenetic programming of early progenitor cells during development may lead to many, perhaps all, human diseases later in life (Feinberg et al. 2006; Feinberg 2008). This is a very attractive model for which there is accumulating evidence, but there is still much more to learn about this potentially unifying disease model. Further, how does the environment (reviewed in Jirtle and Skinner (2007)), both prenatal and postnatal, potentially alter our epigenome? If supported through future research, then the task becomes how to (or should we) intervene in this early embryonic process? Currently, we have very little control over whatever genetic or epigenetic marks are laid down in our developing genomes prior to birth. Beyond scientific issues, interventions through processes such as in vitro fertilization will lead to considerable discussions about the ethical, legal, and social impacts of epigenetic manipulations.

Many important epigenetic questions remain unanswered. For instance, in normal tissues, what are the tissue-specific epigenetic maps and do they change during normal aging, or only with disease development? Do they differ between closely related individuals, as suggested by the twin studies? Are they the same in different ethnic and racial populations? Just as normal somatic cells can be influenced to de-differentiate into induced stem cells, can age-related epigenetic changes be re-directed in such a manner as to reverse aging and/or disease development? How faithful are the male and female germ line epigenomes replicated between generations? What can positively or negatively influence the epigenome of a developing neonate (dietary, environmental, etc)? Are there epigenetic biomarkers of risk for disease development, and can they be altered?

A very recent study of two separate cohorts of families over time demonstrated that indeed there are inter-individual differences in DNA methylation patterns between individuals, that the methylation levels (both global and gene-specific) change over time, and that the amount of changes may be under genetic control (Bjornsson et al. 2008). This is a fascinating concept based on two distant cohorts: one from Iceland and the other from Utah. Like many good studies, they raise many new questions that will be addressed in future studies.

Personalized, Epigenetic-Targeted Pharmacologic Agents

In the future, development of disease should not be seen as a failure of medical science. While we currently do not yet know enough to prevent manifestations of aging or all diseases, future research may allow us to develop newer pharmacologic targets that may include epigenetic modifiers that target or slow the manifestations of aging itself or may be targeted to cure diseases without harmful side effects, thus maximizing the quality of human life. The "Holy Grail" is a personalized medicine that can benefit everyone. Perhaps being able to overlay a deep understanding of epigenetic mechanisms on the findings from the Human Genome Project and its many offshoots will lead us closer.

For now, we do have certain therapeutic agents available that are based on epigenetic modifications. The first generation of these includes demethylating agents such as Vidaza™ (azacytidine; Celgene) and Dacogen™ (decitabine; MGI PHARMA), as well as histone deacetylase inhibitors including valproic acid, Zolenza® (Vorinostat, Merck) and others. Early clinical trials have shown some benefits from these (and other) pharmacologic approaches, but none has thus far been a complete success in terms of treating serious diseases such as cancer. That being said, there are many forms of cancer, and some seem to respond better than others to selected epigenetic therapies. Considerable future studies will be needed to determine optimal dosing routes, timing, and doses of individual and combination therapies. Some of these questions are already being addressed in selected disease conditions (Shaker et al. 2003; Issa et al. 2005; Issa 2005; Lemaire et al. 2005; Momparler 2005; Garcia-Manero et al. 2006; Plimack et al. 2007; Soriano et al. 2007; Raynal et al. 2008; Lemaire et al. 2008a, b). As with historical use of chemotherapeutic agents, it is likely that we will need to learn not only optimization strategies for individual agents, but also the best combinations for specific medical problems. We will also have more advanced forms of pharmacologic agents that target components of epigenetics without the attendant side effects that are currently associated with some of these.

Health promotion, aging, and disease prevention are also currently being addressed, mainly through grass-roots efforts involving individual and complex mixtures of materials that are though to modify epigenetic processes in beneficial ways. Human health is affected by many factors including diet and genetics, epigenetics, and individual metabolism likely playing important roles. Food constituents found in fruits and vegetables (phytochemicals) may interact directly with DNA via methylation reactions. Many studies (reviewed in Ovesna et al. (2008)) have shown that food constituents may affect human health and some constituent interactions may delay or prevent the onset of diseases. Currently, it is common to purchase a variety of plant-derived food supplements from health food and grocery stores. Many of these supplements are faithfully used for a variety of reasons. Only recently are controlled studies being pursued that will help define the real value of such approaches to health promotion.

MicroRNAs (miRNAs), an abundant class of short noncoding RNAs, are widely expressed in mammalian cells and are important in post-translational gene regulation, cell proliferation, apoptosis, and differentiation processes, as well as in cancer, and their expression patterns serve as signatures of different cancer subtypes. Recent evidence (reviewed in Davis and Ross (2008)) suggests that dietary components as diverse as folate, retinoids, and curcumin exert cancer-protective effects through modulation of miRNA expression. Thus, a better understanding of how these foods can contribute to cancer prevention or nutritional status will be enormously helpful, particularly if they can be shown to help prevent development of important human diseases. Several plant compound mixtures are currently in clinical trials of certain cancers and other diseases (Espin et al. 2007). These can be found by running a query to the clinical trials web site at the National Institutes of Health (NIH) (http://clinicaltrials.gov/ct2/results?term=plants).

The Roadmap Epigenomics Program

A multi-layered approach that involves public deposit of data important to study the epigenome is now being used to develop further the Roadmap Epigenomics Program through the US National Institutes of Health (http://nihroadmap.nih.gov/). The main goals of the initiative are to (1) create an international committee; (2) develop standardized platforms, procedures, and reagents for epigenomics research; (3) conduct demonstration projects to evaluate how epigenomes change; (4) develop new technologies for single-cell epigenomic analysis and in vivo imaging of epigenetic activity; and (5) create a public data resource to accelerate the application of epigenomics approaches (http://nihroadmap.nih.gov/epigenomics/).

Targeted research programs will very clearly help move our understanding of the impacts of epigenetics on aging, health, and disease. Program initiatives currently include (1) establishment of Reference Epigenome Mapping Centers that will serve as a resource and be utilized by the scientific community to identify potential therapeutic targets, enhance understanding of disease mechanisms, provide additional insights to genetic susceptibility of disease, pursue therapeutic opportunities in stem cell-based and tissue regeneration strategies, and to understand normal differentiation, development, and aging/senescence; (2) development of an Epigenomics Data Analysis and Coordination Center (EDACC), which will provide data analysis and coordination for all of the Reference Epigenome Mapping Centers, as well as import all other Roadmap Epigenomics Program data generated outside of the mapping centers. The EDACC will be responsible for coordinating with the National Center for Biotechnology Information (NCBI) to develop and implement a data pipeline for transferring and tracking standardized data to NCBI for banking and public utility; (3) an initiative in Technology Development in Epigenomics that will support technology development in epigenetic profiling and/or whole epigenome studies and development of methods to enable in vivo imaging of epigenetic changes in cells, tissues, and eventually intact organisms; (4) a program for Discovery of Novel Epigenetic Marks in Mammalian Cells will support research to identify stable, long-term changes in epigenetic processes and establish the utility of these marks in mammalian cells that could quickly be translated to global epigenome mapping in human cells (conducted by the Reference Epigenome Mapping Centers); (5) a program designed to address Epigenomics of Human Health and Disease that will support research on fundamental epigenomic changes or mechanisms underlying specific diseases, conditions of development or aging, or response to exposures (physical, chemical, behavioral, and social factors).

In summary, there will be enormous research opportunities into the foreseeable future that will lead us closer to the improvements in the quality and quantity of our lives. Just as the Human Genome Project provided a unique foundation for biomedical research, a Human Epigenome Project will likely take us to an even greater understanding of human biology, aging, and medical care.

Acknowledgments Grant Support: National Cancer Institute grants CA 100055 and CA097880 (C.W.C.), CA123018 (H.S.), and the CRC Missouri Chair in Cancer Research (CWC).

References

Albert, T. J., Molla, M. N., Muzny, D. M., Nazareth, L., Wheeler, D., Song, X., Richmond, T. A., Middle, C. M., Rodesch, M. J., Packard, C. J., Weinstock, G. M. and Gibbs, R. A. 2007. Direct selection of human genomic loci by microarray hybridization. Nat Methods. 4:903–905.

Barski, A., Cuddapah, S., Cui, K., Roh, T. Y., Schones, D. E., Wang, Z., Wei, G., Chepelev, I., and Zhao, K. 2007. High-resolution profiling of histone methylations in the human genome. Cell 129:823–837.

Bernstein, B. E., Meissner, A., and Lander, E. S. 2007. The mammalian epigenome. Cell 128: 669–681.

Bernstein, B. E., Mikkelsen, T. S., Xie, X., Kamal, M., Huebert, D. J., Cuff, J., Fry, B., Meissner, A., Wernig, M., Plath, K., Jaenisch, R., Wagschal, A., Feil, R., Schreiber, S. L., and Lander, E. S. 2006. A bivalent chromatin structure marks key developmental genes in embryonic stem cells. Cell 125:315–326.

Bjornsson, H. T., Sigurdsson, M. I., Fallin, M. D., Irizarry, R. A., Aspelund, T., Cui, H., Yu, W., Rongione, M. A., Ekstrom, T. J., Harris, T. B., Launer, L. J., Eiriksdottir, G., Leppert, M. F., Sapienza, C., Gudnason, V., and Feinberg, A. P. 2008. Intra-individual change over time in DNA methylation with familial clustering. JAMA 299:2877–2883.

Bock, C. and Lengauer, T. 2008. Computational epigenetics. Bioinformatics. 24:1–10.

Brena, R. M., Huang, T. H., and Plass, C. 2006. Toward a human epigenome. Nat. Genet. 38: 1359–1360.

Chen, C. M., Chen, H. L., Hsiau, T. H., Hsiau, A. H., Shi, H., Brock, G. J., Wei, S. H., Caldwell, C. W., Yan, P. S., and Huang, T. H. 2003. Methylation target array for rapid analysis of CpG island hypermethylation in multiple tissue genomes. Am. J. Pathol. 163:37–45.

Davis, C. D. and Ross, S. A. 2008. Evidence for dietary regulation of microRNA expression in cancer cells. Nutr. Rev. 66:477–482.

Dahl, F., Stenberg, J., Fredriksson, S., Welch, K., Zhang, M., Nilsson, M., Bicknell, D., Bodmer, W. F., Davis, R. W., and Ji, H. 2007. Multigene amplification and massively parallel sequencing for cancer mutation discovery. Proc. Natl. Acad. Sci. USA 104:9387–9392.

Eckhardt, F., Lewin, J., Cortese, R., Rakyan, V. K., Attwood, J., Burger, M., Burton, J., Cox, T. V., Davies, R., Down, T. A., Haefliger, C., Horton, R., Howe, K., Jackson, D. K., Kunde, J., Koenig, C., Liddle, J., Niblett, D., Otto, T., Pettett, R., Seemann, S., Thompson, C., West, T., Rogers, J., Olek, A., Berlin, K., and Beck, S. 2006. DNA methylation profiling of human chromosomes 6, 20 and 22. Nat. Genet. 38:1378–1385.

Espin, J. C., Garcia-Conesa, M. T., and Tomas-Barberan, F. A. 2007. Nutraceuticals: facts and fiction. Phytochemistry 68:2986–3008.

Esteller, M. 2006. The necessity of a human epigenome project. Carcinogenesis 27:1121–1125.

European Union, Network of Excellence, Scientific Advisory Board, 2008. Moving AHEAD with an international human epigenome project. Nature 454:711–715.

Feinberg, A. P. 2008. Epigenetics at the epicenter of modern medicine. JAMA 299:1345–1350.

Feinberg, A. P., Ohlsson, R., and Henikoff, S. 2006. The epigenetic progenitor origin of human cancer. Nat. Rev. Genet. 7:21–33.

Fraga, M. F. and Esteller, M. 2007. Epigenetics and aging: the targets and the marks. Trends Genet. 23:413–418.
Fredriksson, S., Banér, J., Dahl, F., Chu, A., Ji, H., Welch, K., and Davis, R. W. 2007. Multiplex amplification sequences within 10 cancer genes by Gene-Collector. Nucleic Acids Res. 35:e47.
Gal-Yam, E. N., Saito, Y., Egger, G., and Jones, P. A. 2008. Cancer epigenetics: modifications, screening, and therapy. Annu. Rev. Med. 59:267–280.
Garcia-Manero, G., Kantarjian, H. M., Sanchez-Gonzalez, B., Yang, H., Rosner, G., Verstovsek, S., Rytting, M., Wierda, W. G., Ravandi, F., Koller, C., Xiao, L., Faderl, S., Estrov, Z., Cortes, J., O'Brien, S., Estey, E., Bueso-Ramos, C., Fiorentino, J., Jabbour, E., and Issa, J. P. 2006. Phase 1/2 study of the combination of 5-aza-2′-deoxycytidine with valproic acid in patients with leukemia. Blood 108:3271–3279.
Gebhard, C., Schwarzfischer, L., Pham, T. H., Schilling, E., Klug, M., Andreesen, R., and Rehli, M. 2006. Genome-wide profiling of CpG methylation identifies novel targets of aberrant hypermethylation in myeloid leukemia. Cancer Res. 66:6118–6128.
Hashimshony, T., Zhang, J., Keshet, I., Bustin, M., and Cedar, H. 2003. The role of DNA methylation in setting up chromatin structure during development. Nat. Genet. 34:187–192.
Huang, T. H., Perry, M. R., and Laux, D. E. 1999. Methylation profiling of CpG islands in human breast cancer cells. Hum. Mol. Genet. 8:459–470.
Issa, J. P. 2005. Optimizing therapy with methylation inhibitors in myelodysplastic syndromes: dose, duration, and patient selection. Nat. Clin. Pract. Oncol. 2 Suppl 1:S24–S29.
Issa, J. P., Gharibyan, V., Cortes, J., Jelinek, J., Morris, G., Verstovsek, S., Talpaz, M., Garcia-Manero, G., and Kantarjian, H. M. 2005. Phase II study of low-dose decitabine in patients with chronic myelogenous leukemia resistant to imatinib mesylate. J. Clin. Oncol. 23:3948–3956.
Jirtle, R. L. and Skinner, M. K. 2007. Environmental epigenomics and disease susceptibility. Nat. Rev. Genet. 8:253–262.
Jones, P. A. and Martienssen, R. 2005. A blueprint for a human epigenome project: the AACR human epigenome workshop. Cancer Res. 65:11241–11246.
Keshet, I., Schlesinger, Y., Farkash, S., Rand, E., Hecht, M., Segal, E., Pikarski, E., Young, R. A., Niveleau, A., Cedar, H., and Simon, I. 2006. Evidence for an instructive mechanism of de novo methylation in cancer cells. Nat. Genet. 38:149–153.
Kirmizis, A. and Farnham, P. J. 2004. Genomic approaches that aid in the identification of transcription factor target genes. Exp. Biol. Med. (Maywood.) 229:705–721.
Kuang, S. Q., Tong, W. G., Yang, H., Lin, W., Lee, M. K., Fang, Z. H., Wei, Y., Jelinek, J., Issa, J. P., and Garcia-Manero, G. 2008. Genome-wide identification of aberrantly methylated promoter associated CpG islands in acute lymphocytic leukemia. Leukemia. 22:1529–1538.
Laird, P. W. 2005. Cancer epigenetics. Hum. Mol. Genet. 14 Spec No 1:R65–R76.
Lemaire, M., Chabot, G. G., Raynal, N. J., Momparler, L. F., Hurtubise, A., Bernstein, M. L., and Momparler, R. L. 2008a. Importance of dose-schedule of 5-aza-2′-deoxycytidine for epigenetic therapy of cancer. BMC. Cancer 8:128.
Lemaire, M., Momparler, L. F., Bernstein, M. L., Marquez, V. E., and Momparler, R. L. 2005. Enhancement of antineoplastic action of 5-aza-2′-deoxycytidine by zebularine on L1210 leukemia. Anticancer Drugs 16:301–308.
Lemaire, M., Momparler, L. F., Raynal, N. J., Bernstein, M. L., and Momparler, R. L. 2008b. Inhibition of cytidine deaminase by zebularine enhances the antineoplastic action of 5-aza-2′-deoxycytidine. Cancer Chemother. Pharmacol.
Meissner, A., Mikkelsen, T. S., Gu, H., Wernig, M., Hanna, J., Sivachenko, A., Zhang, X., Bernstein, B. E., Nusbaum, C., Jaffe, D. B., Gnirke, A., Jaenisch, R., and Lander, E. S. 2008. Genome-scale DNA methylation maps of pluripotent and differentiated cells. Nature.
Mikkelsen, T. S., Ku, M., Jaffe, D. B., Issac, B., Lieberman, E., Giannoukos, G., Alvarez, P., Brockman, W., Kim, T. K., Koche, R. P., Lee, W., Mendenhall, E., O'Donovan, A., Presser, A., Russ, C., Xie, X., Meissner, A., Wernig, M., Jaenisch, R., Nusbaum, C., Lander, E. S.,

and Bernstein, B. E. 2007. Genome-wide maps of chromatin state in pluripotent and lineage-committed cells. Nature 448:553–560.

Momparler, R. L. 2005. Epigenetic therapy of cancer with 5-aza-2′-deoxycytidine (decitabine). Semin. Oncol. 32:443–451.

Momparler, R. L. 2003. Cancer epigenetics. Oncogene 22:6479–6483.

Oberley, M. J., Tsao, J., Yau, P., and Farnham, P. J. 2004. High-throughput screening of chromatin immunoprecipitates using CpG-island microarrays. Methods Enzymol. 376:315–334.

Okou, D. T., Steinberg, K. M., Middle, C., Cutler, D. J., Albert, T. J., and Zwick, M. E. 2007. Microarray-based genomic selection for high-throughput resequencing. Nat. Methods 4:907–909.

Ovesna, J., Slaby, O., Toussaint, O., Kodicek, M., Marsik, P., Pouchova, V., and Vanek, T. 2008. High throughput 'omics' approaches to assess the effects of phytochemicals in human health studies. Br. J. Nutr. 99 E Suppl 1:ES127–ES134.

Plimack, E. R., Stewart, D. J., and Issa, J. P. 2007. Combining epigenetic and cytotoxic therapy in the treatment of solid tumors. J. Clin. Oncol. 25:4519–4521.

Porreca, G. J., Zhang, K., Li, J. B., Xie, B., Austin, D., Vassallo, S. L., LeProust, E. M., Peck, B. J., Emig, C. J., Dahl, F., Gao, Y., Church, G. M., and Shendure, J. 2007. Multiplex amplification of large sets of human exons. Nat Methods 4:931–936.

Raynal, N. J., Momparler, L., Charbonneau, M., and Momparler, R. L. 2008. Antileukemic activity of genistein, a major isoflavone present in soy products. J. Nat. Prod. 71:3–7.

Robertson, G., Hirst, M., Bainbridge, M., Bilenky, M., Zhao, Y., Zeng, T., Euskirchen, G., Bernier, B., Varhol, R., Delaney, A., Thiessen, N., Griffith, O. L., He, A., Marra, M., Snyder, M., and Jones, S. 2007. Genome-wide profiles of STAT1 DNA association using chromatin immunoprecipitation and massively parallel sequencing. Nat. Methods 4:651–657.

Rodenhiser, D. and Mann, M. 2006. Epigenetics and human disease: translating basic biology into clinical applications. CMAJ. 174:341–348.

Rollins, R. A., Haghighi, F., Edwards, J. R., Das, R., Zhang, M. Q., Ju, J., and Bestor, T. H. 2006. Large-scale structure of genomic methylation patterns. Genome Res. 16:157–163.

Saxonov, S., Berg, P., and Brutlag, D. L. 2006. A genome-wide analysis of CpG dinucleotides in the human genome distinguishes two distinct classes of promoters. Proc. Natl. Acad. Sci. USA 103:1412–1417.

Schlesinger, Y., Straussman, R., Keshet, I., Farkash, S., Hecht, M., Zimmerman, J., Eden, E., Yakhini, Z., Ben-Shushan, E., Reubinoff, B. E., Bergman, Y., Simon, I., and Cedar, H. 2007. Polycomb-mediated methylation on Lys27 of histone H3 pre-marks genes for de novo methylation in cancer. Nat. Genet. 39:232–236.

Schumacher, A., Kapranov, P., Kaminsky, Z., Flanagan, J., Assadzadeh, A., Yau, P., Virtanen, C., Winegarden, N., Cheng, J., Gingeras, T., and Petronis, A. 2006. Microarray-based DNA methylation profiling: technology and applications. Nucleic Acids Res. 34:528–542.

Shaker, S., Bernstein, M., Momparler, L. F., and Momparler, R. L. 2003. Preclinical evaluation of antineoplastic activity of inhibitors of DNA methylation (5-aza-2′-deoxycytidine) and histone deacetylation (trichostatin A, depsipeptide) in combination against myeloid leukemic cells. Leuk. Res. 27:437–444.

Shi, H., Maier, S., Nimmrich, I., Yan, P. S., Caldwell, C. W., Olek, A., and Huang, T. H. 2003. Oligonucleotide-based microarray for DNA methylation analysis: principles and applications. J. Cell Biochem. 88:138–143.

Shi, H., Yan, P. S., Chen, C. M., Rahmatpanah, F., Lofton-Day, C., Caldwell, C. W., and Huang, T. H. 2002. Expressed CpG island sequence tag microarray for dual screening of DNA hypermethylation and gene silencing in cancer cells. Cancer Res. 62:3214–3220.

Soriano, A. O., Yang, H., Faderl, S., Estrov, Z., Giles, F., Ravandi, F., Cortes, J., Wierda, W. G., Ouzounian, S., Quezada, A., Pierce, S., Estey, E. H., Issa, J. P., Kantarjian, H. M., and Garcia-Manero, G. 2007. Safety and clinical activity of the combination of 5-azacytidine, valproic acid, and all-trans retinoic acid in acute myeloid leukemia and myelodysplastic syndrome. Blood 110:2302–2308.

Squazzo, S. L., O'Geen, H., Komashko, V. M., Krig, S. R., Jin, V. X., Jang, S. W., Margueron, R., Reinberg, D., Green, R., and Farnham, P. J. 2006. Suz12 binds to silenced regions of the genome in a cell-type-specific manner. Genome Res. 16:890–900.

Taylor, K. H., Kramer, R. S., Davis, J. W., Guo, J., Duff, D. J., Xu, D., Caldwell, C. W., and Shi, H. 2007. Ultradeep bisulfite sequencing analysis of DNA methylation patterns in multiple gene promoters by 454 sequencing. Cancer Res. 67:8511–8518.

Tomazou, E. M., Rakyan, V. K., Lefebvre, G., Andrews, R., Ellis, P., Jackson, D. K., Langford, C., Francis, M. D., Backdahl, L., Miretti, M., Coggill, P., Ottaviani, D., Sheer, D., Murrell, A., and Beck, S. 2008. Generation of a genomic tiling array of the human Major Histocompatibility Complex (MHC) and its application for DNA methylation analysis. BMC. Med. Genomics 1:19.

Weinmann, A. S., Yan, P. S., Oberley, M. J., Huang, T. H., and Farnham, P. J. 2002. Isolating human transcription factor targets by coupling chromatin immunoprecipitation and CpG island microarray analysis. Genes Dev. 16:235–244.

Yan, P. S., Perry, M. R., Laux, D. E., Asare, A. L., Caldwell, C. W., and Huang, T. H. 2000. CpG island arrays: an application toward deciphering epigenetic signatures of breast cancer. Clin. Cancer Res. 6:1432–1438.

Perspectives in Aging and Epigenetics

Robin Holliday

Abstract A narrow definition of epigenetics is the inheritance of information that is not based on changes in base sequence. A wider definition is that epigenetics is the sum of all those mechanisms necessary for the unfolding of the genetic programme for development. The fact that identical twins develop to the indistinguishable phenotypes of young adults shows that epigenetic controls are highly accurate. However, as aging sets in their phenotypes often become distinguishable; nor do they have the same lifespans. The same is true of inbred mice kept in a uniform environment. This strongly suggests that stochastic events are an important component of aging, and many of these events could be epigenetic. Already there is evidence that patterns of DNA methylation and histone modification are subject to variation during aging. This can be referred to as epigenetic drift.

Keywords DNA · RNA · Proteins · DNA methylation · Histone modification

Introduction: Epigenetics

The word epigenetics is used in at least two ways. A narrow definition is the inheritance of information which is not based on changes in DNA sequence. A broad definition is that it is the sum of all those mechanisms necessary to unfold the genetic program for development. Thus, epigenetic mechanisms determine that the fertilized egg will initiate a series of orderly events that give rise to the adult. The adult is maintained for a given period of time during which it can reproduce. After this period, other changes begin to occur, which will give rise to the process or processes of aging. These changes are far from orderly, as it is very clear that stochastic events are important. This is well illustrated by monozygotic twins, because development gives rise to identical adults, but the aging of these twins is not the same, as will be discussed later. Similarly, inbred lines of mice are genetically identical,

R. Holliday (✉)
Australian Academy of Science, Canberra, Australia
e-mail: randl.holliday@bigpond.com

T.O. Tollefsbol (ed.), *Epigenetics of Aging*, DOI 10.1007/978-1-4419-0639-7_25,

but their life spans in a constant environment vary considerably. For example, laboratory male CBA mice have a maximum life span of 34–35 months, but the steep fall in the survival curve begins at about 21 months, which is about one-third of the longest life span (see Zurcher et al. 1982; Holliday 1995). It could be said that the epigenetic processes which give rise to the constant phenotype of the adult are slowly eroded by aging, and death follows.

The Multiple Causes of Aging

There is a very large literature documenting changes that occur during aging, much of which has been comprehensively reviewed by Finch (1990). Taken as a whole, the evidence shows that aging is a multicausal event. The changes in DNA include mutations and chromosome abnormalities, attrition of telomeres in dividing cells, and deletions in mitochondrial DNA. Lindahl (1993) pointed out that there may be minor lesions in DNA that are not recognized by repair enzymes, and these would be expected to accumulate with time. Proteins are subject to many changes during aging. The situation in healthy cells is that abnormal molecules are recognized and degraded by proteases. This is not the case in aging cells. There are many studies documenting changes in proteins. There are inactive molecules recognized by specific antibodies to normal molecules (cross-reacting material). There are denatured molecules, or those that have abnormal chemical modifications. The glycation of proteins is one such modification, and this can lead to the formation of AGEs (advanced glycation end products). The "age pigment" lipofuscin contains protein aggregates. There is cross-linking between long-lived molecules, such as collagen. There may be errors in the primary synthesis of proteins. Changes in RNA are less well documented, but in view of the recent studies demonstrating the regulatory role of noncoding RNAs and small RNAs, it would be surprising if alterations of the normal populations did not occur. This could include abnormalities in the splicing of RNA transcripts.

Other well-documented changes during aging include the accumulation of secondary lysosomes or other granular materials. Protein cross-linking leads to hardening of the arteries and increased blood pressure. This and other changes can lead to cerebrovascular disease. There are defenses against the damage that can be caused by reactive oxygen species (ROS), but it is widely believed that such damage is an important component of aging. The immune system is known to lose its efficiency during aging, and there may be a failure to distinguish self from nonself antigens. This can give rise to damaging levels of autoimmunity. Hormonal levels are liable to change, for example, abnormalities in the regulation of insulin can give rise to late-onset diabetes. More generally, it can be said that there is a loss of physiological homeostasis, which would have multiple effects on cells, tissues, and organs. At the molecular level there is the likelihood of a decline in the accuracy of information transfer between macromolecules.

Although many causes of aging can be documented, what is lacking is the significance of each in determining the onset of senescence and its further progression to the aged phenotype. One approach is not to itemize all the different events that can occur, but instead include them all as a global cause of aging, such as a "progressive loss of molecular fidelity," as Hayflick (2007) has done.

Individual causes of aging can often be linked to a particular theory of aging. Thus, there is a somatic mutation theory, a mitochondrial theory, a free radical theory, an immunological theory, and so on. The newest important theory of aging is that it has an epigenetic foundation. So the question arises: how important are epigenetic changes in the general context of aging? Obviously that is the topic addressed by this book.

Biological Reasons for Aging

Aging can be understood only in terms of its evolutionary origins. Apart from a few of the simplest forms, all animals age; therefore, it is of ancient origin. Aging is seen in animals that live in protected environments, such as zoos, under domestication, or in laboratories. In natural environments they rarely reach old age, because there is high mortality from predators, disease, starvation, and drought. It is therefore advantageous for an animal to develop to an adult, reproduce, but not to invest resources into long-term survival after reproduction. The resources saved can instead be channeled into reproduction. It is evident that the strategies for survival vary among mammalian species. Rapid development to adulthood and reproduction are associated with short life spans, as seen in small ground-living animals, whereas slow development and slow reproduction are associated with long life spans, as seen in whales, pachyderms, humans, and the other higher primates.

The adult body, or soma, depends on many repair and maintenance mechanisms, and it can be said that aging is brought about by the eventual failure of these mechanisms (Holliday 1995, 2004, 2006). In mammals at least 12 maintenance mechanisms can be identified, the more important ones being DNA repair, protein turnover, the immune system, the defenses against free radical damage, and the detoxification of toxic chemicals in food. Many comparative studies have now shown that maintenance mechanisms are more efficient in long-lived species than in short-lived ones. Also, events known to be associated with aging occur much more quickly in short-lived animals than in long-lived ones. Examples are the cross-linking of collagen, the rates of somatic mutation, and the frequencies of mitochondrial deletions. The study of each maintenance mechanism comprises a scientific discipline in its own right. Taken together, this is a substantial proportion of all biological research. Also, maintenance mechanisms depend on the activities of many different genes. Although it can be concluded that aging is no longer an unsolved problem of biology (Holliday 2006; Hayflick 2007), there is a great deal more to uncover. Not least, a much better understanding of epigenetic changes during aging is needed, and also the maintenance of the epigenetic controls in the normal adult.

Epigenetics, Development, and Aging

The broad definition of epigenetics is that it plays an essential role in unfolding the genetic program for development. This means that there is precision and accuracy that determines every stage of development until the adult is formed. Given a normal genome and a normal environment, development ensures that every male adult and every female adult of a given species are functionally the same, apart from the relatively minor differences attributable to normal genetic variation. This means that the epigenetic signals and pathways are tightly regulated and controlled. This is most strikingly demonstrated by identical twins, which in most cases are indistinguishable as children or adults. In other words, the genes and the sum of all epigenetic processes give rise to a very exact phenotype.

Aging has little relationship to development, although it has sometimes been referred to as a running down of development (Kanungo 1975). Aging is a gradual process with a gradual change in phenotype. There are many changes in molecules, cells, tissues, and organs, but these are by no means uniform. What happens in one individual may not happen in another, which is the same as stating that stochastic events are important. This is seen in identical twins as they age. Adults which are indistinguishable often become distinguishable as they age, and their life spans are not the same (although more similar than sibs of the same sex). We would expect that some epigenetic controls would be unchanged, whereas others would be altered or lost. Such changes are the antithesis of development. Instead of accuracy and precision, we would have heterogeneity and random changes. There is now remarkable evidence for what has been correctly referred to as "epigenetic drift." In a definitive and comprehensive study of identical twins by Fraga et al. (2005) (and see Martin 2005 and the Chapter "Epigenetic Drift and Aging"), it was shown that there are far more significant differences in DNA methylation and histone modification in aged twins than in young ones. Older identical twins that had different lifestyles showed more differences than those that had lived in a common environment. Nevertheless, the latter also showed significant differences. This is a key set of observation that essentially proves that epigenetic changes occur during aging. These can in part be attributed to environmental influences. However, it must be emphasized that inbred animals in a constant laboratory environment have a range of life spans. In view of the results by Fraga et al. (2005), stochastic epigenetic changes are a likely cause of this variability.

At the cellular level, age changes are heterogeneous. This makes it harder to measure such changes, because what affects one cell may not affect its neighbors. What is needed is a sensitive screen of individual cells. For example, fluorescent antibodies could detect abnormal ectopic gene expression, that is, the expression of a particular gene that is normally silent in that particular population of cells. Any technique that can examine and quantitate differences in individual cells could be profitably used in studies of aging.

DNA Methylation

DNA can be modified by the methylation of cytosine. In mammalian species, about 5% of cytosine is methylated. It was proposed in 1975 that this modification may play a role in controlling gene expression (Holliday and Pugh 1975; Riggs 1975), but there was no direct evidence for this at that time. It was not proposed that all methylated cytosines were important, but only a subset and possibly a small minority. When evidence began to accumulate, it was suggested that DNA methylation should be regarded as an epigenetic mechanism (Holliday 1987). Subsequently the new field of epigenetics emerged as an important discipline in its own right.

DNA methylation occurs predominantly in CpG doublets and it has been known for a long time that these are depleted in genomic DNA. The very important exception is CpG islands which are not depleted in this doublet. Such islands are commonly associated with the promoters of structural genes. There is now abundant evidence that when the island CpGs are methylated, the gene is silent, and when it is unmethylated, it is expressed. There is also much evidence that the demethylation of CpG islands by 5-azacytine leads to gene activation. The term *epimutation* refers to a change in gene expression by methylation or demethylation and it was established that epimutations are heritable. However, it is not clear whether changes in one or a few sites constitute an epimutation or whether the methylation/demethylation of a whole island is necessary. Methylation outside CpG islands may have other functions, for example, the silencing of repetitive sequences, including transposable elements. This causes a problem when total methylation is measured during aging, because it cannot be known whether or not documented changes in methylation would affect the phenotype of individual cells. If epimutations are important during aging, another problem arises, because it would be expected that such changes would be random in the genome. In a population of cells, different changes in methylation, or epimutations, would be heterogeneous and therefore hard to detect.

Model systems, such as CHO cells, demonstrate the reality of heritable epimutations. In cultured mammalian cells it is possible to select a null phenotype, such as the absence of a specific enzyme. From these it is possible to select reactivation of the enzyme. In both cases, accurate measurements of the frequency of silencing and reactivation can be made. It is also known that these changes in phenotype are due to methylation and demethylation, so by definition are epimutations (Paulin et al. 1998).

Such studies are not possible in uniform cell populations in vivo, but one approach is to search for ectopic changes in gene expression. Using modern technologies, it would be possible to screen very large numbers of cells in order to detect an occasional one which may be demethylated and express a gene which is normally silent. Then cell populations from young and old animals would be compared. The silencing of a gene by hypermethylation would be harder to detect.

It is well known that the incidence of carcinomas increases steeply with age, and several chapters refer to the extensive evidence that tumor suppressor genes become methylated and inactivated. This is a case of hypermethylation contributing to an age-associated change, but it could be an uncommon event. It is also evident that its likelihood varies between species. In both rats and humans the formation of carcinomas depends on several successive events, which is why their formation is age-related. It is evident that their frequency is enormously higher in rats, which have a life span of about 3 years, than in humans (Ames et al. 1985). This means that human cells are much more resistant to the changes which will give rise to a tumor, which includes hypermethylation. The same thing is seen in cultured human and cultured rodent cells. The former can be completely resistant to the transformation to immortal lines, whereas the spontaneous transformation of mouse and rat cells is very well documented.

The pattern of DNA methylation is heritable through the activity of maintenance methylases. These recognize hemimethylated DNA at the replication fork and methylate the new strand. The question arises about the accuracy of maintenance methylases. This was examined by Wilson and Jones (1983) using cultured cells throughout their life span in vitro. Total methylation declined rapidly in mouse cells, which have a short life span, at an intermediate level in Syrian hamster cells, which have a longer life span, and more slowly in human cells, which have the longest life span. There were only four data points for human cells, but subsequent studies have confirmed the result (Fairweather et al. 1987). In human cells, maintenance methylation is about 99.5% accurate. This slow decline can be regarded as a molecular clock in normal cells (Holliday 2001), although much more attention has been paid to the molecular clock based on the gradual loss of telomeric DNA (reviewed by Reddel 1998). In transformed lines, neither clock operates.

Epigenetics and Cell–Cell Signaling

During normal development there are many interactions between cells. Cells may be in contact with each other or separated. In the latter case, signaling may depend on diffusible hormones, growth factors, or morphogens. These may interact with a cell surface receptor on the cell which is receiving the signal. An attractive model is that the interaction leads to a change or changes at the DNA level which are epigenetic and can be inherited. The loss or gain of DNA methylation is an obrious possibility. Such a mechanism has a Lamarckian dimension, because an external signal gives rise to a change in the epigenotype which is heritable (Jablonka and Lamb 1995).

After development has been completed, there are still many cell–cell interactions in the adult organism. For example, stem cells may be stimulated and programmed to produce differentiated cells. However, the interactions may well become less precise during aging. There could be changes in the production of a signal and in the cell surface receptors. Signal transduction to the nucleus and an epigenetic switch at the DNA level could also be affected. All these changes could contribute to the loss of physiological homeostasis, which is a major feature of aging.

X Chromosome Reactivation

In female mammals, one of the two X chromosomes becomes inactivated early in development. This inactivation is random, but once it has occurred it is very stable in dividing and nondividing cells. The active and inactive chromosomes share a common cellular environment, so the difference between them is necessarily epigenetic. It has been shown that CpG islands are methylated in the inactive X chromosome, but not in the active one. In mice it has been shown that the inactive X chromosome can become reactivated during aging. In Cattanach's translocation, part of an autosome is inserted into one arm of the X chromosome and this contains the tyrosinase gene. The addition of Searle's translocation to the genotype ensures that the same chromosome is always inactivated (because inactivation of the other produces an unbalanced lethal genotype). Thus, inactivation of the tyrosinase gene produces an albino phenotype. It was shown that when these animals age, they become more and more pigmented as X chromosome reactivation occurs (Cattanach 1974). In another study the X-linked gene for ornithine carbamoyltransferase was used. The activity of this gene can be demonstrated by a histological staining procedure, and Searle's translocation abolishes mosaicism in female animals. Therefore, if the active X chromosome contains a mutant gene and the inactive X has a wild-type gene, then cells do not stain for enzymatic activity. It was found that there are many stained cells in the liver tissue cells in old mice and few in young mice, thus demonstrating the reactivation of the X chromosome during aging (Wareham et al. 1987).

Attempts have been made to demonstrate the reactivation of the human inactive X chromosome in old females, but the methods used are much less sensitive than those available in mice. The reason for this is that any X-linked mutant heterozygote will have a mixture of mutant and normal phenotypes, thus making it almost impossible to detect reactivated wild-type cells. The genetic manipulation procedures available in mice obviously cannot be applied to humans. Nevertheless, the results so far obtained show that the reactivation of the inactive X chromosome is not a common event (Migeon et al. 1988; Pagani et al. 1990). This shows that the epigenetic controls are much tighter in human cells than in mouse cells, just as are the controls that prevent human cell transformation.

Epigenetic and Nonepigenetic Events During Aging

The study of aging has a long history, and a very large literature has accumulated. For most of the twentieth century there were both a huge body of information, which was usually hard to interpret, and many competing theories of aging. By the end of the century, many facts fell into place and it could be said that the biological basis for aging became very much clearer (Hayflick 1994, 2007; Holliday 1995, 2004, 2006 Austad, 1997). It was not a scientific discovery in the usual sense, because it was based on a series of interconnected insights over quite a long period of time. It was also clear that aging is a very complex process or series of processes affecting molecules, cells, tissues, and organs.

The science of epigenetics, in spite of its history, is a relatively new field dating from the 1990s. It is therefore not surprising that it has not been thoroughly studied in relation to aging. One of the most significant advances is "epigenetic drift" (Fraga et al. 2005; Martin 2005). This shows that identity of epigenetic controls during development to the adult, and maintenance of the normal adult, is not sustained into senescence and aging. Instead, differences appear, presumably because the normal controls of epigenetic signaling are gradually eroded. Even on the basis of limited evidence, a strong case can be made that these changes provide an explanation for the stochastic events that determine final life span. The fact that the variation in life span in inbred strains of mice and identical twins in humans has a stochastic basis has been known for many decades, and it cannot be said that this has previously been explained. If epigenetic events provide the explanation, then it follows that such events are of considerable significance. Other evidence that epigenetic events may be very important comes from cloned animals. In this case, imprinting signals are likely to be important. A nucleus from a somatic cell may have an identical DNA base sequence to that in the fertilized egg, but may have lost or changed the normal pattern of imprinting. Without the correct reprogramming, the cloned animal will have epigenetic defects that strongly affect its phenotype and the life span.

Many of the chapters in this book review epigenetic mechanisms which might change during aging. In most cases, experimental evidence is not available, so it could be said that they are candidates for age-associated change. Table 1 lists possible epigenetic changes and also documented nonepigenetic changes based on a very large research base. The narrow definition of epigenetics is that it is the inheritance of information which is not based on changes in DNA sequence. The broad definition is that it is the sum of all those changes required for the unfolding of the genetic program for development. Therefore, it is appropriate to include all mechanisms that affect gene expression. These include alternative splicing of transcripts, small RNAs that influence gene expression, and DNA rearrangements in cells that produce antibodies.

Table 1 Nonepigenetic events and epigenetic events during aging

Changes that are not epigenetic	Possible epigenetic changes
Mutations	DNA methylation
Chromosome abnormalities	Histone modification
Unrepaired DNA damage	Gene silencing
Mitochondrial deletions	Ectopic gene expression
Loss of telomeric DNA	X chromosome reactivation
Damage by ROS	Small RNAs
Secondary lysosomes	Noncoding RNAs
Protein cross-linking	Alternative transcript splicing
Protein aggregation and AGEs	Altered polycomb proteins
Protein modifications and errors	Cell–cell interactions
Membranes and receptors	
Cell loss	
Autoimmunity	

In the future the widely used narrow definition of epigenetics will gradually change to the broad definition. Epigenetics cannot be fully understood until development itself is understood, and this will involve many future advances, including the discovery of new mechanisms. Although it is established that there are multiple causes of aging, the relative effects of each and their relationship to age-associated disease raise many unanswered questions. Much more data are required before the importance of epigenetic defects during aging can be properly assessed.

References

Ames, B.N., Saul, R.L., Schwiers, E., Adelman, R., and Cathcart. R. 1985. Oxidative damage as related to cancer and aging. The assay of thymine glycol, thymidine glycol and hydroxymethyl uracil in human and rat urine. In Molecular Biology of Aging, eds. R.S. Sohal, L.S. Birnbaum, and R.G. Cutler, pp. 137–144. New York, Raven Press.

Austad, S.N. 1997. Why We Age. New York, John Wiley.

Cattanach, B.L. 1974. Position effect variation in the mouse. Genet. Res. 23: 291–306.

Fairweather, S., Fox, M., and Margison, G.P. 1987.The in vitro lifespan of MRC-5 cells is shortened by 5-azacytidine induced demethylation. Exp. Cell Res. 168: 153–159.

Finch, C.E. 1990. Longevity, Senescence and the Genome. Chicago, The University of Chicago Press.

Fraga, M.F., Ballestar, E., Paz, M.F. et al. 2005. Epigenetic differences arise during the lifetime of monozygotic twins. Proc. Nat. Acad. Sci. USA 102: 10604–10609.

Hayflick, L. 1994, 1996. How and Why We Age. New York, Ballantine Books.

Hayflick, L. 2007. Biological aging is no longer an unsolved problem. Ann. NY Acad. Sci. 1100: 1–13.

Holliday, R. 1987. The inheritance of epigenetic defects. Science 238: 163–170.

Holliday, R. 1995. Understanding Ageing. Cambridge, Cambridge University Press.

Holliday, R. 2001. Senescence of dividing somatic cells. In Stem Cell Biology, eds. D.R. Marshak, R.L. Gardner, and D. Gottleib. pp. 95–109. New York, Cold Spring Harbor Press.

Holliday. R. 2004. The multiple and irreversible causes of aging. J. Geront. Biol. Sci. 59A: 568–572.

Holliday, R. 2006. Aging is no longer an unsolved problem in biology. Ann. NY Acad. Sci. 1067: 1–9.

Holliday, R. and Pugh, J.E. 1975. DNA modification mechanisms and gene activity during development. Science 187: 226–232.

Jablonka, E. and Lamb, M.J. 1995. Epigenetic Inheritance and Evolution. Oxford, Oxford University Press.

Kanungo, M.S. 1975. A model for ageing. J. Theoret. Biol. 53: 253–261.

Lindahl, T. 1993. Instability and decay of the primary structure of DNA. Nature 362: 709–715.

Martin, G.M. 2005. Epigenetic drift in aging identical twins. Proc Acad. Nat. Sci. USA 102: 10413–10414.

Migeon, B.R., Axelman, J., and Beggs, A.H. 1988. Effect of ageing on reactivation of the human X-linked HPRT locus. Nature 335: 93–96.

Pagani, F., Toniolo, D., and Vergani, C. 1990. Stability of DNA methylation of X-chromosome genes during aging. Somatic Cell Mol. Genet. 16: 79–84.

Paulin, R.P., Ho, T., Balzer, H.J., and Holliday, R. 1998. Gene silencing by DNA methylation and dual inheritance in Chinese hamster ovary cells. Genetics 149: 1081–1088.

Reddel, R. 1998. A reassessment of the telomere hypothesis of senescence. BioEssays 20: 977–984.

Riggs, A.D. 1975. X inactivation, differentiation and DNA methylation. Cytogenet. Cell Genet. 14: 9–25.

Wareham, K.A., Lyon, M.F., Glenister P.H., and Williams, E.D. 1987. Age related reactivation of an X linked gene. Nature, 327: 725–727.

Wilson, V.L. and Jones, P.A. 1983. DNA methylation decreases in aging but not immortal cells. Science 220: 1055–1057.

Zurcher, C., van Zwieten, M.J., Sooleveld, H.A., and Hollander, C.F. 1982. Ageing research. In The Mouse in Biomedical Research. Vol. IV, eds. H.L. Foster, J.D. Small, and J.G. Fox, pp. 11–35. New York, Academic Press.

ERRATUM

Epigenetic Silencing of Progeroid Syndromes

Ruben Agrelo

Research Institute of Molecular Pathology (IMP),
Dr. Bohr-Gasse 7
1030 Vienna, Austria
agrelo@imp.univie.ac.at, ragrelo@adinet.com.uy

T.O. Tollefsbol (ed.), *Epigenetics of Aging*, pp. 345–369, 2010

DOI 10.1007/978-1-4419-0639-7_26

Chapter 19, page 357, line 7 should read: "We can therefore predict that loss of A-type lamins..."

The online version of the original chapter can be found at
http://dx.doi.org/10.1007/978-1-4419-0639-7_19

Index

T.O. Tollefsbol (ed.), *Epigenetics of Aging*, DOI 10.1007/978-1-4419-0639-7,

GPSR Compliance

The European Union's (EU) General Product Safety Regulation (GPSR) is a set of rules that requires consumer products to be safe and our obligations to ensure this.

If you have any concerns about our products, you can contact us on ProductSafety@springernature.com

In case Publisher is established outside the EU, the EU authorized representative is:

Springer Nature Customer Service Center GmbH
Europaplatz 3
69115 Heidelberg, Germany

Batch number: 10755461

Printed by Printforce, the Netherlands